BROADCASTING / CABLE
AND BEYOND
AN INTRODUCTION TO MODERN ELECTRONIC MEDIA

McGRAW-HILL SERIES IN MASS COMMUNICATION

CONSULTING EDITOR

Barry L. Sherman

Anderson: Communication Research: Issues and Methods
Dominick: The Dynamics of Mass Communication
Dominick, Sherman, and Copeland: Broadcasting/Cable and Beyond: An Introduction to Modern Electronic Media
Dordick: Understanding Modern Telecommunications
Hickman: Television Directing
Richardson: Corporate and Organizational Video
Sherman: Telecommunications Management: The Broadcast and Cable Industries
Walters: Broadcast Writing: Principles and Practices
Whetmore: American Electric: Introduction to Telecommunications and Electronic Media
Wilson: Mass Media/Mass Culture: An Introduction
Wurtzel and Acker: Television Production

BROADCASTING / CABLE
AND BEYOND
AN INTRODUCTION TO MODERN ELECTRONIC MEDIA

SECOND EDITION

Joseph R. Dominick

College of Journalism and Mass Communication
University of Georgia

Barry L. Sherman

College of Journalism and Mass Communication
University of Georgia

Gary A. Copeland

University of Alabama

McGraw-Hill, Inc.
New York   St. Louis   San Francisco   Auckland   Bogotá   Caracas
Lisbon   London   Madrid   Mexico   Milan   Montreal   New Delhi
Paris   San Juan   Singapore   Sydney   Tokyo   Toronto

BROADCASTING/CABLE AND BEYOND
An Introduction to Modern Electronic Media

Copyright © 1993, 1990 by McGraw-Hill, Inc. All rights reserved. Printed in the United States of America. Except as permitted under the United States Copyright Act of 1976, no part of this publication may be reproduced or distributed in any form or by any means, or stored in a data base or retrieval system, without the prior written permission of the publisher.

1 2 3 4 5 6 7 8 9 0   DOW   DOW   9 0 9 8 7 6 5 4 3 2

ISBN 0-07-017817-8

This book was set in Century Oldstyle by Better Graphics, Inc.
The editors were Hilary Jackson and David Dunham.
The production supervisor was Annette Mayeski.
The cover was designed by Circa '86.
The photo editor was Elyse Rieder.
Project supervision was done by The Total Book.
R. R. Donnelley & Sons Company was printer and binder.

Library of Congress Cataloging-in-Publication Data

Dominick, Joseph R.
   Broadcasting/cable and beyond: an introduction to modern electronic media / Joseph Dominick, Barry L. Sherman, Gary A. Copeland. — 2nd ed.
      p.   cm. — (McGraw-Hill series in mass communication)
   Includes index.
   ISBN 0-07-017817-8
      1. Broadcasting—United States.   2. Cable television—United States.   3. Telecommunication—United States.   I. Sherman, Barry L. II. Copeland, Gary.   III. Title.   IV. Series.
HE8689.8.D66   1993
384.54'0973—dc20   92-30610

ABOUT THE AUTHORS

Joseph R. Dominick received his undergraduate degree from the University of Illinois and his Ph.D. from Michigan State University in 1970. He taught for four years at Queens College, City University of New York, before coming to the College of Journalism and Mass Communication at the University of Georgia where from 1980 to 1985, he served as head of the Radio-TV-Film Sequence. He currently serves as the College's director of graduate studies. Dr. Dominick is the author of three books in addition to *Broadcasting/Cable and Beyond* and has published more than thirty articles in scholarly journals. From 1976 to 1980, Dr. Dominick served as editor of the *Journal of Broadcasting*. He has received research grants from the National Association of Broadcasters and from the American Broadcasting Company and has served as media consultant for such organizations as the Robert Wood Johnson Foundation and the American Chemical Society.

Barry L. Sherman is director of the George Foster Peabody Awards program and professor of telecommunications at the College of Journalism and Mass Communication, University of Georgia. Chairman of the department and associate director of the Peabody Awards from 1986 to 1991, he is also director of the Dowden Center for Telecommunication Studies. Dr. Sherman is a graduate of the City University of New York (B.A. 1974; M.A. 1975), and Penn State (Ph.D. 1979). In addition to *Broadcasting/Cable and Beyond*, he is the author of *Telecommunications Management: The Broadcast and Cable Industries*. His writings have appeared in a variety of professional and trade publications, including *Journal of Communication, Journal of Broadcasting and Electronic Media, Communication Education, Journalism Quarterly,* and *Channels*.

Gary A. Copeland is associate professor of telecommunication and film at The University of Alabama. He received his B.A. and M.A. from California State University, Fresno and his Ph.D. from The Pennsylvania State University. He has coauthored the book, *Negative Political Advertising: A Coming of Age*. His research has been published in *Critical Studies in Mass Communication, Journal of Broadcasting and Electronic Media,* and *Journalism Quarterly,* among others. Dr. Copeland is also the senior research associate for Impression Management Consulting, a political media consulting organization specializing in federal and statewide campaigns.

For all our children

CONTENTS

Preface xiii

PART ONE: INTRODUCTION

1 Three Days in Moscow 3
Day One 4
Day Two 5
Day Three 7
Aftermath 8
Implications and Previews 9
Suggestions for Further Reading 10

PART TWO: HOW IT HAPPENED

2 History of Radio 13
The Inventors 15
Boardrooms and Courtrooms 19
Broadcasting's Beginnings 22
Fast Times 24
Radio Days, Radio Nights 29
Adjustment 36
Fragments 39
Radio in the Video Age 39
Radio in the 1990s 43

Summary 44
Suggestions for Further Reading 44

3 History of TV and Cable 45
Roots 46
Postwar TV: 1945–1952 52
Growth Curve: 1953–1962 57
Placid Times: 1963–1975 62
Changes: 1975 to the Present 67
Summary 73
Suggestions for Further Reading 73

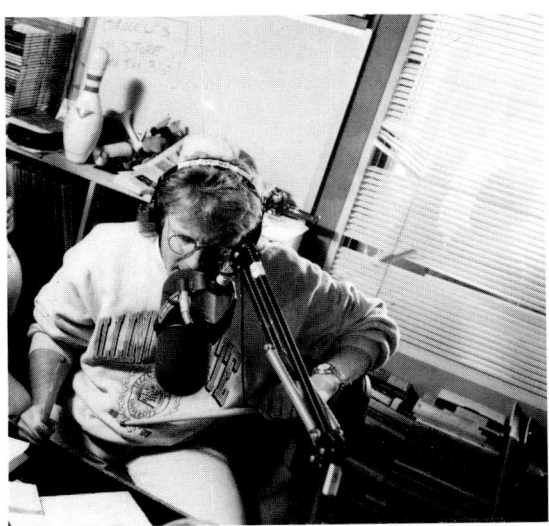

PART THREE: HOW IT IS

4 Radio Today 77
Competition in Today's Radio Business 79
Radio Advertising 80
Ownership: The Big Groups 85
Radio Programming Today 86
Promotions 92
Economics of Radio Today 94

Public Radio Today 96
Radio Station Organization 98
Getting a Job: Radio Employment Today 99
Summary 101
Suggestions for Further Reading 101

5 Television Today 103
Television Now 104
Types of Television Stations 104
The Television Business 106
The Network Television Business 110
Economics of Local Television 112
Station Ownership 114
Station Organization 117
The Job Outlook 120
Summary 122
Suggestions for Further Reading 123

6 Cable and Home Video Today 124
Cable Programming Today 126
Selling Cable Services 131
Cable Economics 131
Cable Finance 135
Cable Personnel 139
Alternatives to Cable 140
Home Video Today 141
The Future of Home Video 147
Summary 147
Suggestions for Further Reading 148

7 Corporate and Organizational Video Today 149
Defining the Field 150
History of Corporate TV 150

Users 152
Applications 153
Organization of Corporate Video 159
Video Networks 161
Videoconferencing 162
Producing Programs: CV versus Broadcast TV 163
Careers in Corporate Video 165
Summary 167
Suggestions for Further Reading 167

8 The International Scene 168
International Electronic Media Systems: A Historical Perspective 170
International Radio Broadcasters 173
International Video 178
Comparative Electronic Media Systems 180
New World Information Order 191
Differences in Electronic Media Systems: A Model 192
Summary 194
Suggestions for Further Reading 194

PART FOUR: HOW IT'S DONE

9 Radio Programming 199
Radio Regulation and Format Design 200
A Matrix of Radio Programming 201
Modes of Radio Production 202
The "Format Hole" 203
Radio and the Music Business 205

News and Talk Formatting 217
The Rebound of Radio Networks 220
Public Radio Programming 222
Summary 225
Suggestions for Further Reading 225

10 Television News Programming 226
The Rise of Broadcast News 227
Television News in the Turbulent Sixties 230
Television News in the Tortuous Seventies 231
Television News in the High-Profit, High Tech Eighties 233
Back to Reality: Television News in the Nineties 236
The Television News Team: Television News Command Structure 238
The Job Outlook in Television News 243
Summing Up: The Public's Appetite for Television News 243
Summary 245
Suggestions for Further Reading 245

11 Entertainment Programming 246
Network Television: The Big Three, Plus One 247
The World of TV Syndication 259
Local Television Programming 266
Programming Stategies 268

Summary 270
Suggestions for Further Reading 270

PART FIVE: HOW IT WORKS

12 Audio Technology 273
Technical Aspects of Broadcasting: Some Basics 274
Steps in Signal Processing 278
A Final Word 299
Summary 300
Suggestions for Further Reading 300

13 Video Technology 301
Basic Principles in Video Processing 302
Video Signal Generation 305
Video Amplification and Processing 311
Video Signal Transmission 312
Video Reception inside the TV Set 323
Video Storage Devices 325
Summary 329
Suggestions for Further Reading 330

PART SIX: WHAT IT DOES: AUDIENCE IMPACT AND EFFECTS

14 The Ratings: Estimating Audiences 333
Uses for Ratings Information 334
Early Attempts at Audience Measurement 336
Rating Services 336
The Ratings Process 343
Summary 353
Suggestions for Further Reading 353

15 Beyond Ratings: Other Audience Research 354
  Music Research 355
  Market Research 357
  Physiological Measures 364
  Consultants 366
  Summary 367
  Suggestions for Further Reading 368

16 The Effects of Broadcasting and Cable 369
  Studying the Effects of Broadcasting and Cable 370
  Theories of Media Effects 372
  Video Violence 374
  A Mean and Scary World? Perceptions of Reality 381
  Sex-Role Stereotyping 383
  Broadcasting and Politics 385
  Television and Learning 387
  Impact of Television Ads 389
  Uses and Gratifications Research 391
  Television and Prosocial Behavior 392
  Summary 394
  Suggestions for Further Reading 395

PART SEVEN: HOW IT'S CONTROLLED: RULES, REGULATION, DEREGULATION

17 The Regulatory Framework 399
  Rationale 400
  History 401
  Regulatory Forces 405
  The Regulatory Nexus 418
  Summary 419
  Suggestions for Further Reading 419

18 Broadcasting and Cable Law 420
  The Communications Act of 1934 420
  The Role of the FCC 420
  Other Federal Laws Covering Broadcasting and Cable 429
  The Law and Broadcast Journalism 437
  Regulating Advertising 445
  International Obligations 448
  Summary 449
  Suggestions for Further Reading 450

19 Self-Regulation and Ethics 451
  Self-Regulation in Broadcasting and Cable 451
  Ethics 462
  Summary 468
  Suggestions for Further Reading 468

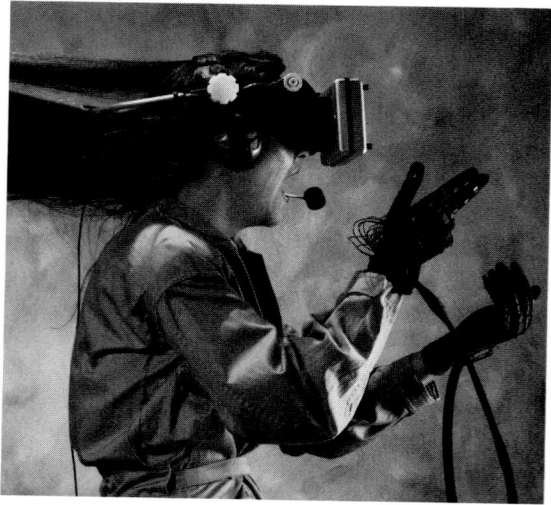

PART EIGHT: WHERE IT'S GOING

20 The Future 471
  The Industry Outlook 472
  Emerging Technologies 473
  Virtual Reality 474
  Holography 476
  Social Concerns 477
  Summary 478
  Suggestions for Further Reading 478
  Glossary 479
  Index 487

PREFACE

Our thanks to our colleagues who were kind enough to adopt the first edition of *Broadcasting/Cable and Beyond* and special thanks to those who were gracious enough to provide us with helpful suggestions for improvement. We have incorporated many of these recommendations in the second edition.

We have retained the same basic goals that made the first edition successful. One of those goals was to convey the excitement of working in the industry. Unfortunately, the exhilaration, immediacy, and opportunity of the field are hard to capture in textbooks. Something significant seems to get lost between the boom box, the shining screen, and the printed page. However, as was the case with the first edition, this book attempts to survey the field of modern electronic media in a way that captures its excitement.

Moreover, broadcasting/cable and related industries have changed radically in the last few years. We have set out to assess these radical developments with a book that takes a fresh look at the new world of the electronic media. With that goal in mind, we have made it a point to keep the second edition contemporary and up to date.

Another important objective of the first edition was to create a book that would be both interesting reading for students and easy for teachers to use. The second edition attempts to fulfill that objective by continuing to use a conversational writing style, modern design, and ancillary teaching materials. We have included easy-to-understand figures, tables, and photos that amplify and illustrate points in the text. Dozens of new boxed inserts that present extended examples or interesting snapshots of broadcasting and its leaders in action are interspersed throughout the book. These boxed inserts are now arranged thematically in order to highlight their connection to topics covered in the text. Each chapter concludes with a summary and a list of readings to make it easier for students to seek out additional information on the subjects discussed in the chapter. Since the specialized language of the industry is extensive and sometimes opaque, we have included a glossary of key terms. Finally, an innovative Instructor's Manual and a Study Guide are available to supplement the text.

NEW TO THIS EDITION

The most obvious modification is the book's organization. The two chapters describing the development of radio and television have been moved to the beginning to respond to the needs of those instructors who prefer to have a strong historical base upon which to build. In addition, we have provided a more detailed treatment of the growth, financing, programming, and current problems of public broadcasting. Since the business climate and financial considerations continue to play a major role in the development of the electronic media, we have also enlarged and made more detailed our description of the economics of radio, television, home video, and cable. Lastly, but perhaps most importantly, this edition tries to keep up with the rapidly changing pace of current events and the increased globalization of the telecommunications industry. New governments, new philosophies, and new worldwide marketing plans have significantly revamped modern telecommunications. Consequently, the chapter on international radio and television received a major overhaul and the opening chapter now focuses on an event of international significance—the failed coup in Russia.

Furthermore, the last two years or so have seen significant changes in the telecommunications realm: Communications companies are becoming more elaborate; new technologies, such as multimedia and virtual reality, have arrived; and social scientists have identified a growing list of issues, including information overload and escapism, that must be confronted. In that connection, Chapter 20, "The Future," has been significantly recast to better reflect changing industry structure, technology, and emerging critical communication issues.

PLAN OF THE BOOK

Broadcasting/Cable and Beyond is divided into eight main parts. Part One illustrates the workings of today's electronic media by examining the role played by television and radio during the unsuccessful coup attempt in Russia. Part Two (Chapters 2 and 3) is a review of significant events in the development of radio and television. The emphasis in these chapters is less on names and dates and more on how history enhances our understanding of current developments.

Part Three is an overview of the current status of the major media industries. Chapters 4 through 6 focus on the core electronic media enterprises: radio, TV, cable, and home video. Each of these chapters pays special attention to the current economic and societal forces that shape the medium. Each chapter surveys the business aspects and the issues affecting the particular industry and includes a discussion of career prospects. Chapter 7 examines the increasingly important field of corporate communications. Part Three closes with Chapter 8, an update on international telecommunications. Although less than three years have elapsed between editions, this is an area that has undergone tremendous change. The new political climate in Europe, a weak international economy, and advances in technology have combined to produce a new picture of global communications characterized by a worldwide flow of content, increasing private ownership of the airways and expanded opportunities for advertising.

Part Four concentrates on the programming process: how what we hear and see gets to our radio and TV sets. Chapter 9 centers on radio programming, including the interplay between performing, recording, and airplay on a local radio station. Chapter 10 deals with television news programming, while Chapter 11 unlocks the secrets of TV entertainment programming from dealmaking in Hollywood to selling shows in the syndication market.

Part Five (Chapters 12 and 13) explains the technical nature of the audio and video signal from the pickup point (a microphone or camera) to its storage on tape, disc or other media.

The audience is examined in Part Six. Chapter 14 answers the question "Who's listening and watching?"—the important business of ratings research. Chapter 15 examines why we listen and watch: beyond ratings to motivations and "psychographics." Chapter 16 speculates on what all this listening and watching may be doing to us—the experimental and survey research conducted at universities and elsewhere to determine the impact and effects of telecommunications.

Media control is the underpinning of Part Seven: telecommunications regulation. Chapter 17 examines the structure and function of the FCC and other forces that shape how the electronic media operate. Chapter 18 looks at the process of regulation in action, including a review of issues regarding program content, political broadcasting, and advertising claims. Chapter 19 looks at media self regulation and the response of the industry, community groups, and individuals to current ethical and regulatory issues.

The text closes with a view of the long-range future of electronic media. In Chapter 20, we examine the major industry trends likely to typify telecommunications in the future—vertical integration, globalization, and privatization—as well as examining the developing technologies of multimedia, virtual reality and holography. The chapter closes with a consideration of some of the social issues likely to be important in the future.

ACKNOWLEDGMENTS

We would like to acknowledge the help of all those who assisted with the second edition. On the industry side, especially helpful were James Duncan, Duncan's American Radio; Rick Ducey, Mark Fratrik, and Brenda Helregel of the National Associations of Broadcasters; Louisa Nielsen of the Broadcast Education Association; Steve Smith, Gannett Broadcasting; Val Carolin, CBS Radio Sales; Dwight Douglas, Burkhart/Douglas and Associates; William Killen, Cox Enterprises; and Tom Dowden of Dowden Communication Investors.

In addition a number of our colleagues at the University of Georgia and the University of Alabama offered helpful comments. Bill Lee, Tom Russell, David Hazinski, Dean Krugman and Allan Smithee are deserving of special thanks. We are also grateful to the help provided by all our students especially, LiQing Zhang, Chunmin Ren, Kristen Smith, Alyson Burner, Patricia Ready, Penny Walls, Denise DeLorme, and David McMahon. Administrative, organization and word processing tasks were ably performed by Myrna Powell, Sonia Davis, and Leslie Hopkins.

We would like to thank those reviewers who offered us insightful comments on how to improve the second edition: Marvin Diskin, Purdue University; Ernest A. Hakanen, Marquette University; Tom Lindlof, University of Kentucky; Ken Nagleburg, Louisiana State University; David Ostroff, University of Florida, Gainesville; Alfred Owens, Youngstown State University; Francine Podenski, City College of San Francisco; Paul Prince, Kansas State University; and Mary Ann Watson, Eastern Michigan University.

We would also like to thank the instructors who responded to our questionnaire and provided valuable insight into the strengths of the first edition and how we could create an even stronger second edition. They are: Virginia Bacheler, SUNY—Brockport; Thomas R. Berg, Creighton University; Richard Boyd, Black Hills State University; Lee R. Goodman, Purdue University—Calumet; Lionel Grady, Southern Utah University; Gordon L. Gray, Temple University; Philip A. Harding, Boston University; Wayne A. Helpler, University of Southern Indiana; Carol Koehler, University of Missouri—Kansas City; Gwen Kuenzli, University of Findlay; Dennis O'Neal, Middle Tennessee State University; April Orcutt, College of San Mateo; Ronald Pesha, Adirondack Community College; Kevin Sauter, University of St. Thomas; Mary F. Sorrentino, San Diego City College; and Joseph Straubhauer, Michigan State University.

Finally, as with the first edition, we offer our thanks to those significant others at McGraw-Hill who teamed up with us to produce a new and improved edition: editors Hilary Jackson and Fran Marino; Kate Scheinman, editing supervisor and Elaine Honig, our ever-observant copy editor. Lastly, we salute our families for standing by us and encouraging us as we went through the revision. To Candy, Eric and Jessica; Susan, Jason, Josh, and Morgan; and Joan and Meaghan—we dedicate this volume.

Joseph R. Dominick
Barry L. Sherman
Gary A. Copeland

PART ONE

INTRODUCTION

CHAPTER 1: THREE DAYS IN MOSCOW

PART 1: INTRODUCTION

In 1975, musician-poet Gil Scott-Heron recorded a song entitled "The Revolution Will Not Be Televised." Sixteen years later the title of that song was proven wrong by events in Russia. And although the events transpired thousands of miles from the United States, they illustrated vividly some of the characteristics of modern electronic media. Let's take a brief look at the way the media covered the attempted coup and then see what implications we can find for our exploration of modern broadcasting, cable, and newer media.

DAY ONE

The new relationship between cable and the broadcast networks in broadcast journalism was illustrated by the coverage. At 11:27 P.M., on August 18, 1991, CNN reported in a dispatch from the Reuters wire service that Mikhail Gorbachev, leader of the Soviet Union, had been replaced by Genady Yenayev. Twenty-nine minutes later, CNN correspondent Steve Hurst was on the air from Moscow with the first live report on the new crisis. Shortly before 5:00 A.M., CNN transmitted live pictures of tanks rolling through the streets of Moscow.

Once again, much like the situation that occurred some seven months earlier during the outbreak of the Persian Gulf War, the three broadcast networks were scrambling to keep up with CNN. The cable network booked twenty-four hours of satellite time each day for a week and settled into a vigil. The broadcast networks, smarting from budget cuts in the news divisions and decreasing profits, were unable to devote as much time and personnel to this breaking story as did CNN. The networks did the best they could with available resources. CBS managed to air a tape of citizens climbing on tanks as the tanks surrounded the Russian Parliament building. NBC's "Today" show, faced with a lack of live pictures from Moscow, presented a long report on Gorbachev's career anchored by Tom Brokaw. The report had been prepared earlier as a "canned" obituary and was sitting on the shelf. Producers ordered it updated and rewritten for the morning show. ABC's anchor, Peter Jennings, was on vacation and "Nightline's" Ted Koppel hosted the news on ABC's "Good Morning America." Also hampered by the lack of live video from the scene, ABC made do with telephone reports and still pictures of their reporters.

In an effort to save money and to operate more efficiently, the major news reporting organizations made use of sharing arrangements with organizations in other countries. CBS News shared its coverage with the Tokyo Broadcasting System. NBC sent a camera crew from Poland to cover the story. The network figured it was less expensive than sending a crew from the United States and the Polish crew also spoke Russian. In addition, NBC and the BBC shared coverage. NBC also took material from Visnews, a company partly owned by NBC that provides video to news reporting agencies all over the world. Similarly, ABC made use of the coverage provided by Worldwide TV News (of which ABC is a majority owner) and shared information with both Germany and Japan. CNN picked up "Vremya," the nightly news show of the Soviet Union, supplied its own translation, and aired the show to subscribers across the globe.

Local television stations supplanted the network coverage by interviewing local experts on the Soviet Union and Russian citizens who had emigrated to the United States. In Atlanta, Georgia, one local station was fortunate enough to have a Russian TV journalist working at the station through an exchange program with the Republic of Georgia, U.S.S.R. The Russian reporter provided unique insights and background as the situation unfolded. Other stations highlighted any local implications of the breaking story in the U.S.S.R.

In addition to CNN, other cable networks altered their programming because of the coup. C-SPAN devoted large blocks of time to expert discussions of the implications of the event. The Weather Channel expanded its coverage of Russia. HBO was scheduled to begin production of an original made-for-TV movie called "Stalin." Plans were put on hold pending the outcome of the coup. The cable network VH-1 had to re-edit an episode of "My Generation," filmed a few weeks earlier, which explored how much things had changed in the Soviet Union.

Although most people in the United States got most of their information about the coup from TV, the coup coverage demonstrated the immediacy of radio. Since much of the initial story broke during morning drive time, radio was able to reach a large part of the population with the news. Radio also demonstrated that it could cover the news efficiently and economically. NBC Radio had only two correspondents in Moscow on an open phone line practically for the entire duration of the coup. CBS Radio was immersed in covering Hurricane Bob when the story broke but quickly flew in reporters from neighboring countries. National Public Radio had two reporters who could speak Russian on the streets of Moscow during the three days of the unrest.

DAY TWO

Within twenty-four hours, the three networks regrouped and were supplying coverage rivaling that of CNN. Since the new Soviet leaders failed to clamp down on international news coverage of the coup, network reporters and crews were able to obtain entry visas and quickly assembled in Moscow. On August 20, 1991, ABC's Diane Sawyer got an exclusive interview with the president of Russia, Boris Yeltsin. On that same day, CBS aired a Dan Rather phone interview with a Soviet journalist inside the surrounded Russian Parliament building. NBC's big scoop came the next day when the coup fell apart by broadcasting shots of the fleeing coup leaders.

The new leaders, however, clamped down on the Soviet media. On Monday, Radio Moscow, the international short-wave radio station, was surrounded by armored personnel carriers and soldiers entered its studios. The station was ordered to carry reports of the decrees that the new State of Emergency Committee was issuing regularly. A similar scenario was played out at Moscow's TV facilities. Instead of the Russian equivalent of "Good Morning America" normally seen on the four TV channels available in Moscow, viewers found test patterns on three of the channels. An announcer on the fourth channel was solemnly reading an announcement to the Soviet people that Gorbachev had to step down because of ill health. Radio Russia and Russian Television, the two broadcasting services of the Russian Republic, were shut down. Two new, dour-faced anchors suddenly appeared on "Vremya," the nightly newscast of the Soviet Union and reported that all Soviet TV and radio stations would be on the air part-time and all independent radio stations would be closed. In the Baltic Republics troops seized the main TV and radio stations.

Despite the crackdown, some voices of resistance were broadcast. The Moscow Echo, an independent radio station, went on the air five times during the coup with defiant messages. Despite the best efforts of the new hard-line leaders, the Moscow Echo stayed on the air until the end of the coup.

The leaders of the coup apparently did not realize that the nature of modern mass communication had changed drastically in the last decade or so. First, mass communication has become an international enterprise, with electronic links blanketing the world.

International radio broadcasting services, particularly the BBC, the Voice of America (VOA), Radio Liberty, and Radio Free Europe, played a major role in the coup. When reporters for the VOA were trapped by the barricade of tanks surrounding the Russian Parliament, they simply took out cellular phones (made in Finland) and called VOA's Moscow Bureau. The bureau in turn transmitted the reports by satellite back to VOA headquarters in Washington, D.C., which then broadcast them back by shortwave to the people of Russia. A similar arrangement was used to relay the speeches of Boris Yeltsin to the Russian people.

The VOA's Russian language service began broadcasting twenty-four-hour newscasts when

CNN covers the violence in Moscow during the second day of the coup attempt. CNN's ratings tripled during the coup attempt.

CHAPTER 1: THREE DAYS IN MOSCOW

the coup started in an attempt to fill the information void caused by the restrictions on the Soviet domestic media. The new Soviet authorities made no attempt to jam the VOA or other Western signals, nor did they attempt censorship or expel the foreign press. As a result, newscasts from the VOA and the BBC were immensely influential. In fact, many of the soldiers surrounding the Parliament building were listening to the BBC to find out what was going on. When the coup finally fell apart, workers in the Russian Federation building facing the U.S. embassy building in Moscow printed a thank-you note on the side of their building: "Thank you, VOA," it read, "for the correct information."

Other enterprising Muscovites picked up CNN newscasts, usually limited to such places as hotels and the offices of foreign businesses, by rigging up specially constructed TV antennas and pointing them at the broadcasting tower in downtown Moscow. The reception was snowy but viewable. The pictures of the resistance in the streets of Moscow helped mobilize the protesters and increased public opposition to the coup, thus hastening its collapse.

Several hundred miles to the south of Moscow, on the Crimean peninsula, Soviet President Mikhail Gorbachev, the primary target of the coup, had also turned to international broadcasting. Under house arrest, Gorbachev and his aides found some old radio sets in the basement and hooked them up. The old sets worked surprisingly well and picked up the VOA, Radio Liberty, and the BBC best of all. The reports of the coup's collapse made it easier for Gorbachev to return to power. During a news conference after the coup Gorbachev praised the BBC for its coverage and asked if the BBC correspondent was in the room so that he could offer personal thanks. Unfortunately, the BBC reporter wasn't present; she was already on the phone filing a report for an upcoming newscast.

Second, modern mass communication has moved beyond the traditional means of broadcasting and the press. New technologies, such as fax machines, electronic bulletin boards, electronic mail, and cellular telephones have opened up channels of communication that are highly difficult to control among people in different countries. Although the coup leaders seized control of the traditional media, they made no attempt to restrict alternative media. Overseas phone links were left open. One Soviet student heard about the coup from an American friend who called her from Nashville, Tennessee. Electronic mail flowed through a satellite link between New York and Moscow. An executive in Virginia learned that violence had broken out on Tuesday evening even before CNN and the three networks when a friend in Moscow sent the E-mail message, "The shooting has begun." Interfax, an independent Soviet news service, sent 300 pages of news to the United States during the coup. Interfax reported that the other Russian republics were not going along with the coup. This news was then relayed back to the Soviet Union by short-wave broadcasting and apparently helped to strengthen the growing resistance movement. Soviet journalists whose papers had been closed or censored resorted to posting faxes on the sides of buildings. Boris Yeltsin was also busy sending fax messages to a friend in Washington, D.C., which were then made available to the U.S. government and the international media. Subscribers to CompuServe, an information network for personal computers, could call up messages from Moscow written by Americans and some Russian citizens. As former NBC News President Larry Grossman summed it up: "No one can operate in the dark of night anymore." As it turned out, the power of the electronic media proved to be more potent than tanks.

At the same time that the international flow of information was undermining their efforts, the coup plotters tried to manipulate public opinion by holding a televised news conference in an attempt to gain legitimacy. Unfortunately, the press conference backfired. No one believed Vice President Genady Yenayev's explanation that Gorbachev was replaced because of bad health. Further, Yenayev's hands shook while he spoke, a telling detail picked up by the TV camera, which did little to inspire confidence in his leadership abilities.

Even American advertising and marketing had its impact during the coup. While holed up in the Parliament building, Yeltsin ordered 260 pizzas and 20 cases of Pepsi Cola from the Moscow Pizza Hut restaurant. Despite the blockade of tanks, the order was delivered without incident. Because of the circumstances, Pizza Hut didn't charge for the food.

DAY THREE

Changes in the media also heralded the coup's dissolution. On Wednesday, August 21, Radio Russia came back on the air with a report that the army troops were leaving the city. At 3:00 P.M. the Soviet-run TV newscast "Vestnik" suddenly changed its tone and started quoting opponents of the coup. On "Vremya" the two solemn-faced anchors who had reported on the actions of the State Committee on the Emergency Situation disappeared and were replaced by two new newscasters who read stories denouncing the coup. It was also announced that TV and radio programming would return to normal the next morning.

The news of the collapse was welcomed by the three networks—ABC, CBS, and NBC. The nets were understandably concerned about the costs in covering what could have been a long and expensive story. Declines in profits, a weak economy, and the $150 million in lost advertising revenue and production costs from covering the Gulf War, made network executives more aware than ever of the bottom-line consequences of news coverage. On Tuesday, the nets tried their best to stick with their regular coverage but most of them ran special reports throughout the day. On Wednesday, the day that the coup collapsed, the three nets went back to regular programming by noon. Luckily the short duration of the coup, only three days—about the same as a prime-time miniseries—saved the networks from making tough decisions regarding news versus commercials.

In the competitive world of broadcast journalism the networks also watched their ratings with care. As the coup unfolded, news audiences increased. ABC's "World News Tonight" and "Good Morning America" showed the biggest jumps during the coup. As the coup foundered, so did viewer interest. Newscast ratings slipped back toward their normal range on Wednesday and Thursday and prime-time news specials were soundly beaten by entertainment shows. CBS's "48 Hours," devoted to the coup, had ratings 25 percent below those of the previous noncoup week. The numbers made it easier for the networks to justify their return to normal programming—and normal commercial revenue.

Even home video played a role in events. In Crimea, Gorbachev's aides found a camcorder and the Soviet president taped a message for the world, proclaiming his good health and denouncing the coup. In Moscow, Yeltsin also relied on

Russian President Boris Yeltsin and his supporters celebrate the failure of the coup outside the Parliament building. Note the TV cameras on the balcony with Yeltsin.

One of Russia's most serious home videos. Mikhail Gorbachev's son-in-law taped this video of the Soviet President while Gorbachev and his staff were under house arrest in the Crimea.

home video, producing a cassette containing a defiant speech. His aides clandestinely distributed the cassettes to foreign journalists. Back on the home front, CNN Video, formed after CNN sold an amazing 400,000 copies of a home video based on their Gulf War coverage, released *The Soviet Crisis: Three Days in August.* Sales were weak, however.

Hollywood also entered the picture. Shortly after the coup failed, an international consortium of movie producers, financed in part by U.S. filmmakers, announced plans to film a motion picture about Boris Yeltsin's role in defying the takeover. The film, tentatively titled *Heroes of August* with Anthony Quinn starring as Yeltsin, was budgeted at $20 million and scheduled for a 1994 release.

When it was all over and the coup plotters had been arrested, the new cable network, Court TV, requested permission from Russian leaders to provide gavel-to-gavel coverage of the trial of those who plotted the coup. The new network thought such an event could significantly help it to get new subscribers. Russian authorities said they would study the request.

AFTERMATH

Although the coup failed, it set forces in motion that would eventually spell the end of the Soviet Union. In place of the Soviet Union was a commonwealth of republics based on a more democratic model that supported the free marketplace philosophy. Western companies were quick to take advantage of the new economic climate. One new business endeavor was a joint venture among the Moscow Telecommunications Company, U.S. West International, and Bell Atlantic International to provide cellular phone service to Moscow. As is the case for many firms who have started to do business in Russia, a significant portion of the training for the new service will be done by video.

Finally, when Mikhail Gorbachev was preparing to sign the official decree that would formally dissolve the U.S.S.R., he discovered that he did not have a pen. Gorbachev borrowed one from Tom Johnson, President of CNN, who was in Moscow trying to arrange an interview. CNN then unexpectedly became a part of the history it was reporting.

Although most news coverage focused on events in Moscow, 400 miles to the north, in what was then called Leningrad and has now reverted to its former name of St. Petersburg, citizens in the city square cheered when they heard the BBC report that the takeover had failed. It was only fitting that the news of this failed revolution should come over a radio link between this city and Great Britain. Some eighty-nine years earlier, not too far from downtown St. Petersburg, a young Italian had inaugurated radio service between Russia and England. Guglielmo Marconi had arrived in St. Petersburg harbor on the *Carlo Alberta,* a yacht especially equipped for wireless. Czar Nicholas II of Russia visited Marconi and witnessed the young man receiving wireless messages from a transmitter in Great Britain. Impressed, the Czar actually operated the wireless equipment himself and sent a message back to England.

The Czar did not survive long enough to see Marconi's invention reach its full potential; Nicholas would be overthrown and later killed as a result of the 1917 Russian Revolution—the same revolution that brought Lenin and communism to Russia and a new name to St. Petersburg. The unsuccessful coup attempt in 1991, whose demise as reported by the BBC brought cheers, was a vain attempt to reinstitute the system that replaced the Czar. As for Marconi, he helped to start another kind of revolution in electronic communication, one that, in its way, had an impact as profound as that which toppled the Czar. This, however, is a story for another chapter—Chapter 2 to be precise.

IMPLICATIONS AND PREVIEWS

The coup, of course, was a news event and most of the participants represented the news divisions of the various organizations involved. An entertainment event, such as a rock concert, would draw a slightly different cast of characters. Nonetheless, the coup coverage represents in microcosm the world of modern-day electronic media. Further, our brief chronology illustrates some important characteristics of the current broadcasting and cable industries and also highlights some of the topics we shall discuss in subsequent chapters. Some examples follow.

First, broadcasting has a long history of providing extraordinary coverage in times of political turmoil. In 1936 the abdication speech of King Edward VIII of England was covered extensively by radio. The coronation of his successor, King George VI, was broadcast live by the BBC and received across the world. Also in that same year radio correspondents covered the Spanish Civil War. The impending war in Europe was the subject of remarkable coverage as NBC alone provided more than 450 special broadcasts during the eighteen days of the Munich crisis in 1938. More recently, TV came into maturity during its coverage of the assassination of President John Kennedy in 1963 and provided unforgettable images of the uprising in Tiananmen Square in Beijing, China, in 1989. As you might have gathered by now, the next section, Part Two, of this book provides the historical context for the modern electronic media.

Second, almost all of the major participants in today's broadcasting and cable industries figured in the coup coverage and its fallout: the major TV networks, local stations, cable networks, international broadcasters, radio networks, radio stations, private video, home video, and even a movie production company. The coup coverage illustrated how the balance of power among these major forces in broadcasting has shifted over the past few years. The once powerful networks have lost viewers to local stations, basic cable networks such as CNN, pay cable such as HBO, and the VCR. Functions once served exclusively by the networks are now shared with international news reporting organizations such as Visnews and Worldwide TV News and their domestic counterparts such as Conus TV.

Also as exemplified by the coup coverage, modern broadcasting is a global enterprise. Reports of the coup and its aftermath were carried on radio and TV services throughout the entire world. The chant, "The whole world is watching," first shouted at demonstrations during the Democratic Convention in Chicago in August 1968, was equally true concerning the demonstrations in Moscow in August 1991.

Accordingly, the next section, Part Three, of the book examines the major forces in broadcasting and cable. Chapter 4 looks at radio; Chapter 5 examines modern TV. The fast-changing world of cable and home video is discussed in Chapter 6. Private and corporate video is the subject of Chapter 7. Finally, Chapter 8 embraces a global focus and examines international and comparative broadcasting.

Third, the coup coverage is a case study in modern TV and radio production. As an unexpected news event, it posed special problems for producers and programmers. It also highlighted the continuing friction between news and entertainment. Part Four of this book focuses on the production of radio and TV programs. Chapter 9 analyzes modern radio production and formatting as practiced at both talk and music stations. Chapter 10 considers the world of news programming, taking into account the TV and radio networks, local stations, CNN, and other specialized networks like C-SPAN. The coup coverage, of course, was surrounded by and often competed against entertainment programs. Chapter 11 examines the production, economics, and scheduling of TV entertainment.

The events in Moscow also demonstrated the use of electronic technology—satellite uplinks, computer graphics, minicams. Since technology continues to have such an important impact on modern broadcasting and cable, Chapters 12 and 13 discuss, in a nontechnical way, audio and video signal processing. In addition, these chapters look beyond traditional broadcasting and cable and examine some of the innovations waiting just over the horizon.

The importance of ratings to modern radio and

TV was illustrated by network executives' close scrutiny of the Nielsen numbers and their decisions to return to regular programming when the numbers started to decline. Since the ratings and other forms of audience research are influential in determining the future of radio and TV shows, and since the ratings are directly related to the bottom line, Chapters 14 and 15 consider the various forms of audience research used by the electronic media.

Further, the social and political effects of the coup coverage are just some of many social phenomena related to broadcasting and cable that are studied by academic and industry researchers. Chapter 16 summarizes the current state of knowledge about the social consequences of the electronic media.

The coup strikingly demonstrated the interplay between the government and the media. Media rules and regulations within a country are directly related to its political system and philosophy. Chapters 17 and 18 examine this important aspect, particularly as it relates to the U.S. electronic media. Chapter 19 looks at a related area—the efforts of the electronic media in self-regulation, an increasingly important area given the recent trend toward less regulation.

Finally, the coup confirmed that new technologies have forever altered the traditional world of broadcasting and cable. Computer bulletin boards, fax machines, and E-Mail have blurred the boundary between mass and interpersonal communication and have raised new debate over the impact of the media. The last chapter in the book looks at the future and what it holds for broadcasting, cable, and society.

SUGGESTIONS FOR FURTHER READING

All of the chapters just previewed end with a section that lists books you can read for additional information. This chapter is different: it ends by suggesting why you should keep reading this book. In other words, you should study electronic media for several reasons:

- Many of you will have careers in the electronic media. This book presents a foundation of information for you to build on as you pursue your professional goals.
- For those of you who do not intend to become media professionals a knowledge of the electronic media will help you become intelligent consumers and informed critics of radio and TV. The electronic media are so pervasive in modern life that everyone should know how they are structured and what they do.
- Finally, all of you will spend the rest of your lives in the information age, an age in which the creation, distribution, and application of information will be the most important industry. At the center of the information age will be the electronic media. Everybody benefits from the scholarly study of an industry that will be a crucial part of business, education, art, politics, and culture. The authors hope that what you learn in this book will be liberating in the traditional sense of a liberal education: we hope that the knowledge you acquire will free and empower you within this era.

PART TWO

HOW IT HAPPENED

CHAPTER 2: HISTORY OF RADIO

The "annihilator of space and time," far from describing some character in a syndicated TV cartoon, was a phrase first used to describe the telegraph. As the United States expanded in the middle of the nineteenth century, it was apparent that earlier forms of communication, which relied on messengers to transport messages physically from one point to another, were no longer adequate. For one thing, the long time lag between the sending of a message and its reception sometimes had serious consequences. The bloody battle of New Orleans was actually fought after the War of 1812 was over but before news of the treaty spread south. The government also had trouble communicating with and keeping track of the millions of pioneers who were moving west. Further, slow transmission of information hampered economic development by making national marketing difficult. Consequently, many inventors were searching for some means of rapidly communicating information across long distances.

Advances in the science of electricity and its applications provided a solution. Although many people were involved in the quest for a workable system, it was the American portrait painter Samuel Morse who is credited with the invention of the telegraph (from the Greek, meaning to write at a distance). Morse sent pulses of electromagnetic current through a wire to deflect an electromagnet, which in turn produced coded markings (eventually called Morse code) on a piece of paper. He patented his system and asked for a government grant to perfect his device. After some hemming and hawing the government came up with $30,000, which Morse used to string a line between Washington, D.C., and Baltimore. His famous first message—"What hath God wrought?"—was sent in 1844. A means of communicating instantly over a long distance, of annihilating space and time, was a reality.

Morse tried to get the government to buy his invention, but the government wasn't interested. Instead, telegraphy in the United States would be controlled by the private sector. The same pattern would be followed by radio.

Further, telegraphy would come to be dominated by large corporations, most notably Western Union. In turn Western Union was itself subsumed by an even bigger company, the American Telephone and Telegraph Company (AT&T). AT&T, along with several other large companies, would also play a dominant role in the development of radio broadcasting.

Finally, although it sounds odd to us now, when the telegraph was first introduced it was hailed as a great moral force, linking together humanity and serving to bring about universal peace and harmony. How could hostility persist when ultimately all nations on Earth were bound together by this wondrous wire? As it turned out, the telegraph was to become more of a commercial than a moral force. Business and industry, particularly the newspapers, became the primary users of Morse's invention. Radio, too, was destined to support itself by commercial means.

Thirty-two years after Morse's first message, Alexander Graham Bell further refined the annihilation of space and time. He patented a device that could send the human voice over a wire—the telephone, another of radio's ancestors. With Bell's invention, people no longer needed to know a complicated code to communicate at a distance. Anybody could talk on the telephone. Further, unlike the telegraph, which relied on a central location and a system of messengers, the telephone could be installed in a person's home. Radio, too, would be a medium designed for the individual home that would require no special skills to understand.

Note, however, that both the telegraph and the telephone were *point-to-point* communication devices. They were typically used to send a message from one source to one receiver. Radio, too, would follow this pattern in its early years until people discovered the advantages of *broadcasting*—sending the same message to a large number of different people.

By the 1880s and 1890s the next logical step in the progression of electrical communication was to liberate it from the wire. (Ironically, as the 1990s begin, much of our electronic communication has now gone back to being carried by wire, thanks to cable TV . . . but we're getting ahead of the story.) Several scientists, including James Clerk Maxwell and Heinrich Hertz, led the way in predicting the existence of and detecting electromagnetic waves that radiated through space. Researchers in France, England, and Russia investigated the nature of these mysterious waves. None of them, however, was able to perfect a system of wireless communication. That distinction would belong to the young Italian inventor, Guglielmo Marconi, whom we have already met in Chapter 1. On December 12, 1901, about seven months before he was to demonstrate the wireless to the Czar of Russia, Marconi was straining against the wind as he traipsed across the frozen, barren coast of St. John's, Newfoundland, carrying, of all things, a kite.

THE INVENTORS

Many people contributed to the birth of radio. There are three, however, who deserve special mention.

Young Man with a Kite: Marconi

Guglielmo Marconi came from a wealthy and cultured Anglo-Italian family. His father was a well-off Italian aristocrat but Marconi had ties with the British Empire through his Irish mother, who was part of the Jameson Whiskey manufacturing clan. While a student at the University of Bologna he had seen a demonstration of the mysterious "radio" waves that were then captivating scientists all across Europe. Marconi had read about the possibility of using these newly discovered waves to send telegraph messages across space without using the wires needed for the conventional Morse telegraph. Enthralled, he set up an experimental station in his father's vegetable garden. Drawing on the work of others, particularly the German scientist Heinrich Hertz, Marconi built a primitive sending and receiving set. Before long the young man was transmitting messages without wires several hundred feet across the rows of beans and cabbages.

He continued his experiments and by 1895 his signals were reaching beyond a mile. Marconi was an entrepreneur, a person who was well aware of the commercial potential of his experiments. It might be stretching the comparison a bit, but Marconi was sort of a turn-of-the-century Ted Turner, willing to devote great energy to promote his innovations. Marconi immediately realized that the new device would be most useful in circumstances where telegraphy by wire was impossible: ship-to-ship and ship-to-shore communication. Accordingly, he traveled to the leading maritime country of the period, Great Britain.

Guglielmo Marconi was able to build a lucrative communications empire on his wireless invention. Here a successful Marconi listens to radio signals in the wireless room of his personal yacht, *Electra.*

Using his mother's connections, he was able to get an audience with the head of the telegraph division of the British Post Office, who had also done some wireless experiments. The British, of course, were interested in the young Marconi's invention because it would help them maintain communications with their extensive naval fleet, merchant ships, and overseas colonies.

Marconi impressed the British, and in 1896 he took out the first patent ever granted anywhere in the world for wireless telegraphy. He formed the Marconi Wireless Telegraph Company and, after a slow start, began to make money as shipowners and sailors realized the value of his invention. Meanwhile Marconi and his employees worked on the problem of increasing the range of his transmissions. By 1889 he had exchanged messages across 36 miles; the next year, across 200 miles. Improvements and refinements came so rapidly that by 1901 he was ready to try the ultimate test—beaming a signal across the Atlantic Ocean.

And so it was that the 27-year-old Marconi was walking down the shore at the entrance to St. John's harbor in Newfoundland, carrying a box kite. Attached to the kite was a wire that would serve as the receiving aerial. At a prearranged time a powerful transmitting station in Cornwall, at the extreme western tip of Great Britain, would send a signal in Morse code: dot–dot–dot, the letter *S*. Marconi, his receiving equipment in an old hospital room, his kite soaring above the angry Atlantic, would try to hear the signal through a headset.

The first try on December 11, 1901, failed. The string holding Marconi's kite broke and the kite was lost at sea. The next day, however, brought success. Both Marconi and his chief assistant heard three low-pitched beeps. Wireless had spanned the Atlantic, a distance of more than 2000 nautical miles. The age of radio was dawning.

A Continuous Wave: Fessenden

After Marconi liberated telegraphy from its prison of wires it was only a matter of time before someone did the same for the human voice. Bell had convincingly shown that voice could be superimposed on an electrical current moving through a wire when he invented the telephone. What was needed now was to do the same thing using radio waves moving through space.

It was easier said than done.

An intent Reginald Fessenden pores over plans to improve his alternator. Fessenden did most of his experiments while a professor at the University of Pittsburgh.

Part of the problem had to do with the way that radio waves were generated. Like Hertz before him, Marconi created waves by generating sparks. An induction coil sent high voltage across a gap. When the voltage was high enough, a spark jumped across the gap, creating an electromagnetic disturbance. A rapid succession of sparks gave rise to a radio wave, which could be interrupted to form Morse code—dots and dashes—that could carry a message.

This worked fine for Morse code but the human voice was another matter. To transmit voice or music or other sounds, something else was needed—a continuous radio wave that could be transformed to carry speech. Perhaps an analogy would help. Creating radio waves using the spark gap is like banging both hands on a piano keyboard: lots of different sounds are produced but the ultimate effect is simply noise. What was needed to send voice would be the equivalent of a person continuously playing one note—middle C, let's say. This continuous tone could be modulated or regulated in such a way as to carry voice and music. How to do it, however, was a big question.

The inventor who came up with the answer was Reginald Fessenden, a Canadian-born electrical engineer who had worked in Thomas Edison's labs. Fessenden proposed building an alternator—a piece of rotating machinery much like those used to generate alternating electrical current for household use. To be used as a radio wave generator, however, a new, improved alternator was needed, one that operated at much higher speeds and generated electrical current at a much higher frequency than ordinary generators. In 1900 it was doubtful that such a machine could be built at all, let alone be useful in generating radio waves. Nonetheless, Fessenden went to work. He asked the General Electric Company to put together a high-frequency alternator according to his design. After a couple of false starts and numerous design changes and alterations, Fessenden had his alternator. In 1906 the machine was shipped to his experimental laboratory at Brant Rock, Massachusetts, and connected to his antenna system.

Fessenden chose Christmas Eve, 1906, for the public unveiling of his new device. But who would be able to hear his test broadcast? The likely candidates were the wireless telegraph operators aboard ships off the Atlantic coast; their receiving equipment should be able to detect his transmissions. Accordingly, Fessenden advertised his broadcast by sending out radiotelegraph messages to these seagoing wireless operators three days in advance.

Christmas Eve was cold and clear down much of the eastern seaboard. Early in the evening wireless operators on ships up and down the coast heard in Morse code the call "CQ . . . CQ," which usually preceded a message. The dots and dashes stopped, however, and through their headphones the wireless operators heard the faint voice of a man speaking. It was Fessenden explaining what was going on. Then came phonograph music, a violin solo by the inventor himself, some readings from the Bible, Christmas greetings, and finally a promise to broadcast again on New Year's Eve. Amazed operators up and down the coast, as far away as Norfolk, Virginia, had heard the first radio program. Fessenden had demonstrated that the continuous-wave transmitter was to be the future of radio. (And, at the same time, he invented the radio variety program.) The invention of the continuous-wave transmitter marked a major breakthrough in radio broadcasting. It utilized a totally new method to generate radio signals—a major break from the traditional spark gap technology. As dramatic as the changeover from propellers to jet engines in airplanes or from a typewriter to a word processor in writing, the continuous-wave transmitter ushered in a new age for broadcasting.

Fessenden's achievements were not limited to transmitting radio signals. He also developed a radically different radio receiving set—called the heterodyne receiver—that was able to detect and decode the continuous-wave signals. Unfortunately for Fessenden, his new inventions did not bring him financial success, primarily because nobody was yet equipped to use them. Fessenden was literally ahead of his time. Ironically, to make a living he went back to working on refinements for the old spark gap transmitter, a technology Fessenden knew was at a dead end. It wasn't until many years later that his contributions were recognized.

The Invisible Empire: De Forest

Financial success was no stranger to Lee De Forest. Neither was financial ruin. Throughout his long career as an inventor he made and lost

Lee De Forest compares his first audion with a later, more advanced model. Frustrated by his patent problems, De Forest later experimented with various systems that would add sound to silent motion pictures.

four fortunes. He also narrowly escaped criminal conviction on fraud charges, was told by a judge to stop inventing and get "a common garden-variety type of job," and was engaged in almost continual legal battles with other inventors. De Forest is best known for his invention of the *audion,* a device that brought him great notoriety but not much happiness.

Before De Forest's invention, there were only two basic ways to receive radio signals. The first was to use a device called a "coherer." This was essentially the device that Marconi had used originally and although greatly improved by 1910, it still had drawbacks. It was hard to build and had trouble detecting weak signals. The second method was to use a crystal detector. In 1906 researchers discovered that certain substances such as carborundum, galena, and silicon possessed the ability to detect radio waves. Moving a tiny wire, called a cat's whisker, over a lump of these substances allowed the listener to hear the dots and dashes of wireless telegraphy or the faint sounds of wireless telephony (as radio broadcasts were referred to in those days). The big advantage of the crystal was that it was inexpensive; almost anybody could afford one. The biggest significance of the crystal was that it created an audience that was interested in radio. Even with advances in radio receivers, crystal sets continued to be popular through the 1920s. Crystal sets, however, had one big disadvantage—they could not amplify weak incoming signals. Consequently, the faint sounds detected by these early radio fans required them to wear earphones and to listen closely. If radio was ever to become a mass medium, something better was needed, something that would amplify incoming signals so that people could listen without wearing clumsy and annoying earphones.

De Forest set about investigating this problem. He experimented with a device called the Fleming valve, which looked like an ordinary light bulb. Originally discovered by Thomas Edison and further developed by John Fleming at Marconi's British company, the device had a thin wire and a metal plate encased within a glass bulb. When the wire was heated, electrical current would flow or not flow to the plate depending on the plate's charge. This allowed the device to translate alternating current, such as a radio wave, to direct current, which could be detected by a receiver. What De Forest did was to modify the Fleming valve by inserting a tiny wire grid between the wire and plate. This may not seem like much, but it had tremendous consequences. The grid acted as an amplifier, boosting weak signals until they were easily detected. Hooking two or three such devices together could amplify signals millions of times. De Forest realized the importance of his invention for radio's future. He wrote in his diary, "[I had] discovered an Invisible Empire of the Air."

Now that the Invisible Empire had been revealed, what next? De Forest used publicity stunts to create a market for this invention he dubbed the audion and to encourage investment in the newly formed De Forest Radio Telephone Company. In 1908 he broadcast a program of phonograph records from the Eifel Tower in Paris. The signals were picked up by French military stations within a radius of 25 miles. Two years later De Forest broadcast a live performance of the New York Metropolitan Opera Company featuring the world famous Enrico Caruso. An audience of about fifty people heard the opera clearly, despite problems from interfering signals. Although the audiences for these early experiments were small, they demonstrated that De Forest had anticipated the future of radio broadcasting. It was not to be a substitute for the telegraph or telephone; listening would not be a solitary experience. Radio was to be a public communications medium, bringing its audience entertainment, education, and information.

Although De Forest had correctly seen the future, it didn't help him much with the present. The Marconi Company sued him, claiming that his audion infringed on their patents to the Fleming valve. Angry stockholders in his new company complained of fraud and De Forest was arrested and made to stand trial. He was eventually exonerated but his company was ruined. Facing financial disaster, he sold the rights to his audion to AT&T for a modest $50,000. (AT&T needed the device to amplify signals in long-distance telephone calls.) He also got involved in another patent battle over the audion that dragged on for twenty years before eventually reaching the Supreme Court. De Forest continued to manufacture audions for sale to amateur radio hobbyists and occasionally broadcast programs of music and talk to help sales. During World War I he manufactured wireless sets for military use. After the war his attention turned to other pursuits and he left the broadcasting field.

The audion contributed to improvements in transmission as well as in reception. It was subsequently refined into the vacuum tube and formed the basis for all radio broadcasting and reception until the 1950s, when it was replaced by the transistor and subsequent advances in solid-state electronics.

To repeat, these three inventors were not the only ones who pioneered early radio. Many others could be mentioned. These three, however, seem to symbolize best the early evolution of radio. Marconi proved radio waves could carry messages; Fessenden developed the modern technique of signal generation; and De Forest refined both transmission and reception to make radio listening a public rather than solitary activity.

BOARDROOMS AND COURTROOMS

Now that radio had been successfully demonstrated, the next step was to refine it and make it commercially rewarding. Accordingly, the next phase of broadcasting's evolution is marked by the activities of corporations more than individuals. It's a tangled story, complicated by legal feuds, conflicting claims, politics, and war. It's also an important phase: decisions made during this period permanently shaped radio's future.

Legal Tangles

To begin, let's review the situation as of 1910. Radio was still thought of as a point-to-point communication device—much like the telephone and telegraph. Despite De Forest's demonstrations, broadcasting, as we know it today, did not exist. Radio's chief use was maritime communication: ship-to-ship and ship-to-shore messages. In fact, in 1910 the United States passed a law that made it mandatory for most passenger vessels to be equipped with a wireless set. In addition, as Marconi demonstrated, there was a potential market for transatlantic communication.

Marconi's company (British Marconi) and its U.S. subsidiary (American Marconi) dominated the field. The company controlled many important patents, like the Fleming valve patent that was to give De Forest so much grief. In addition, it controlled most shore facilities and its equipment was installed on the largest number of ships. The company also had a restrictive policy that prohibited it from exchanging messages with competing companies. Clearly the Marconi Company wanted to dominate the world's wireless market. (At least one Marconi employee had other visions, however. In 1916 a young David Sarnoff wrote what has become one of the most famous memos of all time in which he suggested that radio might become a mass medium designed to bring information and entertainment into the home. Unimpressed, the Marconi Company shelved his idea. Undaunted, Sarnoff later made his vision a reality—but we're getting ahead of the story.)

There were other companies interested in radio. General Electric (GE), for example, held patent rights to the alternators used by Fessenden in his experiments. GE had also made research breakthroughs in the refinement of the vacuum tube. Along with GE, AT&T was also involved. They worried that wireless might eventually compete with the conventional telephone and AT&T wanted to be ready. AT&T also knew that radio technology could be helpful in the telephone business. Yet another company was Westinghouse, which would eventually acquire patent rights both to Fessenden's heterodyne receiver and to further advances in vacuum tube technology.

The result of all this was that no company could produce a superior radio sending and receiving system without stepping on the toes of some other company. For example, take De Forest's audion. After a long legal battle, a U.S. District Court in New York ruled in 1916 that the De Forest invention actually infringed on the original Fleming valve design, whose rights were owned by American Marconi. (The court ruled that De Forest had in fact used Fleming's original wire and plate setup.) Therefore, De Forest couldn't sell the audion as a receiving device without the consent of American Marconi. The third element in the tube, the tiny wire grid, was properly protected by De Forest's patent. But De Forest had sold his rights to AT&T. As a result, each company had something that wouldn't work very well without the thing owned by the other company. To make life even more complicated, another young inventor, Edwin Armstrong, had developed and patented a feedback circuit that made the audion an even better receiver of radio waves. To manufacture a topnotch receiver now meant that his patented device must also be used. Radio might never untangle itself from this mass of legal red tape. Something drastic was needed to clear the way.

Radio Goes to War

Something drastic—World War I—was not long in coming. The crucial role that radio was to play in the conflict was recognized early by the British. Without radio the only way to communicate with allies overseas was by undersea cable, which could easily be cut by the enemy. Consequently, the British turned to radio as a kind of insurance against cable cutting. The Marconi Company provided this insurance by building a chain of high-power stations in Britain, the United States, and other places across the world. When war in Europe broke out in 1914, the British Navy took over the high-power Marconi stations in the interest of national defense.

In the United States the Navy also recognized the importance of radio. It operated three dozen coastal stations and all of its warships were equipped with wireless. When the United States actually entered the war, it came as no surprise that the government turned all radio operations, including commercial stations, over to the Navy. This had two important consequences for the future of radio.

First, the Navy was able to cut through the patent mess by the simple expedient of assuming full liability for patent infringement. In effect, this meant that the companies doing radio research in their labs were free to pool their discoveries to improve the radio system. And improve it they did. Better vacuum tubes, better transmitters, and better receivers were all quickly developed. When the war ended in 1918, the technology was vastly improved.

The second event of consequence came about when the Navy took over forty-five coastal commercial stations and eight high-power stations then owned by American Marconi. At the end of the war the Navy was reluctant to return them. The Navy was convinced that such an important function as international radio communication should not be entrusted to a company (American Marconi) controlled by a foreign power (Britain). In fact, the Navy wanted to make its wartime monopoly permanent. Bills were introduced in Congress to give the Navy exclusive control over radio's future. Obviously, some important decisions would have to be made.

Power Play: Birth of RCA

The first important decision was what to do with the Navy's plan. The goal of the U.S. government after the war was to prevent the British from reestablishing their worldwide domination of wireless. The Navy's proposed monopoly over radio would certainly achieve this, but public and government sentiment was running against the idea. Navy control sounded autocratic and militaristic and the United States had just fought a war against autocracy and militarism. Moreover, there was money to be made from radio; American Marconi had proved this. Commercial interests in the United States were opposed to any form of government restriction of free enterprise. The bill to give the Navy control over radio was tabled and never voted on. Unlike many of its allies in World War I, the United States chose not to place radio under government control.

This decision, however, did nothing to prevent a British monopoly. In fact, in 1919 the Marconi company was attempting to purchase from GE exclusive use of new high-powered alternators developed during the war. If the Marconi organization got this machinery, it would dominate the world market. To make matters worse, now that it was out of the picture, the Navy announced it would no longer be responsible for patent infringement suits. Almost immediately the legal status of the radio industry reverted back to the paralyzing patent tangles that existed before the war. A quick solution was needed.

And one was found—simple, direct, and to the point: buy out American Marconi and start a new company. The idea had its origins in the U.S. government, particularly in the Navy Department. If the Navy was not going to control radio, it certainly didn't want the Marconi company to control it either—far better to put the control in American hands.

The government went to GE with its proposal. GE was probably the only corporation with enough financial clout and radio experience to handle the deal. Owen Young, a GE vice president, headed the negotiations. The bargaining was tough. Meetings were held in Washington, D.C., London, and New York. Eventually American Marconi realized that it had little choice. The Navy still occupied its stations and might not give them back; the U.S. government was against the idea of a British company dominating radio. GE was unwilling to sell its new alternators to American Marconi. The management at Marconi bowed to the pressure. They would sell their American subsidiary. On October 17, 1919, a new company, the Radio Corporation of America (RCA), was chartered to take over the American Marconi Company.

The changeover went smoothly. Marconi transferred all of its assets and operations to RCA. Stockholders in American Marconi got stock in RCA and British Marconi got cash from GE. For workers at the Marconi offices in lower Manhattan not much seemed to change. One day they were working for American Marconi; the next day, for RCA.

This is a good place to pause and consider some essential facts about the new RCA:

1 Its orientation was external. RCA planned to make money by sending wireless telegraphy and telephony to Europe, Latin America, and Asia.

2 It expected its best customer to be the government, particularly the Navy.

3 RCA thought of radio as a point-to-point communication device, much like the telephone or telegraph.

4 Despite the earlier suggestions of Sarnoff, nobody at RCA thought radio would be a medium of *mass* communication, intended for public consumption.

Back to our chronology. The next step was to solve the patent puzzle. The creation of RCA was a big step in that direction since the new company entered into a *cross-licensing agreement* with GE, which allowed each company to use the other's discoveries. This agreement had the effect of pooling the patents once held by American Mar-

---

BACKGROUND

THE ORIGIN OF "BROADCASTING"

One of the many legacies the U.S. Navy left to modern radio was the word "broadcasting." Before World War I, messages sent by wireless were sent to a specific receiving station and the terms "radiotelegraphy" and "radiotelephony" were widely used. During the war, however, the Navy found it necessary to send messages simultaneously to several ships at sea. It would be dangerous for each ship to acknowledge reception of the message since the enemy had radio direction finders that could use the return transmission to target each of the ships. Thus at certain predetermined hours a Navy station would simply "broadcast" messages to all vessels without calling them individually or requiring them to acknowledge and log the message, hence today's usage of the term: a message sent to all with no assurance that it was received.

coni with those of GE. But other patent rights were yet to be secured. AT&T, for example, still held some of De Forest's original audion patents. Under commercial and government pressure, AT&T eventually signed a cross-licensing agreement in 1920, and yet another piece fell into place.

A big piece was still outstanding—Westinghouse—and bringing this company into the cross-licensing agreement would take some serious work. Westinghouse was GE's biggest rival. Westinghouse had bought the rights to several important patents and had plenty of money. The company also thought that it might make money selling radio sets to hobbyists—an idea that would have profound implications (see below). When it was obvious that GE would have the dominant role in RCA, Westinghouse started its own competing wireless company, the International Radio Telegraph Company. It quickly became apparent, however, that RCA had sewn up most of the potential market. Therefore, when Owen Young offered Westinghouse a large block of RCA stock in exchange for placing Westinghouse patents in the patent pool, Westinghouse accepted. As of 1921, then, the big shareholders in RCA were GE with about 30 percent, Westinghouse with about 20 percent, and AT&T with about 10 percent, and another 4 percent was held by the United Fruit Company. Now that they shared the patents, the major stockholders divided up the market. GE and Westinghouse would manufacture radio equipment and RCA would sell it. AT&T, through its Western Electric subsidiary, would manufacture transmitters.

Unfortunately, RCA's architects did not realize that the real future of radio was not in sending personal messages but in broadcasting, providing news, education, and entertainment to the general public. Because of this, all of the painstaking agreements worked out by Young and others would shortly fall apart as radio headed in a new direction. (It is interesting to note that as of 1992, GE was back in the network business, thanks to its purchase of RCA and NBC (National Broadcasting Company). Westinghouse still owned broadcasting stations and AT&T had aspirations to provide entertainment and information services to homes over cable and optical fiber. Of the original owners of RCA, only the United Fruit Company had lost interest in broadcasting.)

BROADCASTING'S BEGINNINGS

It's hard to say who invented broadcasting. As we have seen, the idea occurred to Fessenden, De Forest, and Sarnoff. But Fessenden and De Forest never established a true broadcasting station—one with a regular schedule of programs. Their efforts were experimental and designed to generate publicity. Sarnoff's memo gathered dust for many years.

There were, however, some early efforts at establishing stations. In San Jose, California, Charles Herrold regularly broadcast music and news as early as 1909. Other early stations were located in New York and Wisconsin. All of these early stations had one thing in common: they were experimental, lacking a firm basis for support. Big companies still shied away from the idea of broadcasting. GE, for example, thought it was a faddish and frivolous activity.

The war effectively shut down these experimental stations and the early attempts at broadcasting ceased. By 1920, however, the situation was different. A new industry was about to be born.

Radiomania

Radio burst on the scene in the 1920s and soon became a national craze. There were several cogent reasons for the incredible growth of this new medium:

1 An audience of enthusiastic hobbyists, thousands of them trained in radio communication during the war, was available and eager to start tinkering with their crystal sets.

2 Improvements during the war gave radio better reception and greater range.

3 Business realized that broadcasting might make money.

Its beginnings were modest. In April, 1920, an engineer for Westinghouse, Frank Conrad, began experimental broadcasts from his garage. Conrad became bored with the necessity of continually talking during his test transmissions and instead played phonograph records. Surprisingly, Conrad began to get letters from listeners asking him to play particular records or to conduct a program at a special time. Eventually, to satisfy his new au-

The control room and studio of KDKA, Pittsburgh, generally considered to be the oldest broadcasting station in the nation. Note that early broadcasters had to be careful about loose wires.

dience, Conrad began regular broadcasts. Of course, this meant he would need a lot of fresh records. He arranged to borrow them from a local record store and in return to mention the name of the store on the program. (It wasn't long before the store owner noticed that the records played by Conrad were outselling all others. This lesson, however, would be ignored by the record industry for the next twenty-five years.)

Conrad's operation got some local publicity and came to the attention of Horne's Department Store in Pittsburgh. Horne's placed ads in the local paper promoting Conrad's broadcasts and began to sell inexpensive radio receivers on which to pick them up. Horne's did a brisk business and Conrad's audience continued to grow. It wasn't long before Conrad's employer, Westinghouse, got involved. The company saw an opportunity to make money by selling radio receivers. To do this, however, Westinghouse would have to put Conrad's station on a permanent basis to give set buyers something dependable to listen to. Note that Westinghouse envisioned broadcasting only as a means to sell receivers; the company saw no direct way to profit from owning a station. The station simply stimulated demand for sets. In more modern times, it would be like computer manufacturers developing software to make their machines more attractive.

Acting quickly, Westinghouse moved Conrad's station out of the garage and onto the roof of the tallest building in the company's plant in Pittsburgh. The new station was licensed by the Department of Commerce and was given the call letters KDKA. (It's still in operation today.) As Westinghouse had hoped, the station stimulated demand for radios and the company was swamped with orders for its Aeriola, Jr., a small crystal set that sold for $25.

What made Conrad's broadcasts different from those of others who had preceded him? He demonstrated that there was a *market* for radio and Westinghouse moved quickly to take advantage of that market. Westinghouse also realized that if it worked in Pittsburgh, it could work in other cities. The company quickly opened other stations: WBZ in Springfield, Massachusetts; WJZ in Newark, New Jersey (for a studio WJZ first used a curtained-off section of the ladies' lounge at Westinghouse's Newark plant); and KYW in Chicago, which later moved to Philadelphia. The success of these ventures also helped the entry of Westinghouse into the patent pool and got it part ownership of RCA.

RCA soon dusted off David Sarnoff's memo and began development of a "radio music box." RCA also started station WDY in New Jersey, not far from New York City. GE and AT&T did not miss out either. GE opened a station at its factory in Schenectady, New York. The phone company opened WEAF in New York City. Others also opened stations. As 1922 began there were 28 stations actively broadcasting; six months later there were 378; by the end of the year, 570. Receiving sets were selling rapidly; by the end of the decade almost one-half of the homes in America had a working radio.

Stresses and Strains

The tremendous growth brought problems. For the listener, the biggest annoyance was interference. There was only a limited number of frequencies available that gave good reception and many stations were operating on them, causing great difficulty for the listener as they drowned each other out.

Among the big companies the problem was money. When RCA was founded, its prime purpose was to furnish transatlantic wireless. Selling equipment to amateurs was never expected to produce much revenue. In 1923, however, RCA made $3 million from wireless and $11 million from the sale of radio receivers. The original cross-licensing agreement among the major stockholders of RCA had not envisioned such a situation. Consequently, friction developed. AT&T claimed an exclusive right to sell time for profit on their radio facilities (more about this later) and to interconnect radio stations by wire for "chain" broadcasting while others had to use inferior telegraph lines for interconnection (more about this later as well). The others in the patent pool objected and claimed that AT&T, a telephone company, really had no right in the broadcasting business. (Some sixty years later many cable TV system operators were to complain that telephone companies shouldn't be in the cable TV business either.)

From 1922 to 1924 attempts were made to solve this industrial imbroglio. Feelings ran high. AT&T sold its RCA stock (but stayed on in the patent pool). In 1925, tired from all of the squabbling, the parties agreed to binding arbitration. GE, RCA, and Westinghouse came away the winners in this as nearly all of the issues went in their favor. Stung, AT&T announced that the arbitration decision must be interpreted to mean that the original agreements back in 1920 and 1921 were illegal and AT&T would have to withdraw from them. Another long and costly patent fight loomed on the horizon.

Finally, in 1926 an agreement was hammered out that ultimately solved the problem. AT&T would have a monopoly over wire interconnections of stations. In return the phone company agreed not to reenter the broadcasting field. AT&T's flagship station, WEAF, was sold to RCA for $1 million. The phone company's experiment in broadcasting was over. RCA would eventually but not permanently dominate the field.

FAST TIMES

Things happened quickly for radio in the 1920s. In eight years, from 1920 to 1927, radio went from a fad to a major industry and a major social force. We have already noted how big business got involved with early radio. The three other major developments of this period that helped to shape modern radio were the development of radio advertising, the beginning of radio networks, and the evolution of radio regulation.

Advertising

When broadcasting first started, nobody thought too much about how it was supposed to make money. Most of the early broadcasters were radio and electronics manufacturers. For these companies, radio was simply a device to help sell their products. Other businesses that owned large numbers of early radio stations were newspaper publishers and department stores. For these companies, radio was a promotional device; it helped sell more newspapers or attracted people to the store. Nobody envisioned that it might be possible for a radio station to make money.

It wasn't long, however, before this became a real problem. Early radio equipment didn't have to be precise; amateur-built transmitters would serve adequately. Talent wasn't a problem either. The audience was so taken with the novelty of radio that they would listen to almost anything. Friends, relatives, and neighbors supplied the music and talk heard on many stations. Other stations simply played old phonograph records.

In time, however, costs started building up. To be successful, stations required reliable, professionally built equipment that would produce a strong, reliable signal. Special studios were needed. Technicians were necessary to operate and maintain the equipment. Audiences quickly grew more sophisticated and demanded professional talent. In turn professional performers were no longer satisfied with performing on radio for free and demanded payment. Record manufacturers wanted royalty fees for playing their discs over the air. The rising costs forced many stations off the air. Those that remained scurried to find some way to produce revenue.

There was no shortage of suggestions. David Sarnoff, now an executive at RCA, proposed that a nonprofit corporation be established to conduct broadcasting as a public service. Under this plan broadcasting stations would resemble libraries or museums. Sarnoff asked for philanthropists to contribute large sums of money as an endowment to get things going. Nobody answered his call. Some stations appealed directly to their listeners for donations (a technique still used today by many public broadcasting stations), but they were not successful. Overseas, England solved the problem by instituting a yearly license fee on radio receivers. This plan, however, was given little chance of success in the United States.

It was the phone company that found the answer: a phone booth of the air. Conceiving broadcasting in telephone terms, AT&T proposed a system whereby anyone who had a message for the world would come into their station, pay money, address the world, and leave (just as people did in a phone booth when they made a toll call to a single person). AT&T called this new arrangement "toll broadcasting" and WEAF in New York was offered for this purpose. It wasn't long before customers started to line up outside its radio phone booth. The first was the Queensboro Corporation, which paid $100 for the right to make a ten-minute talk touting the advantages of country living. Not surprisingly, the company had a real estate development in the country where apartments were available. The company repeated this first radio commercial for the next five days and was pleased at the response. Other clients, the Tidewater Oil Company and American Express, quickly followed suit and also broadcast commercial announcements.

Although it seems odd today, this early experiment in commercial broadcasting was resented by many listeners. *Radio Broadcast,* the leading radio hobby magazine, soundly condemned it. There was even talk of a bill in Congress to prohibit all advertising over radio. Despite the criticism, AT&T held steady. By March 1923 WEAF had twenty-five advertisers, including Macy's, Colgate, and the Metropolitan Life Insurance Company. AT&T had demonstrated a system of producing revenue at the station level that was quick and efficient and required little additional expense by the broadcaster.

It took a while for the idea to catch on, primarily because stations operated by RCA were not allowed to broadcast commercials since the original agreement bringing AT&T into the RCA ownership group gave AT&T exclusive right to charge advertising tolls. (As we have seen, this problem eventually led to the exit of AT&T from the ownership of RCA.) Other stations, however, soon adopted the practice. At the end of 1924 enough stations had followed WEAF so that it was obvious that the public did not object to radio ads. If anything, the audience liked the improved programs that came with advertising. Nonetheless, by 1925 there were still hundreds of stations that refused to accept advertising. The beginning of broadcast networks, however, decided the issue (see below). Radio was to be supported by advertising.

Early advertising was subtle and unobtrusive. Many times it was limited to the name of the company that sponsored the program. To illustrate, in the Queensboro Corporation's ten-minute talk the name of the company was mentioned only once. The next step was to link the name of the sponsor to the entertainers: listeners were treated to the Lucky Strike (cigarettes) Orchestra, the Ipana (toothpaste) Troubadours, and the A&P (grocery store) Gypsies. The next development incorporated the name of the sponsor into the program title: "The Eveready Hour" (batteries), the "Palmolive Hour" (detergent), and "The Wrigley Review" (gum). Gradually, however, the skill of advertising copywriters became apparent and such innovations as the singing commercial and the dramatic commercial were born. By 1929 advertisers were spending about $20 million on network advertising and the new medium was on solid financial ground.

Radio Networks

There were three main reasons for the development of chain or network broadcasting in the mid-1920s. The first stemmed from the broadcasters themselves and was primarily economic. It was less expensive for one station to produce a program and to have it broadcast simultaneously on three or four stations than it was for each station to produce its own program. The second reason came from the audience. The listeners of local stations in rural areas, far from the talent of New York, Chicago, or Hollywood, wanted better programs. Networking allowed big-name talent to be heard over small-town stations. A third reason was the desire of advertisers to increase the range of their programs beyond the receivers of local stations and thus multiply potential customers. Given the pressure from these three sources, the development of networks was inevitable.

Once again, it was the phone company that led the way. Using long-distance telephone lines, in 1922 AT&T linked together WEAF in New York and WNAC in Boston for a program of saxophone music. By the end of 1923 six stations were hooked up. By the next year a coast-to-coast chain of twenty-six stations, linked together by a web of phone company wires, was in operation. GE, Westinghouse, and RCA, unable to use phone company lines, tried to improvise their own networks over Western Union telegraph lines, but the results were less than satisfactory. Nonetheless, RCA was able to link up WJZ (Newark), WGY (Schenectady), and WRC (Washington, D.C.). The desire of the phone company to leave the broadcasting arena solved this problem. As we have seen, AT&T sold their stations to RCA in return for the interconnection rights used in network broadcasting.

At RCA a new company was formed to separate the parent corporation from the broadcasting operation. The National Broadcasting Company (NBC) was to oversee broadcasting on two networks. The "Red" network (the networks were apparently named after the color of the crayons used by RCA's executives to outline them on a map) consisted of the former AT&T New York flagship station, WEAF, WSB (Atlanta), WMC (Memphis), WHAS (Louisville), and others. It would be the more commercially successful net-

November 15, 1926. The first radio program is transmitted on the NBC radio network. Sixty-one years later, NBC sold its network radio operation to Westwood One.

work and would carry most of NBC's popular, mass appeal programming. The other network, NBC "Blue," was made up of stations originally owned by RCA, Westinghouse, and GE and was anchored by WJZ (now moved to New York). The blue network generally carried more "high-brow" programs.

In 1927 NBC had a competitor when United Independent Broadcasters sputtered into operation after a few false starts. This company had little success but it did get one lucky break. Industry insiders were predicting that RCA would take over a major recording firm, the Victor Talking Machine Company, then in trouble because of the competition from radio (see below). Victor's biggest competitor, the Columbia Phonograph Corporation, trying to counter Victor's move into radio, merged with United Independent Broadcasters and changed the network's name to the Columbia Phonograph Broadcasting System, hoping for greater name recognition. The record company didn't do well, though. Losing money quickly, the record company backed out of the deal and took back part of its name, and the network became simply the Columbia Broadcasting System (CBS). Whatever its name, the fledgling network's financial fortunes kept dwindling.

Salvation appeared in the form of 26-year-old William S. Paley, vice president of the Congress Cigar Company. Impressed by the effectiveness of radio advertising, Paley bought controlling interest in CBS and set out to make it profitable. Paley acquired a station in New York, strengthened the network's credit, and by 1929 had it turning a profit. There would be no monopoly in network broadcasting.

Local stations scrambled to join the networks. By the end of 1927 NBC had forty-eight stations in its Red and Blue networks; by 1933 it had eighty-eight. CBS started with sixteen stations and had ninety-one by 1933. In 1934 another network, the Mutual Broadcasting System (MBS), was started. The radio networks would be the controlling force in broadcasting for the next twenty years.

Rules

Attempts to regulate the new medium can be traced back to 1903 when a series of international conferences were called to discuss the problem of how to deal with wireless communication. The result of these efforts for the United States was the Wireless Ship Act of 1910, requiring certain passenger vessels to carry wireless sets. Two years later the pressure for more regulation increased when amateur wireless operators were interfering with official Navy communications. In the midst of this, the *Titanic* struck an iceberg and sank. Hundreds were saved because of wireless distress signals, but the interference caused by the many operators who went on the air after knowledge of the disaster spread hampered rescue operations. As a result, the public recognized the need to develop legal guidelines for this new medium. The Radio Act of 1912 required sending stations to be licensed by the Secretary of Commerce, who could assign wavelengths and time limits. Ship, amateur, and government transmissions were to be assigned separate places in the spectrum.

This law worked—for a while. The big problem was that the 1912 act envisioned radio as telegraphy or telephony; it did not anticipate broadcasting. As more stations went on the air, the more apparent the limitations of the act became. To illustrate, the Department of Commerce assigned only two frequencies for all of broadcasting. Every station was supposed to use one or the other. In many communities demand quickly exceeded the available frequencies and the airwaves became crowded with static. To make matters more complicated, the courts said that a license could not be denied to a station simply on the grounds of increased interference. The licensing function was to be clerical not regulatory. Frustrated, Secretary of Commerce Herbert Hoover (later to be president) convened a series of radio conferences among broadcasters, civilian radio experts, and government officials to suggest some solutions. In all, four conferences were convened between 1922 and 1925. Suggestions were made concerning technical standards, allocation of frequencies, copyrights, advertising, licensing, and censorship. Several bills outlining regulatory procedures were submitted to Congress as a result of these meetings. By the end of 1925, however, none had gotten anywhere. Of course, while all of this talking was going on the interference problem kept getting worse as more stations went on the air.

Another approach—self-regulation—also got nowhere. Hoover repeatedly urged broadcasters to avoid government interference by voluntarily

PROFILE

WILLIAM PALEY AND CBS

William Paley, longtime leader of CBS, came from an affluent family and, unlike his counterpart David Sarnoff at RCA, did not have to work his way up from the bottom of the business ladder. After his college graduation, Paley became an executive in his family's Congress Cigar Company and after purchasing the struggling radio network Paley assumed the presidency of CBS.

From the beginning Paley recognized the importance of programming. He also believed that stars made successful programs. While on a ship bound for Europe, Paley heard a Bing Crosby record and immediately telegraphed CBS to sign him up. CBS eventually did, and Crosby went on to become one of the biggest of the radio "crooners." Paley was also instrumental in making the Mills Brothers and Kate Smith into major radio stars.

Paley's greatest programming coup, however, came in 1948 when he successfully lured a number of big stars from NBC to CBS—Jack Benny, Edgar Bergen, Red Skelton, and the cast of "Amos 'n' Andy." Paley shrewdly allowed these performers to incorporate themselves while at CBS, which gave the stars a significant tax break. The "talent raid," as it came to be known, gave CBS a lead over NBC in the programming race.

Paley's achievements extended beyond entertainment programming. He hired Ed Klauber from the *New York Times*, who assembled a cast of reporters and correspondents that would begin a tradition of outstanding broadcast journalism at CBS. Although the two would eventually clash and part ways, Paley persuaded Edward R. Murrow, the famous war correspondent and commentator, to take an executive position at CBS and to sit on the board of directors.

Defying the conventional wisdom, which held that radio had killed the recording industry, Paley purchased the Columbia Records Company during the Depression. The record division, named CBS Records, became a reliable contributor to company profits until it was sold to Sony in 1987. On the technical side, Paley encouraged CBS researchers to develop the LP record.

Of course, not everything went right for Paley. CBS got involved in a money-losing deal to manufacture and distribute TV sets. In the color TV area, CBS was usually playing catch-up to RCA.

---

working together. A first step toward self-regulation occurred in 1923 with the formation of the National Association of Broadcasters (NAB). From the start, however, NAB was more worried about external threats than internal cooperation. It was formed to resist pressure from the American Society of Composers, Authors and Publishers (ASCAP) for royalty payments on broadcast music. ASCAP protected the rights of creative people in music composition and publishing. At the NAB's first convention in New York cooperation was discussed, but the fight with ASCAP was the main topic (NAB and ASCAP would tangle again; see below).

By 1926 the interference problem had worsened and, as was obvious to all of those involved, some sort of federal control was needed if radio was to avoid being suffocated by its own growth. The radio broadcasters themselves requested Congress to provide some kind of legislative solution. The federal legislation finally came about the next year—the Radio Act of 1927. The key assumptions underlying the legislation as ultimately written were:

1. The radio spectrum was a national resource. Individuals could not own frequencies but they could be licensed to use them.
2. Licensees would have to operate in "the public interest."
3. Government censorship was forbidden.
4. Radio service was to be equitably distributed among all of the states.

To enforce the new law, a five-member Federal Radio Commission (FRC) was established. Its first job was to straighten out the chaos caused by interference. To begin, it abolished all

Paley reemerged in the 1980s when CBS got into financial hot water. Octogenarian Paley temporarily came out of retirement to help the company through its difficulties. Paley was interested in the company that he helped found until his death in 1990 at the age of 89.

In January of 1929, William Paley, age 27, makes the connection to inaugurate the beginning of coast-to-coast broadcasting over CBS.

existing radio licenses, forcing all licensees to reapply. Second, it defined the standard AM broadcasting band at 550–1500 kilocycles. (We now call the kilocycle the kilohertz in honor of Heinrich Hertz, who first detected radio waves.) It also developed a plan to classify frequencies into clear, regional, and local channels and specified power, frequency, and times of operation to stations applying for new licenses. Within five years the FRC had solved the interference problem and had laid the groundwork for an orderly system of frequency sharing that would obtain maximum benefit for the public.

In 1934, in an attempt by President Franklin Roosevelt to streamline government administration by bringing telegraph, telephone, and broadcasting communication under the same jurisdiction, the Communication Act of 1934 was passed. This replaced the five-member FRC with a seven-member Federal Communications Commission (FCC). The basic philosophy of the 1927 act, however, was not changed.

Thus in one tumultuous decade, the roaring twenties, a powerful new communications industry was formed. By 1930 radio had been firmly capitalized by big business, had adopted advertising as a profitable means of support, had developed local stations and networks as its distribution system, and had a firm regulatory base. The future looked bright . . . and it was.

RADIO DAYS, RADIO NIGHTS

The years from 1930 to 1948 can be termed the radio years. The new medium grew at a phenomenal rate, became an integral part of American life, developed new forms of entertainment and news programs, and ran into a few problems along the way.

PROFILE

DAVID SARNOFF AND THE *TITANIC*

One of the prominent names in the history of broadcasting is David Sarnoff. Young Sarnoff wrote the famous "Music Box Memo," which correctly predicted the ultimate role of radio. He went on to become chief executive officer of RCA and a leading proponent of TV development. The story of Sarnoff's transition from an unknown youngster to a national figure is almost legendary. As told by his official biographer (who was also his cousin), Sarnoff was on duty as a wireless operator for Wanamaker's Department Store in New York when, on the night of April 14, 1912, the luxury liner *Titanic* struck an iceberg in the North Atlantic and sank with the loss of 1517 lives. Sarnoff allegedly received the distress message, alerted the press, and for the next three days and nights, with no food and little sleep, stayed at his post, the only link between the tragedy at sea and the mainland. Because of his heroic vigil, the young Sarnoff was catapulted into the public limelight.

As it turns out, recent research suggests another story. In the first place, the wireless station at Wanamaker's was a short-range station that probably couldn't receive signals directly from the *Titanic*. More likely, Sarnoff was relaying signals originally received by the high-power Marconi station in Newfoundland, the land wireless station closest to the disaster. In fact, news coverage of the event credits the Newfoundland station with handling most of the crucial distress traffic from the *Titanic* and the ships that sped to the accident site. It also seems that Sarnoff was not the only one involved in the event. A New York newspaper reported that two other operators were with Sarnoff during the disaster, making it unlikely that the young man spent seventy-two hours straight over his wireless set. In short, Sarnoff's station was probably one of many Marconi stations that received news of the tragedy. It appears that as Sarnoff became more famous, the passage of time and a certain amount of poetic license by both Sarnoff and his biographer added a little more drama to the incident.

Nonetheless, even though his conduct during the *Titanic* episode may not have been as heroic as once thought, Sarnoff's accomplishments still speak for themselves.

A young David Sarnoff attends to his wireless key in 1908. Four years later his performance as a wireless operator during the sinking of the *Titanic* would bring Sarnoff to prominence.

Growth

Statistics are dry but they are the best way to document the skyrocketing growth of radio. Table 2-1 helps tell the story—from 618 radio stations in 1930 to 765 in 1940 and 2194 in 1948; from 131 network affiliates in 1930 to 1104 in 1948; from $40 million spent on radio advertising in 1930 to $506 million in 1948; from 6000 employees in 1930 to 52,000 in 1948. Keep in mind that this growth occurred despite a worldwide depression and another world war. In 1930 some 46 percent of all households had radios; in 1948 this was 94 percent. Less than 1 percent of automobiles had radios in 1930; some 35 percent had them by 1948. Whatever index is chosen, the conclusion is the same: radio grew at a rapid rate.

With growth came problems. One of the first had to do with questionable radio content, particularly the swindlers, quacks, and cranks who had joined legitimate broadcasters in starting their own stations. Two of the more infamous were John Romulus Brinkley, or Dr. Brinkley, as he liked to be called, who used his station in Kansas to promote his own patent medicines and a questionable sexual rejuvenation operation (see Chapter 17), and Norman "TNT" Baker, who used KTNT in Iowa to sell his homemade cancer cure. With backing from the federal courts, the FRC was able to shut down both operations in the early 1930s on the grounds that these stations did not further the public interest.

A second problem cropped up in 1933 when newspapers' advertising revenues began to be affected by competition with radio. The newspapers countered by persuading the wire services to stop providing news to radio stations and by refusing to run radio program listings in their entertainment columns. The young networks, afraid to offend the powerful newspaper industry, agreed to a restrictive plan dictated by the publishers, called the "Biltmore Program," which limited the networks to only two daily newscasts of five minutes each, one after 9:30 A.M. and one after 9:00 P.M. (to avoid competition with the newspapers' early and late editions). Under no circumstances were newscasts to be sponsored. The terms were so repressive that the truce didn't last long. Independent and local affiliated stations started their own newsgathering operations and the wire services soon realized that they were fighting a losing battle. Within two years the Biltmore agreements were effectively forgotten and radio news was free to grow.

Perhaps inspired by its victory in this so-called press–radio war, the new medium also flexed its muscle against an old nemesis—ASCAP. In 1937 ASCAP proposed a 70 percent jump in its music licensing fees. Broadcasters countered by forming their own licensing company—Broadcast Music Incorporated (BMI)—which attracted a large number of new songwriters who were dissatisfied with their treatment by ASCAP. After a long and costly battle with BMI and the broadcasters ASCAP agreed to a compromise. BMI, however, would remain as a permanent competitor.

Another problem concerned FM (frequency modulation) broadcasting. Perfected by Edwin Armstrong, FM transmission was first publicly demonstrated in 1933. FM had two big advantages over AM (amplitude modulation): FM was less prone to static and had better sound reproduction quality. Armstrong took his invention to his friend David Sarnoff, now chief executive at RCA, who had supported the inventor's work in the past. Unfortunately for Armstrong, the time was not right for FM. Conventional AM radio was doing fine; it didn't need additional competition. In addition, RCA had made a major investment in AM and was more interested in getting a return on this money than in developing a new radio service. Finally, Sarnoff was committed to another developing technology, television, which he saw as more important. Consequently, RCA would not back the new radio technology. Dismayed, Armstrong built his own FM station but World War II intervened and halted its development. Nonetheless, at the end of the war there were about fifty FM stations in operation. The FCC dealt it another blow, however, by moving

TABLE 2-1 Growth of Radio

Year	Number of Stations	Percentage of Homes with Radio	Number of Employees
1930	618	46	6,000
1950	2,867	95	52,000
1970	6,889	99	71,000
1992	11,164	99	150,000

FM to a different spot in the radio spectrum, thus rendering obsolete about 400,000 FM receiving sets. Although its ultimate future would prove to be bright, FM struggled along for the next twenty years.

The last problems concerned NBC's two networks and the FCC. After a lengthy study of monopolistic tendencies in network broadcasting, the FCC ruled that NBC had to divest itself of one of its two networks. NBC appealed the ruling all the way to the Supreme Court, which upheld the FCC in 1943. Forced to sell, NBC searched for a buyer. It found one in Edward Noble, the owner of New York station WMCA, who had made his money selling Life Savers candy. Noble paid RCA $8 million for the old NBC Blue network and changed its name to the American Broadcasting Company (ABC).

Radio Power

Radio by now had become part of the social fabric. It was the number one source of home entertainment and the stars of early radio programs were familiar to the members of virtually every household in America. Radio news reports had a sense of excitement and immediacy about them that set them off from items in the newspaper. Many audience members turned to radio commentators for authoritative and lucid explanations of the complicated issues of the day. People trusted and depended on radio.

The social power of radio during this time was made obvious by President Franklin Roosevelt. Realizing the political potential of radio, he instituted a series of "fireside chats," informal talks that allowed him to discuss the workings of the government with his audience. These programs were well received and credited with helping Roosevelt move his legislative program through Congress.

Radio's power was demonstrated in a more striking way on October 30, 1938. Orson Welles and his Mercury Theatre of the Air put on a radio adaptation of *The War of the Worlds,* in which Martians invaded the United States. The program used a quasi-newscast style in which bulletins and eyewitness accounts interrupted normal programming. The Martians, using death rays and poison gas, advanced through New Jersey and destroyed New York City. People who tuned in late missed the broadcast's opening announcement that it was a dramatic program. At least a million people panicked, many of them fleeing in their cars to escape the Martian invasion. It took a day or so before everything returned to normal. The event illustrated the credence people placed in the new medium.

During World War II popular radio star Kate Smith conducted a marathon sale of war bonds. On the air for eighteen straight hours, she was responsible for $39 million in pledges to buy bonds. Her feat established the power of mass media for mass persuasion.

Fallout

Radio's ascendancy had an impact on other media as well. Radio's share of the advertising dollar went from 2 percent in 1928 to 10 percent in 1933, with a lot of this money coming at the expense of the newspaper. Radio's immediacy in news coverage effectively killed the newspaper's "extra" edition. A 1939 survey in *Fortune* indicated that 70 percent of the respondents relied on the radio for news and 58 percent thought that radio news was more accurate than that in the newspaper. Magazines also lost some revenue to the new medium.

The one medium most affected by radio was sound recording. The phonograph, invented by Thomas Edison in 1877, developed slowly at first, but by 1900 record players were found in the parlors of many American homes. Even the earliest broadcasters demonstrated that the future of the two media would be meshed when they turned to phonograph records as the content of much of early broadcasting. Despite this apparent fact, during their early years radio and sound recording tried their best to ignore one another.

First, it was the recording industry's turn. When broadcasting became popular in the early 1920s, the phonograph and recording companies refused to recognize it as a threat. The record companies even ignored technical advances in electrical recording because they came from radio technology. For many years the recording industry persisted in using mechanical means of reproduction that dated back to Edison. Employees at the Victor Company, the leading phonograph manufacturer, were forbidden to own radios.

The sound quality of the early radio improved quickly. Before long it was better than that of records. Music on the radio didn't sound tinny, muffled, or scratchy; it sounded like real music. Radio quickly became the preferred medium for music. By 1924 record sales had dropped 50 percent from the previous year. Unable to ignore it any longer, recording companies adopted the better quality electrical recording method in 1927 and began to manufacture combination radio-phonographs. These changes helped a little, but by the late 1920s record companies were looking for financial security. In 1929 Victor merged with RCA, with the broadcasting arm of the company clearly dominant. CBS also reacquired Columbia Records. Radio had triumphed.

Now it was radio's turn to ignore the potential of records. As the Depression of the early 1930s worsened, the recording industry nearly died. In 1927 about 104 million records were sold; five years later only 6 million were sold. RCA and CBS were content to let sound recording struggle since their radio operations were doing a good job withstanding the effects of the Depression. Throughout this period radio was thought of as the medium that provided live music for free. The phonograph provided "canned" music that cost extra. Radio helped nourish this image by the networks' refusal to play records and by carrying many live music programs. In turn, during the 1930s, the record industry put a warning label on most of its records: "Not Licensed for Radio Broadcast." The two media did not yet realize that they could help each other.

The record industry survived the Depression for two reasons. First, Decca Records drastically cut prices on its products and began a large-scale advertising campaign to market them. Other companies copied Decca's successful tactics. Second, there was the jukebox, a coin-operated phonograph that was found in most drug stores, restaurants, and bars in the 1930s (the word "juke" may be derived from the African "jook," which means disorderly). Interestingly, record company executives noticed that songs that got exposure on the jukeboxes were also those that sold well. The idea of promoting records through radio airplay, however, was still not accepted.

In any case, by 1941 the record industry and the radio industry seemed to be going their separate ways and both were doing well. The coming of World War II helped radio prosper but hurt sound recording. Shellac, an essential ingredient in records, was declared a material vital to national defense and was no longer available for record making. If that wasn't enough, a musicians' strike from 1942 to 1944 virtually brought recording to a standstill. The industry wouldn't recover until the war was over.

Programs: Music, News, and Entertainment

As we have seen, programs in the early days of radio were amateurish and undependable. With the advent of networks and commercial sponsorship, however, programming achieved regularity and professionalism. The major types of early radio programs were (1) music and variety, (2) comedy and drama, and (3) news and talk.

In the early 1920s music, almost all of it live from radio studios, was a broadcasting mainstay. In 1925 it was estimated that about 70 percent of radio programming on urban stations consisted of some form of music, much of it classical. Eventually music outgrew the studio, and stations started broadcasting "remotes" from local hotels, featuring well-known dance bands. Bandleaders Paul Whiteman, Vincent Lopez, Guy Lombardo, and Lawrence Welk all made their radio debuts during the 1920s.

Vaudeville performers who were forced out of work by the coming of motion pictures flocked happily to radio, bringing with them their routines, which mixed music and comedy; thus the variety program was born. "Roxy's Gang" on WEAF started in 1923 and lasted for fifteen years. Eddie Cantor, Al Jolson, the comedy team of Weber and Fields, and the Marx Brothers took their turns at the microphone as the Depression began.

Classical and popular music continued to be popular through the early 1930s, accounting for about 65 percent of all programming time in 1932. Variety shows, with their emphasis on light, escapist humor, also grew in popularity. Rudy Vallee, who pioneered a singing style known as "crooning," started a variety show in 1929 which would be popular for a decade. Country-and-western variety shows, like the "National Barn Dance," became national favorites, thanks to the networks.

Just before World War II the importance of music as a program source was dropping, but it still accounted for about one-half of all broadcast time. Classical music, however, had declined in importance and was replaced by dance bands that were popular in the big-band era: Benny Good-

Radio stars of the 1930s often crossed over and appeared in motion pictures. Here the comedy team of George Burns and Gracie Allen appear with Jack Okie in *The Big Broadcast of 1936*.

man, Ozzie Nelson, Tommy Dorsey. Many local stations also began to rely more on recorded music despite the problems that came with it.

On the variety front, programming reached the extremes. The amateur hour, featuring unknown performers, became popular. At the same time, comedy-variety shows featuring big-name professionals such as Bob Hope, Jack Benny, Fred Allen, and Jimmy Durante were also crowd pleasers.

After World War II, music's share of the programming day had dropped to 40 percent, with classical music nearly extinct. Variety shows also became less popular as many of the big stars of the day made the transition to a new competitor —television. The role of music and variety programming on radio was about to undergo a dramatic change.

Experiments in radio drama began back in 1923 when GE's station WGY experimented with plays that were adapted especially for the new medium. The networks, however, were the driving force behind drama. During the evening hours, known as prime time, the networks developed a succession of weekly programs that kept audiences glued to their sets. There was mystery: "Charlie Chan," "Sherlock Holmes," "Lights Out," "The Shadow" (at one time starring Orson Welles), among others. There was romance: "First Nighter," "Grand Hotel," and "Grand Central Station." There were dramatic anthologies: "Lux Radio Theater" and "Hollywood Star Theater."

Before the dinner hour, there were fifteen-minute mystery and adventure shows for children, many of them based on comic strip characters: Little Orphan Annie, Dick Tracy, the Green Hornet. During the morning and afternoon, however, a new program form dominated—the radio soap opera. Probably no other form of radio entertainment was more popular or more scorned than the "soaps." Usually set in a small town and focusing on problems that involved marriage, courtship, family, and occupation, the soaps proved to be remarkably durable (they lasted on radio until 1960). Some of the more memorable: "Ma Perkins"—could a kindly old lady help solve everybody's problems? "Helen Trent"—could a woman over the ripe old age of 35 find romance? "Backstage Wife"—could a simple girl find happiness with a Broadway star?

PROFILE: IRNA PHILLIPS

Irna Phillips was the leading creator and writer of daytime soap operas on radio and TV. She is referred to as "Queen of the Soap Opera" because of her prodigious output of daily serials.

Phillips was born in Chicago, the youngest of ten children. Her family owned a small grocery store and lived on the floor above it. As a child Phillips dreamed of becoming an actress, but she graduated from the University of Illinois in 1923 with a degree in education. She went on to get a Master's degree from the University of Wisconsin and then taught drama and speech for several years at a junior college. In 1930 she was asked to write and perform in a "family drama" for radio station WGN in Chicago where she had worked without pay during her vacations. The show, "Painted Dreams," was one of radio's first soap operas. All six characters and sound effects were performed by Phillips and another actress. Phillips then launched two other radio soaps, "The Road of Life" and "The Guiding Light," the longest running soap in broadcast history. Phillips changed the focus of soap operas from humble, simple characters to professional people, which began the proliferation of soap-opera doctors, nurses, and lawyers.

When soap operas moved to TV, Phillips also made the transition. In twenty years her TV credits included "The Guiding Light" (1952), "The Brighter Day" (1954), "The Road of Life" (1954), "As the World Turns" (1956), "Another World" (1964), "Days of Our Lives" (1965), and "Love Is a Many-Spendour'd Thing" (1967).

Phillips continued to write for "As the World Turns" until a year before her death at 72. In 1991, *Broadcasting* magazine named her a charter member of its Broadcasting Hall of Fame.

The first big comedy hits on radio capitalized on ethnic humor. After a few years on WMAQ, Chicago, "Amos 'n' Andy," featuring two whites who imitated two blacks, went on the NBC network in 1929. It became radio's first classic. Although "Amos 'n' Andy" would be considered racist today, the program almost stopped the nation from 7:00 P.M. to 7:15. A second comedy hit was "The Goldbergs," about a Jewish family on the Lower East Side of Manhattan. Written and produced by its star, Gertrude Berg, "The Goldbergs" was on radio from 1929 to 1946. Other popular comedy shows were "Burns and Allen," "Fibber McGee and Molly," and "The Chase and Sandborn Hour," starring Edgar Bergen and Charlie McCarthy. (Bergen was a ventriloquist who was highly successful on radio but had trouble making the jump to movies and TV primarily because his lips moved when he spoke for his dummies.)

The potential of radio as a news medium was apparent from the start. KDKA's first broadcast covered the Harding–Cox presidential election in 1920. Radio news grew slowly, however, as many people felt the new medium was better suited for entertainment. Some newspaper-owned stations broadcast short news stories, but these were designed primarily to promote readership of the paper. A few radio commentators, who did news analysis, also made their debut around 1930. Radio stations did cover special events such as the Scopes trial in Tennessee, which pitted famous lawyer Clarence Darrow against politician and famed orator William Jennings Bryant in a court case over the theory of evolution, and it covered the 1932 kidnapping of the son of aviator Charles Lindbergh. Regular network newscasts, however, did not start until 1930 when Lowell Thomas anchored a fifteen-minute program on NBC. The press–radio war and the Biltmore agreements further hampered the growth of early radio news.

In the late 1930s radio news was given a boost by events in Europe. As that continent moved closer to war, radio reporters and commentators became more important to U.S. audiences as they described and analyzed unfolding events. H. V. Kaltenborn reported on the 1936 Spanish Civil War from a haystack, unruffled as bullets thudded into the hay above him. Kaltenborn later provided marathon coverge of the Munich crisis in which war with Hitler's Germany was temporarily postponed. As war approached, an im-

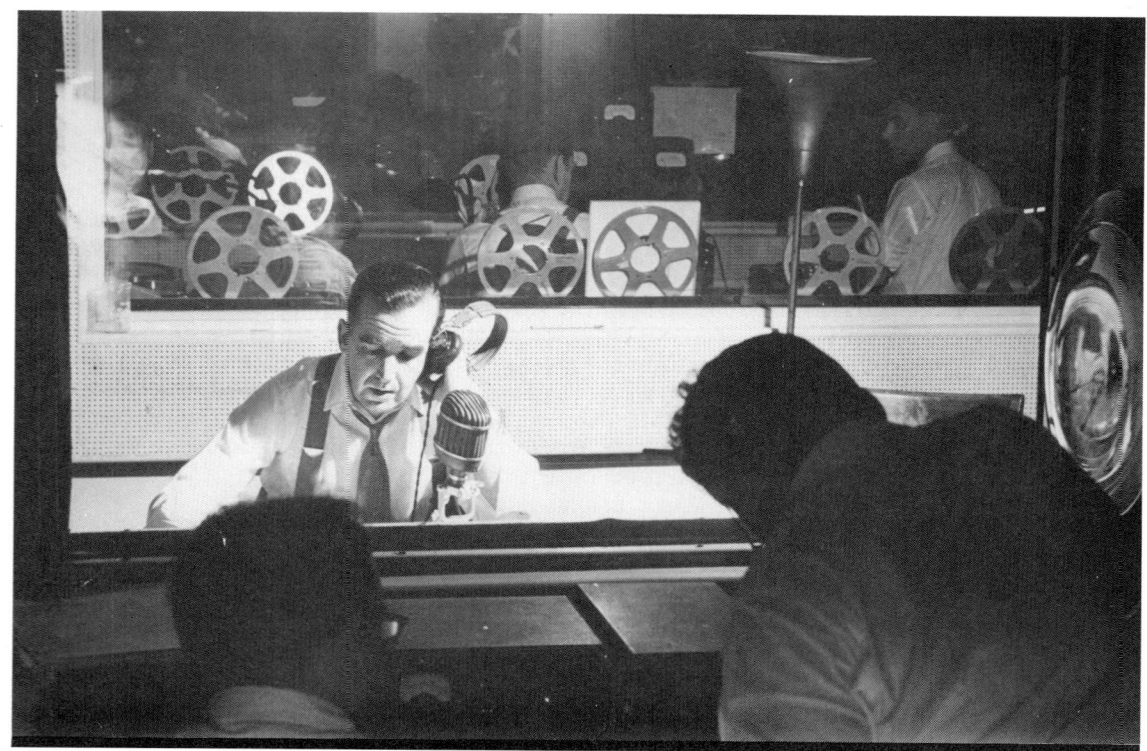

The reports of Edward R. Murrow during World War II established him as one of the greatest radio commentators. He later became one of the pioneers of television journalism.

pressive array of radio correspondents was regularly reporting from Europe. CBS, for example, had Edward R. Murrow in London, Eric Sevareid in Paris, and William L. Shirer in Berlin.

World War II brought out the best in radio journalism. In the United States millions of people huddled around their radios to hear timely reports on the war's progress. The amount of time devoted to news on the four radio networks more than doubled from 1940 to 1945. There were some notable individual achievements. Murrow offered his listeners dramatic accounts of the bombing of London. He once reported from the top of the BBC building in London during an air raid. Listeners could easily hear bombs exploding nearby. Eric Sevareid, on his way to cover China, was forced to jump from a disabled plane high over the Himalayan Mountains. After his rescue he reported his ordeal for his listeners. Larry Lesueur, also of CBS, reported from a British cargo ship that was traveling through mine-infested waters to the Soviet Union. All in all, radio news reached its high point during the war.

The audience came to depend on it for fast, accurate, and thorough reports. Many of the correspondents who covered the war on radio went on to have distinguished careers in TV journalism.

ADJUSTMENT

By the late 1940s and early 1950s, faster than most people imagined, TV became the dominant entertainment and news medium, forcing the powers behind radio to rethink the medium's purpose and goals. Although some people thought radio might go the way of the blacksmith and ice box, it managed to adjust to the changing media environment and to prosper.

At the risk of oversimplification, the coming of TV had the following four effects on radio:

1 Television completely changed network radio broadcasting.
2 Television forced radio to be more dependent on local advertising revenue.

3 Television made radio turn to specialized audiences through the use of particular formats.
4 Television brought the radio and recording industries closer together.

Retrenchment

Large-scale TV programming and TV networking began in 1948. Variety shows starring Milton Berle and Ed Sullivan debuted and were extremely popular. Initially, radio held its own against the new competition. The FCC enacted a freeze on the construction of new TV stations (see Chapter 3), and the shortage of new TV stations kept many national advertisers locked into radio. Network radio advertising declined, but not drastically, down 23 percent from 1948 to 1952. Total radio advertising, thanks to gains in local ads, actually increased by about 13 percent in the same period. Thus the radio industry itself was not in any danger of being made extinct by TV. Nonetheless, it was obvious that network radio was about to face hard times. After the freeze was lifted, mass market advertisers began defecting to TV. By 1955 network radio was taking a financial beating, with revenues dropping by about 58 percent from those in 1952.

Major radio stars like Jack Benny and Bob Hope jumped to TV. The radio networks tried cost-cutting; they broadcast the audio portions of several TV shows (this only served to stimulate the audience to run out and buy TV sets to see what they were missing); and they fired orchestras and writers. Comedians like Groucho Marx and Eddie Cantor were forced to host low-budget radio quiz programs. Faced with low-quality network programs, local affiliates began to jump ship. In 1947 some 97 percent of all radio stations were affiliated with one of the four nets. By 1955 this figure was down to 50 percent. Pretty soon the once-powerful radio networks were reduced to providing ten-minute newscasts on the hour and maybe a few daytime programs. Network radio would take nearly thirty years to recover. At the same time, however, a new radio service was slowly emerging.

All the Hits All the Time: Top 40

The genesis of the new radio came not from the networks but from the local stations. Many current and former affiliates were faced with more time to fill with local programming now that the networks were cutting back. The locals looked for an inexpensive way to fill the time. They found it with the disc jockey (DJ). The idea was not a new one. Despite the network custom of not playing recordings over the air, some independent stations began playing records in the 1930s. Now, in the 1950s, it was an idea whose time had come. Disc jockey shows needed no writers, no actors, no directors, no orchestra—just a glib host and a stack of records to play. Most stations that started out with a DJ format played a little bit of everything, resorting to mainstream appeals, a format that came to be called middle-of-the road (MOR). Other stations began seeking out more specialized audiences. In larger markets some stations developed a rhythm and blues format, designed to attract a black audience. Some rural stations concentrated on country and western music. Many FM stations played classical music.

But the single most influential format developed in this period was started in the middle of the country. The two pioneers generally given credit for it are Todd Storz in Omaha, Nebraska, and Gordon McLendon in Dallas, Texas. Storz noticed that people liked to hear popular songs over and over again; they liked familiarity in their music. Accordingly, Storz checked sales at local record outlets, developed a "hit list," and confined his airplay to songs on the list. McLendon took this basic notion and added a new twist—promotion. He turned his DJs into local celebrities. The new format, dubbed top 40, quickly grew in popularity. About twenty stations used it in 1955 but by the end of the decade hundreds had adopted it.

Top 40's popularity was helped by two factors—one sociological and one musical. The sociological factor had to do with the growth of a "youth culture" during the 1950s. Top 40 was aimed at teens, the fastest growing population segment, who had money to spend and time to listen. Teens were developing their own style of dressing (leather jackets, ducktail haircuts, pony tails), language (daddy-o), and transportation (customized cars). Top-40 radio was a natural rallying point. It drew teens together, emphasized their uniqueness (parents hated top 40), and gave them a common reference point.

The musical factor was the birth of a new form of music that had tremendous appeal among teens—rock and roll. A blend of black rhythm and blues, country, and mainstream white big-band music, rock and roll (later shortened to "rock") became the dominant force in the marketplace. Teenagers flocked to rock at least partly as a means of rebellion (grown-ups didn't approve of it). Early rock stars such as Chuck Berry, Little Richard, Jerry Lee Lewis, and later Elvis Presley seemed unfamiliar and vaguely threatening to the older generation. Young people, however, loved them. In 1956, when Elvis appeared on Ed Sullivan's TV show, an astonishing 83 percent of the TV audience, mostly young, tuned in.

Not surprisingly, rock and top 40 went hand in hand. Rock records were heavy sellers among teens. Their sales put them on the hit lists used by top-40 stations, and this airplay increased their sales even more. Finally, the lesson was apparent for both radio and the record industry. Radio and rock were made for each other. Contrary to past thinking, radio airplay helped to sell records. Moreover, the record companies kept radio's programming costs at a minimum by providing the latest hits free to the station. In turn the radio stations gave the record industry what amounted to free advertising. Record sales nearly tripled from 1954 to 1959; top 40 continued to flourish.

Record sales might have increased even more had the industry not been involved in "the battle of the speeds." Some records were designed to be played at $33\frac{1}{3}$ revolutions per minute, some at 45, and still others at the old standby 78. Radio stations and the record-buying public didn't quite know what to make of it, so many simply stopped buying records until the industry sorted itself out. Spurred on in part by the rock phenomenon, record companies finally agreed that $33\frac{1}{3}$ would be used for long-playing albums, 45 for singles, and 78s would become obsolete. Record promoters quickly supplied DJs at key top-40 stations with the latest releases on 45s.

Payola

Another interesting phenomenon was occurring at top-40 stations. Many DJs were becoming more popular than the music they were playing. In fact, some became big stars in their own right. Allan Freed, generally credited with coining the term rock and roll, became a top-rated DJ in New York and even appeared in a couple of movies. Other celebrity DJs included Wolfman Jack (who appeared in *American Graffiti*), Murray the K, and Bruce "Cousin Brucie" Morrow.

Celebrity or not, the DJ was an extremely important person to the record promoter. In the 1950s the DJ determined what songs he or she would play over the air. Record promoters realized this and from the earliest days of rock and roll they tried to make friends with local DJs. Soon, however, it went way past friendship as promoters gave expensive gifts to influence the DJ's judgment. Ultimately it reached the point where many promoters simply slipped the DJs money under the table or gave them a cut of the sales of a particular record. This practice, called *payola*, reached a climax at a national DJ convention in Miami Beach. Allegedly, record companies had provided a substantial number of prostitutes for the DJs. Local newspapers covering the convention headlined the story "Booze, Broads and Bribes." All of this attention got Congress into the act. Hearings on payola were held the next year, and ultimately the 1934 Communication Act was amended to make payola illegal. (The problem did not go away, however. New rumors of payola surface every few years.)

Alan Freed (seated) was the first disc jockey to become a celebrity. A young Frankie Avalon is second from left.

IMPACT

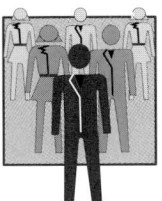

LOUIE, LOUIE

One of the songs that helped make rock and roll an indelible part of the teenage culture was the song "Louie, Louie." Based on a simple four-chord progression, it's a song any amateur band in America could play . . . and did . . . at parties everywhere.

In 1963 an otherwise undistinguished group called the Kingsmen paid $40 to make a quickie record of the song. Not surprisingly, the audio quality wasn't very good. Nobody could understand the lyrics that followed the first line, and many people thought the lyrics were dirty. As a result (maybe), the song sold 12 million copies and has been played countless times on rock radio stations. (Several stations even played the 45 over the air at $33\frac{1}{3}$ revolutions per minute in an attempt to decipher the lyrics. It didn't help; the lyrics were still unintelligible.)

The song became a rock classic. It was featured in the movie *Animal House*. A rock station in California played different versions of "Louie, Louie" for twenty-four hours. In 1985 a Philadelphia DJ started the annual Louie, Louie Day Parade. The 1987 parade drew about 60,000 people, who marched through the streets of Philly humming the song—nobody knew the lyrics.

FRAGMENTS

While adapting to its new environment, radio was still growing. Between 1948 and 1955 more than 1100 new commercial stations were licensed to begin broadcasting. As the number of stations grew, the radio audience was being divided into smaller and smaller fragments. Since the top-40 audience was served by many stations, newcomers looked for other specialized formats that would appeal to other audience segments. As mentioned earlier, black-oriented and country and western formats developed along with top 40. The MOR format was refined and designed to appeal to an older group. The first modern beautiful music station was San Francisco's KABL, started by Gordon McLendon in 1959. This format featured long periods of lush instrumental selections designed to provide unobtrusive background music. San Francisco was also the home of KFAX, the first all-news format, which started in 1960. Other formats that emerged throughout this period were all-talk and jazz. In addition, thanks primarily to the FCC's allocation of space in the FM band, educational radio stations increased from six in 1946 to ninety in 1952. Obviously, radio was turning into a medium with specialized rather than general appeal.

With specialization came localization. Since they were no longer conduits for material produced by the networks, local stations became more responsive to their markets. They could play regional hits; their DJs could make personal appearances at local functions; stations could support local causes; and they could provide more local news. All of this increased radio's attractiveness to local advertisers. In short, radio was redefining its advertising base. In 1945 less than one-third of all radio advertising revenue came from local ads. Ten years later the proportion had nearly doubled. In the future radio would no longer compete with TV or magazines for the national advertising dollar. Instead, it would go head-to-head with the newspapers for local advertising dollars.

To sum up, in about a dozen traumatic years radio made the necessary adjustments that enabled it to prosper in an era that was to be dominated by TV.

RADIO IN THE VIDEO AGE

Since 1960 the TV set has pushed radio from center stage. Television now dominates news and entertainment. The family that used to gather around the radio set in the living room now gathers around the TV. In actuality, this turned out to be a blessing for radio. Liberated from the living room, radio moved to other rooms of the house—bedroom, kitchen, den, playroom—and to the car and finally into people's headbands and showers. It became portable and omnipresent, providing entertainment, information, and companionship to people who weren't near a TV set. Radio found its new role in American life. This concluding section will examine the more significant trends that have occurred in the last three decades or so.

High Tech

Radio became portable during the early 1960s because of the transistor, a tiny device that took the place of the vacuum tube. This made it possible to produce small, lightweight radios that were inexpensive. By 1965 transistor radios were selling at the rate of 12 million per year. Eventually the advent of printed and integrated circuits shrank radio sets even smaller, to the point where they could fit into sunglasses or earphones. In the late 1970s the Sony Corporation introduced the Walkman, a miniature radio-cassette player, which produced high-quality sound through lightweight earphones. Radio was truly a personal medium.

A second technological advance—stereo broadcasting—improved the fortunes of FM radio. The record industry had pioneered the development of stereophonic reproduction. By 1960 most recording companies were issuing discs in the new format and stereo players were selling briskly. The next year the FCC authorized stereo broadcasting by FM stations and fifty-seven of them tried it. The new technique proved popular among listeners. By 1970 there were 688 FM stereo stations, and the number continued to grow. In the early 1980s, in an attempt to revive the sagging fortunes of AM radio, the FCC approved AM stereo, and many AM stations started broadcasting in the new format. Thus far stereo has done little to revive audience interest in the older service.

The recording industry also pioneered another innovation that was to influence radio—the compact disc (CD). Using a computer-assisted technique, digital recording and laser technology, CDs were virtually hiss-free and didn't wear out like conventional discs. Many radio stations, particularly rock stations, replaced their old record libraries with CDs. By the late 1980s almost all radio stations were playing CDs.

Another advance concerned communication satellites. A survey made in the late 1980s disclosed that 85 percent of all stations had at least one satellite dish. The dishes were used to receive feeds from networks and program syndicators.

FM's Rise

Perhaps the most significant event in the radio industry since 1960 has been the way FM has become the preferred service among radio listeners. As we have seen, FM got off to a rocky start and was an unprofitable medium into the 1960s. Then things began to change. The number of FM receiving sets increased by an amazing rate—from 6.5 million in 1960 to 350 million in 1980. The number of FM stations nearly tripled in the same period and FM stations became highly desirable properties. Audiences drastically changed their listening habits. In 1972 FM got only 28 percent of audience listening time and AM got 72 percent. In 1991 the figures were FM, 72 percent and AM, 28 percent.

Why the turnaround? There were at least four main reasons. In the first place, AM stations were hard to come by. Those wishing to go into broadcasting during this time were almost forced to start an FM station since all of the desirable AM frequencies were used up.

The second reason had to do with the growth of stereophonic sound. As both young and old consumers bought stereo sets or components, the infatuation with the new two-channel technique grew. Soon no one spoke any more of "radio-phonograph combinations." Instead, people were buying "sound systems," and an integral part of these new systems was an FM stereo receiver. The enhanced FM sound gave it a distinct advantage over AM among the many audience members who sought improved audio quality.

Third, the FCC passed the nonduplication rule in 1964. This mandated that combination AM–FM operations in markets with more than 100,000 people had to offer separate programming at least one-half the time. Prior to this rule owners of both AM and FM services in the same community mainly used simulcasting, with the FM station simply duplicating the AM station's programming, giving the audience little incentive to listen to FM. Now the FMs would have to come up with original programming for at least one-half of the time. (Ironically, in the 1980s the FCC eliminated the nonduplication rule in part to help struggling AM stations save money by duplicating their FM programming.)

The last reason was a direct result of this new policy. In their quest for original programming FM stations, almost by necessity, developed new formats. One of the first was "progressive rock," typified by the playing of mainly album cuts (many of them too long for the established top-40 format), few commercials, a low-key DJ who kept talk to a minimum, and long music "sets," where several songs were played in succession without interruption.

While all of this was going on, rock music was becoming fractionalized and spinoffs from the top-40 format appeared in the FM dial: jazz rock, hard rock, folk rock, country rock, to name just a few. Aside from rock stations, other FM operators capitalized on FM's good sound quality to offer other formats: beautiful music, easy listening, classical. The listening audience appreciated this new diversity of formats and more and more flocked to FM.

All of this did not bode well for AM. As listenership dropped during the 1970s and 1980s, many AM stations searched for formats to stop their audience loss. Some went to a golden oldies format, playing hits from the fifties and sixties that were originally recorded in monaural sound. Others went to a big-band sound. Still more went to talk formats: all-news and call-in. In the late 1980s, thanks to relaxed FCC rules, a few stations adopted a "raunch" or "shock" format in which a moderator discussed controversial topics. In addition, a concerted effort was made to improve the technical standards of AM radio to enhance its sound quality. Also, as we have seen, AM stereo was adopted by many stations. By 1992 AM stations were still doing well in some large markets but their total share of the listening audience was down and the older service was facing an uphill struggle to regain its prominence.

The Fall and Rise of Network Radio

The 1960s were dark times for network radio. The percentage of radio advertising spent on the networks hovered around 6 percent through most of the decade, a far cry from the 45 to 50 percent the nets had enjoyed during the 1940s. By 1966 the percentage of local stations with network affiliation dropped to 31 percent, compared to more than 90 percent in 1945. Network radio looked for solutions.

ABC was the first radio network to recognize the new nature of radio. Beginning in 1968 ABC split its old network into four separate and specialized networks (Entertainment, Information, Contemporary, and American FM). Local stations with their specialized formats now could find a network with the appropriate content and each of the four ABC networks could have one affiliate in each market. The idea proved to be a good one, and ABC had doubled the number of its affiliates by 1970. CBS, NBC, and Mutual soon followed suit, and by the mid-1980s ABC had seven networks, NBC had three, and CBS had two. The use of satellite transmission made it possible for local stations to pick and choose the exact programs they wanted.

The move by radio to specialized services helped networks survive: by 1980 the number of network affiliates had increased by more than 50 percent. Another sign of the resurgence of network radio was increasing competition. In the early 1980s the following satellite-transmitted networks were in operation: Associated Press, RKO (purchased by United Radio Networks in 1985), CNN Radio Network, Satellite Music Network, National Public Radio (a noncommercial service that started in the late 1960s), and many others. Increased success meant increased revenues and most networks were financially solid. Revenues increased about 10 to 15 percent from 1984 to 1985 and approximately another 15 percent from 1985 to 1986. In 1987 NBC left the network radio business when NBC sold its networks to Westwood One. The nets, however, retained the NBC name.

Coupled with the renaissance in network radio was the growth of radio syndication companies. *Syndication companies,* much like the networks, supply programs to local stations. Some syndicated programming is sent by satellite. In fact, it was becoming difficult to distinguish a traditional network from a satellite-distributed syndication service. Many programs, however, particularly long-form music programs, are still distributed via tape. For example, Bonneville Broadcasting System provides local stations with a complete easy listening format. The station doesn't have to worry about programming the music; for anywhere from $900 to $1500 a month, Bonneville sends the station tapes with all the music it needs. Broadcast Programming Inc. provides the same service for stations with an oldies format. Other syndication companies provide short feature material. The American Comedy Network provides subscribing stations with thirty- to ninety-second comedy bits to spice up their programming.

Fine Tuning

The tremendous growth of radio, with the number of stations on the air tripling from 1958 to 1990, has made radio a highly competitive business. In an effort to stay profitable many stations conduct audience research. Some radio stations do their own research; others call on specialized companies that do it for them. Although this research may take many forms (see Chapter 14), its aim is to make sure that the station's programming reaches its target audience.

Research helps stations to refine their formats or to develop new ones to reach new audience segments. By the late 1980s there had been an explosion in the number of radio station formats. *Broadcasting Yearbook* listed sixty in 1987. In terms of popularity, as of 1991 more than 2400 stations were programming country music and another 2300, the adult contemporary format (a mix of soft rock from the past and present). About 600 stations programmed MOR and another 1000, top 40. There were other formats, however, that were far more specialized. Several stations offered "new age" music. At least one station had adopted an all-comedy format while another was featuring "all-kids" radio. This trend toward increasingly specialized formats seems likely to continue.

Radio and Records

The resurgence of radio has been accompanied by a growing interdependence between the radio and recording industries. The natural union of rock music and radio continued to be beneficial to both media. For example, in the early 1960s, in an attempt to make rock less threatening and more commercially appealing, the record industry introduced a new breed of rock star: white, middle-class, nonthreatening, and wholesome—performers like Annette Funicello, Frankie Avalon, Fabian, Connie Francis, Ricky Nelson. Top-40 stations were eager to play records by these noncontroversial performers. Consequently, record sales increased and top 40 got new listeners as rock went mainstream.

The tremendous success of the Beatles in the mid-1960s paved the way for an influx of groups from Great Britain: Herman's Hermits, the Rolling Stones, the Animals, to name a few. The Beatles brought excitement and innovation to popular music and, interestingly enough, expanded the audience for rock. They were so popular that adults in the 25 to 40 age group began to listen to rock stations to hear their music.

Toward the end of the 1960s and the beginning of the 1970s rock music became associated with the counterculture, those searching for an alternative to traditional American ways and values. "Heavy metal" music went hand-in-hand with the counterculture and many FM stations popularized this new style of music. By the mid-1970s, however, the counterculture had faded. In its place was a new mixture of rock and country and western music. Many radio stations across the country featured this music in new formats called contemporary country or lite country. "Crossover" performers such as Kenny Rogers, Olivia Newton-John, Dolly Parton, and John Denver had hits on both the country and the pop charts. More than 350 stations switched to the new country format between 1978 and 1980. A third popular music form that had great influence in this time period was disco. Prompted by the movie *Saturday Night Fever* (whose soundtrack album sold about 20 million copies), disco enjoyed a brief period of popularity. In 1979 a disco song topped the pop charts for twenty-six out of fifty-two weeks. About a hundred stations, most in the big cities, went to this format. In New York one station switched to disco and in just three months it went from obscurity to the number one station in the market. Disco fizzled out by the mid-1980s, but many stations shifted to an urban contemporary format, which was still popular at the end of the decade.

Perhaps the biggest change in recent years has been the addition of a new element to the radio-sound recording relationship: the music video. Music videos and MTV, the cable network that popularized them, offered a new marketing outlet to record companies. Unlike most radio stations, MTV was willing to take a chance and to air more

R.E.M. performing in their music video, "Losing My Religion." Exposure on MTV and other music-video outlets helped popularize the contemporary hit- and album-oriented rock format at many radio stations.

new releases. This pleased the record companies since it meant more of their products would be promoted, and it also pleased the rock radio stations since they could gauge audience reaction to the videos and add only the most popular ones to their playlist. In short, MTV took some of the risk from radio stations in introducing new releases.

To sum up, the close relationship and interdependence between radio and the record industry has been further cemented in the video age, and apparently it will persist well into the twenty-first century.

RADIO IN THE 1990s

As the decade opened, scarce advertising dollars, increased competition, and rising costs made it more difficult for radio stations to be profitable. Although stations in the top-10 markets were doing well, others were not so lucky. One industry estimate suggested that almost one-half of all radio stations were losing money in 1991, with daytime-only AM stations in those smaller markets hit the hardest. On the national front, the major radio networks were holding their own despite the weakness in national advertising funds. Both ABC and Westwood reconfigured their radio networks for better efficiency.

Radio programming was marked by a shift in the popularity of radio formats. Top 40 or Contemporary Hit Radio was attracting fewer listeners (some critics blamed the lack of memorable or distinctive rock music for this decline) while country music was gaining audiences. Programming strategy continued to center on targeting even narrower demographic segments. Adult contemporary, for example, splintered into at least three different varieties: personality, light, and contemporary.

With regard to technology, the radio industry was awaiting the shift to *digital audio broadcasting* (DAB). This new technology would use the same technique now utilized by compact discs (CDs) to bring high-fidelity sound to radio. The impact of DAB would be significant. In the first place, it would erase the quality differences between AM and FM, perhaps helping many current AM stations to increase their audiences. Second, it would require consumers to purchase new radio receivers in order to hear the improved sound.

SUMMARY

The inventors Marconi, Fessenden, and De Forest were responsible for helping radio take the place of the telegraph. GE, AT&T, and Westinghouse all wanted to have a share of this new medium. Each company owned the rights to different inventions. As a result, many legal battles occurred between inventors and the companies.

World War I was a turning point for radio. The Navy took control of the medium and also took responsibility for preventing patent infringement. After the war the military wanted to keep radio under its jurisdiction. Instead, a new company, RCA, bought out the British-owned company of American Marconi.

Early stations were experimental. In the twenties radio went through a growth period. Frank Conrad started the first radio station, KDKA, in Pittsburgh. Other developments between 1920 and 1927 were radio advertising, radio networks, and radio regulation. However, this growth period of the medium did encounter a few dilemmas, such as the Biltmore Program, the press–radio war, the FM problem, and NBC's ownership of two networks.

Fireside chats by Franklin Roosevelt helped boost the popularity of radio, and the reaction to Orson Welles's "War of the Worlds" broadcast proved that radio was powerful.

Soon TV began to develop, and this changed radio. Local stations began to use DJs, and the top-40 format was conceived. Stations also began to target their audiences.

Technology made many of today's radio luxuries possible: stereo broadcasting, the CD, portable radios, and communication satellites.

The rise of FM can be attributed to the fact that AM stations were difficult for would-be broadcasters to acquire, to the introduction of stereo sound, to the FCC nonduplication ruling, and to new formats.

The interdependence between radio and records ensures that radio has the ability to survive future technological advances. One advance likely to have profound consequences on the industry will be the change to digital audio broadcasting (DAB), expected to occur within the next decade.

SUGGESTIONS FOR FURTHER READING

Aitken, H. (1985). *The continuous wave.* Princeton, N.J.: Princeton University Press.
Archer, G. (1938). *History of radio to 1926.* New York: American Historical Society.
Barnouw, E. (1966). *A tower in Babel.* New York: Oxford University Press.
Barnouw, E. (1968). *The golden web.* New York: Oxford University Press.
Bilby, K. (1986). *The general.* New York: Harper & Row.
Czitrom, D. (1982). *Media and the American mind.* Chapel Hill: University of North Carolina Press.
Douglas, G. H. (1987). *The early days of radio broadcasting.* Jefferson, N.C.: McFarland Publishing.
Douglas, S. (1987). *Inventing American broadcasting.* Baltimore, M.D.: Johns Hopkins University Press.
Eberly, P. (1982). *Music in the air.* New York: Hastings House.
Fornatale, P., & Mills, J. (1980). *Radio in the television age.* Woodstock, N.Y.: Overlook Press.
Gelatt, R. (1977). *The fabulous phonograph. 1877–1971.* New York: Macmillan.
Morris, L. (1949). *Not so long ago.* New York: Random House.
Paper, L. (1987). *Empire: William S. Paley and the making of CBS.* New York: St. Martin's Press.
Schicke, C. (1974). *Revolution in sound.* Boston, Mass.: Little, Brown and Co.
Schubert, P. (1971). *The electric word.* New York: Arno Press.
Settel, I. (1967). *A pictorial history of radio.* New York: Grosset and Dunlap.
Smith, S. (1990). *In all his glory: The life of William S. Paley.* New York: Simon & Schuster.
Sobel, R. (1986). *RCA.* New York: Stein & Day.
Sterling, C., & Kittross, J. (1990). *Stay tuned: A concise history of American broadcasting.* Belmont, Calif.: Wadsworth.
Udelson, J. (1982). *The great television race.* University, Alabama: University of Alabama Press.
White, L. (1947). *The American radio.* Chicago: University of Chicago Press.

CHAPTER 3: HISTORY OF TV AND CABLE

Nine miles from Manhattan, in Flushing Meadow, Queens, a mosquito-infested swamp underwent a magical transformation. At a cost of $156 million the 1939 New York World's Fair was about to open. The theme was "The World of Tomorrow" and many big companies had prepared exhibits that portrayed the wonderful world of the future. There was a 5-acre Town of Tomorrow, a prototype of a planned community. The General Electric exhibit featured Elektro, an electrically powered robot who did all sorts of amazing things—even smoke a cigarette. The General Motors pavilion featured superhighways of the future on which automatic cars were safely and speedily whisked along by remote-control devices. The Borden's exhibit highlighted an invention called the Rotolactor, which, using devices controlled by vacuum tubes, automatically showered, towel-dried, and milked 150 cows. When the Fair opened on April 30, about a half million people attended the opening ceremonies.

Of more relevance to us, however, were the several thousand people who witnessed the opening ceremonies, even though they weren't there. RCA, NBC's parent company, had chosen this event to make a public demonstration of its latest technological marvel—television. NBC had set up a primitive TV camera on a platform about 50 feet from the speakers' podium where it was virtually hidden by dozens of newsreel film cameras. At 12:30 the first TV pictures showed the fair's symbols—the Trylon and the Perisphere. New York Mayor Fiorello LaGuardia was the first politician to be shown in close-up when, curious, he walked up to the TV camera and peered directly into its lens. Franklin Roosevelt became the first president to be televised when his remarks opening the fair were broadcast. A little later David Sarnoff, the head of RCA and the champion of TV's development, called the invention "a new art so important in its implications that it is bound to affect all society."

There were only about 200 TV sets in the New York metro area that day, most of them owned by NBC executives and curious rich people, but there were monitors set up in the lobby at RCA's Manhattan headquarters, in a few department store windows, and at the RCA exhibit inside the fairgrounds. All of the people who saw that first program were impressed with the clarity of the pictures. RCA had TV sets on sale the next day.

Elektro the robot is gone now. Cars have yet to be controlled by sensors hidden in superhighways and farms are yet to be equipped with the Rotolactor. Television, however, is another story. As Sarnoff predicted, it has affected all society. Let's examine how TV traveled the long road of development that eventually led to its inauguration at the World's Fair.

David Sarnoff standing in front of a TV camera as he opens the RCA pavilion at the 1939 New York World's Fair. This was the first time a major news event was covered by television.

ROOTS

The public debut of TV was long awaited. In fact, the *idea* of TV dates back to the 1870s. After the invention of the telephone it seemed natural to add the image to accompany the spoken word. In 1879 the English magazine *Punch* carried an illustration of a couple in front of a fireplace watching a tennis match on a flat TV screen mounted above the mantle. Three years later, a French artist, Albert Robida, drew a series of pictures that eerily predicted the future. Families were shown watching a distant war on a screen in their living rooms. Robida's illustrations also showed families taking academic courses and shopping at home via the screen.

Early TV pioneers also knew the way to go about bringing these predictions to reality. The theory was not too difficult. A black and white photograph in a newspaper is made up of a series of black dots, which can be seen through a magnifying glass. If the dots in one area of the picture are densely packed together, that area appears black. If the concentration is less dense, the area is gray. If there are no dots, the area is white. If a visual image could be broken down into tiny electrical signals (just like the black dots), transmitted, and then reassembled in a receiver, then you would have TV. Sounds simple . . . but it took a while to translate the theory into practice.

One thing that helped was the discovery of a material called selenium. When exposed to light, this material displayed a lowered electrical resistance, thus making it the forerunner of today's photoelectric cell, in which light controls the strength of an electric current. Researchers quickly realized that a huge bank of selenium cells, arranged like the interior of the human eye and wired individually to a bank of lamps, would recreate a crude replica of an image. Unfortunately, to send a picture of any acceptable quality required about a quarter of a million tiny lamps with a quarter of a million attached tiny wires. What was needed was a simpler method whereby the image was *scanned* sequentially, element by element, instead of all at once.

In 1884 Paul Nipkow, a German researcher, came up with a solution. He devised a rotating disk, with perforations arranged in a spiral pattern. As the disk spun around, a different portion of a picture was "scanned" by each perforation, similar to picking up the picture bit by bit through peepholes. In today's terminology (see Chapter 13) each perforation would represent one scanning line. Light that traveled through the perforation then fell on a selenium cell. A few feet away, at the viewing end, another disk, connected by a belt, rotated in sync with the scanning disk, regulating the amount of light that fell on a photoelectrically sensitive screen. Nipkow patented this device but was never able to build a working model. Nonetheless, others kept trying.

In France inventors moved the light source from in front of the disk to behind it, creating what was known as the "flying spot" method of scanning. The advantage of this method was that the subject didn't have to be bathed in high-intensity light as was the case with Nipkow's disk.

In the imperial city of St. Petersburg, Russia, a physics professor, Boris Rosing, studied alternatives to the spinning disk. Using a new device called a *cathode ray tube*, Rosing was able to scan a scene electronically using a stream of electrons steered across the photosensitive screen by magnetism. Rosing took out a patent on this device, which he dubbed the "electric eye," in 1907. Four years later he was able to send a crude picture using his system. Unfortunately for Rosing, his work was interrupted first by World War I and second by the Russian Revolution. In the ensuing turmoil he was arrested and exiled. At the same time, in one of those extraordinary coincidences that sometimes go along with discovery, an English inventor, A.A. Campbell Swinton, had independently developed a similar device. His invention never quite worked properly, but the Englishman's design was adapted by subsequent, more successful pioneers.

Thus at the end of World War I the theoretical principles behind TV had been spelled out and electronic demonstrations of these principles, although not satisfactory, had been achieved. In fact, the actual word "television" (from the Greek, "to see at a distance") was appearing in print, along with many other names for this innovation (see box).

False Starts and Successes

The 1920s were an important time for TV and radio. The decade saw the following developments in TV:

1 A practical but short-lived TV system using mechanical scanning was invented.

2 The search for an improved TV system moved from the workshops of individual inventors to the laboratories of big companies like RCA and GE.

3 The future controlling force of TV was determined. Neither Hollywood nor Broadway nor Washington, D.C., worked to perfect TV. Instead, the task was taken on by the radio industry, thus guaranteeing that a future TV system would be created in the image and likeness of the existing radio system.

PART 2:
HOW IT HAPPENED

BACKGROUND

NAMES THAT DIDN'T CATCH ON

We call it television but when it first started not everybody knew exactly what its final name would be. The following list includes some of the names for TV that didn't make it, as compiled in a 1986 article in *Journalism Quarterly*.

Photoradio	Radiosight
Radiovision	Illustrated Ratio
Eyelids of Radio	Televiewing
Pictorial Radio	Sightseeing by Radio
Tele-eyes	Electrical Imagery
Videocasting	Ethereal Projections
Illustrated Wavelengths	

PROGRAMMING

THE FIRST TV SUPERSTAR: FELIX THE CAT

One of the first TV experiments performed by NBC involved Felix the Cat, an early cartoon feline. In order to have something to televise that wouldn't melt under the bright lights, NBC technicians put a wooden Felix doll on a phonograph turntable and rotated it in front of the camera to give them a moving object to televise. The primitive sixty-line picture was extremely fuzzy, but Felix's likeness was easily recognized.

A few days later NBC received a letter from a woman in New Jersey who had picked up the telecast on an experimental TV set. She complained that since Felix didn't do anything but spin around, the show wasn't very interesting to watch. Her comments weren't profound but she qualifies as the first TV critic.

The indefatigable Felix the Cat appears before a TV pick-up device in the late 1920s.

An American inventor, Charles Francis Jenkins, was a man of many accomplishments. Among his inventions were an improved motion picture projector, an automobile with an engine in front rather than under the seat, and the sightseeing bus. At the turn of the century his interest turned to television or, as he called it, "radio vision." In 1925, using a newly developed scanning system, he transmitted a TV signal 5 miles to a primitive receiver. (This first TV show was not very exciting; it consisted of ten minutes of Dutch windmills in motion.)

At about the same time in Britain, John Baird publicly demonstrated something he called a "televisor" in the window of a London department store. Using the spinning Nipkow disk, he presented three shows daily for three weeks. The results were encouraging enough to gain financial backing for Baird from several investors who thought that they could make money selling TV sets.

This idea had a certain appeal back in the United States as well. Jenkins started his own firm to manufacture both TV transmitters and TV receiving sets. By 1929 eighteen "visual broadcast" stations had been licensed by the Federal Radio Commission (FRC). The audiences for these experimental telecasts consisted mainly of electrical engineers and others adventurous enough to invest money in one of the early TV receivers. The General Electric station in Schenectady, New York, even broadcast the first TV drama, a play called "The Queen's Messenger," in 1928 and followed this with a series of regularly televised programs. Other stations in New York, Boston, and Chicago followed suit. Most of these early stations were started by electronics manufacturers hoping to increase sales. (Note how this parallels the early impetus behind radio broadcasting, as discussed in Chapter 2.)

In a few years this early boom in mechanical TV was over. The Depression, which started in 1929, made it difficult to get the money necessary to support further TV development. Moreover, the FRC had its hands full regulating radio. It did little to establish TV standards or to find it a permanent home in the electromagnetic spectrum. Finally, the picture quality of early mechanical TV wasn't very good. The screens were only a few inches square, the picture itself looked orangish or pinkish, and the whole system was too bulky, too noisy, too dim, and too fuzzy.

Mechanical TV development came to an end. (Well, not quite. In the 1950s CBS, as we shall see, came up with a spinning three-disk system to televise color. Further, a much improved version of this system was used in the 1970s to transmit color TV from the moon to Earth.)

The real future of TV would be tied to electronic scanning.

Zworykin and Farnsworth

The two individuals who were most responsible for the development of the American system of electronic TV were Vladimir Zworykin and Philo Farnsworth. First, let's look at the achievements of Zworykin.

Vladimir Zworykin was a student of Boris Rosing in St. Petersburg. After World War I he emigrated to America and joined the research staff at Westinghouse. While there he started work on what he called the "iconoscope," the eye of an electronic TV camera. In 1929 Zworykin convinced RCA's David Sarnoff, also a Russian immigrant, of the advantages of the electronic system. Sarnoff, confident that TV was the wave of the

One of the two American inventors of electronic television, Vladimir Zworykin, with his iconoscope. Zworykin later pioneered the application of electronics engineering to medicine and helped develop the electron microscope.

The other important American inventor of electronic TV, Philo Farnsworth, is shown with his "image dissector." After a long and complicated patent battle, RCA paid Farnsworth $1 million for the rights to use his inventions.

future, brought Zworykin to RCA's research labs, where Zworykin continued to develop TV pickup tubes and receivers. As the 1930s began, Zworykin, aided by RCA's vast resources, was making real progress.

Philo Farnsworth grew up in Rigby, Idaho, far from the big companies in the east. While in high school he astounded his science teacher by diagraming on the chalkboard a system for electronic TV. Farnsworth moved to Salt Lake City, Utah, in 1926 and convinced private investors to back his research into TV. The young inventor soon developed the "image dissector," which accomplished the same thing as Zworykin's iconoscope but used a different design. (There were, however, enough similarities to spark a patent battle with RCA, as discussed later.) The Depression forced Farnsworth and his backers to spend more time looking for money to support further research. Farnsworth eventually negotiated a contract with Philco, a leading radio set manufacturer and rival of RCA. Along with Westinghouse, RCA, and Philco, other companies that also experimented with early TV were GE, AT&T, and CBS. By 1930 it was clear that corporate America would be solidly behind future TV development.

It was also becoming clear that the existing radio industry would be the major force in shaping the new medium. By the end of the 1920s, organizations that could supply programming for TV—NBC (RCA's subsidiary), CBS, and the Don Lee System (a regional network on the west coast)—were taking the lead in promoting the new medium, and TV set manufacturers were content to follow their lead. Consequently, TV came into being with an organized pattern (networks and local stations) and a support system (commercials) already in place.

Experiments and War

Throughout the 1930s the quest to improve TV continued. Zworykin perfected his all-electronic system in the 1930s. NBC was operating experimental TV station W2XBS (later to become WNBC, New York) and CBS had its own station as well, W2XAB (later WCBS). Other stations soon went on the air in the midwest and west.

The biggest supporter of TV during the Depression years was David Sarnoff of RCA. At a meeting of RCA stockholders in 1935 Sarnoff announced a million dollar plan to take TV out of the lab and into the public arena. The first step was to improve picture quality. Zworykin's new all-electronic system increased the number of scanning lines in the early sets from 120 to 240 (the more scanning lines, the clearer the picture). By 1937 TV was using 441 scanning lines (close to today's standard of 525 lines).

Another step forward in the evolution of TV technology occurred late in the decade when a complicated patent suit between RCA and Farnsworth was settled in Farnsworth's favor. RCA agreed to a cross-licensing agreement with the inventor and was now free to use some of his ideas in the design of their products. Farnsworth then devoted his energies to manufacturing TV receivers, but ill health forced him to curtail his research. Thus in just ten years TV had changed from a thing of whirling disks and belts and flickering lights to an all-electronic system without a single moving mechanical part.

In 1937, connected by cables leased from the phone company, stations in New York and Philadelphia shared programs, thus demonstrating the system that was the precursor of network broad-

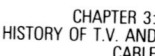

PROGRAMMING

PRIME-TIME TV IN 1931

W2XAB, the experimental TV station licensed to CBS, went on the air in July 1931. If you had been one of the lucky few to have a TV set in this era, here's what you could have watched during prime time that year.

8:00–8:30	At-home party with Alvin Hauser
8:30–8:45	The Television Mystics with Richard Kenney
8:45–9:00	Doris Sharp, the television crooner
9:00–9:30	Education feature: piano lesson
9:30–9:45	Julya Mahony, soprano
9:45–10:00	The art of bookbinding
10:00–10:15	The Kaye Faye Show
10:15–10:30	Roger Kinney, baritone
10:30–10:45	Kathryn Parsons, "The Girl O' Yesterday"
10:45–11:00	The Singing Vagabond

casting. Keep in mind that the early TV stations were licensed as experimental stations. This prohibited them from broadcasting ads, but it allowed them to televise any play or film without having to worry about commercial authorization. Consequently, Broadway plays, vaudeville acts, motion pictures, and original drama were televised. NBC also put into operation two mobile TV vans, which roamed about New York City sending back signals to the transmitter via microwave. Most of the time these remote broadcasts were pretty mundane—planes landing at the airport, traffic on the expressway, pedestrians walking by—but occasionally a newsworthy event was televised. The mobile units covered several fires and once even inadvertently telecast a suicide. (A camera happened to be pointed at an eleventh-floor window of a building when a young woman jumped.) This experimental programming generated some public curiosity and excitement but it didn't exactly cause a national craze.

As we have seen, RCA publicly demonstrated TV in 1939 and shortly thereafter NBC started televising a regular program schedule of ten hours a week, consisting mainly of sports and old films. Set sales moved slowly. By the end of 1939 only 1000 had been sold in New York, even with hefty price cuts. Nonetheless, RCA had competitors lining up to join it in the new field. CBS and Philco were planning to manufacture sets and CBS opened a large TV studio in Grand Central Station. Allen DuMont, a young inventor and manufacturer of cathode ray tubes, marketed his own brand of TV sets and bought TV stations in anticipation of starting another network.

In an effort to stimulate lagging set sales both CBS and RCA presented competing plans to the FCC to introduce color. The FCC in turn held long hearings trying to determine the technical standards for TV, finally culminating in a 1941 statement of policy. The FCC postponed the color issue, saying it was too early to make a decision (the problem would resurface later). Also in 1941 the FCC agreed to authorize licenses for full-time commercial TV stations. On July 1 of that year NBC's station in New York carried the first TV ad ever broadcast. It was a test pattern in the form of a Bulova clock face. The image was on screen for a full minute while the second hand completed its sweep. Bulova paid $4 for the ad. By December 1941 thirty-two commercial TV stations were licensed. Then, just when it seemed that TV was poised to take off, the United States entered World War II. Early TV stations responded with a burst of enthusiasm. On the day Pearl Harbor was attacked it was rumored that Nazi planes were crossing the Atlantic to bomb New York City. At CBS a camera crew laboriously hauled one of the bulky early TV cameras to a top-floor window to televise the bombers as they arrived. The crew then realized that any arriving bombers could use their TV signal as a beacon and head right for them. The crew hurriedly hauled the heavy device back down the stairs. Luckily, the raid was just a rumor.

The development of a commercial TV system was interrupted by the war. Station construction was halted and all but a handful went off the air. On another front, however, World War II accelerated the technology behind TV. Scientists involved in perfecting TV went into the military and studied high-frequency electronics. Their work greatly improved the U.S. system of radar and also advanced the technical side of TV. As the war neared its end, it was apparent that TV would be back, stronger than ever.

POSTWAR TV: 1945–1952

When the war ended, the broadcasting industry made immediate preparations to shift its emphasis from radio to TV. Assembly lines that had been used to turn out war materials were retooled to produce tubes and TV sets. Returning soldiers, skilled in radar operation, were hired by many stations that were eager to use their electronics knowledge. Set manufacturers made plans to advertise their new, improved products.

In 1945 the FCC made a series of decisions that helped set off the postwar TV rush. First, it moved the FM service to a different part of the spectrum, giving priority to TV (and hampering FM development, as discussed in Chapter 2). Second, it lifted a wartime ban on TV station construction. Third, it decided to continue the development of postwar TV according to prewar technical standards. This last decision was welcomed by RCA and most of the set manufacturing industry since they had retooled and were ready to go. On the other hand, the decision caught CBS, RCA's main rival, off-guard. CBS had argued that the FCC should postpone any decision about TV's future until it decided the unresolved issues relating to color TV standards. As a result, RCA got a head start. The U.S. decision to stick by its 1941 standards made the United States one of the first countries to spell out the technical guidelines for a TV system. Many European countries, however, waited until after the war to adopt standards and, as a result, opted for a plan that gave their domestic systems better picture resolution than that chosen by the United States. The search for standards would resurface some forty years later during the quest for a high-definition TV system.

Now that they were ready to go, early TV broadcasters faced the problem of how to fill all of those empty hours. They turned first to sports. Boxing and wrestling were naturals; they were easy to televise. Football followed, then the roller derby, then pro baseball. By 1947 the combined network schedules included twenty-nine hours of sports per week, about 60 percent of the total air time. Television even got its first soap opera in 1947. Entitled "A Woman to Remember," it was quickly forgotten. Cooking shows, travelogs, parlor games, discussion shows, and newsreels rounded out the early schedules. These proved no match for the high-quality entertainment people had grown used to on radio.

The situation changed, however, in 1948. First, network TV programming was introduced. Second, popular radio shows made the transition to TV. The first was "The Original Amateur Hour" on the DuMont Network. NBC moved its "Texaco Star Theater" to TV in the same year, and it quickly became the most popular show on TV. The camera work was excellent, the budget allowed for lavish sets, costumes, and guest stars; and finally, there was Milton Berle, soon to be known as Mr. Television. The avuncular Berle was TV's first superstar. Unlike his predecessors who simply transferred a radio, nightclub, or vaudeville act to TV, Berle designed his show for the small screen. When his program came on Tuesday nights at 8:00 P.M. it seemed as though the whole country stopped to watch. Once about 95 percent of all the sets in use were tuned to his show. CBS countered with its own variety show, Ed Sullivan's "Toast of the Town." Sullivan, a newspaper columnist with no discernible TV talent, introduced a succession of comedians, singers, jugglers, animal acts, ballet dancers, puppets, and acrobats. Sullivan's show quickly followed Berle to the top of the ratings charts. Other variety shows followed; nineteen were on the fall 1948 schedule.

People loved it. After a slow start TV took off. By 1948 there were 29 TV stations on the air, another 80 were authorized, and the FCC had applications for about 300 more. Television set sales also picked up. Only 8000 homes had sets in 1946; two years later 172,000 homes, mostly in the northeast, were equipped with the latest model sets. There were four networks in operation: NBC, CBS, DuMont, and ABC, which was formed during World War II when NBC had to divest itself of one of its two radio networks.

Elvis Presley appeared live on the "Steve Allen Show" in 1956. As you can probably infer from the photo, he sang "Hound Dog."

Freeze

The growth of TV during 1948 was phenomenal. Set manufacturers couldn't keep up with the demand. More stations signed on and license applications were piling up; it seemed that every city wanted at least one TV station. The FCC was bogged down by paperwork and was worried that so many new stations might cause interference problems, much like what had occurred in the development of early radio. The FCC finally acted in September 1948, when it placed a freeze on licensing new TV stations. The freeze would last three and a half years. During this time the FCC studied the future of TV.

The TV industry, however, was not in a state of suspended animation during the freeze. Several significant things happened. First, stations whose applications had been approved before the freeze were allowed to go on the air. By 1950, in the middle of the freeze, 105 stations were on the air. Most cities had only one station; only twenty-four had two or more. Some places like Austin, Texas, and Little Rock, Arkansas, had no TV stations at all. This meant that the stations which had a monopoly on TV in their community had time to build an image and cultivate audience goodwill. These prefreeze stations would be the leaders in the coming years.

Second, the freeze effectively shut out the struggling DuMont and ABC TV networks. In those markets that were lucky enough to have two stations, the stations preferred to join the better established NBC or CBS network. DuMont and ABC could not get their programs into those markets where no station was available to carry them.

The networks continued to wire up stations using AT&T's cable. Before the entire country was connected, programs were recorded on *kinescope*, a film of a TV screen, and sent from station to station. By 1951 a coast-to-coast hookup of fifty-two cities was completed, using coaxial cable, and live network broadcasts could reach about 95 percent of the American homes then equipped with TV. On one of the first shows carried nationwide, CBS's "See It Now," Edward R. Murrow opened the show with monitors televising live shots of both the Atlantic and Pacific oceans. Despite the freeze, TV had grown to nationwide proportions.

Blacklist

Unfortunately, there was no freeze on suspicion or cowardice; the period following World War II marked the beginning of one of broadcasting's less auspicious eras. This period, known as the "Cold War," saw relations between the United States and the Soviet Union decline. At home, people began to suspect that many domestic problems were caused by Communists and people who sympathized with them. Because of its special prominence and power, the entertainment industry got special scrutiny.

A group of three former FBI agents started a newsletter called *Counterattack* in 1947. This publication, usually with little evidence, reported alleged Communist influences and named alleged Communist sympathizers in various industries. Eventually it came to focus on the broadcasting industry. In 1950 the same group published *Red Channels*. Sporting a cover illustration of a red hand about to grasp a microphone, this book listed 151 people who were allegedly working on behalf of the Communist party. The book shocked the entertainment industry since the people named were among the most prominent and successful in motion pictures and broadcasting. Based on the questionable information contained in *Counterattack* and *Red Channels,* broadcasting executives, fearful of offending sponsors, drew up an informal *blacklist,* a listing of people whose loyalty to the United States was suspect. Anyone unfortunate enough to be on this list had to go through a humiliating twelve-step process of "rehabilitation" or be barred from working in broadcasting and films. As a result, several prominent actors, reporters, directors, and writers had their careers cut short.

Note that the impetus for the blacklist came from private citizens, not from the government. For its part, however, the government did little to discourage its application. FBI Director J. Edgar Hoover supplied information from anonymous sources to the FCC about a potentially suspicious license applicant in California. Senator Joseph McCarthy and the House Un-American Activities Committee were conducting widely publicized investigations about Communist infiltration of the State Department and other government branches. The House Committee even investigated the alleged Communist sympathies of Lucille Ball.

Inspired by *Red Channels,* another group dedicated to eradicating the "Communist conspiracy," Aware, Inc., sprang up in 1953. Aware had ties to a Syracuse, New York, supermarket chain that threatened to boycott the products advertised on TV shows that featured Communist sympathizers. As was the case with *Red Channels,* Aware's labeling of someone as a sympathizer was usually based on flimsy, incomplete, and inaccurate evidence. In the climate of the times, broadcasting executives took Aware's threatened boycott quite seriously and more performers found themselves out of work because they had aroused the ire of Aware. One such performer was a CBS talk-show host named John Henry Faulk, who sued Aware for libel. Aided by Edward R. Murrow, Faulk eventually won a record $3.5 million award (he was able to collect only a small part of it because Aware had few assets and little money). Murrow also figured in another event that helped end the blacklist. In 1954 he used his "See It Now" TV program to expose the questionable tactics of Joe McCarthy, who had made many unsubstantiated allegations about people's loyalty. After Murrow's revelations the senator lost most of his influence.

Live from New York or Filmed in Los Angeles

Programming developed and flowered during the freeze period. This was the golden era of TV drama as high-quality plays were performed live over TV almost every night of the week. Most of the early set owners were affluent and lived in the northeastern part of the country. Accordingly, they were used to seeing high-caliber plays. Television responded to this rather specialized audience with a series of prestige productions. In 1948 "Studio One" presented a televised adaptation of Shakespeare's *Julius Caesar* with a young Charlton Heston in the cast. Other notable productions from the early 1950s included Paddy Chayefsky's "Marty," Rod Serling's "Patterns," and Reginald Rose's "Twelve Angry Men." Of course, not everything on TV measured up to this level. CBS introduced TV versions of its long-running radio soap operas in 1951, and wrestling and roller derby still accounted for six and a half hours of weekly network prime time.

New forms of programming cropped up. CBS experimented with a science fiction anthology series, a forerunner of "The Twilight Zone," called "Out There" in 1951. NBC, under the guidance of programming expert Sylvester "Pat" Weaver, premiered "Today" in 1952; as of this writing it is still going strong. NBC also demonstrated the viability of late-night TV in 1951 with "Broadway Open House," a comedy-variety talk show that was the forerunner of "The Tonight Show." "Mr. Wizard," designed to make science entertaining to children, debuted in 1951. Big radio stars like Jack Benny, Burns and Allen, and Groucho Marx also crossed over to TV.

Perhaps the most significant premiere during the freeze was "I Love Lucy" in 1951. In the opening episode Fred and Ricky want to go to the fights but Lucy and Ethel want to go to a nightclub. After several zany schemes, disguises, and misunderstandings, the boys win and they wind up at ringside. (This particular formula with this particular cast would be repeated for 180 episodes. Lucille Ball would appear on TV, in one format or another, in four different decades.) This program was important not only for its durability

Early productions on live television were often quite ambitious. This is a scene from the 1949 Studio One production of "Mary Poppins."

but also because Lucille Ball and Desi Arnaz formed their own production company, Desilu, and filmed their show in Los Angeles. This marked the beginning of Hollywood involvement in TV production (Hollywood had basically ignored the new medium up to this point) and presaged the era when independent production companies would control much of the supply of TV programs.

Lucille Ball also had the foresight to realize that live TV was inefficient. Recording her shows on film meant that they would last for many years and give her an edge when, in subsequent years, a syndication market would create an increased need for programming.

Sixth Report and Order

The freeze thawed in 1952 when the FCC produced a document called the Sixth Report and Order, which cleared up several problems.

First, after a couple of false starts, the commission came to a final decision about color TV standards. RCA's version was ultimately chosen over the CBS version because the RCA system was all-electronic and compatible with existing black and white sets. The CBS version used a variation of the spinning Nipkow disk, which made the commission somewhat uneasy because it was outdated and could not be received on existing sets. (CBS wasn't that disturbed about losing this battle to RCA since early CBS color TV marketing efforts were failures and the company still had doubts about color TV's ultimate profitability.)

Second, the FCC opened up new spectrum space in the ultra-high-frequency band (UHF channels 14 to 69) to accommodate the hundreds of applicants seeking stations. Unfortunately, because of their wavelength, UHF signals didn't travel as far as those in the older VHF (very high frequency) band (channels 2 to 13) and most TV sets couldn't get UHF signals. Therefore, UHF stations started off with a technical disadvantage. This disadvantage was further intensified because

PROFILE

LUCY AND THE RED SCARE

The Communist scare even touched Lucille Ball, the top female star then appearing on TV. The comedienne was called before the House Un-American Activities Committee in 1952 and was asked to explain why, sixteen years earlier, she had registered to vote for the Communist party ticket in the upcoming election. Lucy explained to the committee that she had in fact expressed her intention to vote Communist but had done it only to placate her aging grandfather, who had become somewhat radical in his later years. She had never actually voted and had forgotten about the whole episode. Lucy thought that was the end of it. Naturally, she was surprised to learn a year later that the committee had once again called her to testify about the 1936 voting registration incident. Lucy repeated her story and the committee seemed satisfied. She was assured her testimony would remain secret.

Two days later a popular radio commentator announced that "the top television comedienne," no doubt meaning Lucy, was a member of the Communist party. The next day the newspapers picked up the story. Lucy's entire career was threatened. Indeed, the careers of several others were ended for much less. She was then making a movie with MGM as well as doing her extremely popular "I Love Lucy" series. Both projects could easily be ruined by this kind of damaging publicity.

Ironically, the story reached its height on the night that the first "I Love Lucy" show of the new season was about to be filmed before a live audience. Both Lucy and her husband, Desi Arnaz, were unsure how the audience would react. They were afraid that they would be booed off stage. Desi made an emotional speech to the crowd, declaring Lucy to be 100 percent American and denouncing the newspaper stories as "bunk." The audience roared its approval. When Lucy was introduced, she got a standing ovation. The public clearly loved Lucy.

A representative from the House committee quickly called a news conference to announce that Lucy was not under suspicion. Newspaper headlines exonerated her. Even the radio commentator who broke the story retracted it and apologized. As Desi Arnaz summed it up, "The only thing red about Lucy is her hair color and even that's not real." Lucy's great popularity had conquered. Others were not so lucky.

CBS Chairman William S. Paley with one of the network's early color TV cameras. The CBS color system was ultimately dropped in favor of one developed by rival RCA.

VHF stations had been able to establish themselves during the freeze. In addition, the FCC adopted a policy called intermixture in which both UHF and VHF stations were assigned to most of the top markets. Given the inherent problems with UHF, it came as no surprise that these stations could not compete effectively in the intermixed markets.

Third, the problem of stations interfering with each other was solved by increasing the geographic separation and by limiting the maximum power of stations using the same channel.

Fourth, the FCC worked out a table of channel assignments that would provide TV service to all parts of the United States. The commission decided to choose a policy of local service rather than regional service. Unlike radio, where several clear channels were set aside to assure wide geographic coverage, all of the TV channels were assigned to local communities.

Finally, thanks to the efforts of the first woman FCC commissioner, Frieda Hennock, 242 channels were set aside for noncommercial TV stations.

GROWTH CURVE: 1953–1962

The next ten years saw an incredible surge in the fortunes of TV (see Table 3-1). Statistics tell the story. In 1952 some 34 percent of all households had TV. Ten years later 90 percent were equipped. There were 108 stations broadcasting in 1952. Ten years later there were 541. In the same time period the amount of money spent on TV advertising more than quadrupled. By any yardstick, TV was booming.

New Wrinkles

On the technology side, the Ampex Corporation introduced videotape recording in 1956. This new technique solved the problem of how to store TV shows. Prior to this, live TV shows could be preserved only by a kinescope recording. The quality of videotape was so good that programs recorded on tape were hard to distinguish from live broadcasts. Moreover, it made program production cheaper and would ultimately replace film in the production of many situation comedies. Finally, it helped kill off live drama on TV. With tape, the show could be shot and edited long before air time; mistakes could be corrected. Many critics claimed that shows recorded on tape lacked the excitement and the intensity of live TV.

TABLE 3-1 Growth of Television

Year	Number of Stations	Percentage of Households with TV Sets	Number of Employees
1950	98	9	9,000
1970	862	95	58,400
1992	1693	99	150,000

On another front, RCA began to market color TV sets in 1954 and NBC presented several programs in color (such as the World Series and "Peter Pan") to help spur set sales. The first color sets had $12\frac{1}{2}$-inch screens and were priced at $1000 (about $4000 in today's dollars). Not only were they expensive; they were also hard to tune and frequently broke down. Consumer resistance was immediate. Only 5000 sets had been sold by 1955. On one of the color variety specials NBC presented to promote the new service, Bob Hope quipped, "There's a tremendous audience watching us in color tonight—Mr. Sarnoff and his wife." RCA, having then spent some $70 million in developing color, toughed it out. New and improved sets came off the production line, prices came down, and NBC presented nearly 700 hours of color programming in 1959. By 1962 about a half million color sets were sold and RCA finally announced that it had made a profit from their sales. The future of color TV would continue to improve.

The FCC also realized that UHF's technical disadvantage was creating problems for those stations. First of all, not many sets could pick up the UHF signals (channels 14 to 69). Regular production model TV sets were sold with a tuner built in for the VHF channels, but consumers had to buy a special converter to get UHF. Moreover, since its signal strength was weaker, in any market where there were both VHF and UHF stations, advertisers naturally preferred the stronger VHF outlets. Similarly, networks were more likely to sign up VHF stations as affiliates. As a result, many UHF stations were having trouble staying in business. One of the solutions that the FCC considered was "deintermixture." In markets where there were both UHF and VHF stations it was proposed that the VHF stations be moved into the UHF band. Naturally, the existing VHF stations were not thrilled with this solution and it went nowhere. Finally, in 1961 the FCC persuaded Congress to pass the All-Channel Receiver Bill, which required all newly manufactured TV sets to be able to receive all TV channels, not just VHF. Even so, UHF TV grew slowly.

Two other technological milestones occurred during this period that were to prove highly significant for the future of TV. The significance of the first was immediately apparent. A rocket roared off Cape Canaveral in July of 1962 carrying *Telstar,* the first active communication satellite capable of relaying signals across the Atlantic Ocean. Only twenty-five years after his death, wireless signals were crossing the Atlantic in a way Marconi had never imagined.

The ultimate significance of the second event was overlooked by virtually everybody. A new idea, introduced in the early 1950s, was gaining ground. People who lived in mountainous areas could not get good over-the-air TV reception. Residents of these areas hit upon a novel solution. They would put an antenna on top of one of the tall peaks and run wires down to the homes in the valley. The residents would pay a fee to receive their programs over a cable. This system was called *community antenna TV,* or *CATV.* Later CATV would come to stand for cable TV. Conventional broadcasters thought that transmitting signals by cable was a clever idea but that it would have little general application. They were mistaken.

Hollywood

Not all was going well for the networks. The DuMont Network, shut out of many markets because the existing stations were affiliated with NBC or CBS and lacking its own stations in the major cities, finally went out of business in 1958. ABC was also having problems. Salvation appeared in 1953 when ABC merged with Hollywood-based United Paramount Theaters. The theater company brought a fresh influx of cash to the struggling network, which enabled it to develop new programs that were competitive with those of CBS and NBC.

The merger also marked the beginning of another important trend: a closer relationship between TV and the movie industry. At first, Hollywood basically ignored the upstart medium, hoping it would go away. When it was clear that TV was here to stay, Hollywood turned hostile. Movie stars were prohibited from appearing on the small screen. No major studio would allow its films to be televised. In retrospect, it's apparent that the film industry missed a chance to have a key role in TV's early development. The TV industry needed a production facility for its programs. Hollywood was the natural place, but the major film studios saw TV as competition and would have nothing to do with it. Eventually TV networks turned to advertising agencies, which packaged programs for them. (As we shall see, this arrangement had its good and bad points.) The agencies got out of the business when program production costs got too high and when a system of participating "spot" advertisers replaced the single sponsor.

After a few years it became clear that the movie industry was losing its audience to TV. Worried, the Hollywood studios tried several things to win back its popularity—bigger screens, Cinerama, 3-D, mature themes, big-budget films—but nothing seemed to work. Ultimately the movies came to terms with the new medium—thanks, in part, to Mickey Mouse. ABC, with its Hollywood connection, made the breakthrough. It signed a contract with Walt Disney studios for a one-hour weekly series, "Disneyland," to premiere in 1954. Although not a major studio, Disney was still an established, mainline Hollywood operation, and its willingness to deal with TV signaled a new philosophy among motion picture producers. ABC in turn was in dire need of popular programs. "Disneyland" was a big hit, giving ABC one of its few top-10 rated programs of the decade. The success of "Disneyland" also prompted ABC to team up with Disney to produce "The Mickey Mouse Club," which premiered in 1955 and became one of the most popular children's shows of all time. NBC and CBS made arrangements with several studios for programs especially produced for TV. ABC in turn signed with Warner Brothers to produce a traditional western series for TV called "Cheyenne." In 1955, in another example of the growing Hollywood–TV alliance, about 2000 theatrical films made before 1948 were released for TV.

By 1957 Hollywood had taken over much of the general TV program production. The amount of network prime-time programming that originated on the west coast in that year was nearly 71 percent of the entire schedule. The marriage of Hollywood and TV became complete in the next few years when post-1948 movies were released to the new medium and when the major motion picture studios began to produce movies for TV.

Quiz Shows: The Big Money

The idea of winning prizes for answering questions was not a new one. Both radio and early TV shows gave away modest prizes to their successful contestants. But the quiz program that premiered on CBS on June 7, 1955 was unprecedented. Its title said it all: "The $64,000 Question." Contestants started by answering easy questions for small sums of money. Eventually, if they answered correctly, they progressed to the four big-money questions: $8000, $16,000, $32,000, and finally $64,000.

The show went to great lengths to demonstrate its honesty. Armed guards watched the safety deposit box that held the questions and answers. A bank official appeared and testified that nobody had access to that box "except the editors," a small but significant exception. Contestants answered questions from an isolation booth.

Ordinary people with whom the public could identify were preferred as contestants. A 28-year-old Marine whose category was cooking became the first to win the big money.

"The $64,000 Question" was successful beyond all imagination. It quickly displaced "I Love Lucy" as the number one show. The sponsor of the program, Revlon, was ecstatic as sales skyrocketed. At one point the company couldn't make its lipstick fast enough to satisfy consumer demand. When the quiz show aired on Tuesday nights, sometimes 80 percent of the sets in use were tuned in.

The other networks scrambled to catch up. NBC unveiled "The Big Surprise," where contestants could win $100,000. CBS replied with "The $64,000 Challenge," which had a top prize of $128,000. "Break the Bank" offered a $250,000 top prize. NBC finally did the inevitable. It trotted out "Twenty-One," where contestants might win an unlimited amount of money. One of the contestants on "Twenty-One" was Charles Van Doren, a charismatic, 30-year-old Columbia University English instructor. Van Doren quickly became an audience favorite as his winnings mounted to $143,000.

Not everything, however, was going smoothly. Ugly rumors were circulating. It was revealed that quiz show producers tried to keep popular contestants on their shows by "controlling" the questions they were asked. For example, on "The $64,000 Question" an Italian shoemaker whose category was opera was asked questions only about Italian opera, his specialty. French and German operas, which he knew less about, never came up in his questions. Even uglier rumors surfaced to the effect that contestants who drew big audiences were given the answers to their questions and received acting lessons on how to look tense and nervous while answering.

The New York district attorney investigated the possible rigging. He found one disgruntled "Twenty-One" loser who charged that he had been told to "take a dive" against Van Doren. Van Doren responded by issuing a denial and stated that the quiz shows were honest. Even as he was making this pronouncement, a grand jury was impaneled to hear mounting evidence, and several quiz shows were quietly canceled.

Finally, a losing contestant on "Twenty-One" offered convincing proof of rigging. He composed three letters that contained the questions and answers from upcoming shows and mailed the letters to himself by registered mail before the shows aired. The unopened envelopes were then presented to the grand jury as evidence. Other contestants saw that the lid was off and admitted that they, too, were given answers. In 1959 the House of Representatives convened hearings on the topic. One of its key witnesses was Charles Van Doren. He read a long, anguished statement admitting his complicity. He had been regularly coached and given answers.

By 1960 all of the big-money quiz shows were off the air. Public trust in TV was shaken to such an extent that the networks took several steps to rebuild their tarnished image. First, they assumed more active control over program development, no longer giving as much power to producers and sponsors. Second, the next few years were notable for a boom in network documentary production, such as "CBS Reports," which was designed in part to enhance network credibility. Finally, in order to discourage a recurrence of the problem, quiz show tampering was made illegal.

PROGRAMMING

"STRIKE IT RICH"

Not all was golden in the golden age of TV. In fact, some programs, particularly quiz and game shows, were positively exploitative. Take, for example, "Strike It Rich," one of the most criticized shows on TV during the 1950s. Critics referred to it as "commercial television gone berserk," "a disgusting spectacle and national disgrace," and a video version of "kick the cripple."

Every week the show asked those in its TV audience to mail in letters detailing why they needed help. About 5000 letters a week poured in from desperate people, telling their sob stories to the show's staff. People wrote for money to buy hearing aids, for medical treatments, to buy caskets for departed loved ones, for clothing that had been lost in a fire, for new eyeglasses, for false teeth, for artificial limbs, for lost dogs, for Christmas gifts for their children. The most deserving of these letter writers (actually the most miserable) were invited to come on the show. They were given a chance to win the things they desperately needed by answering the five easy questions. But first they had to relate their tear-jerking tales of woe to the national TV audience. Even if they missed questions, the truly needy wouldn't go away disappointed. At the end of the show, the "Heart Line," a regular phone in the middle of a heart-shaped piece of scenery, was opened up so that viewers from across the nation could volunteer help. A viewer might call and offer a job to a contestant who was unemployed and could not feed his family. Another call might be from a doctor offering to perform a free tonsillectomy needed by a contestant's little child. Another caller might offer furniture to replace what a contestant lost in a flood. And so on.

Eventually the MC of the show resorted to picking needy people from the show's studio audience. Thousands of people traveled to New York in the hopes of being chosen as contestants. They kept coming even after the show broadcast warnings not to come. Many of the unsuccessful ones wound up on the welfare rolls of New York City. The city's welfare commissioner demanded that the show be shut down since it was in reality providing welfare without a license. Despite all the furor (or perhaps because of it), the show spent four years on prime-time TV, finally running out of misery in 1955. It was revived in syndication in 1973 but it didn't last. Apparently the viewing audience had suffered enough.

Kids, Westerns, and News

Quizzes weren't the only programs to enjoy popularity. Encouraged by the success of "Disneyland," the networks catered to the growing child audience by introducing child-oriented prime-time series such as "Lassie," "Rin Tin Tin," and "The Lone Ranger." These joined such daytime staples as "Captain Kangaroo," "Howdy Doody," and "The Mickey Mouse Club." The type of programming that TV provided for children would be a future source of controversy.

The biggest programming trend, however, was aimed at adults. Around the mid-1950s the image of the western as entertainment geared only to kids changed, thanks to the success of such serious theatrical films as *High Noon* and *Shane*. Not surprisingly, ABC, with its Hollywood contacts, was the first network to translate this trend successfully to TV. "The Life and Legend of Wyatt Earp" premiered in 1955 and concentrated on more mature story lines, earning it the title of "adult western." Viewers liked it and other networks quickly ventured into the adult western format. CBS transformed a radio program into a western that would run for twenty years, "Gunsmoke." Three more westerns premiered the next season; ten more in 1957, including such classics as "Maverick" and "Have Gun, Will Travel." In 1958 and 1959 twenty-seven westerns dominated prime time. Finally, the craze hit the saturation point and by 1962 only a handful were left on the air.

Television news grew slowly during this period. Nightly network newscasts were only fifteen minutes long. There were, however, prime-time programs that dealt with current affairs. The best of these was Edward R. Murrow's "See It Now" on CBS. During its six-year run from 1952 to 1958 the program dealt with such controversial

CHAPTER 3: HISTORY OF T.V. AND CABLE

One of the early adult westerns, "Gunsmoke" ran for twenty years on CBS. The first choice for the lead role of Marshall Matt Dillon was John Wayne. Wayne turned down the role but suggested a friend of his, James Arness, for the part. Needless to say, Arness was grateful for the recommendation.

PROGRAMMING

ADULT WESTERNS AND GIMMICKS

With the increasing numbers of adult westerns on TV in the late 1950s and early 1960s, it became harder and harder for viewers to tell them apart. To combat this problem, producers and writers came up with gimmicks so that their hero would have something distinctive to make him stand out from the crowd. Some obvious examples: "The Adventures of Jim Bowie" starred a character who, naturally enough, used a Bowie knife instead of a gun to fight off the bad guys. The hero of "Yancy Derringer" carried a derringer (what else?) in his fancy hat. Bat Masterson, hero of the series of the same name, batted his opponents with a gold-tipped cane. Wyatt Earp carried two Buntline specials—oversize .45-caliber pistols whose extra range enabled him to drop desperadoes clear across the county. In "Wanted Dead or Alive" bounty hunter Josh Randall (played by Steve McQueen) carried a 30–40 sawed-off carbine. Perhaps the most unusual gimmick was that of a character called Sundance, the star of a series called "Hotel De Paree." His trademark was a string of brightly polished silver disks that he wore around the base of his cowboy hat. When he got into a gunfight, he would maneuver the sun's reflection off the disks into his opponent's eyes, giving Sundance a decided edge. (He was never seen in a gunfight on a *cloudy* day.)

issues as the future of nuclear technology, the war in Korea, and the relationship between cigarette smoking and cancer. In the documentary area, NBC produced the award-winning twenty-six-week series "Victory at Sea." In 1962 the same network filmed the efforts of several West Germans as they dug an escape tunnel under the Berlin Wall. "The Tunnel" became a global news event and was seen by a huge audience.

PLACID TIMES: 1963–1975

The next dozen years or so marked a fairly stable period in TV history. Television continued to thrive. The networks were the dominant source of entertainment and news and enjoyed more or less steady financial growth. In any given minute of prime time about 90 percent of the sets in use were tuned to network shows. Local independent stations offered little competition. VHF stations dominated UHF stations in the quest for audiences. Competition from cable TV and pay TV was not serious. Programming, while sometimes innovative, basically followed the formats set down in the 1950s. True, there were some events within the industry that ruined the placid mood, but compared to what was to come, these years seem tranquil.

Statistics document TV's continuing growth. About 96 percent of all homes had TV by 1975; 70 percent were equipped with color. From 1963 to 1975 more than 325 new stations signed on, a growth of more than 50 percent. Advertising revenue more than doubled.

Technology

On the technological side, the slow but steady growth of cable TV (CATV) began to capture industry attention. As we have noted, CATV began as a means to bring better reception of existing TV channels to mountainous areas. The new service grew slowly. By 1966 there were only 1570 CATV systems in the entire country and only 3 percent of U.S. homes had cable. At the same time, however, the concept behind CATV was changing. In addition to carrying local stations, CATV began to import the signals of distant stations, to which the system did not previously have access. This, of course, meant additional, unforseen competition for traditional broadcasters and they were not amused by this turn of events.

Particularly disturbed were the owners of UHF stations, who were afraid the new signals would force their operations, at best marginally successful, out of business. The FCC responded in 1965 to 1966 with a set of restrictive rules that protected over-the-air broadcasting. (The prevailing philosophy at the FCC at this time envisioned cable as merely an extension of traditional TV signals. Thus it is not surprising that any challenge to broadcasting by cable's development would be curtailed by regulation.) Cable systems had to carry all TV stations within 60 miles and couldn't carry shows from distant stations that duplicated those offered by local stations. In 1968 the commission ruled that CATV systems in the top 100 markets had to get specific approval before they could import the signals of distant stations. Taken together, these rules effectively inhibited the growth of CATV and made sure that any growth would be limited to smaller communities. While all of these rules were being made, CATV systems were quietly improving their technology so that by the late 1960s many systems could carry as many as twenty different channels.

In 1972 the FCC, pressured by both traditional broadcasters and the cable industry, issued yet another set of rules. Among other things these rules specified:

1 who was to regulate cable (local communities, states, and the FCC)
2 twenty-channel minimums for new systems
3 carriage of all local stations
4 more rules on the importation of distant signals, including the nonduplication provision mentioned earlier and
5 the approval of pay cable services.

Once again, the major impact of these rules was to discourage the growth of cable in urban areas. Cable was insignificant to city residents, who already received good reception from a number of local stations. Cable system operators were not encouraged to bring service to urban areas, since the stringing and installation of cable was expensive in densely populated areas, and once cable was installed the operators would have a big job making sure that no imported signal duplicated local programs. To top things off, a major cable company nearly went bankrupt. Cable did, however, grow in midsized markets and by 1974 cable had penetrated a little more than 10 percent of all TV homes. On balance, its future did not look promising. As we shall see in the next section, however, things changed.

At the same time, UHF TV stations were still struggling, but there were hopeful signs. First of all, the number of TV sets in use that could receive UHF signals, thanks to the All-Channel Receiver Bill, increased from 10 percent in 1963 to 89 percent in 1974. Second, UHF stations got a boost from CATV. Cable systems were required to carry all local stations, and this included UHF. Once on the cable the signal strength disadvantage of UHF no longer mattered, putting UHF and VHF stations on an equal footing. These two factors helped UHF increase its audience reach, but overall VHF stations were still far more profitable.

Communication satellites became more important to TV. Those that followed *Telstar* were placed in a synchronous orbit about 22,300 miles up, which meant the satellite maintained its position relative to a point on earth and could serve as a convenient relay for ground stations. These geostationary satellites had many advantages over lower orbit satellites: not as many were required, there was no loss of signal since the higher orbit satellites never went out of sight, and less elaborate tracking systems were needed. Networks looked at the possibility of shifting from coaxial cable to satellites to interconnect their stations.

Public TV

Noncommercial TV progressed slowly during the early 1960s, growing to 100 stations in 1965. Early attempts to organize these stations into a network arrangement amounted to little. Finally, in 1963 an organization called National Educational Television (NET) was formed. NET was a "bicycle" network—programs were not distributed electronically but were physically shipped on film or tape to member stations. Along with most early efforts, the NET arrangements suffered from too little money.

A major development occurred in 1967 when a study sponsored by the Carnegie Foundation recommended a plan for what it called "public" TV. The term "public broadcasting" came to include a wide range of station owners: universities, school boards, state governments, school systems, and community organizations. Indeed, some of the problems that were to plague public broadcasting were in part caused by its heterogeneous nature. In any case, one of the important things about this report was a shift in philosophy: noncommercial TV would no longer be limited to programs stressing formal instruction and education. Instead, it would provide an alternative to commercial programming. With amazing rapidity (just eight months later), Congress created the Corporation for Public Broadcasting (CPB). The main function of CPB was to channel money into programming and station development. Two years later CPB created the Public Broadcasting Service (PBS) to manage the network interconnection between the public stations. Local public station managers soon dominated the leadership of PBS, thus setting the stage for what was to become a continuing source of friction between the two organizations: PBS resented the concentrated bureaucracy of CPB and preferred decentralized power; CPB preferred strong, centralized management.

It wasn't long before CPB and PBS tangled over two related problems: money and politics. In the first place, Congress never provided enough funds for public TV, making it difficult to establish any long-range plans. In the second place, the feeling of President Richard Nixon and his administration was that public TV was dominated by liberals. In an attempt to weaken the power of the CPB Nixon proposed giving money directly to the local stations. For its part, CPB got into its own internal political squabble with the PBS—the organization that managed the actual network operation—over who exactly was in charge. As a result, public TV was continually short on funds, fuzzy in its goals, and on shaky political ground. Nonetheless, the service presented some award-winning programs: "Masterpiece Theater," "Upstairs, Downstairs," "Black Journal," "Sesame Street," and "The Electric Company." As the mid-1970s came to a close, however, its future was somewhat cloudy.

Impressionable Audience?

As TV became an established part of the social fabric, many concerned citizens worried over what it was doing to its audience. Especially troubling was the impact of TV violence on children.

Although "The Man from UNCLE" was a spy spoof that featured comic book style violence, it was criticized as one of the shows that put an undue emphasis on aggression.

This problem received congressional attention in the early 1950s when a Senate committee held hearings on the relationship between antisocial TV shows and juvenile delinquency. The committee had little in the way of research on which to base its recommendations and suggested that the topic receive further study.

The problem resurfaced in the early 1960s thanks to the success of an ABC action-adventure program, "The Untouchables." Focusing on the gang-warfare days of Chicago mobster Al Capone, the series was popular and inspired several violent imitators. All of this mayhem caught the eye of another Senate committee, which again held hearings off and on from 1961 to 1964 on the possible connection between TV violence and antisocial behavior. This committee did have research to examine, some of which suggested a connection between watching violent TV shows and viewer aggression. The committee never came to a firm conclusion and again called for more research.

The issue was examined again a few years later by the National Commission on the Causes and Prevention of Violence and yet again in 1969 by the Surgeon General's Advisory Committee on Television and Social Behavior. Both of the groups ultimately concluded that although the relationship was relatively weak, TV violence was a cause of aggression in real life for some members of the audience (see Chapter 16).

The other area that caused concern was TV advertising aimed at young children. In the early 1970s a citizens' group, Action for Children's Television (ACT), called attention to the potentially harmful effects of TV ads for sugary cereals, candy, toys, and fast food on children who were not able to distinguish program content from commercials. Supported by research indicating that kids were particularly vulnerable to TV ads, ACT was able to convince a consumer-oriented FCC to begin hearings on this issue. Ultimately, under the FCC's pressure, the TV industry agreed to reduce the amount of advertising on Saturday morning kids' shows, prohibit popular and credible hosts from selling products on their shows, place a "separator" (a few seconds of black or a billboard saying "We'll be right back after these messages") between the program and the commercial, and eliminate drug and vitamin ads from kids' shows because these products tasted too much like candy and kids might overdose on them.

Regulations and Economics

There were two other rules passed by the FCC in this time span that had an impact on the way the TV industry earned its money. The first of these, like the changes in children's TV, came about because of a citizens' group.

A government report linking cigarette smoking and lung cancer appeared in 1964. Almost immediately there was political pressure to put health warnings on cigarette packs or, more extreme, to ban all cigarette advertising completely. This, quite naturally, created controversy. Broadcasting stations, under FCC rules, have special obligations concerning controversial issues. The fairness doctrine (largely repealed by the FCC in 1987; see Chapter 18) required that broadcasters present all sides of a controversy. Late in 1966 a New York lawyer named John Banzhaf IV requested time from a local station for antismoking commercials. Banzhaf and his group, Action for Smoking and Health (ASH), argued that since the station carried ads that encouraged smoking, and

PROFILE

JOAN GANZ COONEY

Joan Ganz Cooney is the founder of "Sesame Street" and created the Children's Television Workshop.

Cooney graduated from the University of Arizona in 1951 with a degree in education. She worked as a reporter for two years before moving to New York to become a publicity writer for NBC. From 1962 to 1967 she produced public affairs documentaries for New York's public broadcasting station, WNET.

In 1966 Cooney was asked by the Carnegie Corporation to do a study of the possible uses of public TV in preschool education. The Children's Television Workshop (CTW) was founded in 1968 on the strength of her report, which found that in homes with children the TV set was on as many as sixty hours a week. The idea behind the CTW was to use TV to give poor children the same preparation for school that most middle-class children were getting. CTW's first effort, "Sesame Street," became a huge success, winning Peabody awards in 1970 and again in 1989.

Inducted into the Academy of Television Arts and Sciences' Hall of Fame in 1990, Cooney continues to search for innovative ways to use TV in education.

---

since smoking was indeed controversial, the station had an obligation to air messages that told the antismoking side. The station said no and Banzhaf went to the FCC. Most observers thought that Banzhaf was an imaginative guy but that his request would be seen as amusing and immediately denied. Much to everyone's surprise, the FCC sided with Banzhaf and decreed that antismoking spots must be carried for free. Health organizations immediately offered a number of creative and effective spots.

The tobacco industry became nervous that these antismoking spots would cut their sales and wanted to stop all broadcast ads immediately. The broadcasters didn't like this idea since 10 percent of their revenue came from tobacco ads. Instead, the broadcasters wanted to air the smoking ads when few children were watching. An act of Congress finally settled the issue. After January 2, 1971 no cigarette ads would be allowed on TV. (This date was chosen so that the big cigarette companies could advertise on New Year's day football games one last time.) The broadcasters complained that the annual $200 million they lost from this law would cause hardships. The networks, however, quickly made up their loss by replacing cigarette ads with those from other sponsors.

The second event that caused some financial upheaval was instigated by the FCC itself. Concerned that the networks were dominating TV, the commission, espousing the principle of diversity, announced the prime-time access rule (PTAR) in 1970. This rule, in effect, gave the 7:30 to 8:00 P.M. (E.S.T.) time period back to the local stations. The FCC wanted to open up the program production market to non-Hollywood companies and evidently hoped that more locally originated shows dealing with community issues would find their way to the screen. Instead, most stations opted for low-cost game shows or nature programs. Rather than encouraging local production, the rule had two unanticipated effects. First, it made several syndication companies rich and heralded the importance of the TV syndication market in the future. Second, it helped ABC become more competitive. Over the years ABC usually ran third in the three-network race for the ratings. Consequently, the network always had financial problems. Now, thanks to PTAR, ABC no longer had to program (and lose money on) the 7:30 to 8:00 time period. The network used its savings to develop new programs that enabled it to challenge the two older networks. In fact, in the late 1970s ABC became, for a short time at least, the number one rated network.

Hillbillies, Vulcans, and Cops

There were several overlapping trends in TV programs in these years. The first started in the early 1960s and was due, in part at least, to a change in the demographic makeup of the audience. When TV first emerged, most of the audience lived in urban areas. As set prices declined and more local stations came on the air the audience became more rural. CBS was the first network to notice this and unveiled a number of programs designed to appeal to this new market segment. The first of these, "The Andy Griffith Show," featured the comic adventures of a kindly sheriff in Mayberry, North Carolina. The quintessential rural show debuted in 1962. "The Beverly Hillbillies" was about the Clampetts, a family of hillbillies who struck oil on their property and moved to posh Beverly Hills, California. Although panned by critics (but loved by fans), "The Beverly Hillbillies" contained amid the corn some biting satire on modern society. In fact, some modern writers argue that the show was ahead of its time. (In fashion too—its star, Jed Clampett, wore a sport coat and T-shirt long before the style was popularized by Don Johnson on "Miami Vice.")

Naturally, the success of these two shows inspired an abundant harvest of imitators, most of them on CBS, the network that excelled at turning out rural comedies. There were "Petticoat Junction" (a mother and her three daughters run a hotel in the backwoods town of Hooterville); "No Time for Sergeants" (two country types join the Air Force); "Gomer Pyle" (country type joins the Marines); and "Green Acres" (a reversal of "The Beverly Hillbillies"—two city types move to the country). The rural comedy craze would last for several seasons.

A second programming trend appeared to be connected with the social climate. Beginning with the assassination of President John Kennedy in 1963, the midsixties were marked by political unrest and violence. Perhaps as a reaction to the dismal nature of the real world at the time, TV programmers introduced a world of escapist fantasy and the audience gravitated to such programs as "My Favorite Martian" (an avuncular-looking Martian pals up with an earthling); "My Living Doll" (a voluptuous female robot pals around with her creator); "The Munsters" and "The Adams Family" (implausible families of monsters coping with suburbia); "Lost in Space" (the Swiss Family Robinson in orbit); "I Dream of Jeannie" (a girl from a lamp pals around with an astronaut); "Voyage to the Bottom of the Sea" (a nuclear sub battles sea monsters); and "Star Trek" (most people already know the premise behind this one). Never in TV's history has there been such a concentration of fantasy shows as there was in the middle to late 1960s.

As the decade ended, the "youth culture" had gained prominence and the watchword in society was "relevance." Again influenced by the social climate, TV scurried to present relevant programs about young people. In 1970 the networks treated their audiences to shows about young social workers, young teachers, young lawyers (two shows), young doctors, young cops, and young revolutionaries (in the 1776 American Revolution, of course). The audience wasn't buying, however, and all but one of the relevant shows were canceled before the year's end.

The early years of the 1970s were characterized by a shift toward a more conservative political climate. "Relevance" was replaced with "law and order." Again mirroring society, TV's next trend was shows that featured law enforcement officers as their main characters. A check of the three networks' fall schedules from 1970 to 1974 disclosed that thirty-eight series featuring cops and crime premiered during this period.

The last trend in entertainment programming was the maturation of the situation comedy. Pioneered by CBS, the first of these new sitcoms was "The Mary Tyler Moore Show." Although it began as a conventional sitcom, this show went on to combine more contemporary attitudes and outlooks with the comedy premise. Another groundbreaker was "All in the Family," a show whose lead character wasn't the totally lovable stereotype usually found on situation comedies. In fact, Archie Bunker was a bigot. The show's scripts explored previous taboo areas of language, sex, and politics. "All in the Family" injected a new style of realism into the situation comedy genre. More realistic comedies followed: "Maude," the adventures of a female liberal—the opposite of Archie Bunker; "M*A*S*H," wise-cracking doctors in the Korean War whose experiences also had relevance for the Vietnam War; "Good Times," a black family trying to make it in a Chicago housing project; and "Chico and the Man," a Mexican-American works for the conservative owner of a run-down garage. Interestingly, the situation comedy genre succeeded in bringing relevancy to TV where the dramatic form failed.

Finally, this period also marked the coming of age of TV news. The networks expanded their nightly newscasts from fifteen to thirty minutes in 1963. Walter Cronkite of CBS and the team of Chet Huntley and David Brinkley became well-known anchors. A few months after going to their expanded formats, the networks gained national praise for their coverage of the Kennedy assassination and funeral. Television news also showed its unique capacity for live coverage of news stories. The networks offered gavel-to-gavel coverage of political conventions, live TV pictures of the first steps on the moon, coverage of the civil rights movement, and disturbing pictures of the war in southeast Asia.

Audiences for both local and network newscasts grew steadily. As a result, regularly scheduled news programs turned a profit. At many local stations advertising revenue from news programming constituted the station's single most important source of income. Many big-city local stations increased the length of their early evening newscasts to one hour or more. Advances in electronic news gathering meant that live pictures could be sent directly back to the station from the scene of a news event. Influential documentaries such as "The Selling of the Pentagon" showed that the networks were not afraid to tackle controversial issues. All in all, TV journalism continued to grow.

CHANGES: 1975 TO THE PRESENT

The period from 1975 to the beginning of the 1990s heralded tremendous change in the TV industry. New TV technologies emerged to compete with traditional TV, increased competition lessened network domination of audience viewing, and the industry itself was reshaped by changes in the economic and business climate.

Growth of Cable

There were two basic reasons for the explosive growth of cable in the late 1970s, one technological, the other regulatory. In 1975, a then little-known company in the pay-TV business, Home Box Office (HBO), rented a transponder on the communications satellite *Satcom* I and announced plans for a satellite-interconnected cable programming network. Cable systems could set up their own receiving dish and HBO would transmit to them first-run movies, which the operators could then sell to their subscribers for an additional fee. Although pay TV was not a new idea (it had been tried before on a limited basis but never really caught on), HBO's new arrangement meant wider coverage of cable systems at a lower cost. Further, the new programming service provided a reason for people in urban and suburban areas to subscribe to cable. Now the big attraction was no longer better reception of conventional channels but content that was not available to regular TV viewers. In a few years other cable-only channels were also distributed by satellite—Showtime, The Movie Channel, Christian Broadcasting Network—as well as independent local stations, dubbed "superstations," such as WTBS in Atlanta and WGN in Chicago. Other specialized cable networks—ESPN (sports), CNN (news), and MTV (music videos)—soon followed. Cable now had a lot more features to attract customers.

The second reason came from the FCC. By the mid-1970s the commission realized, with some help from the courts, that its 1972 rules were stifling cable's growth. Consequently, the FCC postponed or canceled the implementation of many of its earlier pronouncements and changed its philosophy: it would henceforth encourage competition between cable and traditional TV. Eventually, as the Reagan administration advanced its deregulation policies, the FCC dropped most of its rules concerning cable. In 1984 Congress passed the Cable Communications Policy

The outspoken Ted Turner, maverick station owner, who attracted some of the networks' audiences by starting superstation WTBS, CNN, TNT, and The Cartoon Channel.

Act. The law, which was incorporated into the 1934 Communications Act, endorsed localism and set up a system of community regulation tempered by federal oversight. The FCC was given definite but limited authority over cable. The local community was the major force in cable regulation, which it exercised through the franchising process. The act gave cable operators, among other things, greater freedom in setting their rates and released them from most rules covering their program services.

Taken together, these two factors caused a spurt of cable growth, which attracted the interest of large media companies such as TCI, which in turn invested in cable. This in turn caused cable to grow even faster. In fact, the growth was so great that many cable companies in a rush to get exclusive franchises in particular communities promised too much and had to cut back on the size and sophistication of their systems. Nonetheless, although the growth rate tapered off a bit in the mid-1980s, the statistics are still impressive (Table 3-2). From 1975 to 1987 the number of operating cable systems nearly tripled. The percentage of homes with cable went from about 14 percent in 1975 to 50 percent in 1987. Even the urban areas shared this growth and at least parts of many big cities were finally wired for cable.

By 1988 the cable industry had become dominated by large multiple-system operators (MSOs). The era of a locally owned "mom-and-pop" cable system was over. The top-10 MSOs controlled more than 54 percent of the nation's subscribers. The largest MSO, TCI, Inc., alone had more than 10 million subscribers.

Network Audience Erosion

The increased number of cable channels meant increased competition for the three traditional networks. Their audience shares declined as more people opted for other viewing alternatives. But new cable channels weren't the only source of competition for the networks. From 1975 to 1983 about seventy new independent stations started broadcasting, most of them UHF. Many of these were carried by local cable systems, received wide coverage, and siphoned off more of the networks' viewers. Finally, videocassette recorders (VCRs) became popular. From only a handful in operation in 1978 VCRs were in 50 per-

TABLE 3-2 Growth of Cable and VCRs

Year	Cable Penetration (%)	VCR Penetration (%)
1975	13	0
1980	23	5
1985	46	30
1990	55	70

cent of U.S. homes by the end of 1987. Many people used VCRs to record and play back network shows, but many also used them to watch rented theatrical movies on tape, which ate into the three major networks' audience. If this wasn't enough, a fourth network, the Fox Broadcasting Company, premiered a limited program schedule in 1987. Although many of its shows got off to a rough start, Fox divided the available audience into even smaller segments. By the late 1980s the three major networks' share of the audience had declined from 90 percent to less than 70 percent.

Evening Soaps, "Warmedies," and News

One of the more interesting trends in programming during the late 1970s and early 1980s was the emergence of the continuing-episode prime-time series (also called prime-time soap operas). Culturally, this period was called the "me generation" with emphasis on individual fulfillment and personal achievement. The prime-time soaps focused on rich people who had a continuing array of problems, indicating perhaps that although a lot of money can bring a lot of fun, material success was no guarantee of happiness. One of the first prime-time soaps grew out of a book and successful miniseries. "Rich Man, Poor Man, Book II" premiered in 1976 and was followed by "Dallas" (rich people in Texas); "Dynasty" (rich people in Colorado); "Flamingo Road" (rich people in the South); "Falcon Crest" (rich people in California); and "Dynasty II: The Colbys" (more rich people in California).

By the mid-1980s the most significant trend was a shift back to the warm and wholesome family situation comedy. These shows were called "warmedies." Although part of the reason behind this trend may be societal (the country was rediscovering the values of the traditional family during this time), much of it was economic.

Bart and the rest of the Simpson family attend the opera. "The Simpsons" helped the Fox network become competitive in the early 1990s.

Family-oriented situation comedies were in demand in the syndication market; they did much better than action-adventure programs. "The Cosby Show" is the best example of this trend. Premiering in 1984, the program quickly went to the top of the ratings and regularly attracted an audience of more than one-half the sets in use. It inspired many imitators. By 1986 there were sixteen warm, wholesome family sitcoms on the three major networks.

News programming continued to expand. Ted Turner began an around-the-clock news service, the Cable News Network, in 1980. It was quickly joined by a related service, Headline News. The three networks countered by expanding their own news programs with late-night and early morning news programs. By mid-1982 the three networks had added a total of forty hours of news every week. Although some of this programming was ultimately cut back, it signaled the importance of news as a major part of TV programming.

The end of the decade saw the emergence of what was labeled "trash TV" or "tabloid TV." This genre consisted of talk shows and other "reality-based" programs that featured confrontation, sensationalism, or titillation to attract an audience. Presided over by a new breed of TV hosts, dubbed the "news punks," this genre included such offerings as "Geraldo," "The Morton Downey Jr. Show," and "A Current Affair." All of these programs had an undercurrent of violence that sometimes rose to the surface, as happened when host Geraldo Rivera had his nose broken in an on-camera melee during a show about race relations. (Geraldo got into a similar fight in 1992.)

Programming innovations on the cable networks were few. The pay channels (HBO, Showtime, etc.) offered mainly movies with an occasional music, sports, or comedy special. Series produced exclusively for cable generally resembled their network prime-time counterparts. Some cable channels, like the USA network, CBN, and Lifetime, relied heavily on old network shows and copies of the network interview format. The Weather Channel presented twenty-four hours of weather (what else?) and ESPN broadcast twenty-four hours of sports, but both were familiar formats to viewers. Several home-shopping cable channels premiered during the mid-1980s, but their programming was simply a long series of commercials. Perhaps the only innovative forms of programming were MTV, which broadcast primarily music videos, and C-SPAN, which carried live coverage of Congress. Otherwise, cable programming was basically more of the same.

The overall impact of cable on the TV industry has been considerable. As we mentioned, the growth of cable has played a large part in the continuing erosion of network audiences. In addition, cable has affected the standards of acceptability on broadcast TV. Movies and programs on cable are more explicit and deal with more adult themes than programs on over-the-air TV. For example, "Brothers," a series aired on Showtime, featured gay characters in leading roles and dealt with the topic of homosexuality in a way that would make the series unacceptable to prime-time network TV. As a consequence, broadcast TV networks and stations have relaxed many of their standards in an effort to compete. The sexual frankness of the 1988 miniseries "Favorite Son" or the raw humor of Fox's "Married . . . with Children" are two obvious examples of this trend.

Further, cable has encouraged TV to become more like radio in its increasing reliance on tightly defined formats. The greater number of available TV services are dividing up the audience, particularly in major cities, into definable segments. MTV, for example, is aimed at the 15- to 24-year-olds and VH-1, MTV's soft rock counterpart, goes after the 25- to 44-year-old segment, just as the contemporary hit radio format and the adult contemporary format are geared toward similar age groups. In like manner, the Nashville Network tries to attract a country-western audience and ESPN goes after a largely male audience by programming sports.

Mergers and Cuts

There were important business developments as well. Starting in the mid-1980s, American industry experienced merger mania, as many firms were bought by other companies. The broadcasting industry was no exception. In 1985 Capital Cities Broadcasting acquired ABC, an unusual merger in that the smaller company took over the larger one. And there were examples of history coming full circle. GE, one of its original owners in 1919, reacquired RCA (and NBC) in 1986 for $6.3 billion. In addition, after fending off a takeover attempt by Turner Broadcasting, William Paley, the founder of CBS, reemerged in 1986 to become the company's leader. Eventually Laurence Tisch, from the Loew's Corporation, was tapped by Paley as chief executive officer.

These new owners exemplified another significant trend in the industry. Under an FCC that increasingly endorsed the philosophy of "let the marketplace decide," the test of a successful broadcast operation became how much of a profit it made as opposed to how well it served the public interest. The new breed of network owners was led by individuals who were "bottom-line" oriented and devoted their skills to the entrepreneurial side of broadcasting.

Not surprisingly, one of the first tasks faced by these new owners was to improve the networks' financial status, hurt by declining audiences and rising program costs. All of the networks tightened their belts but ABC and CBS were particularly affected. Both networks laid off large numbers of employees and cut back expenses. The news departments were especially hard hit. It appeared that the era of easy money was over for the networks.

Mergers affected other segments of the electronic media as well. Sony bought CBS records in 1987 and a West German company, Bertelsmann A.G., acquired the record division of RCA. The biggest deal, however, occurred in 1989 when Time Inc. and Warner Communication announced plans to merge into Time Warner, Inc., thus becoming the world's largest media conglomerate.

Public TV: Searching for a Mission

Public TV's recent history has been marked by two main themes: a general lack of money and a search for a purpose. A second Carnegie Commission Report, dubbed Carnegie II, was released in 1979. The report reviewed the problems that had plagued PBS from its inception: a lack of long-range funding from Congress, the lack of insulation of public TV from political squabbles, a clumsy managerial structure, and a need to define its mission. Carnegie II called for more federal funding of public TV and recommended replacement of the CPB hierarchy. The report had little impact, partly because it was issued during a recession and partly because of a change in political direction.

When the Reagan administration took office in 1980, as part of its overall emphasis on reducing government involvement in creative enterprises, it called for drastically reduced funding of public broadcasting. State governments, a second important source of funding, also cut back on their funding dollars. As a result, public TV stations began to look for more support from both businesses and viewer contributions.

All of this was taking place at the same time that the basic mission on public broadcasting was being debated anew. Should it return to its original purpose during its inception in the 1950s and become an educational service? Or, should it provide more general appeal programming that was an alternative to commercial broadcasting? But, what exactly was an alternative? "High-brow" programming for the culturally elite or programs such as "Austin City Limits"? Should it compete with the major networks or become a service for minority interests? Which minorities? Such questions are still being asked today.

To make matters more complicated, cable networks were providing some of the material that had previously been the province of public TV and public TV was beginning to look more like a commercial network. The Discovery Channel, the Arts and Entertainment Network, and the Learning Channel presented educational and prestige programming that siphoned away some PBS viewers. Moreover, public TV began rerunning episodes of series that had first appeared on commercial TV, such as "The Paper Chase," "Lassie," and "Lawrence Welk." Some programs that first appeared on PBS, such as the National

Geographic Specials and "Siskel and Ebert at the Movies" made the jump over to commercial TV. Finally, public TV stations, in an attempt to raise more money, adopted a policy of "enhanced underwriting," announcements from companies that help underwrite the costs for programming, which sounded suspiciously like commercials.

Despite its problems, PBS still produced programming that garnered critical praise and loyal fans. "Sesame Street" celebrated its twentieth year on the air in 1989 with its second Peabody Award. "Viet Nam: A TV History," "Nova," and "Cosmos" also won numerous awards. Regularly scheduled news and public affairs programs, such as "The MacNeil/Lehrer News Hour," "Washington Week in Review" and "Wall Street Week" score steady if not spectacular numbers in the ratings.

TV and Hollywood

In recent years the relationship between the movie industry and the TV industry has grown even closer. Hollywood studios regularly turn out a large number of made-for-TV movies every year. Further, virtually all of prime-time network program production is done by the TV divisions of Hollywood studios. (Many Hollywood-based studios, however, are now producing programs in Canada to cut costs.) The same is true for syndication, where major motion picture companies control more than one-half of the market.

Another notable trend is the growing importance of home video and pay cable as new markets for theatrical film. At the same time, the importance of broadcast TV in the profit equation of a film has diminished. As it currently stands, many films are released first to theaters, then to the videocassette market, then to cable, and only afterward to networks and independent stations. In the early 1990s a typical motion picture made about as much money from TV as it did from theatrical release. Revenue from the pay cable channels (HBO, Showtime, etc.) accounted for much of this, but the tremendous growth of the VCR sparked the formation of a new market for movies. More than 30 million prerecorded cassettes—mostly movies—were sold in 1990. More than 20,000 cassettes were on the market and video rental outlets were cropping up all over. From just a handful in 1980 the number of rental stores had grown to more than 17,000 in 1990. Competition also increased. Most video rental stores had at least one competitor within a five-block radius.

Just as they do at the box office, the major studios dominate the videocassette market. The ten major Hollywood studios accounted for more than two-thirds of all revenue from the sale of videocassettes in 1990. (Keep in mind that the studios make their money only from the sale of videocassettes to consumers and to stores that rent cassettes. As it currently stands, the studios get no revenue from the actual rental of videocassettes.) Finally, to complete the circle, by 1986 all of the major networks had, with varying success, produced films that were released to movie theaters. As the 1990s began, the relationship between the two industries had never been closer.

Technology

In the 1970s TV production equipment became smaller and easy to carry. One of the results of this was the development of electronic news gathering (ENG), which revolutionized TV coverage. Using portable cameras and tape recorders, reporters no longer had to wait for film to be developed. In addition, ENG equipment was frequently linked to microwave transmission, which allowed live coverage of breaking news.

The 1980s saw the development of satellite news gathering (SNG). Vans equipped with satellite uplinks made it possible for reporters to travel virtually anywhere on Earth and to send back a report. Local stations sent their own correspondents to breaking news events in Europe and Asia for live reports. Stations also formed satellite interconnection services that swapped news footage and feeds. SNG profoundly altered the relationship between affiliates and networks. Many local stations, since they no longer had to depend on the networks for international or national news pictures, questioned the relevance of a network newscast.

Some new technologies didn't do very well. *Subscription TV (STV),* a method of receiving pay TV from conventional TV stations, couldn't compete with cable and was virtually defunct by the end of the 1980s. Low-power TV stations, with an effective coverage radius of only 10 to 15 miles, did not catch on as quickly as many had expected. *Videotex* and *teletext*, two techniques for sending electronic pages of information and graphics over the TV set, have not yet achieved rapid acceptance among the general audience (they did better in Europe).

Another promising development, *direct broadcasting by satellite (DBS),* is also off to a slow start. DBS uses high-power communication satellites to send original programming direct to umbrella-sized receiving dishes mounted on rooftops. (DBS, however, is different from a backyard satellite dish. The backyard dish is much bigger and can pick up broadcasts from weaker communication satellites. The backyard dish can pick up only programs sent by other sources, such as cable channels, whereas a DBS system would have its own original programs.) The startup costs of such a system, however, were substantial and only a few companies had the resources to experiment with it.

Further, for many people living in rural areas the backyard satellite dish, or TVRO (for TV reception only), may have become an adequate substitute for DBS. After a period of impressive growth TVRO sales leveled off after many program services scrambled their signals. Currently, TVRO owners must buy a descrambler and pay monthly fees to receive the scrambled channels. Nonetheless, thanks to cable companies and other firms that are making it easier for TVRO owners to subscribe, the industry is enjoying a modest comeback.

Television in the 1990s

The erosion of network audiences continued during the early years of the decade but the rate of decline seemed to be slowing down. A weak economy and a general shortage of advertising dollars resulted in a fairly bleak profit picture for both networks and local stations. There were no mergers and/or acquisitions in the industry as there were in the 1980s. Several reports in the trade press, however, speculated that one network, perhaps NBC, was an acquisition target of a major film studio, a conjecture that was supported by the departure of NBC programming executive Brandon Tartikoff to Paramount Pictures in 1991.

Cable subscribership remained flat as a proliferation of new cable networks further frac-

Maverick producer David Lynch brought the short-lived "Twin Peaks" to ABC in 1990.

tionalized the TV audience. Some industry executives were using the word "slivercasting" to describe the highly focused appeal of some of the new services such as The Cowboy Channel and the Science Fiction Channel. The premium cable channels such as HBO and Showtime saw their subscribership remain flat or slightly decline, and top management was worried about the increasing competition from PPV services. On the regulatory front, Congress was reexamining the Cable Communications Act of 1984 and most in the industry thought that regulations governing cable would be strengthened.

Network entertainment programming was marked by a return to the familiar in 1991 when established TV stars such as James Garner, Carol Burnett, and Robert Guilliame returned in TV series. All of their series, however, were quickly canceled. The Fox network continued to score programming successes as two of its shows, "In Living Color" and "Beverly Hills, 90210" became popular among young audiences. In the news area, CNN showed that it had become a major force to be reckoned with in broadcast journalism when it scooped the three networks with live coverage of the outbreak of the Persian Gulf War.

Both broadcasters and cablecasters watched with great interest as decisions by the FCC and the federal courts gave an initial approval to AT&T and the regional phone companies to provide limited video information services. If these companies are finally permitted to provide a full range of programming, the traditional competitive structure of American broadcasting and cable would be drastically altered.

Public TV still suffered from chronic funding problems but did streamline its management system. PBS centralized its programming management in 1990 by appointing an executive vice president for programming, Jennifer Lawson, with authority to develop and schedule programs. Lawson decided to lead off the 1990–1991 season with "The Civil War," which garnered record ratings (and additional audience support) for PBS. The outlook for increased federal funding, however, was gloomy.

On the technological front, the key word was "digital." This technique of sending TV signals encoded as a series of "0s" and "1s" allowed for a whole new generation of special effects. It also made possible signal compression, a method that would allow most cable systems to be upgraded to 150 channels. Engineers were also using digital signals to minimize the spectrum space used by high-definition TV (HDTV), an innovation allowing the United States to catch up with Japan and several European countries in the race to develop a workable standard for HDTV.

SUMMARY

There were two forms of early TV: mechanical and electronic. In the 1920s the development of mechanical TV was moving along at a steady pace. However, the boom ended when the Depression cut off sources of research funds; the FRC, busy regulating radio, ignored the new medium; and people realized TV was too bulky, too noisy, too dim, and too fuzzy. In the meantime researchers like Zworykin and Farnsworth continued to make advancements in electronic TV. After World War II the FCC stepped in to promote TV.

In 1948 four events took place: network programming began, radio shows went to TV, demand for TV was high, and the FCC ordered a "freeze" while it determined standards. Eventually the FCC's Sixth Report and Order ended the freeze.

After 1952 there was a great period of growth in TV. There was the advent of videotape recording; the TV and movie industry began to work together; quiz shows became popular (until they were plagued by scandal); and kids' shows, westerns, and news shows were generating a following.

UHF was at a disadvantage during this time, but eventually the use of cable TV would help UHF become more competitive. Public TV was established as an alternative to commercial TV.

In recent years cable growth caused network audiences to decrease. Other current trends in TV have been evening soaps, comedy shows about families, longer newscasts, and tabloid TV.

SUGGESTIONS FOR FURTHER READING

Abramson, A. (1987). *The history of television: 1880–1941*. Jefferson, N.C.: McFarland & Co.
Baldwin, T., & McVoy, D. S. (1983). *Cable communication*. Englewood Cliffs, N.J.: Prentice-Hall.
Barnouw, E. (1975). *Tube of plenty*. New York: Oxford University Press.
Castleman, H., & Podrazik, W. (1982). *Watching TV*. New York: McGraw-Hill.
Castleman, H., & Podrazik, W. (1984). *The TV schedule book*. New York: McGraw-Hill.
Fabe, M. (1979). *TV game shows*. Garden City, N.Y.: Doubleday.
Henderson, A. (1988). *On the air: Pioneers of American broadcasting*. Washington, D.C.: Smithsonian Institution Press.
Shulman, A., & Youman, R. (1973). *The television years*. New York: Popular Library.
Singleton, L. (1986). *Telecommunications in the information age* (2nd ed.). Cambridge, Mass.: Ballinger.
Sterling, C., & Kittross, J. (1990). *Stay tuned*. Belmont, Calif.: Wadsworth.
Udelson, J. (1982). *The great television race*. University, Ala.: University of Alabama Press.

PART THREE

HOW IT IS

CHAPTER 4: RADIO TODAY

*My name is RADIO! My influence shall abide!
I, Magic Box, am something years ago
The wizards dreamed of in Arabian Nights.
Science has conceived and brought to birth
More wondrous far than legends' figments
    wrought
By the ingenious bards of long ago. . . .*

*I feel like a spirit medium that can bring
The listener what'er he wishes from the void.
Do you want multitudes of thoughts, all types?
Full measure comes with the revolving dial;
The masters wait to pour out symphonies
That rock the world and set your soul on fire. . . .*

Robert West, "My Name Is Radio!" (1941)

We'll be looking for caller number ten but first you've got a lock on a thirty-minute block of rock direct from stereo compact disc on the hot new Z-93 . . . Hot . . . hot . . . hot . . . hot.

DJ on large-market FM station (1992)

Satellites! VCRs! high-definition TV!—amid the furor of today's communications explosion it's easy to overlook persistent, enterprising, unassuming radio. If nothing else, radio is resilient. It has withstood frontal attacks from an array of new media services, each promising to sound the death knell of the radio business. But in every instance, from the introduction of sound pictures in the late 1920s to the arrival of TV in the 1940s and the birth of music TV in the 1980s, radio has rebounded, reformulated, and, most important, remained.

Radio today is as vital as ever. It is chameleonlike in form: from the supertiny Walkman to the suitcase-sized "boom box." It is omnipresent: most households have five radios or more. It fills the air: there are more than 11,000 stations on the air in the United States alone. And it seems to meet our needs: somewhere on that dial there can be found almost every form of music, all kinds of advice, hundreds of ball games and special

The two extremes of radio today: the very personal stereo headset and the very public oversized boom box.

events. Perhaps most important of all, in the face of an unprecedented flow of competition from within and outside the industry, radio remains economically viable.

Sometimes in our crush to credit TV for almost everything good or ill in our culture, there is a tendency to overlook radio. Let's not make that mistake. To borrow from some of its supporters, let's remember: radio is red hot! In fact, "heat" is the appropriate metaphor. Everything about radio—from its competitive policies to its flamboyant personalities—tends to be intense. Let's don our insulated gloves and delve into the radio business.

COMPETITION IN TODAY'S RADIO BUSINESS

Any discussion of the radio business must begin with one word: competition. Radio is arguably the most competitive of contemporary media. By almost every criterion there is more "radio" than anything else. There are more radios than there are TVs (about three times as many). There are five times as many radio stations as there are daily newspapers and nearly ten times as many radio stations as TV stations. About the only thing there is less of in radio is revenue. To understand commercial radio today, we must begin with its economics.

For some reason many people have difficulty with economic terminology but few fail to understand the intricacies of pizza. So imagine you and a group of friends have just been served two pizzas at your favorite restaurant. As usual, the pies have been sliced. Let's see where the cuts are.

Advertising Revenue

Pie 1 (Figure 4-1) is apportioned on the basis of 1991 advertising revenue. In 1991 advertisers spent over $125 billion trying to convince the American public to buy their products. The bulk of that spending, over $33 billion, went to the newspaper business. Television got the second biggest slice, about 22 percent, or $26 billion. A lot went to direct mail and magazines; nearly a quarter went to other media, like billboards, bumper stickers, and blimps. Look at the paltry radio slice: less than 7 percent of total advertising

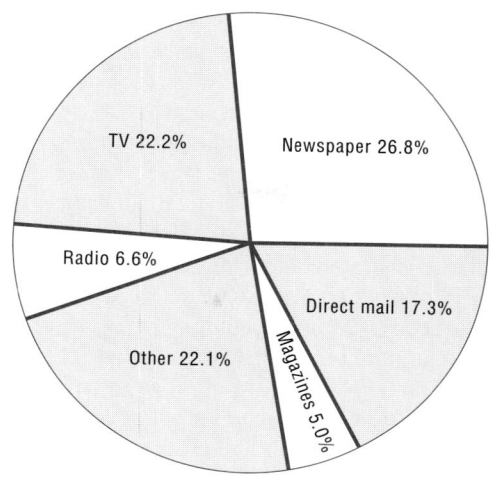

FIGURE 4-1 Advertising revenue by medium. (Source: McCann-Erickson, Cox Research.)

expenditures, representing about $8.5 billion. Lesson number one, then, for radio, is that if this pizza party were given in honor of America's leading advertising vehicles, you would leave comparatively hungry.

The Station Universe

Pie 2 (Figure 4-2) is divided on the basis of a radio station type. There are over 11,000 radio stations on the air. Just under one-half of these (47 percent) are commercial AM stations. Although many of these are powerful stations emanating from large cities, the majority of AM stations are relatively low-powered, local operations serving the small and midsized communities that com-

FIGURE 4-2 Types of radio stations, 1992.

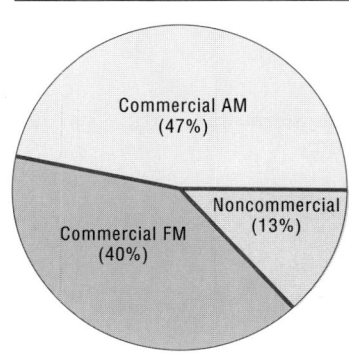

prise the essence of "middle America." Some (about 14 percent) broadcast in stereo; most offer the stable but static-filled AM signal people got used to hearing in their cars and at their beaches in the 1950s and early 1960s.

Four in ten radio stations are commercial FM operations. Because of their relatively late arrival on the scene and their technical requirements (described in detail in Chapter 12), the majority of the FMs are currently allocated to midsized and large-sized cities. Recently, however, the Federal Communications Commission (FCC) allocated around 700 new FM stations to small communities. Although reception over distance is problematic, the majority of commercial FMs broadcast in crisp, balanced high-fidelity stereo.

The remainder of radio stations, about 1400, or 12 percent, are designated noncommercial stations. Most noncommercial stations are FMs. There are at least four types of these stations. The dominant ones, at least in terms of federal and local money available, are those affiliated with National Public Radio, or NPR. College radio includes some stations that meet NPR criteria but many that do not; these operate under the direction of student clubs and activities. Religious FM stations represent another one-quarter of the noncommercial band, a segment that has been growing in recent years. And there is a small core of "community" educational FMs, which rely mainly on support from local listeners.

The point of this pizza is that radio today is largely a locally based medium. Most stations are commercial AMs, trying to operate on the advertising revenue available from local advertisers in the smaller markets. Large cities are increasingly being dominated by FM commercial stations, where major advertisers call the shots. Within the noncommercial environment the pressure is on to find funding: from the federal government, from religious organizations, from foundations, from university administrations or student fees, or from individual listeners.

Share of Audience

Figure 4-3 is simple but profound. Today more than 70 percent of radio listening is to FM stations. In many of the nation's largest markets the figure is even higher; for example, in Atlanta it's 80 percent. FM's better technical quality is drawing more listeners, particularly those who like to listen to music. Since advertising revenue normally follows the share of listening, the majority of FM stations make money, and major-market FMs lead the pack with profit margins higher than 20 percent. (A *profit margin* is calculated by dividing a station's profit by its revenue. For example, if a station has a profit of $10,000 on total revenues of $100,000, its profit margin would be 10 percent.) On the other hand, most AM stations today lose money. Profitable AM stations tend to be those in major markets, programming news and talk, or longtime leaders in America's smaller towns, which are "full-service" stations of music, news, and local sports, much the same in the 1990s as they were in the 1940s.

The recession that struck the United States in the early 1990s had a profound effect on radio. While most commercial FMs in major markets remained profitable, many small-market stations struggled for survival. In 1990 alone, according to the National Association of Broadcasters, over 300 radio stations went out of business. Of those that "went dark," in industry parlance, about two-thirds were AMs, the remainder was FMs.

RADIO ADVERTISING

A quick reading of the foregoing discussion might suggest that commercial radio is in trouble: too

---

TECHNOLOGY

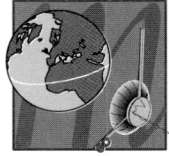

RADIO LEADS THE WAY

Radio is found in more American homes than any other consumer audio product according to a recent survey by the Electronic Industries Association. Radios are in 98 percent of all U.S. homes. In second place are audio systems, both portable and component, that include radio, phonograph, and tape and CD players in 90 percent of homes. Answering machines are in 25 percent, cordless phones in 20 percent, and home security alarms (classified as an audio product since they emit a sound) are in 10 percent of homes.

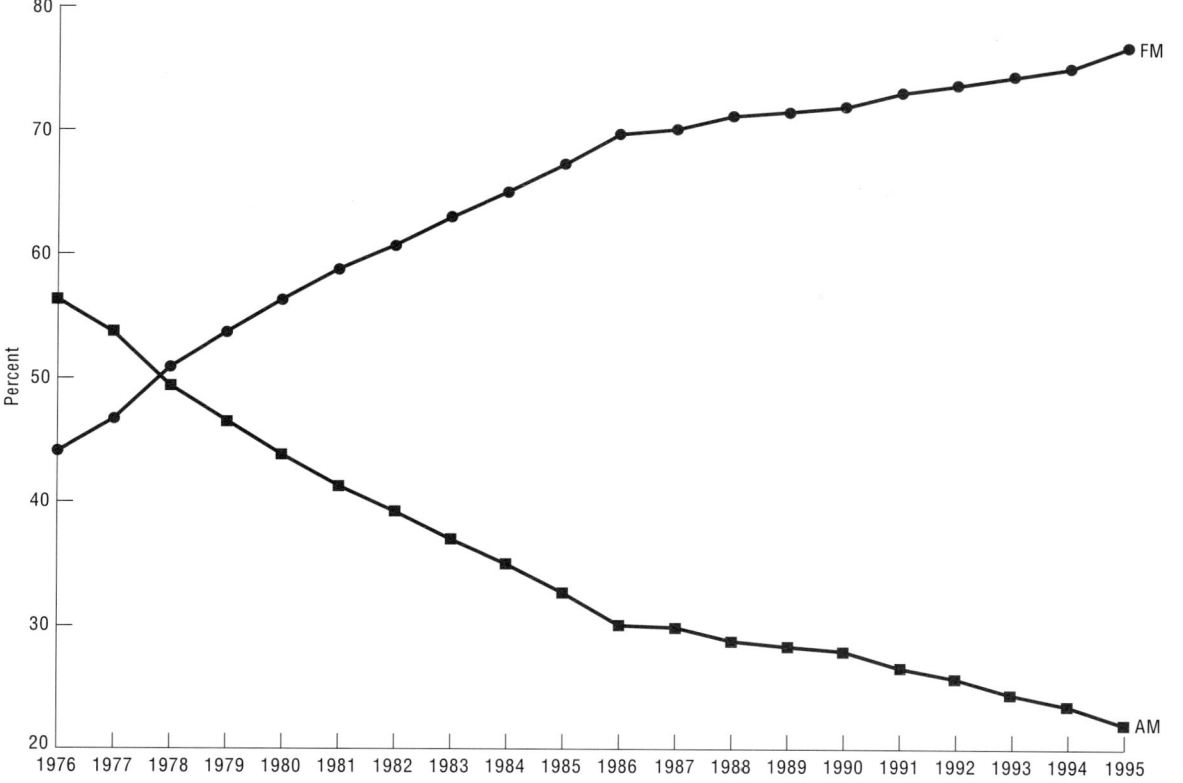

FIGURE 4-3 FM/AM share of the audience. (Source: James H. Duncan, *American Radio,* Fall, 1990. Projection to 1995 by the authors.)

many stations, too few advertisers. The fact is, however, that while the percentage of advertising expenditures on radio remained relatively constant in the 1980s, the volume of radio advertising increased substantially. Take a look at Table 4-1. Advertisers spent $1.3 billion on radio in 1970; by 1980 that figure had almost tripled to $3.7 billion. Radio advertising in 1990 approached $9 billion, almost another threefold increase in ten years! To place the figures in perspective, the number of stations in 1970 was about 6000, a figure that did not even double by 1990. Thus, while competition has certainly increased, so have the stakes. As we shall see, today's leading stations can reap phenomenal profits through advertising revenue.

The Search for Spots

Just like the TV business, there are three types of advertising purchases made in radio. The time segments available for commercials in radio and TV are called *spots,* and this term is also used to refer to the commercials themselves. Local spot sales, known as *local,* is the sale of commercial advertising by stations to advertisers in their im-

TABLE 4-1 Radio Advertising Volume, 1965–1990 (millions of dollars)

Year	Net	National Spot	Local	Total
1965	$ 60	$ 275	$ 582	917
	7%*	30%*	63%*	
1970	56	371	881	1308
	4	28	67	
1975	83	436	1461	1980
	4	22	74	
1980	183	779	2740	3702
	5	21	74	
1985	365	1335	4790	6490
	6	21	74	
1990	433	1626	6780	8839
	5	18	77	

* Percentages may not equal 100 due to rounding error.

SOURCE: McCann-Erickson, Inc., New York. Reported in *Statistical Abstracts of the United States, 1987,* p. 538; Radio Advertising Bureau Reports, reported in *Marketing News,* October 1, 1990.

mediate service area. Auto dealers, appliance stores, and restaurants are frequent local advertisers. Each station has a group of salespeople who call on local businesses and attempt to sell them ad time. An important selling tool for radio stations is the *rate card,* a summary of what the station will charge for an ad. Figure 4-4 is a typical rate card.

Station ad rates are pegged to the *share of the audience* that is listening to the station at a given time. Shares are determined by Arbitron (see Chapter 14), which publishes radio ratings reports. The larger the share is, the more money a station can charge for its commercial spots. Most stations give discounts if an advertiser buys a large number of spots (called a *package*) and if the advertiser commits to buying spots that will run over an extended period of time. Ads cost the most during morning and afternoon drive time, when the largest number of people is in the audience.

Radio stations also make use of *cooperative advertising,* or simply *co-op.* Many local retail stores sell items made by national manufacturers. In a co-op arrangement the national firm will share the cost of advertising with the local business. Thus the Maytag Company might pay part of the cost of local radio time purchased by Green's Appliance Store. Local retailers like co-op because, in addition to helping them pay for the ads, it can allow them to use ads produced by national ad agencies and to tie their local businesses in with a national campaign. In fact, at some stations one or more members of the sales staff are assigned exclusively to deal with co-op plans.

National spot sales refers to the sale of commercial radio time to major national and regional advertisers. For example, Ford and General Motors buy national spots so that their commercials are heard all over the country but at different times on different stations. The local Smith Ford Dealership or the Jones GM dealer would buy local spots. National spot sales are normally made in behalf of local stations by station *representation firms,* or *reps.* Reps maintain offices in the nation's leading cities, like New York, Los Angeles, Dallas, and Atlanta, which are also home to the nation's leading advertisers and their agencies.

The third type of radio advertising is *network sales.* This refers to the sale of commercial advertising by regular and special radio networks, like ABC, Mutual, and NBC, or a network that carries a local college football game, to be aired on their programs. These spots run on each station that carries the network; local stations receive no revenue for their sale. Instead, the network might offer the station some available time within the programs for local sale.

As Table 4-1 reveals, in addition to a change in amount, there has been a subtle shift in the type of advertising expenditures made by ad agencies in radio. In 1970 about 30 percent of all radio buys was of the national spot variety. Local sales accounted for about two-thirds of all revenue. Network radio was in substantial decline at this time, with advertisers spending under 4 percent of their radio budgets on network programs. Today nearly 80 percent of all radio sales is local. National spot sales accounts for under $1 in $5 of radio advertising; the bulk of sales goes to the top-rated stations in the biggest markets. And despite predictions of doom, network radio has rebounded. The success of personalities like Larry King and Paul Harvey, the entrepreneurship of Norman Pattiz, and the rise of satellite-distributed services have renewed advertiser interest in radio networks. (We shall trace this development in the programming chapter.)

In daily business practice, what these figures mean is that radio advertising today is a fiercely competitive battleground (there's that violent, competitive metaphor again). At the local level, stations use every weapon to lure advertisers. At the network level, there is fierce competition among stations to align with the highest rated performers or news personalities, from Paul Harvey to Rush Limbaugh to Charles Osgood. But the battle for advertising dollars is perhaps most visible in the station rep business as the stations struggle for a share of the national spot market.

Examine Figure 4-5, perhaps the most important illustration in the chapter thus far, which tracks station rankings in the nation's top-10 radio markets since 1976. As in the TV business, national spot advertising is usually sold on the basis of station popularity, as measured by ratings. Almost without exception the highest rated stations will get the majority of advertising expenditures. The people placing the advertising, known as *media buyers,* usually go with these higher ranking stations, leaving the other stations out of

GRID RATE CARD #89-1

KSON AM/FM COMBINATION

Contact your KSON AM/FM Account Executive for prevailing grid levels.

97.3 FM • 1240 AM

CLASS AAA

Monday-Sunday 5 a.m.-10 a.m.

GRID LEVELS	UNIT COST (60s or 30s)
I	500
II	430
III	370
IV	320
V	270

CLASS AA

Monday-Sunday 10 a.m.-3 p.m.

GRID LEVELS	UNIT COST (60s or 30s)
I	350
II	300
III	265
IV	230
V	200

CLASS A

Monday-Sunday 3 p.m.-9 p.m.

GRID LEVELS	UNIT COST (60s or 30s)
I	400
II	340
III	290
IV	250
V	220

CLASS B

Monday-Sunday 9 p.m.-12 midnight

GRID LEVELS	UNIT COST (60s or 30s)
I	170
II	150
III	130
IV	120
V	110

BEST TIMES AVAILABLE

Monday-Sunday 5 a.m.-12 midnight

GRID LEVELS	UNIT COST (60s or 30s)
I	320
II	280
III	250
IV	220
V	190

▲ Wednesday through Saturday schedules—next highest grid level.
▲ No discount for AM or FM only.

SPECIAL FEATURES

▲ Rollin' Radio Show
▲ Sponsorship of Local News
▲ Live Personality Ad-Libs
▲ Special Program Sponsorships

FIGURE 4-4 A radio station rate card. (Courtesy: Jefferson-Pilot Broadcasting Company.)

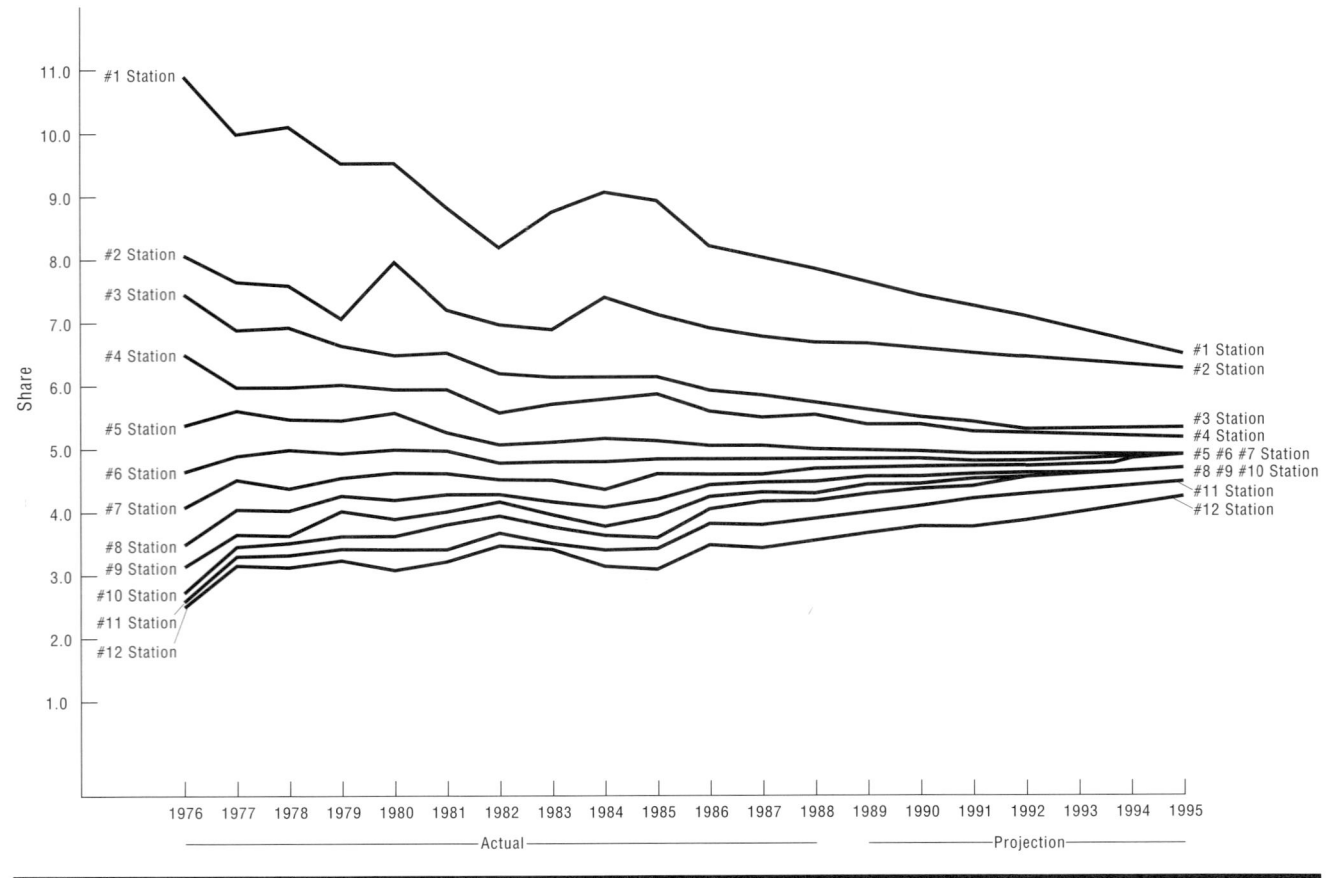

FIGURE 4-5 Actual and projected audience shares for persons over the age of 12 in the top-ten radio markets. (Source: Katz Radio Group.)

the huge advertising expenditure. Stations ranking considerably above their competition thus outbill and outearn their competition, usually by a wider margin than their ratings advantage. Consequently, a station's share of the audience is an extremely important sales tool for a media rep. The higher ranking stations are much easier to sell to media buyers.

In 1976 the top-rated station earned an 11 percent share of the audience, a full three points ahead of the second-ranked station, with an 8 percent share. The tenth-place station in a big market averaged about a 3 percent share of audience. In 1981 the leading station earned less than 10 percent of the total audience, now only a percentage point or so ahead of the second-ranked station. Today a 7 percent share of audience can be a market leader, and as many as three or four other radio stations can be within a percentage point or two of the top. Indeed, the tenth-ranked station, with "only" 3 or 4 percent of the audience, is still within striking distance of the top. With the right programming or promotional strategy, it might trade places with the leader in the next ratings period. If it's able to do this, the station will see a significant increase in its revenue since its media rep is better able to sell spots to national advertisers.

If current trends hold, by the middle of the 1990s the top station in a major market will attract about 6 percent of the total audience; the number ten station will attract about one-half of that (between 2 and 3 percent). Stations will become even more reliant on promotional efforts and targeted programming to maintain a competitive edge.

The Rise of the "Megarep"

As we have seen, the volume of spot radio advertising has grown steadily since 1976, from just under $400 million to over $1.5 billion. Over the same period an unprecedented series of mergers caused the number of radio rep firms to spiral downward. In 1976 there were over twenty-five major national rep firms owned by twenty-two different companies. In 1987 there were only fourteen major rep firms, owned by just five companies!

Two rep conglomerates account for the major share of spot advertising expenditures today. The Katz Radio Group includes such firms as Banner and Republic, independent firms until they merged with Katz. The other major rep firm is Interep, which owns Group W Radio Sales and Major Market Radio, among others.

OWNERSHIP: THE BIG GROUPS

The increasingly competitive nature of the radio business is reflected in the kinds of companies now involved in the medium. For nearly one-half a century radio was generally a small business. Owners were also operators and lifelong residents of the community in which the station operated. Many owned other businesses in the area, such as automobile dealerships, restaurants, even local newspapers. Reflecting this ownership pattern and the "homespun" environment in which they operated, such stations were commonly referred to as "mom and pop" stations.

Although many stations, particularly in small towns, are still locally owned and operated, radio in large cities has become a corporate battleground. Today the business is dominated by powerful groups of stations. These groups have the large operating budgets, specialized personnel, and management acumen to compete vigorously for listeners and advertisers. In many radio markets the groups own both an AM and an FM station (known as combos). Table 4-2 ranks the nation's largest radio group owners. Some of the groups (particularly those at the top of the list) may be familiar to you, as their holdings also include TV stations, cable interests, even movie studios. Examples of groups of this type include Capital Cities/ABC, which owns the ABC TV network; CBS, also an owner of a TV network; and Shamrock, the radio arm of the Disney empire, which includes Buena Vista Films, Touchstone Films, a local TV station, The Disney Channel, and of course Disneyland and Disney World.

However, many of the nation's leading groups tend to specialize in radio. Examples of this type include Infinity, Emmis, and Jacor, which rank fourth, sixth, and eleventh, respectively.

Just as one encounters a McDonald's, a Wendy's, and a Burger King in every city, the radio one finds is likely to be "franchised" as well. Critics claim that this is one reason many stations sound exactly the same from city to city, featuring the same programming, promotions, even personalities. If all-news KYW in Philadelphia is reminiscent of KFWB in Los Angeles and WINS in New York, it may in part be due to the fact that all three are owned by Westinghouse. Similarly, country stations in Wichita, Shreveport, Omaha, Denver, and Springfield (Missouri) may all drawl alike since Great American Broadcasting owns country combos in all of these cities.

Jeff Smulyan, chairman of the board, Emmis Broadcasting.

TABLE 4-2   America's Leading Radio Groups

Rank	Group	Revenue	Stations	Revenue per Station
1.	Capital Cities/ABC	$232,200,000	18	$12,400,000
2.	CBS	216,900,000	19	11,415,000
3.	Westinghouse	200,000,000	17	11,765,000
4.	Infinity	131,500,000	14	9,393,000
5.	Cox	103,900,000	12	8,658,000
6.	Emmis	101,900,000	10	10,190,000
7.	Bonneville	96,200,000	10	9,620,000
8.	Gannett	91,200,000	12	7,600,000
9.	Viacom	90,000,000	10	9,000,000
10.	Great American	87,200,000	17	5,129,000
11.	Jacor	75,800,000	12	6,317,000
12.	Noble	73,700,000	14	5,264,000
13.	Malrite	73,500,000	10	7,350,000
14.	Susquehanna	70,900,000	15	4,727,000
15.	Greater Media	67,400,000	12	5,617,000
16.	EZ	62,900,000	12	5,242,000
17.	Shamrock (Disney)	58,800,000	12	4,900,000
18.	Booth/Genesis	55,000,000	14	3,929,000
19.	Tribune Co.	50,000,000	3	16,667,000
20.	Nationwide	50,000,000	13	3,846,000
21.	Beasley	50,000,000	14	3,571,000
22.	Cook Inlet	47,600,000	7	6,800,000
23.	Summit	46,600,000	9	5,178,000
24.	NewCity	46,400,000	11	4,218,000
25.	Evergreen	45,600,000	6	7,600,000

* 1990, estimated.

SOURCE: *Duncan's Radio Market Guide,* 1991. Used with permission.

Advocates of this trend maintain that group ownership allows for economies of scale (more efficient programming, better news gathering) to keep big-city radio exciting and interesting. Critics charge that group owners lack the sensitivity to community concerns that "mom and pops" have. If you're planning a radio career and favor local independent ownership, you may wish to consider small-market radio. If you're heading for New York, Chicago, or Los Angeles, the "corporate culture" is no doubt in your future.

RADIO PROGRAMMING TODAY

At the risk of sounding stereotypical and sexist, a snapshot of radio listeners today might include a teenager "zoning in" to a personal stereo on a school bus, a middle-aged executive listening in her luxury automobile, a city youth listening to his oversized "boom box," a trucker listening while traveling the highway in his eighteen-wheel rig, and a secretary listening to a desktop radio while at her computer workstation. It is highly unlikely that each is listening to the same type of station.

With more than 11,000 stations on the air, with more than 25 stations typically available for most listeners, and with radio competing for audiences against tape and CD players, TV, movies—even live entertainment like concerts and theater—radio has become a focused and highly targeted medium. That is, rather than programming to meet the broadest tastes of the largest numbers of people in their listening areas, most stations today cater to a narrow market segment, the core of listeners who prefer a certain type of programming. The two key components of this trend are target audience and format.

The concept of *target audience* emerges from advertising research, which shows that the majority of sales of a given product are made by a minority of the public: the target market for that product. For example, the overwhelming majority of beer is purchased by men between the ages of 18 and 49, teenagers account for most movies attended, and adults over 45 take the most European trips. In developing their campaigns, advertisers try to identify the target market for a product and then to develop appeals that meet the needs of this group.

Commercial radio today, particularly in the largest cities, is programmed in precisely the same fashion. Reflecting this trend, in fact, cities themselves are known in the business as "markets." Management identifies a target audience by its age, sex, music preferences, lifestyle, and other information, and it develops a program strategy to satisfy that group. The program strategy is known as the radio station's *format*. A successful radio station consistently delivers its intended target audience, in both aggregate size (quantity) and lifestyle preferences (quality). Its listeners are an identifiable subgroup, largely similar in age, sex, habits, leisure pursuits, and other characteristics. This makes the station attractive to advertisers: the name of the game in commercial radio.

Figure 4-6 lists today's most common radio formats and their typical target audiences. Chapter 9 will detail how a format evolves from inception to execution. Until then, let's examine the key components of format radio today.

Adult Contemporary

The most popular radio format today is adult contemporary, or AC. Depending on the station's interpretation and the type of music played, AC stations may also be known as "soft rock," "oldies," or "mellow rock" stations. AC stations have a number of distinguishing characteristics. In music the emphasis is on soft, nonmetallic rock and roll, strong on melody and familiarity. The songs played are by rock artists familiar to adult listeners, from the Beach Boys to the Beatles. Oldies played are likely to have been top-10 hits in their heyday.

Announcers are generally pleasant, friendly, innocuous, and noncontroversial. In fact, many listeners would be hard put to name the announcers at their favorite AC station since the main reason they listen is for the music.

AC stations are popular because they tend to attract the audience most in demand by advertisers and marketers: women between the ages of 25 and 54. The appeal of AC is wide: from urban areas to rural, from college educated to grade school, from upper income to the poverty line. But AC is particularly strong "where it counts" to many advertisers: among middle- and upper-income housewives and working women in urban and suburban areas. Today one in five FM stations plays a version of AC. AC is popular on AM as well, particularly in the pure "oldies" format, concentrating on hits from the heyday of AM radio in the 1950s and early 1960s.

Country

Country radio stations are second in listenership among the major formats, and there are more stations with this format than any other. The

PROGRAMMING

NEW FORMATS FOR AM RADIO

Some of the most interesting experimentation in radio today is taking place on the AM band, as managers try to develop new formats to attract audiences. There has been an "all-Beatles" radio station, KBTL, in the Houston area. A Cincinnati station, capitalizing on rumors that "The King" was still alive, went to an all-Elvis format. WFAN in New York is dedicated to the hard-core sports enthusiast, with a relentless barrage of play-by-play, sports talk, and interviews. K-PAL in Little Rock was devoted exclusively to children, with a unique mix of educational and entertainment features for the under-12 set. Sadly, the station quickly went out of business.

As FM share continues to increase, look for AM to try more alternative formats. In development is an "all-motivation" service devoted to nutrition and health, stress reduction, and personal improvement programming. A natural extension of all-news is the all-financial format, featuring stock and business reports and investment information.

What's next? Based on recent trends in popular culture, what about all-wrestling? all-therapy? all-Madonna? The mind boggles. . . .

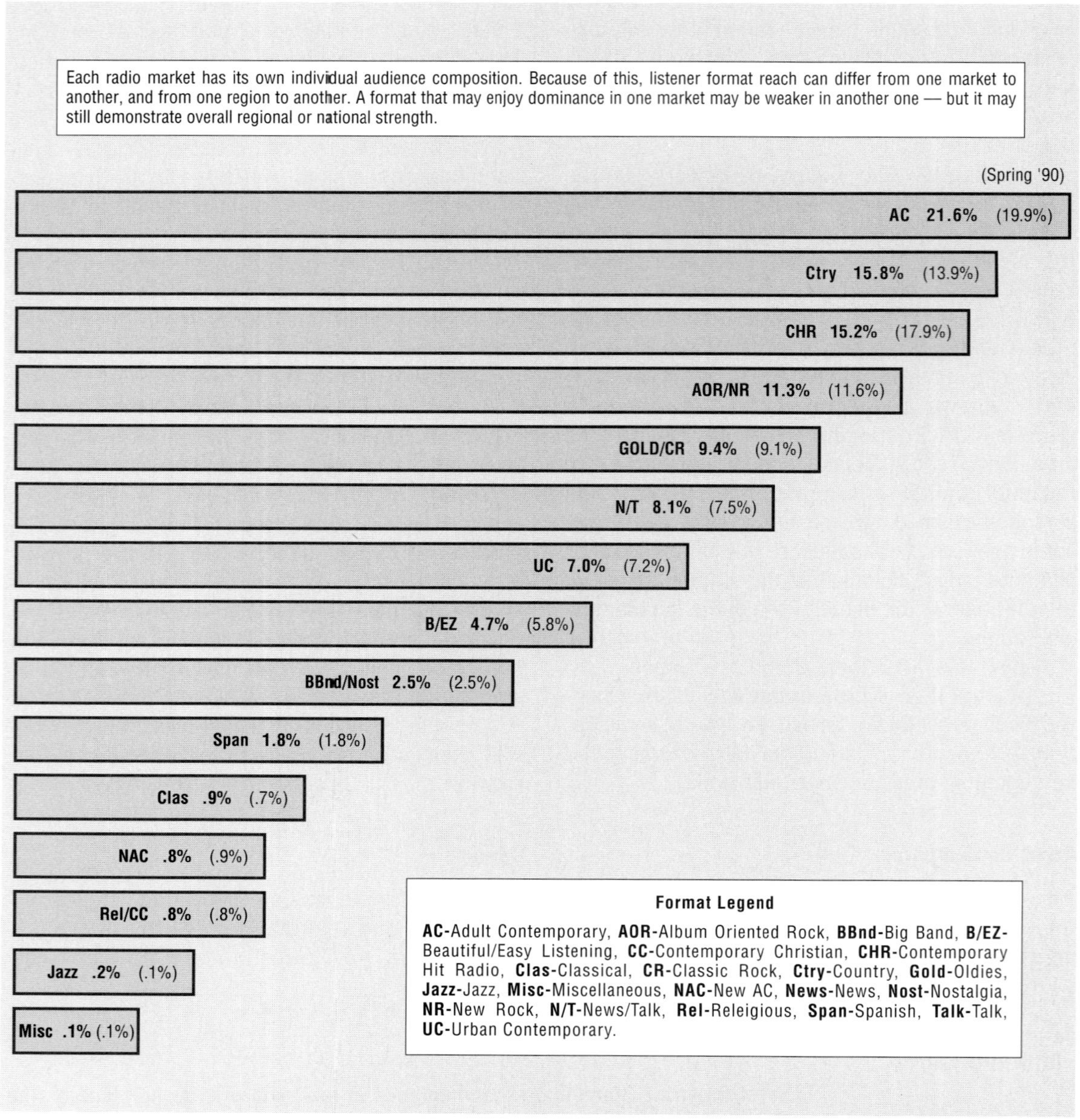

Each radio market has its own individual audience composition. Because of this, listener format reach can differ from one market to another, and from one region to another. A format that may enjoy dominance in one market may be weaker in another one — but it may still demonstrate overall regional or national strength.

(Spring '90)

AC 21.6% (19.9%)
Ctry 15.8% (13.9%)
CHR 15.2% (17.9%)
AOR/NR 11.3% (11.6%)
GOLD/CR 9.4% (9.1%)
N/T 8.1% (7.5%)
UC 7.0% (7.2%)
B/EZ 4.7% (5.8%)
BBnd/Nost 2.5% (2.5%)
Span 1.8% (1.8%)
Clas .9% (.7%)
NAC .8% (.9%)
Rel/CC .8% (.8%)
Jazz .2% (.1%)
Misc .1% (.1%)

Format Legend
AC-Adult Contemporary, **AOR**-Album Oriented Rock, **BBnd**-Big Band, **B/EZ**-Beautiful/Easy Listening, **CC**-Contemporary Christian, **CHR**-Contemporary Hit Radio, **Clas**-Classical, **CR**-Classic Rock, **Ctry**-Country, **Gold**-Oldies, **Jazz**-Jazz, **Misc**-Miscellaneous, **NAC**-New AC, **News**-News, **Nost**-Nostalgia, **NR**-New Rock, **N/T**-News/Talk, **Rel**-Releigious, **Span**-Spanish, **Talk**-Talk, **UC**-Urban Contemporary.

FIGURE 4-6 Format reach chart showing the percentage of listeners nationwide who listen to each format. (Source: *R&R Ratings Report,* 1991.)

reason is clear: these stations predominate in the rural areas and smaller cities of America, particularly in the south and west. This by no means suggests that country is "hick" or "backwards" in nature. Many of the nation's most successful stations play country music, even in such "sophisticated" markets as Detroit, Phoenix, and Washington, D.C.

As a radio format, country music has decided advantages. Its appeal is broad. People of all ages listen to country; unlike AOR (see below), for example, it's a cradle-to-grave experience. Like most of the preferred formats, it delivers more women than men. Even so, large audiences of both sexes respond to the music. Country fans are loyal: they tend to listen to a single favorite

Two of the better known DJs in America: Mark and Brian of KLOS, Los Angeles. In addition to being the top-ranked morning drive time duo, Mark and Brian were stars of a short-lived TV series on ABC in 1991.

station for long periods of time, making these stations a prime target for advertisers.

The growth in popularity of country in the 1970s and 1980s led to the development of various derivatives. "Traditional" or "classic" country stations consider themselves the purest players of the format. The emphasis in music is on the country and western standards of twenty or thirty years ago. Current country songs, particularly those that might receive airplay on CHR or AC stations, are avoided.

"Contemporary" or "modern" country stations follow the rules of CHR. They concentrate on current hits on the country charts, particularly the most up-tempo or upbeat tunes. Strong airplay is usually given to the "superstars" of country, like Garth Brooks, Randy Travis, and Reba McIntyre.

Regardless of the music orientation, most country stations are "full-service" operations. Announcers tend to be friendly and helpful, directly involved in community events. Unlike some of the other contemporary formats, most country stations provide news, weather, and other information to their listeners. Remote broadcasts are common, from concerts and fairs to shopping malls and drive-ins.

Part of the continuing appeal of country music is its success on both AM and FM. Country is heard nearly evenly on the two bands. This means that the format has participated in the FM boom but also remains viable in AM. Like rock and roll, country radio is here to stay.

Contemporary Hit Radio

Targeting a younger audience than AC is the contemporary hit radio format, or CHR. Contemporary hit radio may also be referred to as top-40 or current-hit radio. Contemporary hit radio is like an audio jukebox. The emphasis is placed on the most current music, the songs leading the charts in record sales. The music played is almost always bright or up-tempo. Slow songs, long songs, and oldies (even those only a few months old) are avoided. Songs play again quickly (a program strategy known as *fast rotation*) and are removed from the playlist as soon as there is evidence that their popularity is declining. Top 40 is a misnomer: today's CHR stations may play as few as twenty or thirty songs.

"Screamers" are common on CHR stations. Disc jockeys (DJs) tend to be assertive, high-energy personalities who sprinkle their shifts with humor, sound effects, and gimmickry. Contemporary hit radio stations sound "busy" compared to ACs: the air is filled with contests, jingles, jokes, buzzers, whistles, and, above all, hits. Currently about 15 percent of all stations play this format, the majority being FM stations in large markets. Contemporary hit radio is most popular with the age group that buys single records and hot albums: preteen, teen, and young adult women. This is the group in recent years that has made mammoth stars of Paula Abdul, Janet Jackson, and Michael Bolton.

Album-Oriented Rock

The teens and young adults who wouldn't be caught dead with a Michael Bolton record are likely to be listening to album-oriented rock (AOR) stations. We mean, of course, preteen, teen, and young adult males. In fact, of all the major formats, AOR is most heavily targeted, or *skewed,* to male listeners. Over 60 percent of the AOR audience is male.

In many ways AOR is the opposite (some fans say the antidote) of AC and CHR. Short, tight playlists are avoided in favor of long musical segments, or *sweeps*. The music is different, too. Songs played are "harder" or "heavier," ranging from the classic groups like Led Zeppelin and the Doors to contemporary superstars like Guns 'n Roses and The Black Crowes.

Album-oriented rock personalities are as diverse as the music. On one end of the scale is the sultry female DJ. On the other end is the bizarre world of raucous male announcers who liberally sprinkle their talk with double entendres, innuendos, and ethnic barbs.

With the "baby boomers" of the 1950s hurtling toward middle age, some AOR stations have begun to target older males who may have abandoned the format for AC and other radio stations. Today "classic rock" AOR stations sprinkle their playlists with psychedelic sounds of the 1960s and 1970s, from Dylan to the Grateful Dead, in search of more so-called mature (or at least older) males. This mix has apparently been successful, and the classic rock format is becoming increasingly popular.

Other AOR stations have targeted younger listeners, who often see themselves as trend-setters. This version of the format is known as "new rock," or "modern rock," and has made superstars of such groups as R.E.M., Jesus Jones, U2, and Nirvana.

In addition to being a predominantly male-oriented format, AOR is overwhelmingly an FM phenomenon. Roughly two of every ten FM stations can be labeled AOR or a derivative. Album-oriented rock is virtually unheard of on AM—probably because AOR listeners, with their home component systems, car stereos, and portable CD players, prefer the superior fidelity of the stereo FM band.

News and Talk

News and talk is a broad radio format. At one end of the scale are "all-news" operations, usually AM stations in major metropolitan areas that program twenty-four hours a day of news, sports, weather, and traffic information. Examples include WINS and WCBS in New York, WBBM in Chicago, and KNX and KFWB in Los Angeles. At the other end of the scale are "all-talk" stations, which rotate hosts and invite listeners to call in on a range of topics—current affairs, sports, auto mechanics, counseling in every area, even sex. Examples are New York's WOR, WIOD in Miami, and KSTP-AM in Minneapolis. Between these extremes there are many "news and talk" stations, some of which mix play-by-play sports and occasional musical segments into the format. However, even the "hybrids" center their programming on information services.

Whatever its unique interpretation, talk radio is a growth format, particularly on the AM band. In fact, almost half of all AM listening today is to news/talk stations. Part of the appeal of news/talk is its audience composition. The news format attracts "big numbers," particularly during important drive time in the morning and afternoon. It follows that people who listen for news, traffic, and weather information are on their way to jobs. This makes them an ideal target for advertisers.

Black/Urban Contemporary

The black or urban contemporary (UC) format refers to the percussive, up-tempo sounds of the stations in America's major cities. Black is actually a misnomer for the format: people of all racial and ethnic backgrounds enjoy the music. Urban contemporary arose out of the "disco craze" of the mid-1970s, when a gyrating John Travolta captivated the culture and sent a new generation to the dance floors. By the 1990s break dancing had merged with rap music to create a new force in UC. Today variations of the format range from the rap, "hip-hop," and "house" music of Salt 'n Peppa, Hammer, and Bell Biv Devo to the more traditional rhythm and blues sounds of Whitney Houston and Natalie Cole.

Beautiful Music

Beautiful music, or "easy listening" (B/EZ) refers to the stations that program "wall-to-wall," "background," or "elevator" music. One of the stalwarts of FM (which stood for "fine music" in its early days), easy listening is an evolving format. Where once the "rules" called for only instrumental music (Mantovani, 101 Strings, Tony Matola, etc.), today it is common to hear Billy Joel, Joni Mitchell, Neil Diamond, and a range of other well-known vocalists on B/EZ air.

The primary reason for this change is the aging audience for the format. Beautiful music appeals

The urban contemporary radio format has been doing well lately thanks to the increasing popularity of rap music and rap artists such as Hammer.

mainly to older listeners, those above 45 years of age. Today's mature audiences grew up after the dawning of the age of rock and roll. They are as likely to relax and unwind to pop tunes as they are to lush orchestrations.

Beautiful music stations tend to have unique program elements. Music is generally played continuously; breaks for commercial announcements and news segments, if used, are kept to a minimum. Announcers have pleasant, low-key styles. They will never shout at you. Contests and other aggressive promotions are eschewed in favor of "image-enhancing" station events (sponsorship of appropriate music performances, for example).

The success of the format is not based on audience size, although in many markets beautiful stations boast big audiences. Rather, these stations attract a high-quality audience of professionals and managers. In addition to being high earners, this audience tends to listen to the radio for long periods of time and to be loyal to a single station or to very few stations. Interestingly, despite a vast gap overall in household income and education, in effect the audience for beautiful music shares much with that of country and UC. They are dedicated, loyal, and highly identifiable to advertisers.

Other Formats

Together, the formats just described account for nearly 90 percent of radio listening in the United States. The additional 10 percent is filled out by a number of other formats.

A format that targets the older audience is nostalgia, sometimes known as "big band." The music featured is the sound of the 1940s and early 1950s when bands like those of Benny Goodman, Gene Krupa, and Duke Ellington were popular. The bands were often "fronted" by a solo performer, many of whom had solid singing careers on their own. Frank Sinatra, Helen O'Connell, Doris Day, and Nat "King" Cole (Natalie's dad) fall into this class.

Appealing to the vastly growing population of immigrants from Central and South America is the Hispanic format. By the early 1990s, nearly 300 stations were programming in the Spanish language; most of these were on AM, many in large cities in the northeast and southwest.

Religious stations appeal to a variety of faiths, but Christian stations are most plentiful. In addition to delivering inspirational talks, many religious stations include music in their format. Those targeting the black audience are generally known as gospel stations. Religious stations tend to be most popular among older women (above the age of 50). The majority of religious stations (60 percent) can be found on the AM dial, usually at the high end of the dial where local stations are most plentiful.

While classical music and jazz are the backbone of public radio stations (see below), some fifty or so commercial operations play these formats. Classical stations attract a very upscale listener: between 35 and 54 years of age, typically college-educated and professional. The jazz enthusiast shares a similar profile. Although the number of classical and jazz aficionados is small (compared, say, to the audience for adult contemporary), the "high quality" of their listeners makes these stations potentially attractive advertising vehicles. In the late 1980s a new format emerged that seemed to merge classical and jazz, with elements of rock and pop included. "New age" [also known as "new adult contemporary (NAC)] music, featuring pianists like Liz Story and ensembles like Mannheim Steamroller and Shadowfax, began to find airplay in some cities. New age has a light, "airy" sound (some critics call it the sound of junk food). By the early 1990s about a dozen commercial stations played this type of music, led by KTWV-FM, "the Wave," in Los Angeles.

It sometimes appears that there are almost as many other formats as there are radio stations. Filling out the dial there are Portuguese, Greek, Polish, German, and other foreign-language stations. Over twenty stations program exclusively to American Indians and Eskimos. Clearly radio today is in an era of format diversity.

PROMOTIONS

As competition for listeners soars, promoting the station increases in importance. With so many stations to choose from (more than fifty in Los Angeles, for example), and so much format duplication (New York has four AOR stations, Chicago has five CHR stations), stations rely on promotions to convince listeners to tune in. Promotions help differentiate one station from another and give greater appeal to a particular station. Differentiation and increased appeal, it is hoped, lead to higher ratings and higher advertising rates. In fact, most radio stations schedule promotions to coincide with ratings periods (see Chapter 14). Consequently, promotion is directly related to a station's profitability.

There are three major forms of promotions used by radio stations. The first is advertising. Many stations, especially in the big cities, rent billboards, take ads in the newspapers, or buy TV spots to publicize a DJ, format change, or new image. The second form consists of promoting special events, such as rock concerts or road races, that give the station increased visibility among its audience members. If these events benefit the community, as a food drive would, then the station also profits from improved public relations. Frequent on-air spots of ten, twenty, or thirty seconds publicize the upcoming event and stress the station's connection with it. The third type of promotion—contests and giveaways—have been taken to new heights by radio stations. The typical contest requires the listener to do something special to win, such as identifying

PROGRAMMING

WARNING: BEING A DJ CAN BE HAZARDOUS TO YOUR HEALTH

You would think that sitting in a studio all day playing records would be a pretty safe job, right? Wrong. In this age of high-power radio competition, a lot of DJs have been involved in high-risk station promotions. For example:

- An FM station in Sacramento, California, ran a Smash the Zoo promotion in which all five members of the station's controversial morning "Zoo Crew" sat in an old car that had all the window glass removed. Listeners were given sledgehammers and allowed to pound on the car. The car was supposed to rest on a platform so that people swinging the sledgehammers wouldn't be able to reach the open windows. Unfortunately, the platform never arrived. With the DJs in easy reach, members of the crowd started flailing away with 10-pound sledgehammers. Luckily, the promotion was stopped before anybody got hurt.
- The sports director on a Nashville station was camped out on a billboard protesting a losing streak by the University of Tennessee football team. Unluckily, he fell off and suffered a severe gash on his leg.
- A morning DJ for a Phoenix station pogo-sticked the entire 26-mile route of the Boston Marathon—it took him two and half days—and wound up with severe knee damage.
- An Abilene, Texas DJ tried to raise money for the United Way by spending eighty-one straight hours on a Ferris wheel. One night, while adjusting his sleeping bag, he fell off. Luckily for him, he was at the bottom of the wheel at the time. Unluckily, though, the Ferris wheel attendant who immediately rushed to his aid was hit by the wheel and suffered a concussion.

In the competitive world of radio, promoting the radio station is important. Here are some of the promotional devices used by jazz station **WBGO-FM** in Newark, New Jersey.

the titles of songs by listening to short excerpts, remembering the names of the station's DJs, finding a station's "lost" call letters, or solving a puzzle by listening to broadcast clues. On the other hand, a giveaway simply requires the listener to be the tenth or twentieth caller to the station or to answer the phone with "WWWW is my radio station" or to call the station within a specified time after hearing his or her name read on the air. The listener knows instantly if he or she is a winner.

Since the 1970s the prizes connected with contests and giveaways have gotten bigger and bigger as stations entered into fierce promotional rivalries. For example, a Houston station put a listener in a bank vault containing a million dollars and gave the contestant ninety-three seconds (the station's frequency) to grab as much as possible. The listener wound up with about $125,000. A Los Angeles station specializes in giving away expensive automobiles. In the past few years one lucky listener won a Porsche 944 and another listener won a Rolls-Royce. Along with cash, other stations give away shopping sprees, vacations, concert tickets, stereo equipment, and other merchandise. One Nashville station even built a name for itself based on the unusual prizes it gives. A breast enlargement operation went to one winner while another was covered in honey, allowed to roll around in a swimming pool filled with cash, and got to keep the bills that stuck.

Some station managers point out that promotions really amount to "buying an audience" for a station, which results in fickle listeners who simply tune to the station that's offering the most. Others note promotions can get out of hand and cost the station more money than they can bring in. But most agree that promotions are a permanent fixture in modern radio.

ECONOMICS OF RADIO TODAY

One of the distinguishing features of contemporary radio is its focus on financial performance. Employees and managers at today's radio stations, both public and commercial, are under intense pressure to operate in an efficient and cost-effective manner. An examination of the economics of the radio industry underscores this and other financial trends in radio today.

Life on the Bottom Line: Commercial Radio

As we discussed earlier, there are three main sources of income for commercial radio stations: network advertising, national/regional spot advertising, and local advertising. On average, today's radio stations earn three-quarters of their income from local sales, just under a quarter of every dollar from national/regional spot advertising, and the remaining two cents or so from payments from radio networks carried by the stations.

On the expense side, about a third of every dollar spent goes to general and administrative costs (rent, utilities, office supplies, and other overhead items); another third goes to sales expenses (including advertising and promotion, research costs, travel, and client lunches). For each dollar spent about a quarter goes to news and entertainment programming and production. The remaining amount (under a dime) is spent on engineering: keeping the transmitter, tubes, and transistors humming. By themselves these percentages may seem somewhat abstract and may obscure the different economic picture for AM, FM, and AM–FM combo stations. To clarify these comparisons, examine Table 4-3.

Table 4-3 presents revenue and expenses for a typical midsized American city with a population between 250,000 and 500,000 people. We could be in Toledo, Knoxville, Rochester, or any of dozens of similar cities. The table contrasts the varying economic climate for a full-time AM station, an AM–FM combo, and a full-time FM facility.

First, let's inspect the income disparities. Since the AM–FM combo is actually two stations in one, not surprisingly it rakes in the most advertising dollars, about $1.5 million. The FM stand-alone takes in about $1.1 million a year, and the AM takes just under $600,000 or about one-half the revenue of the other types of stations.

Note that the FM station nearly doubles the local advertising revenue of the AM station and that the AM–FM combo nearly triples the revenue of the stand-alone AM station. It is no wonder that today many AM stations are in perilous financial straits.

On the expense side, note that operating an AM–FM combo and an FM stand-alone costs more than twice as much as running an AM station. The difference is not much in engineering and programming. The overhead expenses at an

TABLE 4-3 Revenue and Expenses for Typical AM, AM–FM Combos, and FM Stations

Station Revenue & Expense Items	AM ($)	AM (%)	AM–FM Combo ($)	AM–FM Combo (%)	FM ($)	FM (%)
Network compensation	12,114	2.1	22,728	1.5	9,505	0.8
Nat'l/Reg'l advertising	137,107	23.3	341,186	22.0	259,245	23.1
Local advertising	438,397	74.6	1,189,219	76.6	852,733	76.0
Total time sales	587,618	100.0	1,553,133	100.0	1,121,483	100.0
Agency & rep. commissions	50,292		149,766		133,092	
Other Revenue	33,648		51,822		25,666	
Total Net Revenue	570,973		1,455,188		1,014,057	
Departmental expenses						
Engineering	29,879	5.1	67,843	4.7	38,860	3.7
Program & promotion	124,110	21.2	285,470	19.8	198,655	19.1
News	46,941	8.0	50,740	3.5	20,813	2.0
Sales	111,663	19.1	304,673	21.1	226,287	21.8
Advertising & promotion	35,478	6.1	100,441	7.0	77,380	7.5
General & administration	236,464	40.5	633,359	43.9	476,158	45.9
Total expenses	584,535	100.0	1,442,526	100.0	1,038,154	100.0
Pretax profit	(13,561)		12,662		(24,097)	
Employment						
Full-time	13		23		15	
Part-time	5		9		5	

SOURCE: *1990 NAB/BCFM Radio Financial Report*; used with permission.

FM facility and AM–FM combo are much higher, as are sales and promotion expenses.

The bottom-line figures are illustrative of trends discussed throughout the chapter. Both the typical AM and FM stations reported operating at a loss in 1990. The AM–FM combo earned only $13,000—a profit margin of less than 1 percent! Thus the "big money" was concentrated in the hands of the major group owners in the major markets. In fact, the typical FM station in a major market (population over 1 million) reported a profit margin exceeding 15 percent. Stations in smaller cities and towns are struggling today to stay afloat in a difficult economy.

The Trickling Pipeline: Economics of Noncommercial Radio

The 1980s will be remembered for many things. When it comes to public broadcasting, the decade will be epitomized by one word: accountability. This was the era in which public stations saw a decreasing flow of support from federal and local governments and increasing competition for funds from private sources, such as local business and charitable foundations. The result for the 1990s is a cost-conscious, belt-tightening atmosphere at America's public radio outlets, which some observers claim is making them more and more like their commercial counterparts.

To illustrate, note the differences in funding sources in the period between 1975 and 1995, illustrated in Figure 4-7.

In 1975 federal, state, and local governments provided about 67 cents of every dollar going to public broadcasting. Listeners contributed about 12 cents and foundations and businesses contributed about 8 cents each. The remaining amount came from other sources. Today the government pays less than one-half the bill. Through subscribership and membership programs listeners now account for nearly a quarter of the budget. Business involvement has more than doubled, promoting some critics to argue that the educational band has become too commercial. Foundation grants are on the decline, dipping from 8 to 4 percent of funding. And other unconventional income sources have increased. Included in this "grab bag" are such items as listener magazines, merchandise (tote bags and umbrellas are particularly popular), even toys and games.

By 1995 listeners may provide a greater share of support for public broadcasting, as much as 30 percent of the budget. State and federal funding will together comprise another third. Corporate and business underwriting may increase to the point where it accounts for at least one-fourth of the income base. An era of "social Darwinism" may be dawning in public radio: survival of the fittest (in terms of ability to be self-supporting) may become the rule.

PUBLIC RADIO TODAY

The bulk of America's radio stations seek profits through advertising sales, but about one in six stations do not. Into this class fall approximately 1500 noncommercial radio stations. There are three main types of noncommercial stations: community, college, and public.

Community stations are those that are licensed to civic groups, nonprofit foundations, local school boards, or religious organizations. There are about 350 of these stations, which operate in the FM band (between 88 and 92 megahertz), providing a range of services from coverage of local issues, study-at-home classes in conjunction with local schools, religious services, and other "home-based" activities.

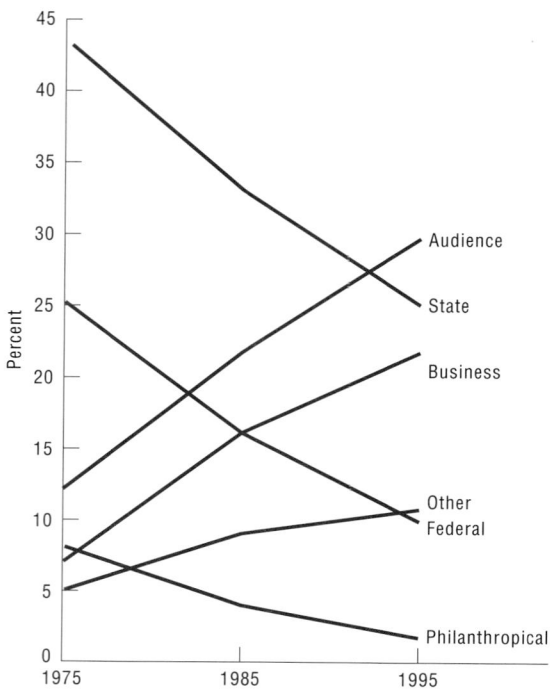

FIGURE 4-7 Public broadcast funding, 1975–1995. (Source: National Public Radio. 1995 estimate by the authors.)

IMPACT

THE POWER OF PUBLIC RADIO

While public radio has faced considerable financial hardship in recent years, and its listening levels fail to match those of commercial radio stations, the medium has enormous political power and influence. This power is strongest in and around NPR's headquarters in Washington, D.C.

Over the years aggressive reporting on "Morning Edition" and "All Things Considered" has had considerable impact on the political process. In the early 1970s NPR vigorously covered the Watergate scandal, which ultimately led to the resignation of President Nixon. More recently, the network was at the heart of coverage of the Iran-Contra scandal, the Persian Gulf War, the nomination of Robert Gates to head the CIA, and other matters of national and international scope.

NPR gained some additional attention in late 1991 when Nina Totenberg, a reporter for "All Things Considered," helped break the story alleging sexual harassment by Judge Clarence Thomas while he was the head of the Equal Employment Opportunity Commission. Ms. Totenberg made national headlines after an appearance on ABC's "Nightline" with Senator Alan K. Simpson of Wyoming. According to witnesses, in a parking lot following the program Senator Simpson waved copies of the journalistic code of ethics at Ms. Totenberg. Ms. Totenberg responded with a burst of verbal gunfire, including some language that is not normally heard on the radio.

"I am not saying I didn't curse," said Ms. Totenberg. "I should have just kept my mouth shut and left."

College radio is a broad category comprising about 800 stations licensed to universities and some secondary schools. About 650 of these stations are members of the Intercollegiate Broadcasting Society (IBS). College stations are a diverse group. However, most of them share a similar programming pattern. The musical mix is eclectic and "progressive," featuring program blocks of new wave, new age, reggae, folk, jazz, and other alternatives to standard formats. Many college stations operate as training sites for students planning broadcast careers. Thus in addition to announcing, staffers gain experience in news, play-by-play sports, public affairs, and promotions.

Public radio stations are also known as *CPB-qualified stations*. These stations meet criteria established by the Corporation for Public Broadcasting (CPB), enabling them to qualify for federal funds. Criteria for CPB support include full-time staffs of at least five, annual budgets exceeding $150,000, and a minimum operating schedule of eighteen hours per day. These stations are the backbone of the National Public Radio (NPR) service, well known for its national news programs "All Things Considered" and "Morning Edition" and for entertainment offerings. By 1992 NPR had nearly 400 member stations.

Diversity is the byword for public radio programming. Figure 4-8 illustrates the typical programming mix in public radio. About two-thirds of the air time at public radio stations is devoted to music. Classical music is the preferred musical type, followed by jazz, folk, contemporary, and opera. News and public affairs programs amount to about one in five programming hours, a much higher proportion than that found in commercial broadcasting (excluding news/talk stations, of course). Other informational programming, ranging from cultural and arts reviews to history and biography, comprises about 7 percent of public programming. About 4 of every 100 hours is devoted to spoken-word performance pieces, such as radio drama and poetic readings. Interestingly, for a medium sometimes known as "educational" radio there is very little direct instruction going on. Today less than 1 percent of public radio air time is devoted to classroom-type instructional programming.

As one might expect, the audiences for public radio tend to be comparatively smaller than those

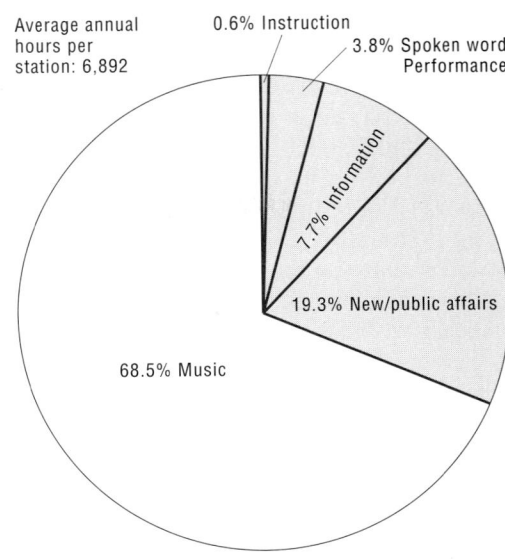

FIGURE 4-8 Public radio programming hours by content category. (Source: National Public Radio.)

for commercial stations. Nationwide, public radio averages about a 2 percent share of audience at any given time. This compares with shares of 8 to 20 percent for highly ranked commercial radio outlets. However, in some cities public radio attracts sizable audiences.

The public radio audience is somewhat highbrow. Studies have shown that the majority of listeners to public radio stations have college degrees and many listeners have advanced and professional degrees. You may recall that this profile overlaps the target audience of some commercial radio formats, such as advertiser-supported classical and jazz stations and beautiful music operations. In some cities, in fact, there is considerable competition between commercial and noncommercial stations for audiences. In Cleveland, for example, both commercial WKSU-FM and noncommercial WCLV-FM attract about 100,000 classical listeners per week. In Los Angeles, jazz enthusiasts are attracted to commercial KKGO-FM and noncommercial KLON-FM. While the competition is tame in comparison to commercial radio "wars," it may heat up as public radio struggles for survival in an era of budget cutbacks and dwindling federal support.

Noncommercial and commercial stations also compete, in a limited way, for advertising dollars.

This may sound contradictory since the term "noncommercial station" seems to suggest one that does not run ads. Nonetheless, public radio stations are allowed to accept corporate *underwriting*. Under this arrangement a company agrees to pay for all or part of the costs of a program and in return gets a mention at the beginning and end of the show. There are some restrictions on what the underwriter can say during these mentions that differentiate them from traditional advertising. An underwriter, for example, can mention a product but cannot claim that the product is superior to its competitors. Despite these limitations, many companies are eager to underwrite programs primarily because of the prestige involved. Texaco, for example, which spends relatively little on radio advertising, is a prominent underwriter of public radio programs.

RADIO STATION ORGANIZATION

Regardless of their size, commercial radio stations tend to share an organizational pattern. Figure 4-9 illustrates the flow of managerial control common to radio stations. The illustration is typical of a radio station in a midsized market, such as those described earlier in the section on radio economics.

There are typically four "core departments" at each station: operations, programming, sales, and engineering. *Operations,* also known as the *traffic department,* has the responsibility of placing advertising on the station in accordance with the contracts signed with advertisers. This is a difficult task: at any given time dozens of different contracts are in force, each with varying schedules for air time, length, position, and so on. For this reason many radio stations have automated their traffic departments to some degree with computer systems. The operations director or traffic manager heads this important department.

The *program department,* headed by the program director, has overall responsibility for the sound of the station, including music, news, and public affairs. Stations with a music format may also employ a music director to oversee the development and implementation of the format; news/talk stations may appoint a news director to handle the logistics of news and public affairs coverage.

The *sales department* is very important. Sales personnel are fond of pointing out that theirs is the only department that makes money; all of the others spend it. Led by the sales manager, this department is responsible for the sale of commercial time to local, regional, and national advertisers. Depending on the size of the station (and

FIGURE 4-9 A radio station table of organization.

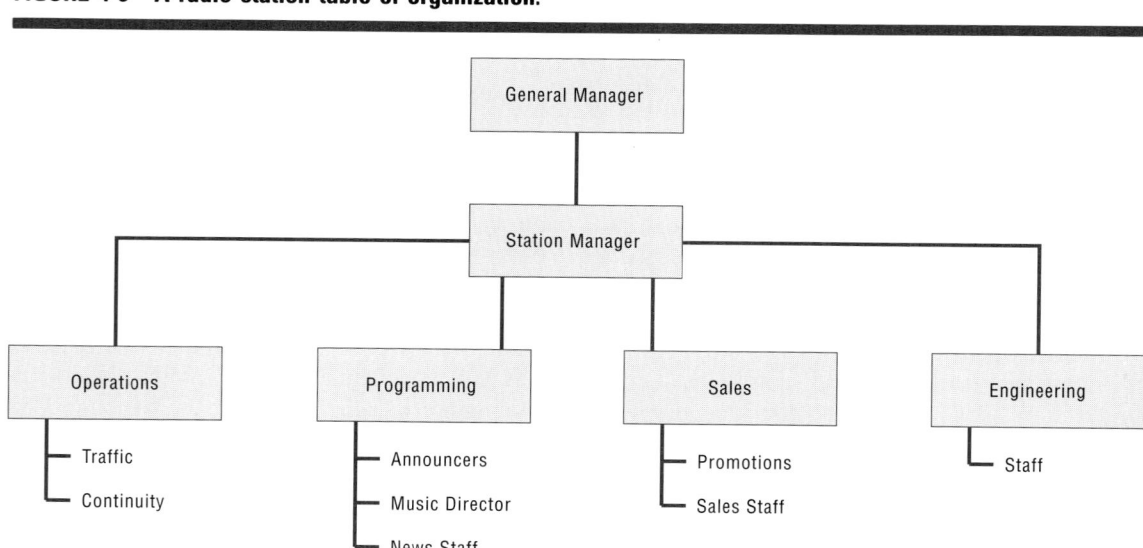

the size of its commercial client list), stations may employ both a local sales manager to oversee local sales and a national sales manager to handle spot advertising accounts. Today's commitment to research and promotion makes these important functions of this department, prompting many stations to value highly a promotions director and research manager.

The *engineering department,* headed by the chief engineer, basically has one function: to keep the station on the air with the best signal possible. Many radio stations once had large engineering staffs, with as many as five to ten full-time employees, but relaxed federal regulations, improved electronic equipment, and competition from other businesses for engineering talent led to the streamlining of engineering departments. Some stations retain the services of a consulting engineer, who works part-time, as needed, to keep the station in prime operating condition.

Top-level management of and responsibility for a radio station is in the hands of the general manager (GM) or station manager. The GM is responsible for business and financial matters, including station revenues and expenses, short- and long-term planning, budgeting, forecasting, and profitability. The GM must run the station in accord with local, state, and federal regulations. The GM is responsible for maintaining and representing the station's image in the community. General managers also hire the major department heads, establish their goals, and monitor their performance.

Some large stations, particularly those in large markets and those owned by station groups, have both a GM and a station manager. In this case, the station manager has responsibility for the day-to-day operations of the radio station, such as hiring and firing, making sure the bills are paid, and keeping up employee morale; the GM reports to the "home office," representing the station to its corporate ownership, to the community, and to federal, state, and local regulatory bodies.

Traditionally, the route up the corporate ladder into the management of radio stations starts in sales; most radio GMs have a background as account executives, promotion directors, or research managers. However, it is not unheard of for GMs—even station group owners—to come from the music or announcing ranks. At a radio management meeting one will hear many well-modulated announcer voices; a very high percentage of owners and operators are former DJs.

GETTING A JOB: RADIO EMPLOYMENT TODAY

We know that there are about 11,000 radio stations in the United States. From an employment standpoint that's good news, but there is a downside. Most radio stations today play recorded music, are increasingly committed to cutting costs, and are turning to automation and satellite syndication. Thus despite the industry's seemingly large size, the radio work force is actually quite small. The FCC estimates that about 100,000 people work in the radio business. The majority (about 60,000) work at commercial radio stations. Noncommercial radio employs about 3500 people. The remainder of radio jobs are at corporate headquarters and networks.

Opportunities for minorities and women have been increasing in recent years. For example, in 1977 the percentage of women in radio jobs was 28 and ethnic minorities accounted for only 10 percent of employees. By 1990 almost 40 percent of the radio work force was female and the percentage of minorities had risen to 15. Most of the new opportunities for minorities and women had occurred in sales and announcing positions. Today the majority of salespeople at many large stations are women. Whereas in 1980 black and Hispanic announcers were virtually unheard of (and unheard) outside of "ethnic" stations, radio today has an increasingly multiethnic and multicultural sound.

The typical commercial radio station employs about seventeen full-time and part-time personnel. On average, eight of the full-timers work in programming, five in sales, three in traffic or office positions, and one in engineering. The part-timers are most likely to be announcers or clerical personnel.

It takes more people to run an AM–FM combo than any other kind of station. Combos employ twenty-four people on average, compared to seventeen at full-time AM or FM operations and only seven at low-power AM daytime-only stations. Stations in larger markets employ more people.

An FM music station in a major city (population greater than 1 million) may employ up to forty people; an all-news operation in the same city may hire over fifty.

Salaries in the radio business are largely a function of the size of the market. Overall, the average radio salary is low—only about $20,000 per year. In smaller markets most people, especially announcers, earn at or slightly above minimum wage. A salary of about $13,000 is typical for most small-market air personalities. However, in larger markets, top personalities—particularly the popular DJs in the lucrative early morning time period—can earn six-figure salaries. Figure 4-10 compares radio salaries by occupation and size of the market.

In radio the better paying jobs are in sales. Account executives in large markets often exceed $100,000 in annual income. The nationwide average is over $110,000 per year for radio sales managers in major markets. This trend is also apparent in smaller markets. For example, a small-market sales manager averages about $27,000 per year in earnings, about twice as much as the average salary for the program director, chief engineer, or news director. In fact, in some markets a successful sales manager might even make more than the GM.

FIGURE 4-10   Radio salaries. (Source: 1990 National Association of Broadcasters/Broadcast Financial Managers Report.)

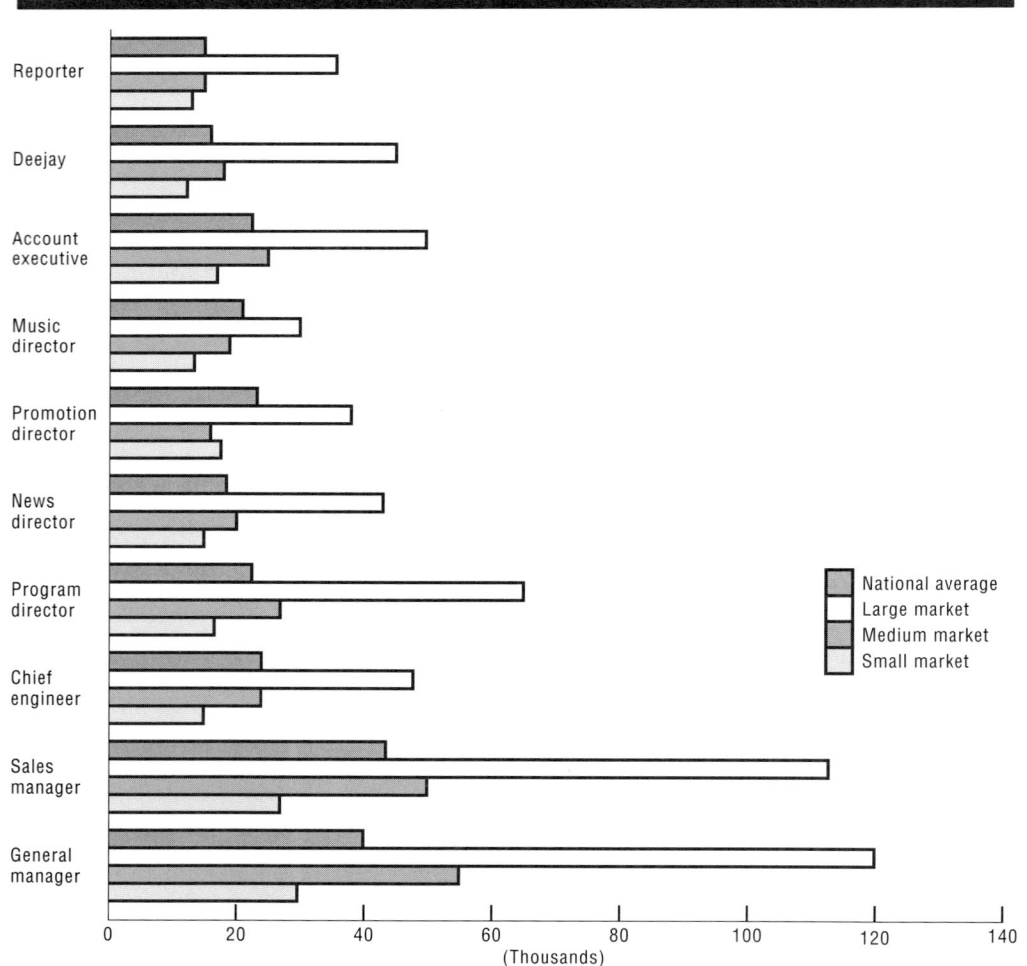

PROFILE

YOU TOO CAN BE A RADIO STAR

You never know when the hand of fate will touch you on the shoulder, changing your life forever. Take the case of Michael Burton. He was a maintenance man for the building in which an L.A. radio station was housed. One day he accidentally walked into the studio during the morning drive show and on a whim the DJs asked him to do the weather. Burton did and the feedback was so favorable that he was kept on the show. His career really blasted off a few weeks later when he started doing a "rap" weathercast. Listeners loved it. He now makes personal appearances and has hired an agent and a publicist.

Then there's Jim Trenton. After failing the bar exam, Trenton started writing restaurant reviews. He bumped into a local DJ at a restaurant one night who asked Trenton to do his food reviews on the air. Trenton became one of the station's best-known personalities.

Then there's Debra Thomas. She was a legal secretary when she won a contest at a local radio station. When she went to the station to claim her prize, Thomas impressed management with her poise and vocal delivery. She soon became a regular on the station's morning show.

Then there's Howard Henkin. Henkin was a New York City cab driver who was driving in a rainstorm when he heard a DJ report that the weather was sunny and mild. Henkin stopped his cab and called the station to complain. He was told if he could do a better job he should come to the studio and do it. Henkin said "Okay" and went to the studio. Impressed by his outspoken manner, the station manager put him on the air. He spent a decade doing weather at the station and now serves as the station's director of community affairs.

Who knows?

SUMMARY

Judging by the number of radio stations on the air and by the number of radios in homes, there is no doubt that the medium has survived the threat of TV. However, in 1990 radio received less than 7 percent of total advertising expenditures. This means that some stations, mainly FMs, are profitable, whereas others, mainly AMs, are losing money.

Funding for the medium is both public and private, and stations are either commercial or noncommercial. Noncommercial stations include NPR member stations, religious channels, college radio, and community educational stations. Most radio stations are becoming more local in their focus.

Drive time is the period when the most people are listening to radio; advertisers therefore pay more money for slots during drive time. Three types of advertising carried on radio are local, spot, and network.

Radio stations, once family-owned operations, now are merged into groups of stations owned by corporations. And because of intense competition, most stations have turned to format radio, targeting their programming toward specific factions of society. Some fear these trends are resulting in the depersonalization of the medium.

Funding of noncommercial radio has changed in recent years. The government now provides less money to these stations, so listeners and businesses are filling in the gaps. This appears to be making noncommercial programming increasingly similar to that of commercial stations, especially in its competitive nature.

Most radio stations have four major departments: operations, programming, sales, and engineering. The GM is responsible for all of the executive decisions at a station.

SUGGESTIONS FOR FURTHER READING

Blume, D. (1983). *Making it in radio*. Hartford, Conn.: Continental Media Company.

Busby, L., & Parker, D. (1984). *The art and science of radio*. Boston, Mass.: Allyn & Bacon.

Duncan, J. H. *American radio* (various years).

Ellis, E. I. (1986). *Opportunities in broadcasting careers*. Lincolnwood, Ill.: VGM Career Horizons—A Division of National Textbook Company.

Fornatale, P., & Mills, J. E. (1980). *Radio in the television age*. Woodstock, N.Y.: Overlook Press.

Hilliard, R. (1985). *Radio broadcasting* (3rd ed.). New York: Longman.

Johnson, J. S., & Jones, K. K. (1978). *Modern radio station practices*. Belmont, Calif.: Wadsworth.

Keith, M. C., & Krause, J. M. (1989). *The radio station* (2nd ed.). Boston: Focal Press.

Lewis, P. (1989). *The invisible medium: public noncommercial and community radio*. Basingstoke, England: MacMillan Education.

O'Donnell, Lewis B., Hausman C., & Benoit, P. (1989). *Audio station operations*. Belmont, Calif.: Wadsworth.

Schulberg, B. (1989). *Radio advertising: the authoritative handbook*. Lincolnwood, Ill.: NTC Business Books.

CHAPTER 5: TELEVISION TODAY

After about thirty years of relative stability, vast changes have recently taken place in the TV industry. As an example, consider what has happened in Omaha, Nebraska.

In 1955, there were only two TV stations to choose from. Since there were three networks, KMTV split its affiliation between CBS and ABC. Most of the time the station stayed with the more popular CBS network. Omaha viewers saw only a few of the programs carried by ABC.

Ten years later (1965) not much had changed. A new TV station affiliated with ABC had signed on. At least now Omaha viewers could get the full programming lineup of all three networks.

Another ten years and still not much changed. A noncommercial station carrying the Public Broadcasting Service had joined the lineup, but everything else was the same.

Now let's fast forward to the 1990s. Cable has come to Omaha. Instead of just four viewing choices, residents of Omaha now have fifty-four. Excuse the cliché, but it's a whole new game (Figure 5-1).

This chapter examines the traditional or over-the-air TV industry, an industry in an age of transition. The next chapter will look at the cause of much of the change: the cable TV industry.

TELEVISION NOW

Watching TV is one of America's favorite pastimes. More than 98 percent of the homes in America—over 90 million homes—have at least one TV. Of the homes with TV sets 96 percent have a color set. Over one-half of the homes in the United States have more than one set. More than a decade ago the census reported that more American homes had TVs than indoor toilets; that disparity continues to grow.

According to the A. C. Nielsen ratings company, the average American home has the TV set on more than seven hours each day. Over 66 percent of Americans turn to it for their news and 55 percent rate TV as the most believable news source among the various media.

TYPES OF TELEVISION STATIONS

There are over 1500 TV stations in operation today. The various types of stations are depicted in Figure 5-2.

Commercial and Noncommercial Stations

The stations depicted in Figure 5-2 can be divided into two categories: commercial and noncommercial. The primary distinction between the two is the way in which each type of station acquires the funds to stay on the air. Commercial stations—75 percent of the total number of TV stations—make their money by selling time on their stations to advertisers. Noncommercial stations are not allowed to sell advertising. These stations, which were set aside by the FCC for educational, civic, and religious groups, must adhere to the FCC mandate not to sell advertising time. They survive strictly through donations from individuals, businesses, and the government.

VHF and UHF Stations

Another way to categorize stations is by the channels on which they broadcast. Stations that broadcast on channels 2 to 13 are called *VHF*, or *very high frequency*, stations. Those stations that broadcast on channels 14 and above are called *UHF*, or *ultra high frequency*, stations.

VHF frequencies have historically been the preferred channels for broadcasters. During the 1950s and most of the 1960s TV sets often did not have the capability to receive UHF signals. The quality of the UHF signal was also judged to be inferior to that of a VHF signal.

Cable technology has erased many of the distinctions between VHF and UHF broadcasting. Cable TV provides the subscriber pictures of equal quality whether the station is VHF or UHF. Cable has also made selecting a UHF station as easy as selecting a VHF. On some cable systems UHF stations are reassigned to a VHF channel.

Figure 5-2 reveals that the most common type of TV stations today (nearly 600 stations, or 39 percent) are commercial UHF facilities. A

COX Cox Cable Omaha — CHANNEL LINE UP

LIMITED BASIC

#	Channel	Description
1	VIEWER'S CHOICE	Movie Previews & Schedules
2	COX 2 ... For YOU!	
3	TBN	Inspirational programming
4	HEALTH & WELLNESS	
5	CBS - KMTV	Omaha CBS affiliate
6	C-SPAN II - GOV'T. ACCESS	Government coverage
7	PREVUE Guide	Hourly listings of programs
8	NBC - WOWT	Omaha NBC affiliate
9	ABC - KETV	Omaha ABC affiliate
10	KPTM - #42	Omaha's Fox Station
11	PBS - KYNE	Channel #26 - Omaha
12	PBS - KHIN	Channel #36 - Red Oak
13	C-SPAN	Cable Satellite Public Affairs Network
16	EDUCABLE	Programming from Lincoln
17	THE ALPHA CHANNEL	Educational programming
18	T.V. CLASSROOM	Educational programming
19	THE LEARNING CHANNEL	Educational programming
20	PBS - KUON	Channel #12 - Lincoln
21	ECUMENICAL ACCESS	Religious programming
22	VIEWER'S CHOICE	Movie Previews & Schedules INFORMATION SERVICES NETWORK
23	PUBLIC ACCESS	Local Programming

TIER II

#	Channel	Description
24	QVC NETWORK	Shopping Channel
25	WWOR	New York Independent
26	WGN	Chicago Independent
27	TBS	Atlanta Independent
28	LIFETIME	Programs for your lifestyle
29	THE NASHVILLE NETWORK	
30	VH-1	Video Hits #1
31	ESPN	Worldwide Sports
32	USA NETWORK	In touch with today's world

COMPLETE BASIC Continued

#	Channel	Description
33	CNN	Cable News Network
34	NICKELODEON	Entertainment for kids
35	MTV	Music Television
36	TNT	Turner Network Television
37	THE DISCOVERY CHANNEL	
38	ARTS & ENTERTAINMENT	
39	BET	Black Entertainment Television
40	THE FAMILY CHANNEL	All family entertainment
41	ENTERTAINMENT NETWORK	
42	CONSUMER NEWS AND BUSINESS CHANNEL	
43	THE WEATHER CHANNEL	National, regional & local weather
44	CNN HEADLINE NEWS	Comprehensive news coverage
45	AMERICAN MOVIE CLASSICS	
46	COMEDY CENTRAL	
47	PRIME SPORTS	

PREMIUM SERVICES

#	Channel	Description
14	CINEMAX *	Movies, music & comedy programs
15	HBO *	Blockbuster movies & specials
48	THE DISNEY CHANNEL *	All family entertainment Selections from the Disney library
50	THE MOVIE CHANNEL *	The ultimate movie theater
51	VIEWER'S CHOICE *	Cox Cable's Pay-Per-View Channel
52	SHOWTIME *	Programs for every audience
54	BRAVO *	

Premium Channel - Extra Charges   *Effective 9/1/91*

FIGURE 5-1 TV listings in Omaha, 1992.

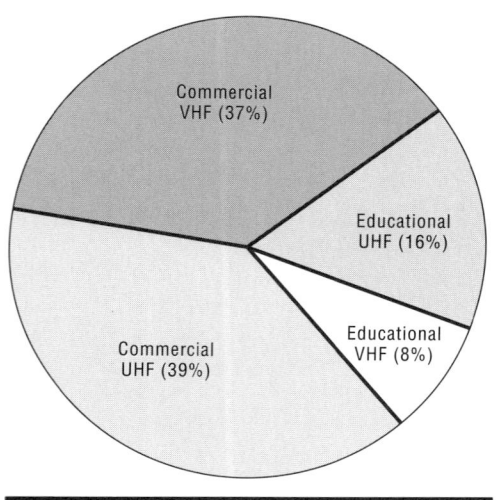

FIGURE 5-2 TV stations on the air, 1992.

nearly equal number of stations (37 percent) are commercial stations operating in the VHF band. There are nearly 250 noncommercial UHF facilities (16 percent) and about one-half that number of noncommercial VHF stations.

Of course, like the situation in radio today, the commercial stations do not share equally in TV revenues. And noncommercial TV faces the same problems of funding that public radio faces. Let's examine the business of commercial TV first.

THE TELEVISION BUSINESS

Because commercial TV is first and foremost a business, to understand TV, one must understand the business aspects: the dollars and cents of TV. As TV becomes more and more a "bottom-line" industry, profits increasingly drive the decisions made in TV.

Television's revenues come primarily from the sale of time to sponsors. Perhaps more accurately, TV stations deliver audiences to sponsors. A TV program can snare the attention of a large number of people. From a TV executive's perspective, programming is the sideshow that lures advertisers into the tent.

The amount of money that a particular station or network can charge sponsors is influenced by several factors, such as the number of people predicted to watch a given program or time period and the number of commercials the advertiser wishes to place with the station. In general, the amount that the station or network will charge is based on the estimated number of people viewing a program, and that estimate is based on the

The television control room is the nerve center for network and local TV production. Seated in the middle is the director—the person who calls the camera shots. To the director's right is the switcher—the person who punches the buttons that determine what picture is sent out over the air. To the director's left is the assistant director—the person in charge of timing the program and making sure the script is followed accurately.

PROFILE

THE CZARINA OF PUBLIC TV

Public TV faces difficult times in the 1990s. The rise of cable networks, many of which feature similar programming, from nature documentaries to dramas with a British accent, has eroded public TV audiences over 10 percent in the past five years. Funding has been in short supply, and the prospects for additional revenue from the government and private sources appear slim.

As if on cue, someone stepped up to seize the day and to pump new vigor into public TV. In October 1990 the PBS Board of Directors revamped the network's organization, giving sweeping new powers to the Director of National Programming. Ms. Jennifer Lawson took the helm shortly thereafter.

In short order the former professor of communications and civil rights advocate took charge. Under her watch the immensely successful "The Civil War" was scheduled. She purchased ads for PBS shows on commercial TV. In development for PBS there are a new situation comedy about life in Russia today, children's game shows, and programs featuring more pop (as opposed to classical and operatic) music.

In a profile in *Time* magazine Ms. Lawson said, "a perfect program to me is one where the viewer never questions the value or importance. Entertainment and intelligence can live well together."

---

ratings. The more people the ratings estimate to be watching at a particular time, the more a station or network will charge.

Television advertising can be broken down into three classifications: network, national spot, and local. Network advertising is defined as the commercials sold by and seen on the programs aired by the three commercial networks: ABC, CBS, and NBC, and the younger Fox network. National spot advertising is comprised of the commercials that national advertisers place on selected stations across the nation. For example, makers of harvesting equipment might wish to purchase ads in a large number of rural markets. Buying time on the network would be inefficient for them since it would be unlikely that many people in New York City or Los Angeles would be shopping for harvesters. Local advertising consists of commercials that are shown on the broadcast station in your area, which feature products and services in the local community served by the TV station.

Trends in commercial TV advertising are depicted in Figure 5-3.

Network Advertising

As you might expect, the networks generate tremendous revenue. In 1990 the total advertising revenue of the three networks was nearly $10 billion. The amount of money the networks take in from advertising will continue to increase into the mid-1990s, although network TV advertising revenue is somewhat flat, as Figure 5-3 demonstrates.

These figures suggest that even while the number of people viewing the networks declines, the money that the networks bring in from advertising will increase. The increase will be caused primarily by increased competition among advertisers to get their messages seen—and the three major networks, despite a declining share of the audience, still provide the largest collection of people at one time. However, the rate of growth will not be so large as it would have been if the networks did not face continuing competition from cable networks and other entertainment options.

National Spot Advertising

An advertiser who wants to promote his or her products or services in selected areas rather than purchasing network time, which usually covers the entire country, will buy time from individual local stations to target the advertiser's messages to a specific locality or region.

Most spot buying is handled by a local station's *national representative,* usually referred to as the *rep.* The rep represents local stations to national and regional buyers. National reps make it easier for buyers to purchase time by being a central contact point for a given station. Usually a na-

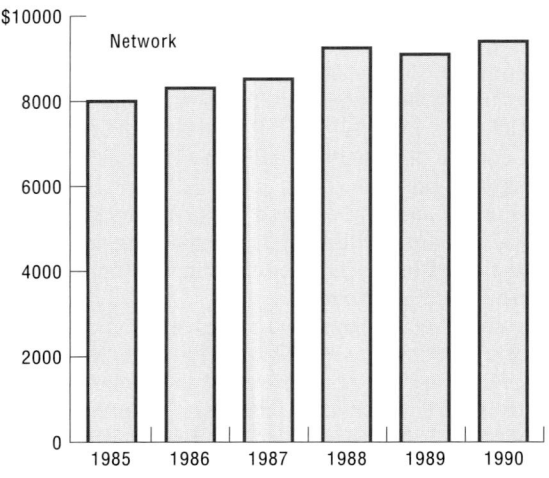
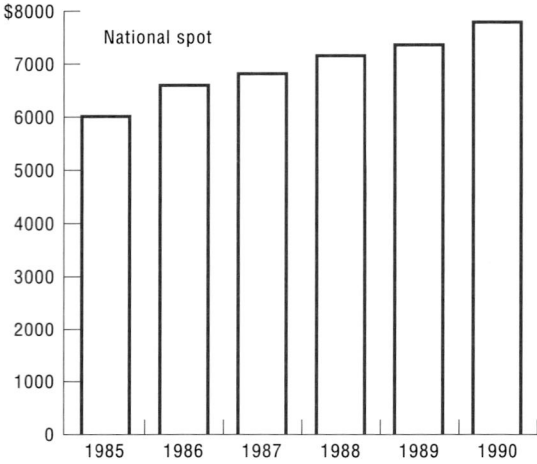
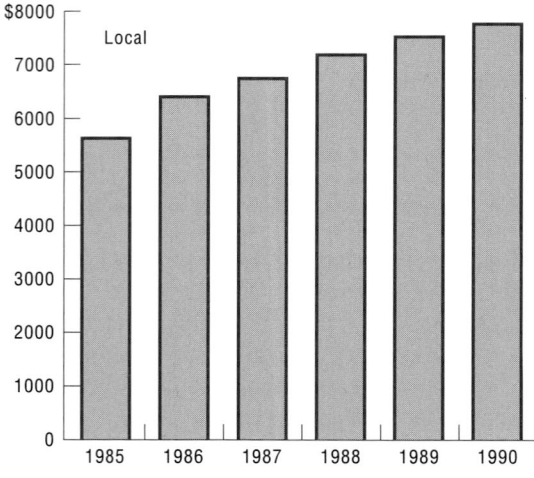

FIGURE 5-3 Television advertising volume (in millions). (Source: McCann-Erickson/Television Bureau of Advertising.)

tional rep will represent only one station in a given market. This prevents conflicts of interest.

Reps will also offer programming advice to help the station increase its number of viewers. This advice is not altruistic, however; more viewers mean better ratings. It is easier to sell the available advertising time at higher rates to national advertisers for stations that have the highest ratings in their areas. More advertising time sold at a higher cost means more profit for the rep, whose fee is a percentage of sales, as well as for the station.

Katz Communications is the nation's largest TV rep firm, representing more than 200 TV stations and generating annual revenues approaching $1 billion. Other leading rep firms are HRP, Blair, and Telerep.

As demonstrated in Figure 5-3, national/regional spot sales have grown steadily in recent years. In 1990 spot business accounted for nearly $8 billion in total TV sales. As the traditional TV networks continue to lose audience, national spot revenues should increase. Potentially this means more money for TV stations and for their rep firm.

Local Advertising

Today an increasing amount of the advertising revenue of a local station comes from the sale of advertising time to local merchants. Car dealers, appliance stores, lawyers, and others seek to inform people about their services and products through advertising on the local station.

Television stations employ salespeople, called *account executives,* to visit potential advertisers and to demonstrate how advertising on their TV station will increase their business. A good account executive will help the businessperson, or *client,* plan an advertising strategy based on the amount of money the advertiser has to spend. The account executive should help the client plan the time of the year to advertise and the time of the broadcast day to advertise and may, particularly in smaller markets, help write commercials. Account executives are usually paid a *commission,* a percentage of the value of advertising they sell.

The primary tools of the account executive are the ratings and the list of prices for advertising, which is called a rate card. The *ratings* show the client approximately how large an audience would

THIRD & FOURTH QUARTER RATES

			Submission Levels-$$
			3rd Qtr underlined
		DAYTIME	and 4th Qtr in bold

M-F	530-6AM	**NBC News @ Sunrise**	150 125 **100** <u>75</u> 50 25
M-F	6-7AM	**11 Alive Today**	500 **450** 400 350 <u>300</u> 250
M-F	7-9AM	**The Today Show**	450 **400** 350 <u>300</u> 250 225
M-F	6-9AM	**Early Morning Rotation**	500 450 **400** 350 <u>300</u> 250
M-F	9-12N	**Morning Rotation**	200 175 150 **125** <u>100</u> 75

The Joan Rivers Show (9-10A)
The Judge (10-1030A)   300 250 **200** <u>175</u> 150
The Judge (1030-11A)   (M-F/10-11A)
Trialwatch (11-1130A)
Closer Look with Faith Daniels (1130-12N)

M-F	12N-1P	**Noonday**	300 275 **250** <u>200</u> 150 125
M-F	1-4PM	**Afternoon Rotation**	500 450 **400** 350 <u>300</u> 250

Program Lineup Includes:
Days of Our Lives (1-2P)
Another World (2-3P)
Santa Barbara (3-4P)

FIGURE 5-4 A television station rate card. Because of the volatile nature of television sales, most rate cards are simply computer printouts. Salespeople use these rates as a starting point for negotiations with clients. (Courtesy Gannett Broadcasting.)

see the commercial and the demographic makeup of that audience. The *rate card* lists the prices for the various parts of the broadcast day. Figure 5-4 is a sample rate card.

Local businesses have increased in importance to local stations in terms of advertising revenues. In 1970 local TV stations received only 35 percent of their advertising revenues from local merchants. In the 1990s advertising from local merchants and service providers supplies, on average, a little more than one-half of the total advertising revenues for local stations.

Today local advertising sales, like national spot, is about an $8 billion dollar annual enterprise. Television is becoming increasingly like radio in that its primary source of revenue is the sale of advertising to local merchants.

Other Announcements

Not all announcements on TV are commercials. Besides advertisements there are promotional and public service announcements. Both types appear at the network and local levels; neither provides direct compensation to the station.

Promotional announcements, or *promos,* are short announcements that publicize a program on the station. These messages remind you of the content of a show: "Tomorrow on 'Geraldo': Satanic transvestite drug-abusing prostitutes who want to adopt." Or the promo may just remind the viewer of an upcoming program: "News tonight at 11."

The second type of announcement is the *public service announcement (PSA).* These announcements, as the term implies, are unpaid advertisements for a charity or philanthropic endeavor. An announcement urging you to donate to the United Way or to give blood to the Red Cross would classify as a PSA.

Like the radio business, commercial TV stations do not participate equally in their share of the billions of dollars available from TV advertising. The next section discusses the medium's winners and losers in the high-stakes world of commercial TV.

THE NETWORK TELEVISION BUSINESS

For many years the TV business (like the U.S. automobile business) was dominated by three giant companies. At the peak of their power in the 1960s and 1970s ABC, CBS, and NBC dominated the viewing habits of the nation. That dominance was especially acute in prime time: the evening hours during which the overwhelming majority of American households was watching TV. On the east coast and in the far west prime time ran from 7:00 P.M. to 11:00 P.M.: in the midwest and certain parts of the Rocky Mountain region, prime time spanned 6:00 P.M. to 10:00 P.M.

The power of the three networks over the years is illustrated in Figure 5-5.

At the height of their prime-time power the three networks commanded more than nine in the ten of all TV homes with the TV set on. Facing competition from new sources—most notably independent stations (see below), cable, and home video (see Chapter 6), the 1980s saw the beginning of a significant decline in the networks' audience share.

Today viewers to ABC, CBS, and NBC comprise about six in ten TV households in prime time. By the dawn of the next century network share may dip below one in two TV homes.

ECONOMICS

THE HIGH COST OF BATTLE

As the network share of the TV audience continues to decline, executives at CBS, NBC, and ABC have come up with a novel sales approach. It seems that TV coverage of wars, natural disasters, tragedies, and other crises leads to increased network viewing. Unfortunately, many advertisers have been reluctant to advertise in these types of programs, fearing a negative association between their products and the particular national crisis (such as "Today's air disaster has been brought to you by . . ."). NBC alone lost more than $50 million in ads that were canceled during the Persian Gulf War.

Recently the Network Television Association, a trade group representing the three networks, has gone on the offensive. It commissioned a research study of American viewers and found that seven in ten supported advertising in war-related specials and more than eight in ten felt it was okay to put ads in regular news reports featuring war and disaster coverage. A sixteen-minute videotape was produced to be shown to potential advertisers. The message of "Advertising During Times of Crisis" is simple: war, crime, personal tragedies, hurricanes, earthquakes, and volcanoes are good for business.

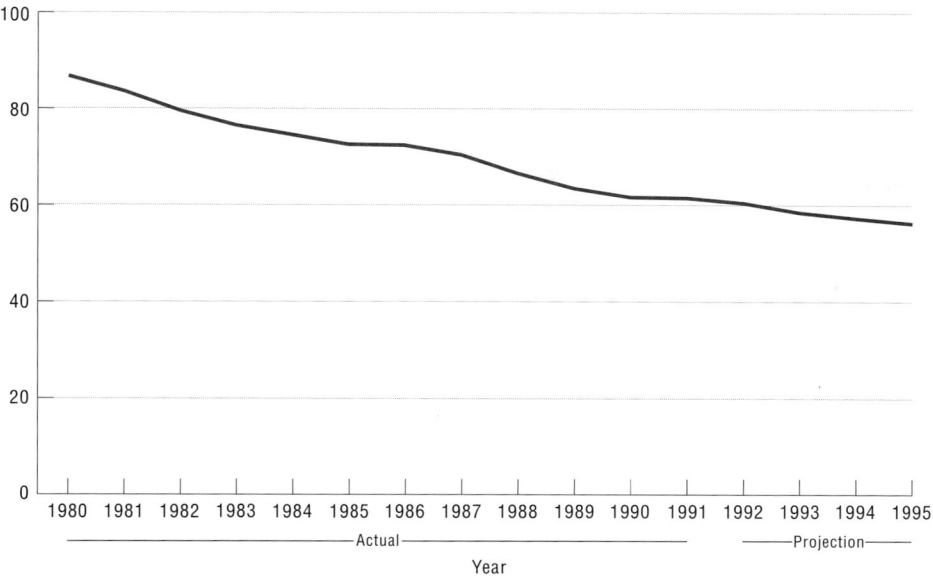

FIGURE 5-5 Network share of audience. (Source: Constructed by the authors.)

Since viewing equates with advertising revenue, the dip in network share has impacted significantly on network advertising revenues. In 1980 advertisers spent just under $5 billion on network TV spots. The three networks collected virtually all of that revenue ($4.827 billion to be exact). The paltry remainder went to a few fledgling cable networks (like WTBS out of Atlanta, Georgia) and some prime-time specials that were not presented on one of the major networks. In 1985 advertisers spent nearly $9 billion on TV commercials. By then the proportion that went to ABC, CBS, and NBC had slipped to about 86 percent, or about $7.5 billion. In 1990 the three networks claimed less than 75 percent of the money spent trying to reach a nationwide audience (about $11 billion). Their gross revenue figure was about $8 billion, an amount almost unchanged from five previous years. Cable networks got a big piece of the pie (about $1.3 billion). But most of the rest ($1.6 billion) went to a comparatively new network.

Fox Broadcasting Company

The Fox Broadcasting Company (FBC) was launched in 1986 by Australian media magnate Rupert Murdoch. Murdoch had purchased the former Metromedia stations and was interested in using his newly acquired TV studio (20th Century Fox) to produce programs for these stations. The decision was made to introduce programming slowly—first, one day of the week (Sunday), then gradually extending it throughout the week.

Today Fox programming is available seven nights per week, with late night, and weekend morning programming on the drawing board. Many successful shows now attract audiences and advertisers to Fox, including "Married with Children," "The Simpsons," and "Beverly Hills

Part of the gang from "Beverly Hills 90210." Fox used this show to draw young viewers from the other three networks.

90210." The impact of Fox on the network TV business has been significant.

In 1988 Fox posted losses of $95 million. Three years later the network billed over $1 billion in advertising revenues and its profits were estimated at over $70 million.

The picture for Fox continues to be bright. By the middle of the decade its gross advertising revenues may match those of the "big three."

ECONOMICS OF LOCAL TELEVISION

The major networks seem to get all the attention in the TV business, even though their star is in decline and they account for only about a third of all the advertising revenue spent on the medium. This section looks at where the rest of the dollars go—to the nearly 1100 commercial TV stations.

Figure 5-6 rates the various types of TV stations in economic terms. The rankings range from "5-star" stations, which traditionally have been the most profitable, to "½-star" stations, which have faced considerable financial hardship and apathy among America's TV households.

Television's Cash Cows: Network Owned-and-Operated Stations

At the top of the rankings of commercial TV stations there are those that are owned outright by the corporate parents of the three established TV networks—ABC, CBS, and NBC. In industry parlance, these are owned-and-operated stations, or "O&O's," for short. These are the 5-star stations in Figure 5-6.

Network O&Os have traditionally been the most profitable of all TV stations. Located in the VHF band, O&Os are considered the flagship stations of their network. Situated in the largest TV markets, they often boast the call letters of their network. In addition to owning NBC, General Electric is the parent of WNBC in New York and KNBC in Los Angeles, plus stations in Chicago, Denver, Miami, and Washington, D.C. CBS owns WCBS in New York and KCBS in Los Angeles, as well as other stations in Minneapolis, Chicago, and San Francisco. ABC O&Os include WABC (New York), KABC (Los Angeles), plus additional TV stations in Chicago, San Francisco, Philadelphia, and Houston.

Ownership by a major network guarantees a steady supply of programming to these stations and a high profile for potential advertisers. Owned-and-operated stations are typically local news leaders in their marketplace. The fact that they emanate from the corporate or regional headquarters of their networks permits economies of scale and access to programming and personalities that other stations can't match.

For these reasons network O&Os have traditionally been the most profitable of all TV stations. Annual profit margins above 50 percent have been commonplace. Even as the ratings of their parent networks decline, O&Os remain "cash cows" for their companies.

Cash Calves? Major Network Affiliates

The second most profitable class of TV facilities has been those facilities affiliated with one of the three "old-guard" networks. In industry parlance, such stations are network-affiliated stations, or *affiliates*, for short. Each of the major networks has about 225 affiliates around the country.

A few network affiliates are in the highband UHF region, but the overwhelming majority of ABC, CBS, and NBC stations is on VHF (channels 2 to 13). Among the nation's oldest and most established stations, affiliates have been very profitable operations. Like O&Os, most affiliates have been leaders in local news in their communities. Many have been in operation since the dawn of TV in the early 1950s, and over the years they have cultivated enormous goodwill among their viewers. Affiliates have often used their sizable revenue stream to purchase the best syndicated programs, from popular quiz and talk shows (for example, "Oprah!" and "Geraldo") to network hits (like "Cosby" and "Designing Women").

FIGURE 5-6 Rating the stations. The more stars, the more the potential for profit.

	VHF	UHF
Network O & O	★★★★★	
Network Affiliate	★★★★	★★★
Fox-Independent	★★★	★★
Non-Fox Independent	★★	★
Low power TV	½ ★	½ ★

DEVELOPMENTS IN PROGRAMMING

RUSH LIMBAUGH ▲
Everybody's talkin' at me. In the early l990s, outspoken radio personality Rush Limbaugh—along with Chevy Chase and Whoopi Goldberg—were several of the performers who were slated for their own syndicated TV talk shows.

◀ RESCUE 911
Reality-based TV, which includes such shows as "Rescue 911," has become a common fixture on network TV. These shows are attractive for at least two reasons: (1) real life supplies a never ending stream of story lines; and (2) they are less expensive. "Rescue 911" costs about $600,000 per half-hour compared to about $700,000 for the average sitcom.

RODNEY KING VIDEO ▲

The video that shook up a nation. The impact of the camcorder revolution was graphically illustrated by amateur photographer George Holliday's taping of the Rodney King incident, which was then aired on national news programs.

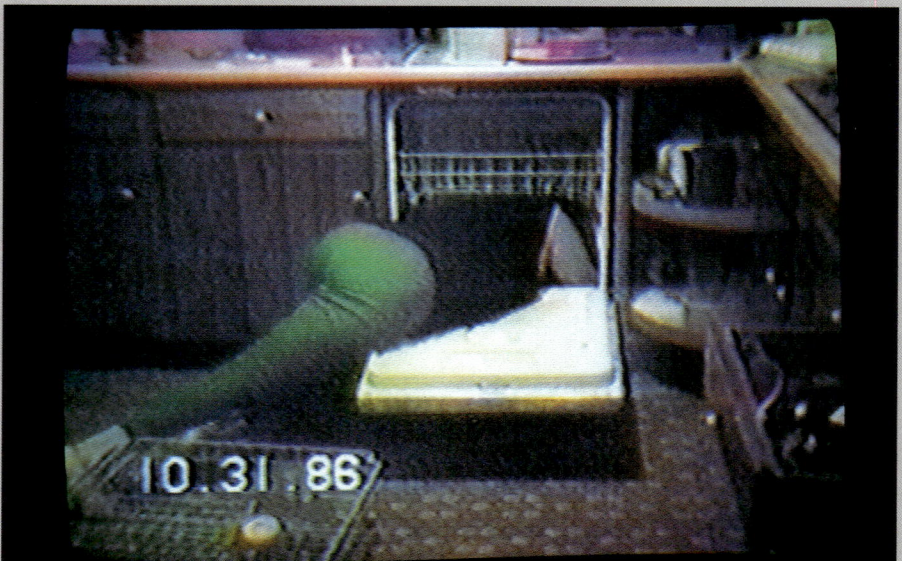

AMERICA'S FUNNIEST HOME VIDEOS ▲

The other extreme of the camcorder revolution: unsuspecting woman finds dishwasher has turned into Jaws. This memorable segment appeared on "America's Funniest Home Videos," one of several programs based on viewer-produced comedy.

STUDS ▲
Syndicated programs such as "Studs" and "The Love Connection" stretch the limits of acceptable taste on TV in an effort to capture viewers. This approach seems to be working for "Studs." Ratings for the show jumped 14 percent from 1990 to 1991.

ENG ▲
Canada is becoming an affordable source of popular television series. Alliances' "ENG" was one of several Canadian-produced shows to enjoy popularity both in the U.S. and Canada.

KATO-CHAN AND KEN-CHAN ▶

Long known as a country that produces video hardware, Japan also produces an impressive assortment of TV programs. "Fun TV with Kato-Chan and Ken-Chan" is a popular show with kids.

EURODISNEY ▲

Bonjour Michel le Souri. The internationalization of one of America's most successful media companies is illustrated by the opening of EuroDisney near Paris. Disney has offered packages of TV programming to Russia and other European countries.

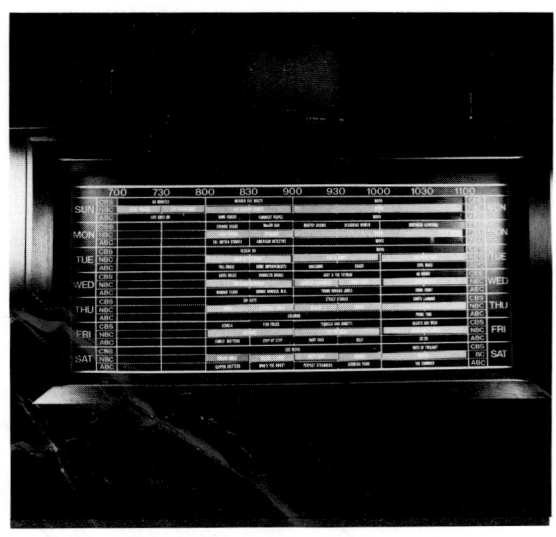

The board that holds the networks' fall schedules.

From an income standpoint, VHF affiliates in the nation's largest markets are the best of this type. According to the National Association of Broadcasters, the typical major-market affiliate in 1990 returned 40 cents in profit for each dollar in sales. This is why in Figure 5-6 we give these stations four stars. Nationwide figures for affiliates, including those in the UHF band, show an average 22 percent return on investment, a figure hard to match by any other industry, particularly in lean economic times.

Independents: Fox-Fired?

The next class of TV stations are those that do not rely on CBS, NBC, and ABC for prime-time and other programming. Such stations are known as independents, or "indies." The leading independents today are those owned by or affiliated with Fox.

As mentioned earlier, Rupert Murdoch's News Corporation is the parent of both the Fox network and the former Metromedia stations. Fox-owned independents include KTTV in Los Angeles, WTTG in Washington, D.C., WNYW in New York, and several others.

Traditionally, independent station profits have been diminished by the high costs of acquiring programming (see Chapter 6). One answer has been to rely on longer-form programs, like movies and local team sports. The attractive programming emanating from Fox has provided additional help.

Fueled by Fox, most major-market independents are profitable enterprises. In 1990 the typical large-market independent reported pretax profit above $5 million, with an annual profit margin of about 15 percent. However, many smaller-market indies, especially those in the UHF band without Fox programming, are operating either close to margin or in the red.

Low Power to the People: LPTV

A new, relatively unknown force in TV is *low-power television (LPTV)* stations. The FCC authorized this new service in 1982 to create openings for minority ownership of TV stations and to increase the number of broadcast offerings in a community. To promote minority investing in these stations, the FCC promulgated rules that would show preference for minority applicants. In theory low-power TV would increase broadcast offerings to communities by increasing the number of TV stations that served those communities.

To restrict coverage to the community to which an LPTV station is licensed, the FCC placed limits on the power of LPTV stations. An LPTV station can transmit at 100 watts VHF and 1000 watts UHF. Regular TV stations can be assigned transmitter powers 1000 times more powerful than these. This low power (hence the name) limits the signal to a very small area. Some owners of LPTV stations are hoping that the FCC will change its rules from limiting the coverage by capping the broadcast power to limiting by barring interference: the station may operate as powerfully as it wishes so long as its signal does not interfere with that of another station on the same channel.

By the early 1990s about 1000 LPTV stations were in operation. Located mainly in rural areas (Alaska has the most LPTV stations), LPTV has faced financial hardship to date. In most cases LPTV operations have been unable to compete with affiliates and full-power independents for attractive programming. Their limited broadcast range has made it difficult for LPTV to interest advertisers in the medium. This is why we place LPTV at the bottom of our rankings of TV stations, with only $\frac{1}{2}$ star (see Figure 5-6).

Profit and Loss in Television Today

To shed light on TV economics, let's examine TV stations in a "typical American city," with a population between 1 and 2 million. Such cities include Salt Lake City, Utah; San Antonio, Texas; Louisville, Kentucky; and Little Rock, Arkansas. Table 5-1 describes the financial performance of the typical network affiliate and independent for cities this size in the early 1990s.

Because of its commitment to news, the typical affiliate employs over 100 people. The independent station operates with less than one-half that number.

On the revenue side, the affiliate receives over $1 million each year from its network, representing nearly 8 percent of total revenues. The indie receives less than a quarter of a million dollars from Fox and/or other networks, including regional sports nets, which is only about 4 percent of total revenue.

All told, the affiliate attracts nearly three times the total advertising revenue of the independent ($13.1 million to $5.4 million). On the expense side, the affiliate puts nearly 15 percent of its expenses, over $1.5 million, into its news. The independent's income goes to purchase programming (38 percent of expenses; over $2.2 million dollars per year).

The bottom-line figures are most revealing. The network affiliate reports profits of $680,000; a profit margin of about 6 percent. The independent, on the other hand, is more than $1 million in the red! It is easy to see why cost-cutting is commonplace at many TV stations today.

STATION OWNERSHIP

Generally, one wouldn't ask "who" owns a TV station, but rather would ask, "what." Television stations are so expensive that few individuals can afford to own them. Instead, most TV stations are owned by companies that own other stations or by investment groups.

Today, a single entity may own twelve TV stations, although an owner's influence is limited by a cap on the potential total audience covered by the TV stations. The FCC decided that no owner could have stations that served more than 25 percent of the nationwide number of TV homes. For example, ownership of twelve VHF TV stations in the twelve largest markets in the United States would be prohibited because the stations' signals would cover more than 25 percent of the nation's TV homes.

New York City contains about 7.8 percent of the homes with TV; Los Angeles has 5.2 percent; Chicago, 3.5; Philadelphia, 3.0; San Francisco, 2.4; and Boston, 2.3. These six cities constitute 25 percent of the nation's viewers. If you owned stations in each of these six areas, you could not own another because of the 25 percent audience cap. Even though you haven't reached the station limit, you have reached the audience limit. Then let's say that you decide to buy stations in Houston (1.7 percent) and in Washington, D.C. (1.8 percent). To do so, you have to sell your Chicago station (3.5 percent) or a station serving a larger number of homes. In other words, a TV station owner is limited by both the number of stations that may be owned and the number of homes the owner may influence.

TABLE 5-1 Revenue and Expenses for a Typical Affiliate and Independent Station

Station Revenue & Expense Items	Affiliate Average ($)	(%)	Independent Average ($)	(%)
Network compensation	1,010,426	7.7	237,435	4.4
Nat'l/Reg'l advertising	5,782,021	44.1	2,637,430	48.6
Local advertising	6,306,971	48.1	2,549,776	47.0
Total time sales	13,099,417	100.0	5,424,641	100.0
Agency & rep. commissions	2,018,703		964,644	
Other revenue	259,065		95,260	
Total net revenue	11,339,779		4,555,257	
Departmental expenses				
Engineering	771,615	7.2	336,197	5.7
Program & production	2,513,441	23.6	2,216,350	37.4
News	1,555,069	14.6	32,762	0.6
Sales	1,055,098	9.9	703,456	11.9
Advertising & promotion	327,646	3.1	289,646	4.9
General & administration	4,435,981	41.6	2,354,430	39.7
Total expenses	10,658,850	100.0	5,932,840	100.0
Pretax Profit	680,929		(1,377,583)	
Profit margin		6%		(−30%)
Employment				
Full-time		98		38
Part-time		9		3

SOURCE: *NAB/BCFM 1990 TV Financial Report*, pp. 39, 68.

IMPACT

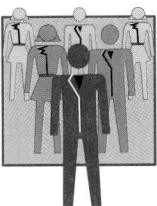

HOW WE PASS THE TIME

In a survey of the American public only 12 percent said that watching TV was their primary activity in the evenings. This study, conducted by Decision Research Corp., found that people do a number of things in the evenings besides just watching TV.

The survey revealed that 70 percent of the people did watch TV at night, but watching TV was usually associated with doing something else: cooking, cleaning, even reading. Other studies have shown that people often "watch" TV while engaged in other activities.

Other activities that were reported include reading, the second most popular activity (22 percent), working (18 percent), and shopping (9 percent). Going to meetings, exercising, and having sex were each tied at 2 percent. These activities turned out to be more popular than attending a movie, which weighed in at only 1 percent.

So while TV is our most popular activity in the evenings, it is not our only one.

At least one person in Denver counts on people watching TV in the evenings. The prime-time prowler breaks into people's homes and robs them while they are lulled by the sound of TV. The prowler has entered as many as six houses in one night through unlocked doors or windows, taking a few easily transported items. Said one policeman, "It's a little bizarre. Most burglars would rather find an empty house than a resident inside."

Portraits such as this were prominently featured in Ken Burns' "The Civil War" on PBS. This series was the highest rated PBS program of all time.

There are two exceptions to this rule. If your company has a financial interest in a station that is financially controlled by minorities, your company need not count this station against the limit. Thus you could have a maximum of twelve plus the minority-controlled station, but the home coverage rule still applies. This exception is meant to encourage minority ownership of stations.

The second exception is based on whether your TV station is a VHF or UHF station. For UHF stations each home covered counts as only one-half a home toward the 25 percent coverage rule. For example, New York City, which has 7.8 percent of the homes with TV, would count as only 3.9 percent for a UHF station. So if you own UHF stations, you can have more stations in larger markets, but the maximum number of stations rule applies.

Table 5-2 ranks America's leading TV groups, using the criterion of total U.S. household coverage.

It should not be surprising that network O&Os lead the pack in terms of TV household coverage in the United States. Together, the stations owned by ABC, CBS, and NBC reach more than one in five TV homes. Fox is right behind, as is another leading owner of independents, the Tribune Company (which owns WGN in Chicago, WPIX in New York, and WGNX in Atlanta, among others).

The telemarketing giant Home Shopping Network owns more than a dozen stations, mostly UHFs in medium to small markets. Its total reach is nearly as great as the three major networks and

TABLE 5-2 Television's Top-20 Station Groups

Company	Total U.S. Household Penetration (%)
CapCities/ABC	24.35
NBC	22.36
CBS	20.78
Fox	19.44
Home Shopping Network	19.05
Tribune	18.69
Gillette	11.05
Univision	10.67
Chris crafts Ind.	10.67
Gannett	9.99
Westinghouse	9.93
Telemundo	9.22
Cox	8.98
MCA	7.72
Scripps Howard	7.57
Hearst	6.78
A.H. Belo	5.65
TVX	5.44
Disney	5.34
Gaylord Broadcasting	5.25

SOURCE: Standard & Poors Industry Survey, 1991.

Fox. Other leading TV groups appeared on our previous list of leading radio groups, including Gannett, Westinghouse, Cox, and Disney (parent of Shamrock radio group).

Television Station Sales

In the fifties and sixties TV stations were rarely bought and sold. Those who wanted to get into broadcasting would simply find an available channel and build a station. Finding stations for sale was unusual and sometimes the sign that something was very wrong with the station. Ownership of a station was perceived to be a long-term investment—one that wouldn't pay off in one or two years. Cashing in on such an investment would require an extended period of ownership.

Owners of TV stations in the fifties and sixties frequently had roots in the community from other enterprises. Even if the station was owned by a distant corporate owner, the station executives felt tied to their communities. Television stations were seen by most in the industry to be profit-making businesses, but they were also considered a part of the community, to which they owed service and allegiance.

At the national level the big-three commercial networks were ranked among the most secure institutions in the United States. True, ABC had sought acquisition deals with other companies in the past. But by the 1970s ABC was just as established a business enterprise as the other two networks. Television at the network and local levels was run by conservative businesspeople who tried to run up as little debt as possible. This situation persisted until the 1980s.

Then the sky fell. In the latter half of the 1980s a remarkable series of events occurred. First, the unimaginable: all three commercial networks changed ownership. ABC was the first to change hands when Capital Cities—a much smaller company than ABC—bought the network.

CBS fell to the leadership of Laurence Tisch when the company floundered after the retirement of William Paley—the man who had built the company—and a threatened buyout from Ted Turner. Tisch was seen as CBS's white knight who would save the network from the likes of Turner.

NBC was the last of the commercial networks to belong to a new master. Ironically, General Electric's purchase of RCA, NBC's parent corporation, gave GE back the ownership of NBC, which it had shared with RCA originally in the early years of radio.

What was happening at the networks was just a much more visible pattern of what was occurring at the local station level. Local TV in the 1980s was also experiencing changes in ownership. Mainstays of the broadcasting industry were selling their TV stations or severely cutting back their TV activities. Companies such as Metromedia, Golden West, and Gulf were prominent, respected broadcasters, each of which owned a number of stations. But corporate heads saw it was time to reap the rewards of their investments, and these companies sold their station holdings.

What was occurring at the top and middle levels of TV ownership was also occurring at the base of the pyramid with smaller local TV owners. Many smaller owners of TV stations saw a chance to make a profit and sold their stations also.

The time arrived when TV stations became another investment commodity, probably best exemplified by the trend to call broadcast stations "broadcast properties."

Regulatory Reasons for Change

These changes were the result of both financial and regulatory decisions. The FCC, which licenses stations to operate, had previously kept the pace of sales of local stations slow. Their rules made it a difficult and lengthy process to transfer the license from one organization to another. The FCC changed its rules in the mid-1980s and made it easier for purchasers of TV stations. The key rule changes that increased the sale of TV stations are:

- The prohibition of the ownership of both newspapers and broadcasting stations in the same community. This meant that newspaper companies that owned broadcast stations in the same town had to decide whether to sell the paper or the stations.
- The increase in the number of stations media companies could own, that is, the previously discussed boost to twelve TV stations.
- The removal of the three-year waiting period before an owner could transfer a station's license.

As a result, the pace of station sales accelerated. From 1986 through 1990 over 400 individual TV stations, and groups of stations, representing over 500 TV properties, changed hands. This means that over three-fourths of the TV stations in the United States changed ownership at least once in recent years.

Money for these purchases came primarily from borrowing. The total debt owed by broadcasters (TV and radio) is estimated at $40 billion. This averages out to a little over $160 for each person in the United States. The companies that hold these debts must pay back the loans with interest. These interest payments have proven to be such a heavy burden on some stations that there have been severe cutbacks in programming and personnel.

There are signs that the buy/sell binge of TV stations is slowing down. The large debt created to purchase stations has combined with the escalating cost of programming to make purchasing TV stations less attractive. When you purchase a station, you agree to pay off the station's outstanding debts for programming, cars used by reporters, and the like. As you will learn in the programming chapter, the cost of programming has created huge debts for some broadcasters.

The new owner must take into account all of these debts when purchasing.

The large debts lessen the attraction of investing in TV stations. As we noted earlier, advertising revenue should continue to grow for local stations, but it is questionable if the rate of increase will be sufficient to finance the large debt and turn a profit for the owners. The result is that the purchasing of local TV stations is becoming less desirable.

STATION ORGANIZATION

The organizational structure of a TV station varies according to the size of the organization. There is no specific way in which stations are organized, but there are some general areas common to most TV stations. Figure 5-7 presents a typical organizational structure for a station in a large community.

At the top of the organizational ladder is the general manager (GM) or station manager—two different names for the same job. This person is ultimately responsible for the operation of the station. If the station is part of a group of stations, the GM usually is a vice president in the parent organization.

Television stations are generally divided into five divisions, each division having its own head who reports directly to the station manager. The five areas are sales, engineering, business, programming, and news. Each of these areas is vital for the efficient operation of a TV station.

Sales

Sales is the most important part of the TV station—at least according to anyone in the sales department. This division of the TV station is headed by a general sales manager. It is her or his job to oversee the sales staff—both local and national. As mentioned before, the salespeople for the station are called account executives or sales representatives.

The sales department is also in charge of *traffic* and *continuity*. Traffic is not the helicopter reports but that part of the station that schedules commercials and verifies that scheduled commercials are aired properly. Traffic departments are responsible for the program logs that tell the people in the control room when each video event

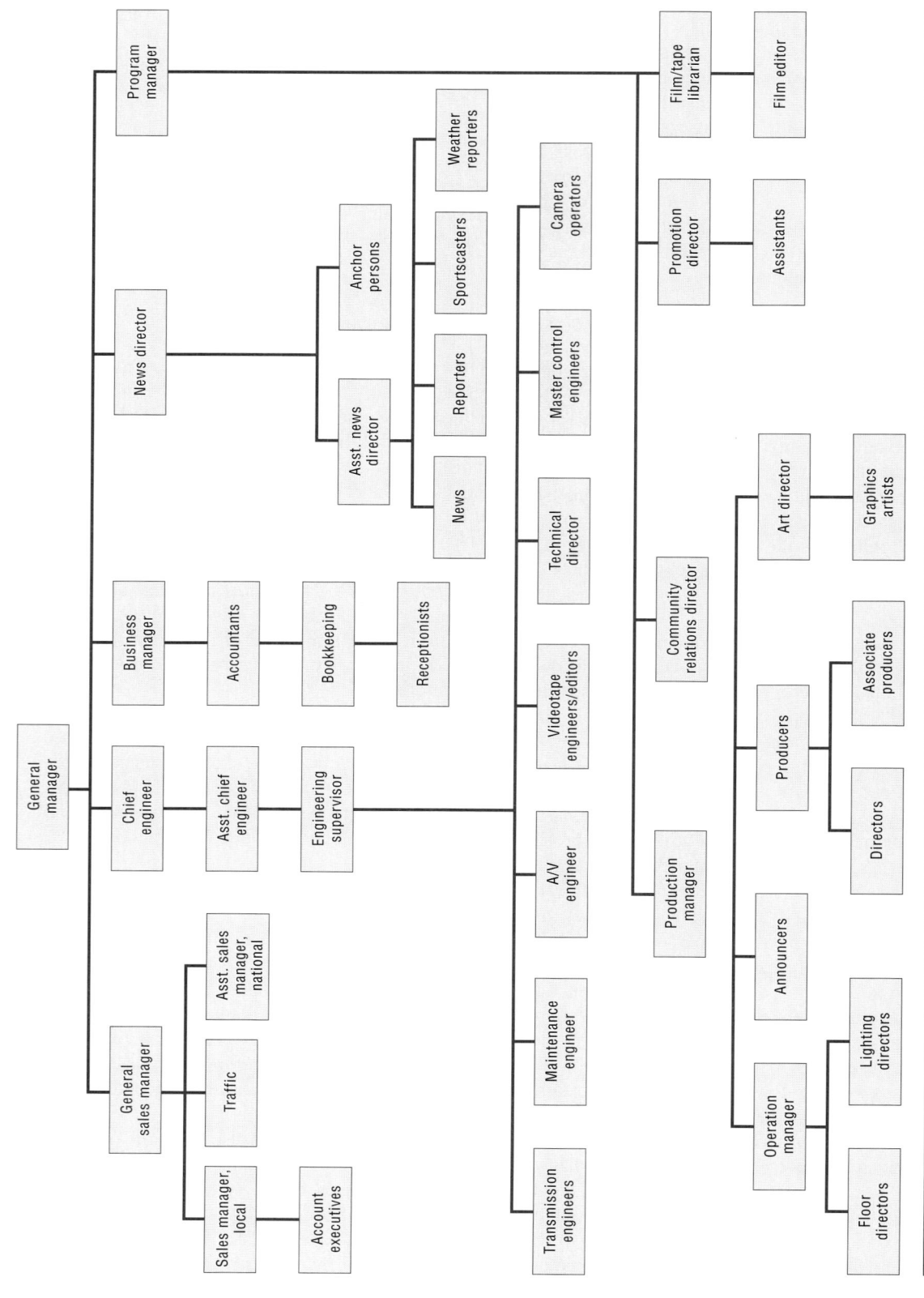

FIGURE 5-7 The organizational structure for a television station in a large community.

is to occur. The continuity department makes sure the station's schedules have no interruptions between commercials and programs.

The verification that an advertisement was played is just as important as scheduling it. If the scheduled commercial doesn't air or only partially airs—say, because of an equipment malfunction—then the sponsor is entitled to a *make-good*—a free commercial in the same time category—to replace the commercial that didn't air. Make-goods are given in lieu of returning the advertiser's money.

Engineering

The second major division in a TV station is engineering. Engineering is the most important part of the TV station—just ask any engineer.

The engineering department is responsible for the maintenance of the equipment, including the transmitter. If there is an equipment failure, it is up to a member of the engineering department to find the problem and correct it.

In sufficiently large or unionized stations engineers run the audio/video equipment. People who load the videotape machines, edit videotape, run the audio board, and push the buttons on the switchers are from the engineering department. Camera operators are also usually from this department. In smaller stations or nonunion shops there may be no hard-and-fast rules about who uses what equipment.

Business

The third area of the TV station is the business division. The business division is usually headed by the business manager. Most business managers feel that they run the most important division of the station. Accounts payable—money owed *by* the station—and accounts receivable—money owed *to* the station—are handled in this division. Everyone who is owed money by the station and everyone who owes money to the station has their paperwork go through this division. Receptionists and secretaries are also a part of the business division.

Programming

The fourth and, according to the people who work there, the most important area of a TV station is programming. The program director, often abbreviated PD, oversees a number of subdivisions and, in consultation with the station manager, is responsible for the purchase of all new programming for the station and the scheduling of the programming during the broadcast day.

Three subdivisions under the program director can be identified at most stations. Usually the largest subdivision is run by the production manager, who oversees studio workers such as floor managers and lighting directors, art directors and cinema/videographers, producers, directors, and production assistants.

The second subdivision is headed by the community relations director and her or his staff. This division is usually in charge of PSAs. The community relations director may also be in charge of TV programs examining minority interests at the station or may be the official spokesperson for the station at community events.

The third subdivision is run by the promotion director, who typically has three duties. First, the promotion director oversees the creation and placement of messages that promote programs, movies, specials, and the station's image. Second, she or he plans and runs activities designed to gain publicity—such as sponsoring a local charity road race. Third, the promotion director is responsible for the purchase of advertising in other media. This person is in charge of the commercial aspects of promoting the station, whereas the community relations director is usually seen as being in charge of the altruistic side of the station.

News

The fifth division is news, which, as any newsperson will tell you, is the most important part of the station. The news division is headed up by the news director, who is in charge of the reporters, news writers, anchors, sportscasters, and weather forecasters. The news division of a station is supposed to operate independently of influences from the other divisions and so is separate from all other divisions. The smaller the station is, the harder it is to maintain this independence. As with the other division heads, the news director reports directly to the station manager.

News has a special place in broadcasting. Stations with active news departments become more important to—and are viewed more favorably by—the communities they serve. The overall

A local TV reporter for WABC in New York covering the Wall Street beat.

quality of a station is often judged by how good its news department is. The value of a station during a sale can be affected by the reputation of the news department. A good news department is an asset to a station.

Departmental Evaluation

As indicated, each department thinks it is the most important department at the station. Which department is really the most important? Are the sales executives right that the station would close its doors if they were not finding sponsors? Are the programming and production people right when they argue that without them there would be nothing to sponsor? Or, do the engineers, who insist that without them there wouldn't be any signal, have the better claim?

The answer is that all of them are important. A station is an interdependent organization. Each division is important in its own right and is important to the other departments. The removal of almost any one of the divisions—with the exception of news—would force the station to close its door and to go dark.

The sales department may be the most "equal" among equals. It is the division that directly generates revenues for the stations. Perhaps, more important, most station managers come from the sales division and may well have a greater affinity for this department.

THE JOB OUTLOOK

About 80,000 people work in TV today—nearly 70,000 in commercial TV, with the remainder in public TV. Reflecting on the financial realities of the business, employment in commercial TV has grown in some areas (like research and marketing), stabilized in some departments (like production and sales), and declined in others (most notably, local TV news and engineering).

The highest earnings tend to be in the northeastern and northern states; the lowest is in the south. Salaries depend on a number of factors: the size of the community in which the station is located, whether the workers at the station are unionized, the dominance of a station in its market, and so on.

Sales executives are usually paid on a commission basis. To survive, they must sell time. If they are not good at selling time, they rarely make enough to live on and so move on to a different field. If they can sell, then TV can be very lucrative. There is always a strong demand for good salespeople in broadcasting, and TV is no exception.

Production crews and those who appear on the air are represented by unions at some stations. Organized, that is, unionized, stations almost without exception pay their employees better than nonunion stations in the same area. There are several different unions representing workers at a station. Generally performers are represented by AFTRA, the American Federation of Television and Radio Artists. Engineers and production people are usually represented by NABET, the National Association of Broadcast Engineers and Technicians, IATSE, the International Alliance of Theatre and Screen Employees, and/or IBEW, the International Brotherhood of Electrical Workers.

Unions usually represent the workers at contract renewal time, when management and union come to agreement on compensation and work responsibilities. Unionized stations usually have fairly strict rules about what tasks each person is allowed to perform. At a Capital Cities station in Fresno, California, 16-millimeter film is put in the projectors by engineers, proud members of IBEW. The 35-millimeter slides are put in the projector by the announcer/directors, dues-paying members of AFTRA. Engineers are not supposed to touch the slide projectors and the announcer/directors are not allowed to touch the film projectors. As we will see, new video and computer technologies are making more and more of these distinctions moot.

Reflecting on national trends, unions have been in decline in TV. Whereas nationwide the number of employees belonging to a union was once as high as one in three, today fewer than one in six workers belongs to a union. This trend has also been felt in TV, as the power of NABET, IATSE, and other organizations is much less today than it was in the 1960s and 1970s.

Historically, like many other industries in the United States, TV has been primarily a white, male-dominated industry. Steps were taken in the late 1960s and early 1970s to begin to rectify the imbalance.

Changes have been slow in coming but there continue to be opportunities. Employment of women has increased significantly in recent years. Women constitute about 37 percent of the employees at commercial TV stations and about 45 percent at noncommercial stations.

The changes in broadcasting can probably best be seen just looking around you in your broadcasting classes. A few years ago broadcasting classes of twenty-five students might include three or four women. If your class is typical, the number of men and women studying broadcasting with you is very close to evenly split.

Most of the positions for women in TV tend to be on the clerical side of the business, although there continues to be a slow expansion of women in all areas of the broadcasting industry, especially sales.

Ethnic minorities have not fared as well as women in increasing their numbers in TV stations. Minority employment at stations nationwide is a little over 19 percent. (The FCC defines blacks, Asians, Hispanics, Native Americans, and Pacific Islanders as ethnic minorities.)

Like the situation for women, inroads into management positions have been hard to come by for ethnic minorities. Most TV jobs for minorities are in the labor and craft fields, from toting cameras to cleaning studios. However, the number of blacks and other minorities in management has increased somewhat (from 10.3 percent in 1985 to 12.2 percent in 1990).

Television Salaries

Like the radio medium, pay in TV is largely determined by the size of the market one works in. Simply put, the larger the city is, the more money you will earn. In addition, on average, VHF stations pay more than UHF stations and network affiliates pay better than do independents. Figure 5-8 illustrates these trends for selected jobs in TV stations.

As is true for most businesses, managers make the most money in TV. Nationwide, the typical GM earns well over $40,000 per year, over $80,000 in the nation's top-50 TV markets. Like radio, sales managers make the most money. In fact, you would be hard-pressed to find a sales executive in a top-10 TV market earning less than $100,000 per year.

Next to sales, salaries in the news department are the highest in TV. In the top-10 markets anchors and reporters can earn six-figure sal-

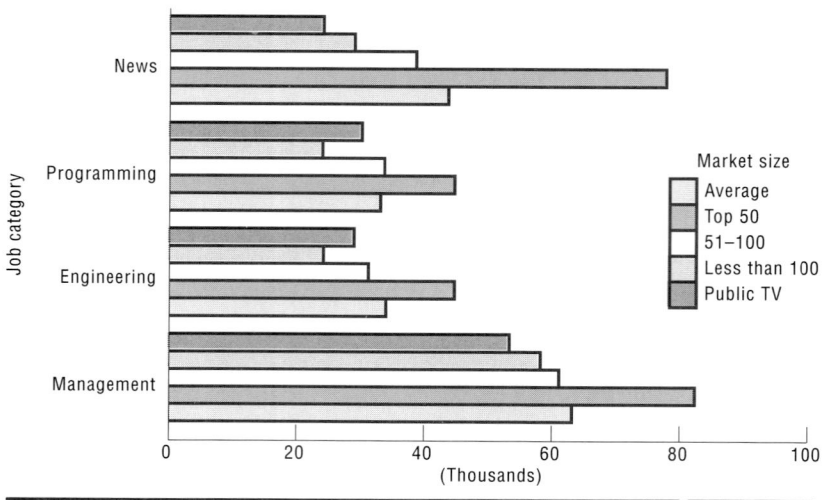

FIGURE 5-8 TV salaries. (Source: 1990 National Association of Broadcasters/Broadcast Financial Managers Report.)

aries. The downside is that salaries slip significantly in smaller markets. In addition, news staffs have been trimmed substantially in recent years to help stations meet sizable debt payments in a troubled economy.

Engineering and programming positions, from the video editor to the camera operator, tend to provide steady employment at stable but lower salaries than high-profile sales and news jobs. There is less volatility from market to market in salaries and employee benefits.

A check of salaries in public TV reveals, not surprisingly, that these positions pay less than their commercial counterparts. Public TV salaries in major markets are similar to those in commercial stations in small markets. It is clear that one seeks a job in public TV more for altruistic or public service reasons than for the quest for "big money."

SUMMARY

Virtually every home in America now has a TV set, and in each average household the set is on more than seven hours per day. Most Americans rate TV as the most credible news source.

There are two types of TV stations: commercial and noncommercial. Commercial stations can be further divided into network affiliates and independents. Some noncommercial stations are affiliated with the Public Broadcasting Service (PBS).

VHF stations broadcast on channels 2 to 13. UHF stations broadcast on channel 14 and above. Historically, VHF stations have been more prosperous than UHF stations, but the carriage of UHF stations on cable may change that situation. The licensing of low-power TV stations (LPTV) is an attempt by the FCC to provide more local service to a community.

Cable has made more programming services available to viewers. This has resulted in fewer viewers for programs on the major TV networks.

Revenue in TV comes from the sale of advertising. Advertising can be divided into several categories: network, national spot, and local. In addition, TV stations carry public service announcements and promos for upcoming programs.

In general, a single organization can own no more than twelve TV stations as long as these stations do not serve more than 25 percent of the nationwide number of TV homes. Sales of stations have increased in the last decade thanks to more liberal FCC regulations and favorable financial conditions.

Stations are typically organized into five departments: sales, engineering, business, programming, and news.

SUGGESTIONS FOR FURTHER READING

Auletta, K. (1991). *Three blind mice: How the TV networks lost their way.* New York; Random House.

Block, A. B. (1990). *Outfoxed: The inside story of America's fourth television network.* New York: St. Martin's Press.

Heighton, E., & Cunningham, D. (1984). *Advertising in the broadcast and cable media.* Belmont, Calif.: Wadsworth.

Hilliard, R. (1989). *Television station operation and management.* Stoneham, Mass.: Focal Press.

Hilsman, H. (1989). *The new electronic media.* Stoneham, Mass.: Focal Press.

Sherman, B. (1987). *Telecommunications management.* New York: McGraw-Hill.

Smith, M. (1984). *Radio, TV and cable.* New York: Holt, Rinehart and Winston.

Verna, T. (1987). *Live TV.* Stoneham, Mass.: Focal Press.

White, Barton, & Satterthwaite, Doyle. (1989). *The selling of broadcast advertising.* Needham Heights, Mass.: Allyn and Bacon.

Williams, H. (1989). *Beyond control: ABC and the fate of the networks.* New York: Atheneum.

CHAPTER 6: CABLE AND HOME VIDEO TODAY

The next time you scan the dials with your remote control or browse the thousands of tapes at the video store, consider this: a few years ago cable TV was available in just a small number of households, mostly those in mountainous areas where "rabbit ears" couldn't pull in a watchable signal. VCRs were owned only by TV networks and big stations; families went out to the movies, they didn't rent them for viewing at home. A quiet revolution in electronic media has taken place. Cable and home video today are an important component of the electronic media landscape. From modest beginnings each has come to occupy a pivotal place in the programming and economics of modern electronic media.

The Growth of Cable TV

Cable TV's rapid growth began in the 1960s and continues today. This growth can be seen in three different indicators: the number of cable systems in operation, the number of households that subscribe to cable, and the percentage of TV homes in the United States that are on the cable. This information is summarized in Table 6-1.

In 1960 about 600 small cable TV systems were in operation, providing service to about 700,000 households. About 1 percent of TV households were "wired." People subscribed to cable because it provided access to a good-quality picture from distant TV signals through a common tall antenna. Cable TV then stood for "community antenna television."

By 1965 little had changed. Although the number of systems had doubled, just over a million households were on the cable, representing a bit over 2 percent of the total TV households in the United States. Virtually all of the new systems were community antenna operations, springing up mainly in the south and west, where many communities gained access to TV for the first time.

A portent of things to come could be noted by 1970. Nearly 2500 systems had sprung up and the number of subscribers tripled to just under 8 percent of American TV homes. By this time a number of systems had begun operating in wealthy suburbs and chic "upper-crust" inner-city areas, from Santa Barbara, California, to the upper eastside of Manhattan. In addition to offering good signals, these systems began to supply special-interest programming to their clientele, ranging from home-team sports to eclectic political and social programming (this was the late sixties, after all).

TABLE 6-1 The Growth of Cable Television

Year	Operating Systems	Subscribers (millions)	TV Homes (%)
1960	640	0.7	1.4%
1965	1,325	1.3	2.4
1970	2,490	4.5	7.6
1975	3,366	9.8	14.3
1980	4,048	15.5	20.5
1985	6,600	37.3	43.7
1990	10,200	54.0	58.0
1995	13,000	60.0	62.0

SOURCE: Compiled by the authors from various industry publications.

The decade of the 1970s began cable TV's "gold rush" period. For many consumers, having cable TV became a status symbol (just as having TV had been a status symbol for people in the early 1950s). The new cable offered first-run, uncut movies, unlike the heavily censored, commercial-ridden movies on "regular" TV. Many cable systems signed contracts bringing baseball, basketball, and hockey games to their service and away from the over-the-air channels in their communities. Many cities required cable systems to provide access channels for community-based programming. Some of the shows on these channels were fascinating: topics ranged from sex (like "Dirty George" in Manhattan, who invited women to disrobe before his cameras); to politics of the left and right; to travel, cooking, and educational programs.

From 1970 to 1980 the number of cable systems grew from about 2500 to over 4000. The number of subscribing households quadrupled, from just over 4 million to nearly 16 million. By 1980 one in five households in the United States was on the cable.

Cable's growth spurt continued in the 1980s, although by the end of the decade the pace of cable expansion had begun to slow. By 1985 some 6600 systems were operating, delivering service to just under one-half of the TV universe. In 1988 cable passed the "magic number": more than one-half of all households with TV in the United States were cable subscribers (about 45 million homes).

Today about 11,000 cable systems deliver cable to over 53 million TV homes, approaching 60 percent of TV homes.

Interestingly, the growth of cable TV proceeded in a different direction from conventional broadcasting. Radio and TV stations sprang up first in large metropolitan areas. Then, with increased power and better transmission equipment, service was expanded to include suburban and rural areas. Cable, on the other hand, began in the most inaccessible rural areas, spread next to affluent suburbs, and, because of technical obstacles, political battles, and bureaucratic problems, had been slow to catch on in major metropolitan areas (see Table 6-2).

Channel Capacity

Another way to appreciate the growth of cable is to consider the number of channels available to viewers. In the cable business this is known as *channel capacity*.

TABLE 6-2 Cable Penetration in Large Cities

Year	New York (%)	Chicago (%)	Los Angeles (%)	National Average (%)
1980	11	1	12	20.5
1984	30	21	29	42
1988	41	39	41	51
1992	58	60	57	60

SOURCE: *Cablevision*, various issues.

Just a few years ago virtually all cable systems were twelve-channel operations. Attaching the TV set to the cable generally filled most of the channels from 2 to 13. In the mid-1970s new systems offered thirty-five channels, requiring most TV homes to have a converter box. By the 1980s many older systems were upgraded and new franchises were awarded to companies providing fifty-four or more channels. Today's cable channel capacity is graphically illustrated in Figure 6-1.

More than one-half of the nation's cable systems, representing two-thirds of all cable viewers, provide over thirty different channels of programming. Nearly one in four cable subscribers can view fifty-four or more cable channels. Only a few systems today are restricted to the six to twelve channels of TV programming that one can receive over the air with "rabbit ears" or rooftop antennas. Clearly cable provides more viewing alternatives. Whether more TV is *better* TV is an important question that we address in some detail in a later chapter.

CABLE PROGRAMMING TODAY

Getting into the TV business today is something like opening a new shopping mall. There may be room for 100 retail stores in the new mall, but how many will actually sign a lease? Which stores will be most popular with shoppers, and thereby take the lion's share of the profits? While some stores will be "mass merchandisers," attempting to bring in lots of customers with discounting

---

TECHNOLOGY

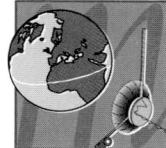

"QUANTUM" LEAP: TIME-WARNER'S 150-CHANNEL CABLE SYSTEM

In December 1991 BQ Cable in Queens, New York, made a new option available to its subscribers. The service is called *Quantum*, billed by BQ cable's parent company, Time-Warner, as the nation's first 150-channel cable system. In addition to carrying virtually all of the available basic and pay services, fifty channels are reserved for PPV events. With Quantum, Time-Warner hopes to lure back some of the customers lost to the video store: over twenty different movies are offered on a PPV basis each day, including many recent hits.

It is too early to tell if Quantum has lept into the hearts of cable subscribers. However, at least one household is happy. The first customer to sign up for the service was a housewife in Queens. To publicize Quantum, Time-Warner made a gift of the service for a year, plus twenty free PPV events. In addition, Time-Warner threw a press party (at her house). Other gifts included a stuffed animal for her small baby (from HBO), a globe (courtesy of the Monitor Channel), and an old-time radio (from the Nostalgia Network).

practices and volume sales, others will be specialty stores or boutiques, aiming at a narrowly selected clientele (like teens and young adults for Benetton and urban professional males for Brookstone). The same situation applies in the quest for cable audiences.

There are now over 100 program services available to cable, with an additional 20 or more in the planning stages (see box). Some are "mass marketers," like Sears or Macy's; others are boutiques, like Banana Republic or The Sharper Image, Ltd.

For convenience, cable programming today can be divided into three broad classifications: basic cable services, pay cable services, and specialty services. Let's examine each in detail.

Basic Cable Services

The backbone of CATV is its lineup of basic services. These are the program services available for the lowest subscription charge, typically from $10.00 to $20.00 per month. There are two main types of basic services: local and regional broadcast signals and advertiser-supported cable services.

From "must carry" to "may carry": Local/regional broadcast signals For years cable systems were obligated to provide space on their systems to retransmit local TV stations within their communities. Such rules, known as "must carry," were declared unconstitutional in 1985. Since then cable operators have been able to retransmit the TV signals of stations in and near their communities without compensating those stations. For obvious reasons broadcasters have fought diligently to force cable operators to pay to carry their stations. As this edition goes to press, broadcasters are seeking to impose a system whereby they will be compensated by cable systems carrying their signals, a set of rules known as *retransmission consent*.

Even if retransmission consent laws are passed by Congress, cable systems are still expected to carry the majority of local broadcast stations. Many subscribers choose cable to improve reception of stations in their area. Most of them expect to see the stations they are accustomed to watching, even after they subscribe to the cable. Thus local and regional stations are likely to remain in the channel lineup of the nation's cable systems.

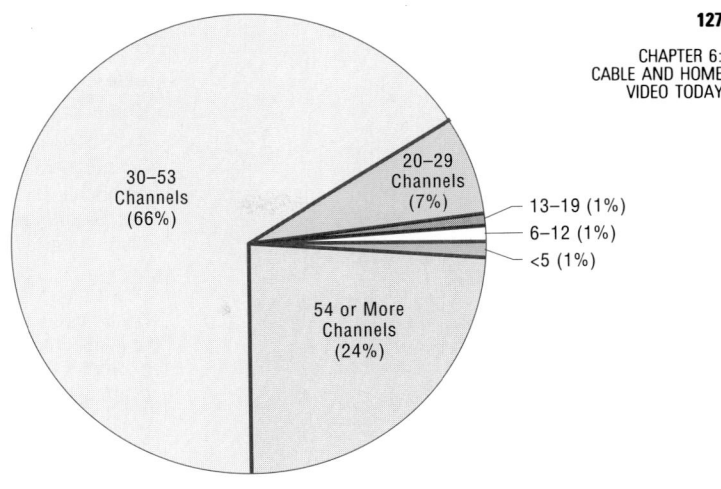

FIGURE 6-1 Cable channel capacity. (Source: A. C. Nielsen, *1990 Report on Television*, p. 2.)

Advertiser-supported basic cable services The second classification of basic cable services is advertiser(ad)-supported cable networks, program services specifically designed to reach cable audiences. Like the broadcast networks (ABC, CBS, NBC, and Fox), they carry national advertising. They also provide opportunities for local cable systems to place their own advertising spots.

The leading ad-supported cable networks include three channels offered by Turner Broadcasting Systems: CNN, Headline News, and Turner Network Television (TNT); the Black Entertainment Network (BET); the Arts and Entertainment Network (A&E); the Nashville Network (TNN); Nickelodeon; the Weather Channel; and MTV. Some basic services are dedicated to health (Lifetime), some education (The Learning Channel). The Cable News and Business Channel (CNBC) caters to consumer and business news; the Courtroom Television Network provides extended coverage of crime and legal issues.

Superstations, which are independent TV stations in large metropolitan areas that are licensed to satellite companies for national distribution, also offer ad-supported basic services. Local cable systems contract with the satellite distributor to carry the station on their cable system. The superstations feature a steady diet of movies and sports. Some even carry local and national news. Because of this, cable viewers in Alaska can avidly follow the Chicago Cubs and viewers in Guam

CABLE PROGRAMMING TODAY

Basic Services	Affiliates	Subs
ACTS Satellite Network	478	9,500,000
American Movie Classics	2,828	35,000,000
America's Disability Channel	238	14,200,000
Arts & Entertainment Network	7,000	51,000,000
Black Entertainment Television	2,407	31,371,550
Bravo	455	6,000,000
CNBC	3,000	43,000,000
Cable News Network	10,877	58,892,000
Channel America	13	429,300
C-SPAN	4,055	54,000,000
C-SPAN II	800	24,500,000
Comedy Central	1,282	18,250,000
Country Music Television	1,974	13,700,000
Courtroom Television Network	N.A.	4,800,000
The Discovery Channel	9,397	56,000,000
E! Entertainment Television	777	18,750,000
ESPN*	23,300	59,195,000
EWTN	774	23,300,000
The Family Channel	9,500	53,500,000
Fox Net	275	1,000,000
Galavision	249	1,750,000
Headline News	5,506	46,856,000
Home Shopping Network	1,502	18,000,000
Home Shopping Network II	400	7,000,000
HSN Infomercial Channel (new)	N.A.	N.A.
The Inspirational Network	850	6,500,000
International Channel	35	2,200,000
KTLA	292	4,800,000
KTVT	481	2,200,000
The Learning Channel	1,196	15,600,000
Lifetime	5,400	53,000,000
Mind Extension University	402	15,000,000
The Monitor Channel	340	3,500,000
MTV	7,430	56,600,000
The Nashville Network	12,259	53,900,000
National College Television	380	5,983,244
Nickelodeon	8,635	55,400,000
Nick At Nite	3,837	50,250,000
North American Television	8	513,000
Nostalgia Television	640	12,300,000
Prevue Guide	835	24,526,075
QVC Fashion Channel (new)	380	6,000,000
QVC Network	3,900	41,000,000
SCOLA/News Of All Nations	35	2,800,000
Silent Network	238	14,200,000
SportsChannel America**	58	2,320,000
TBS Superstation	11,105	57,207,000
Telemundo	36	1,362,036
TNT	6,958	54,190,000
The Travel Channel	735	17,500,000
Trinity Broadcasting Network	1,015	13,300,000
Univision	814	11,062,692
USA Network	10,100	58,000,000
VH-1	3,985	41,800,000
Video Jukebox Network	96	9,050,000
VISN	670	10,200,000

CABLE PROGRAMMING TODAY (Cont.)

Basic Services	Affiliates	Subs
The Weather Channel	4,500	49,063,000
WGN	13,969	34,900,000
WPIX	641	9,200,000
WSBK	73	2,000,000
WWOR	3,013	13,500,000

Pay Services	Affiliates	Subs
Cinemax	5,458	6,400,000
The Disney Channel	7,000	5,665,000
Encore	854	25,000
Home Box Office	8,833	17,300,000
The Movie Channel	3,250	2,800,000
Showtime	6,000	7,400,000
TV-Japan	5	N.A.

Audio Services	Affiliates	Subs
AEI Spectra Network	160	5,000,000
C-SPAN Audio Network	38	2,419,000
C-SPAN Audio Network II	23	1,500,000
Cable Radio Network	71	2,700,000
Digital Cable Radio	14	4,200
Digital Music Express	N.A.	N.A.
Digital Planet	2	N.A.
Japan Cable Radio	7	2,500
KLON	22	760,000
Moody Bible Institute	47	715,533
Satellite Radio Network	237	483,000
Superaudio	481	7,652,000
WFMT	143	1,400,000

Text Services	Affiliates	Subs
AP Business Plus	100	2,000,000
AP News Cable	185	2,500,000
AP News Plus	100	2,300,000
Cable SportsTracker	42	1,389,708
EPG	96	4,537,815
EPG Jr.	548	582,117
Reuters NewsView	127	2,437,000
Story Vision Network	40	2,100,000

Channel Promotion/ Computer Services	Affiliates	Subs
NuStar	840	19,000,000
X*Press	600	14,000

Regional Basic Services	Affiliates	Subs
Arizona Sports	1	310,000
Atlanta Interfaith	3	225,000
Bay Area Religious Channel	6	113,000
Cable TV Network of New Jersey	34	1,500,000
CAL-SPAN	42	2,100,000
The Ecumenical Channel	9	170,000
Empire Sports Network	15	316,000
Florida Tourism Channel	20	754,000
Home Sports Entertainment	475	2,400,000
KBL Sports Network	67	1,200,000
Madison Square Garden Network	16	4,500,000
Meadows Racing Network	17	700,000
Midwest Sports Channel	90	610,000
NewsChannel 8 (new)	8	650,000

CABLE PROGRAMMING TODAY (Cont.)

Regional Basic Services	Affiliates	Subs
News 12 Long Island	4	601,000
Northwest Cable Sports Network	3	700,000
Orange County Cable News	8	350,000
Pennsylvania Cable Network*	28	750,000
Prime Sports/Intermountain West	30	328,000
Prime Sports/Midwest	6	182,000
Prime Sports/Northwest	65	1,100,000
Prime Sports/Rocky Mountain	110	935,000
Prime Sports/Upper Midwest	4	185,000
Prime Ticket	128	4,200,000
SportsChannel Bay Area	35	300,000
SportsChannel Chicago	75	1,980,000
SportsChannel Cincinnati	14	300,541
SportsChannel Florida	74	1,100,000
SportsChannel Ohio	33	904,000
SportsChannel Philadelphia	41	1,300,000
SportSouth	70	1,000,000
Sunshine Network	189	3,078,542

Regional Pay Services	Affiliates	Subs
Home Team Sports	205	2,200,000
New England Sports Network	171	380,000
Prism	87	470,000
Pro-Am Sports	240	750,000
SportsChannel Los Angeles	76	125,000
SportsChannel New England	164	1,300,000
SportsChannel New York	117	1,400,000
SportsChannel Pacific (new)	65	1,700,000

*Formerly Pennarama.
SOURCE: *Cablevision*, October 21, 1991, pp. 70–73.

can decry the nightly news from New York City. The major superstations are Atlanta's WTBS, Chicago's WGN, KTVT from Los Angeles, WWOR (from Secaucus, New Jersey, near New York City), and WPIX (New York).

Pay Services

Pay services became popular in cable in the 1970s as a source of home viewing of theatrical motion pictures, major sporting events, and entertainment specials. They are called pay services since subscribers must pay an additional fee to receive the service. In return, pay services do not generally carry advertising. Their selling feature is original programming that is not available on broadcast TV without the commercial interruptions experienced in "free" TV. The leading pay services are listed in Table 6-3.

The giant of pay cable is Home Box Office, available on nearly 9000 cable systems. Over 17 million homes, about one in five American households, receive the channel. HBO's parent company, Time Warner, Inc., also owns Cinemax, now in just over 6 million homes. Other important pay services include Showtime and the Movie Channel (both owned by Viacom), and the family-oriented Disney Channel.

Specialty Services

This "mixed bag" of cable services share one important characteristic: each specialty service is designed for narrowly focused, or "targeted," audiences. Audio services provide stereo radio-type programming over the cable. Attaching the TV cable to the home stereo receiver virtually guarantees noise-free reception for hi-fi enthusiasts. Some cable audio services today include the Cable Radio Network, Digital Music Express, Satellite Radio Network, Superaudio, and a radio "superstation," WFMT (see box).

TABLE 6-3 The Leading Pay Cable Services

Network	Number of Subscribers	Owner and Headquarters	Launch Date	Content
HBO	17.5 million	Time-Warner Inc., New York	November 1972	Movies, variety, sports, specials, documentaries, children's programming
Showtime	7.5 million	Viacom, New York	July 1980	Movies, variety, comedy specials, Broadway adaptations
Cinemax	6.5 million	Time-Warner Inc., New York	August 1980	Movies, comedy, music specials
The Disney Channel	6 million	Walt Disney Co., Burbank, California	April 1983	Original feature films, specials, series, classic Hollywood films
The Movie Channel	3 million	Viacom, New York	December 1980	Double features, film festivals, movie marathons

SOURCE: *Channels,* December 1988, p. 90; *Cablevision,* October 21, 1991, p. 52. Projected to 1993 by the authors.

A particular class of specialty service is teletext. Text services provide words and pictures on the screen. This is a useful way of presenting news and weather, travel information, financial reports, and the like.

The leading text service today is EPG, the electronic program guide. This service lists the programs that are available on a cable system's channel lineup to help viewers decide which shows to watch. EPG (and its twin, EPG Jr.) is available to over 5 million subscriber households.

Just as the cable can be used to receive stereo music signals, with appropriate hardware and software, cable TV can bring news and business wire services to viewer TV screens and home computers. Leading text services of this type include NuStar, available on over 800 cable systems and 15 million homes, X*Press, AP News Cable, Reuters NewsView, and the Story Vision Network.

SELLING CABLE SERVICES

Cable is marketed to attract different types of cable consumers. The minimum monthly charge for cable TV service generally includes a menu of local broadcast stations, ad-supported cable networks, and a few superstations. To get the full range of cable programming services, subscribers must pay additional monthly fees. Cable operators tend to offer their additional services in packages. That is, for an additional $15, the subscriber may move from twelve to twenty-four

PROGRAMMING

CHICAGO'S WFMT: THE RADIO SUPERSTATION

Most discussion of cable centers on TV. Often overlooked is cable's ability to transmit high-quality audio signals to subscribers. One successful sound operation in cable is WFMT, a fine-arts radio station that emanates from Chicago. In its listening area WFMT is well known as an honored and treasured radio station, one of the few advertiser-supported classical music operations. In 1979 United Video of Tulsa obtained a license to include WFMT in its package of satellite-fed cable services. Today WFMT is fed by the Galaxy I satellite to more than 350 cities in 44 states. Its estimated national distribution is over 3 million.

Recently the station launched a second satellite music service. The Beethoven Satellite Network (BSN) draws from WFMT's library of over 40,000 recordings to provide classical music to radio stations around the country. Despite his genius, it's doubtful the great composer could conceive of his notes traveling 23,000 miles and back every day to listeners throughout the western hemisphere!

Some of the many services available to today's cable TV viewer.

channels of service. Adding another $15 might bring the subscriber to thirty-five channels of service. The "high-end" customer, paying up to $75 or more, may receive fifty-four channels, three or four pay services, as well as the ability to order special events. The process of moving subscribers up the ladder from basic to pay services is known as *tiering*. Let's examine how tiering is practiced in the cable business.

Types of Cable Households

In the cable business not all homes are created equal. Cable companies make clear distinctions between the types of households in their service areas on the basis of which program options they elect. The various types of TV homes in the United States can be viewed as a pyramid, as illustrated in Figure 6-2.

Homes Passed

The base of the pyramid consists of those households in the United States that are in an area served by cable TV. This statistic is known as *homes passed (HP)*. Homes passed are all households that could subscribe to a cable system if they wanted to. The cable literally passes by these households.

In real numbers, there are about 93 million homes with TV in the United States. About 82 million homes are located in an area served by cable, for an HP figure of just under 90 percent.

Cable Households

The next level of the pyramid are those HPs that decide to subscribe to cable TV. This figure can

FIGURE 6-2 The cable subscriber pyramid.

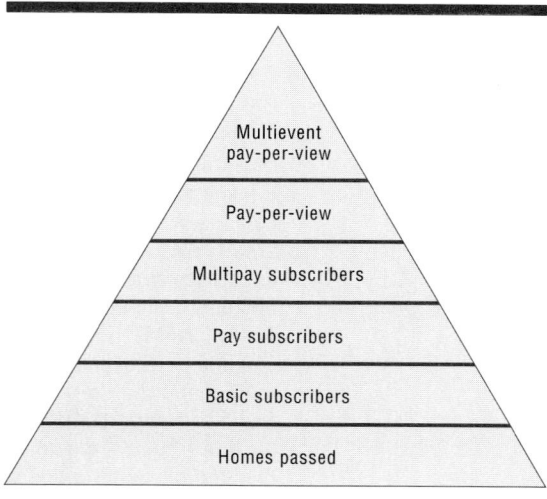

Total U.S. TV households

be calculated as the ratio of subscriber households to HPs. The resulting percentage is sometimes called the cable system's *basic penetration*. For example, if a cable system passes 100 homes and 85 take the cable, the basic penetration rate is 85 percent. Obviously, a cable operator would like all HPs to take cable service, that is, a penetration rate of 100 percent. This ideal world does not exist. The percentage of households in cable areas that elect to subscribe varies widely. Some suburban systems have enough "upscale" consumer households to boast over 90 percent basic penetration. Other systems, including many in poor inner-city and rural areas, report penetration rates below 50 percent. Industrywide, about two-thirds of HPs are cable subscribers, about 55 million homes.

Pay Households

Those cable homes that pay an additional fee for the pay services listed earlier (such as HBO, Disney Channel, or Showtime) are known as *pay households*. Through the period of cable's massive growth—from the mid-1970s to the mid-1980s—as many as three-fourths of all cable subscribers also took a pay service. Faced with competition from new media outlets (mainly video stores), the percentage of pay units began to drop in the late 1980s. Today about 32 million homes have pay cable, representing about one-third of TV homes in the United States (and a bit more than one-half of all cable homes).

Multipay Households

There are two types of pay cable households: those that elect just one pay service and those that subscribe to more than one. The homes that take more than one service are known as *multipay* households. Although the percentage of pay households that buys more than one premium service is relatively low (under 20 percent), this level of the pyramid is highly sought after by cable executives. Multipay subscribers pay the biggest cable bills—$50 per month or more—much of which is additional profit for the cable entrepreneur, since the operator can negotiate reduced rates from pay services to "bundle" their program offerings.

Pay per View

Near the top of the subscriber pyramid are those households that can choose their pay programming selectively by ordering it as desired from the cable company. This is known as *pay per view*, or *PPV*.

PPV requires special cable technology. Pay-per-view homes require cable boxes that can be isolated by the cable company and separately programmed (so that only the home that orders the event will receive it). Such devices are known as *addressable converters*.

About one in three cable homes today is addressable by PPV. Many cable operators are staking their future on PPV. By the middle of the decade addressable technology should be in more than one-half of all cable households.

At the top of the cable pyramid is multievent PPV, also known as impulse PPV. As channel capacity increases and addressability spreads, cable executives expect consumers to buy on impulse, to purchase many different events each month on a PPV basis. This optimism is based on the great success of boxing matches, concerts, and other events offered on a PPV basis in recent years. For example, the farewell tour of the Judds was watched in over 250,000 households, grossing more than $50 million. The Rolling Stones' "Steel Wheels" tour attracted more than 175,000 homes. The success of boxing on PPV led Time-Warner to form TVKO, a PPV service dedicated to major sports events.

A number of national PPV services have sprung up in recent years to provide programming of interest to cable subscribers. The leading PPV services are profiled in Table 6-4.

Movies are the backbone of PPV today. Request Television, Viewer's Choice, and the Cable Video Store allow PPV subscribers to select from a number of recent box office hits. Late at night the lineup often includes adult films (the core of the Spice and Playboy at Night services).

CABLE ECONOMICS

Like the previous chapters on TV and radio, an understanding of cable requires analysis of its economic structure. Who owns cable TV? Where are its revenues? What are its greatest cost fac-

TABLE 6-4 Leading Pay-per-View Services

Network	Launch	Systems	Addressable Subscribers
Action pay-per-view	9/1/90	46	2,000,000
Cable Video Store	4/1/86	105	1,200,000
Playboy at Night	9/1/87	175	5,000,000
Request Television	11/27/85	518	8,600,000
Spice	7/1/89	80	3,400,000
Viewer's Choice	11/26/85	400	9,000,000

SOURCE: *Cablevision,* December 2, 1991, p. 40.

tors? This understanding begins with the way cable systems get built: the franchise process.

Cable Franchising

Cable systems share one important characteristic with radio and TV stations: entrepreneurs can't simply announce they are building a system and start construction. In broadcasting, stations must apply for available frequencies and go through a long, sometimes tortuous licensing process at the FCC. Although entering the broadcast business is a federal matter, cable entrepreneurs enter into negotiations with state and local governments to gain the right to construct their systems. The contract negotiated between a municipality and a cable company is known as the *franchise agreement.*

A stipulation of many such agreements mandates that systems return some of their income to the municipality in the form of franchise fees (see "Cable Expenses," below). Franchise agreements can also require cable systems to provide studio facilities, free drops to schools and government offices, and so forth.

The power of municipalities to govern cable operations through the franchising process was dealt a serious blow in 1984 with the passage by Congress of the Cable Policy Act. Designed to streamline regulation of the cable business just as radio and TV had been deregulated a few years earlier, the act imposed limits on city governments in setting cable rates and franchise fees and in policing cable programming. However, the act reaffirmed cities' rights to franchise systems. The result of the act is that the franchise process continues to exert enormous influence in the relationship among a cable system, its subscribers, and local governments.

As this edition went to press Congress was considering legislation granting additional power to cities in the franchising process. New cable laws would restore the role of city governments in regulating cable rates and would require cable systems to provide minimum levels of basic services to all subscribers.

Cable Ownership

Cable TV is different from broadcasting in one important respect. Whereas there are federal caps on the number of broadcast stations that can be owned by a single entity, there are no such limits imposed on cable ownership. In other words, a cable entrepreneur can own as many systems and boast as many subscribers as can be amassed, subject to antitrust law. Consequently, the cable business is marked by a concentration of ownership by a large number of *multiple-system operators (MSOs).* Owners of only one system are known as *single-system operators (SSOs),* or simply *cable system operators (CSOs).* The MSOs dominate the business, as depicted in Table 6-5.

Each of the leading MSOs operates in excess of 100 systems and serves between 1 million and 9 million subscribers. The largest MSO is Tele-Communications, Inc. (TCI), with nearly 10 million subscriber households. ATC, a subsidiary of Time-Warner, Inc., is the second largest MSO, with nearly 5 million subscribers. Companies with over 2 million homes include UA Entertainment and Continental Cablevision. Another Time-Warner company, Warner Cable, boasts nearly 2 million subscriber households.

The hardware necessary to receive pay-per-view TV. Subscribers can see recently released movies for about $5 and big sporting events, like championship fights for about $30 to $35.

TABLE 6-5 Top-20 Cable MSOs

Rank	Multiple System Operator	Basic Subs
1.	Tele-Communications Inc.	9,493,682
2.	ATC*	4,700,000
3.	UA Entertainment	2,840,000
4.	Continental Cablevision	2,760,000
5.	Warner Cable*	1,919,000
6.	Cox Cable (7)	1,661,277
7.	Comcast (6)	1,661,000
8.	Jones Intercable	1,646,132
9.	Cablevision Systems	1,619,333
10.	Storer Cable Communications	1,605,900
11.	Newhouse Broadcasting	1,267,200
12.	Times Mirror Cable Television	1,135,153
13.	Cablevision Industries	1,122,800
14.	Adelphia Communications	1,095,300
15.	Viacom Cable	1,069,700
16.	Sammons Communications	919,411
17.	Century Communications	884,000
18.	Falcon Cable	873,572
19.	Paragon Communications	838,341
20.	Prime Cable	694,798

* Parent company Time Warner merged ATC and Warner Cable in 1992.
SOURCE: *Cablevision,* November 18, 1991, p. 96.

The biggest individual systems are located in major suburban areas and range from about 200,000 to 500,000 subscriber households (See Table 6-6.). The largest single system is Cablevision Systems' operation in Long Island, New York, with nearly 600,000 basic subscribers. Time-Warner's ATC system in Orlando, Florida, is the second largest, with nearly 500,000 homes. Other large systems are in suburban Seattle, San Diego, Phoenix, Tampa, Honolulu, and Chicago.

The concentration of cable by the MSOs is further evidenced by this fact: although there are over 10,000 operating systems, more than one-half of all cable subscribers in the United States send their checks for monthly service to one of the top fifteen MSOs! In fact, one in five monthly bills for cable service is paid to TCI alone! From an economic standpoint, clearly the MSOs exert enormous influence on the cable industry.

CABLE FINANCE

Running a cable system is considerably different from radio and TV broadcasting. Unlike radio and TV, which distribute their signals by "air" (which is still free in most parts of the world), cable companies must physically link each subscriber household to the headend by wire. This requires cable systems to stretch cable through underground viaducts and across utility poles, into private homes and apartment houses. This is a very costly and time-consuming process, leading to expenses for construction and maintenance that

ESPN covers auto racing. The basic service cable network tries to lure viewers away from major network programming with its coverage of key sporting events.

TABLE 6-6 The Top-20 Cable Systems

Rank	Location	Operator	Basic Subs
1.	Long Island, NY	Cablevision Systems	561,385
2.	Orlando Complex, FL	ATC	455,000
3.	Puget Sound, WA	Viacom	377,900
4.	Brooklyn/Queens, NY	ATC	326,000
5.	San Diego, CA	Cox Cable	323,759
6.	Phoenix, AZ	Times Mirror	289,363
7.	Manhattan, NY	ATC	254,000
8.	San Antonio, TX	KBLCOM	240,000
9.	Tampa/St. Petersburg, FL	Paragon	224,988
10.	East Orange, NJ	Maclean Hunter	224,357
11.	Honolulu, HI	ATC	216,000
12.	Montgomery County, MD/Arlington, VA	Hauser	208,500
13.	Chicago Suburbs, IL	Continental	206,000
14.	Houston, TX	Warner	197,000
15.	Sacramento, CA	Scripps Howard	196,000
16.	Chicago, IL	TCI	191,700
17.	Atlanta, GA	Prime Cable	191,157
18.	Fairfax, VA	Media General	190,848
19.	Wayne, NJ	UA Entertainment	188,046
20.	Hampton Roads, VA	Cox Cable	182,487

are typically much higher than those at TV stations.

Unlike broadcasting, where revenue comes mainly from advertising, the overwhelming majority of cable revenues comes directly from audiences via subscriber fees. And, as we have seen, whereas broadcasters can use the airwaves free of charge, many cable operators must return part of their revenues in the form of franchise fees to local governments. Let's examine cable finance by looking at its primary sources of revenue and expense. These are graphically illustrated in Figure 6-3.

Cable Revenue

In the broadcast business more than 90 percent of revenue comes from advertising and less than 10 percent of income comes from other sources. In cable almost 90 percent of revenue comes from subscriber fees; less than 10 percent comes from advertising and other sources.

Subscriber fees Increased cable penetration, greater channel capacity, the availability of PPV, rate deregulation, and other factors led to a substantial increase in subscriber spending for cable TV in the 1980s. For example, in 1980 subscriber fees, or sub fees, accounted for about $3 billion. By 1985 sub fees totaled $9.5 billion; in 1990 revenue from cable households was estimated at $21.4 billion.

The average monthly cost for basic cable service nationwide today is about $20. Adding a pay service increases the monthly average to about $30. A full menu of pay services can jack the cable bill up to $50 or more. Pay-per-view adds additional fees: one major prizefight alone can nearly double the bill. Installation fees add $10 to $30, and the monthly charge for a cable converter (in systems offering more than twelve channels) can be an additional $5 to $15.

Cable advertising Like subscriber fees, cable advertising experienced spectacular growth in the 1980s. In 1980 the total advertising volume in cable was under $60 million. This figure represented less than 1 percent of total industry revenues. By 1985 advertising revenue had increased more than ten times, to over $800 million. Income from cable advertising in 1990 was over $2.5 billion, with projections for the middle of the decade approaching $5 billion.

The bulk of cable advertising revenue goes to the advertiser-supported cable networks (such as WTBS, CNN, and ESPN). The rest of the advertising income goes to the thousands of local systems.

In 1991 two-thirds of $3 billion spent on cable advertising went to ad-supported basic cable networks. About $800 million went to local/regional cable spot advertising; the remainder was spent on regional sports networks, such as SportsChannel New England and Home Team Sports.

Cable Expenses

Cable is an expensive business to operate. The cable itself must be built and maintained; satellite dishes and microwave horns are needed to receive programming; and studios need to be constructed and staffed. In addition, cable systems must maintain an expensive inventory of converters, trucks, and spare parts. Personnel costs for needed office and field crews are high. And, like broadcasters, cable operators must pay license fees to receive programming.

For these reasons cable finance typically involves a large, long-term debt that is gradually

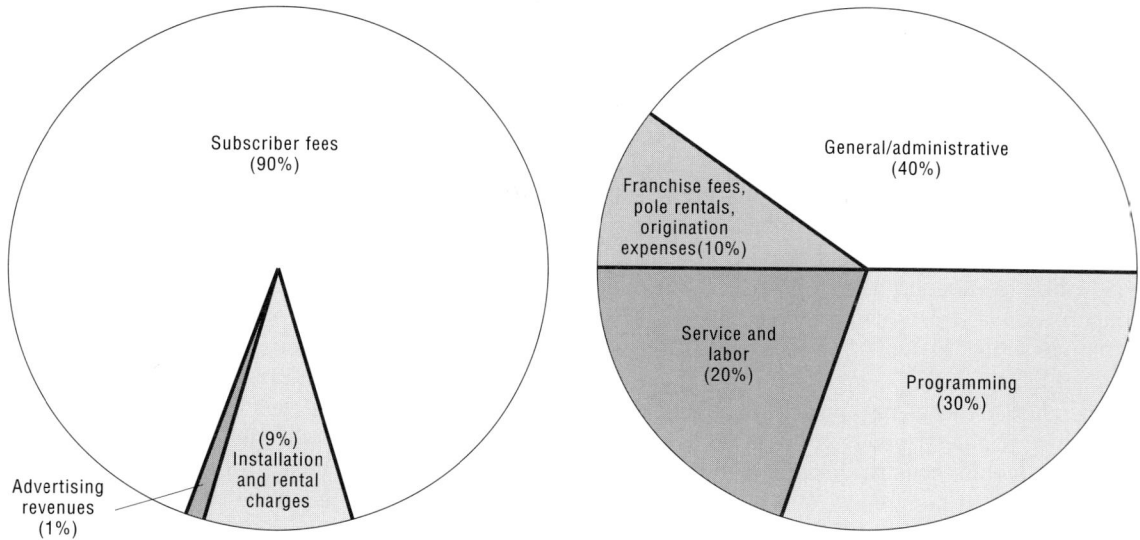

FIGURE 6-3 *Left,* the cable revenue pie. *Right,* the cable expense pie.

paid off as the numbers of HPs and subscribers escalate.

Capitalization expenses The primary cable expense is the cost of capitalization: the physical construction of the cable facilities and the wiring of subscriber households to the system. The capital investment in physical facilities can fluctuate wildly depending on population density and the type of utility configurations in the community. Generally, the industry rule of thumb calculates capital costs at $20,000 to $40,000 per mile for *aerial construction,* wiring a community by using existing telephone and utility poles. *Underground plant,* the process of laying cable in existing underground utility conduits, can range from $50,000 to $250,000 per mile. In the process of wiring the larger cities some cable operators have found conduit space filled by existing telephone and utility services and have had to dig new tunnels. In some cases existing wiring was unacceptable for cable's complex video and audio trans-

ECONOMICS

THE VARYING PRICE OF CABLE TELEVISION

One of the reasons for increased congressional scrutiny of cable TV is the widely varying rates charged for service around the country and the different number of channels available for those rates. A short trip around the country illustrates what we mean.

Today the average cable customer pays about $20 per month for cable and receives about thirty channels of service in return. The most pricey cable systems are in Los Angeles and New York. In those cities the average cable bill approaches $30 per month. In return the consumer receives fifty channels or more of programming. Salt Lake City, on the other hand, seems to be cable-poor—only twenty-five channels are offered. The good news is that monthly bills are below $17. Things are slightly better in Phoenix, where monthly rates average about $22 and the number of channels is forty-two.

The bottom line is that consumers have become increasingly unhappy about the rise in cable rates. One recent survey revealed that while the consumer price index rose about 6 percent each year in the late 1980s, cable rates rose, on average, at a rate of 14 percent or more. No wonder rate regulation is at the heart of much pending cable legislation!

missions, requiring systems to lay new cable. Thus capital costs for such new construction have escalated to as much as $350,000 per mile.

Operational expenses Once the cable system has been constructed, operating expenses fall into distinct categories, depicted in Figure 6-3. Like broadcasting, general and administrative expenses include buildings and grounds, equipment and office personnel, sales expenses, and other expenses relative to operating a media business. This consumes about 40 cents of the cable expense dollar.

Also like broadcasting, programming expenses eat up a sizable proportion of cable revenues. Cable systems acquire programming for fees based on their number of subscribers and the popularity of the network. For example, it might cost a cable operator 25 cents per subscriber per month for ESPN, 15 cents per subscriber for MTV, and 5 cents for CNBC. Operators pay more for the premium services, up to a fifty–fifty split of the costs for HBO or Showtime. For example, the system charging the subscriber $10 per month for HBO may send $5 to HBO for the service.

Program costs have escalated precipitately in recent years, led by price hikes from the pay services and many of the leading basic services, including USA, ESPN, and CNN. Today programming represents up to a third of cable expenses.

Unlike their broadcast counterparts, many cable operators subcontract a sizable part of their operations to outside support services. These include installation and repair crews and telephone answering services. Service and labor costs account for 20 cents of the cable expense dollar.

While not all systems are required to pay franchise fees to local governments, many of the larger systems funnel up to 5 percent (the current ceiling allowed by Congress) of their gross revenues back to their communities in this manner. Other expenses include pole rentals (another 3 percent), origination (studio) expenses, and satellite and microwave costs.

Cable systems have taken a variety of cost-cutting steps. Some have hired their own full-time construction and maintenance crews to streamline costs in the service area. Major MSOs are buying geographically adjacent systems so that operations can be combined, or clustered, for more efficient management. To hold down program costs, the major MSOs negotiate package rates with the pay and basic services. Most MSOs now use centralized billing, processing the bills from all of their local systems at their national headquarters.

Cable Profits

With high up-front and operations costs, it might appear that making money in cable is difficult, if not impossible. But the truth is that cable can be an enormously profitable enterprise. After all, few businesses find subscribers writing a check to them every month, year after year, "just for TV." Add to this phenomenon the increase in revenues from cable advertising, continued decline of network viewing, the rise of PPV, and the result is a rosy outlook for cable economics.

Despite its enormous construction costs, profitability can come quickly to new cable systems. Once systems pass the construction phase, expenses tend to become controllable, if not constant. For example, adding a new household might bring in $25 per month in subscriber fees but might cost only $5 to wire and program. Simple arithmetic indicates a 500 percent profit margin!

Older, "mature" systems have excellent profit potential. The evidence is that coaxial cable lasts fifty years or more. Systems constructed in the growth period of the 1970s will function effectively with relatively low maintenance until well into the next century. Many of these systems were built before the rapid increases in construction and utility costs that have marked recent years. Their debt, relatively small to begin with, is likely to have been reduced or eliminated by now.

For these reasons cable operations have received particularly glowing reports from financial analysts. The increasing value of cable systems is driven home by a key industry indicator: cost per subscriber. When a cable system is sold, cable investors and industry observers divide the sales price by the number of subscriber households to arrive at this figure. It is a good measure of how much a cable system is truly worth.

In 1977 the typical cable system sold for under $400 per subscriber. By 1980 the figure had risen to about $650. By 1985 cable systems were sell-

TECHNOLOGY

"I WANT THREE MTVs": NEW COMPRESSION TECHNOLOGY

One of the main problems facing cable programmers is the shortage of transponder space available on communications satellites. Ever since the *Challenger* explosion in 1986 the demand for satellite channels has outpaced the supply, as it has taken longer and longer to get new communications satellites into orbit.

In the early 1990s a new technological development made it possible to squeeze more and more channels into the available slots on orbiting satellites. Known as video compression, the technique has enabled major cable programmers to offer different versions of their services for different audiences. In late 1991 HBO became the first programmer to use compression, sending three different versions of HBO to its subscribers, and thus offering a choice of nine different movies in the prime evening hours. Other programmers have followed suit, including ESPN and CNBC.

Not to be outdone, MTV announced that it too would offer three different services beginning in 1993. While plans were to maintain a broad appeal rock format on one channel, it was expected that the two other versions of MTV would emphasize different music, such as rap and college/alternative.

---

ing for over $1000 per subscriber household. Today cable systems trade in the $2000 to $3000 range! Few industries can match the pace of this economic growth.

CABLE PERSONNEL

During the era of cable's great expansion, from the late 1970s to the mid-1980s, it was estimated that the cable industry was expanding by an average of 1000 new jobs per month. This is an astonishingly high figure, particularly since this expansion coincided with an era of high national unemployment and a serious economic recession.

Cable employment slowed in the late 1980s and early 1990s, as most systems completed their construction, and many cable networks faced cost-cutting due to increasing competition.

By 1990 the cable industry employed just over 100,000 people, nearly a 30 percent increase over the 1985 tally. About 7 percent of the total cable work force was found at MSO headquarters; the remainder was found at the thousands of operating systems around the nation.

Whereas the size of the staff in a radio and TV station is usually based on the size of the station's home city, in cable, the number of employees is based on the number of subscriber households. That is, systems with more homes hire more people. This makes sense: large systems are more likely to sell advertising and to program their own channels. They also need more customer service representatives, installers, and technicians. A large system (more than 50,000 subscriber households) is about the size of a large TV station, with seventy-five or more full-time employees. Midsized systems (10,000 to 50,000 subscribers) are more like large radio stations, with twenty-five to fifty people on staff. Small cable systems (under 10,000 households) are run like small-market broadcast stations, both radio and TV, with fewer than twenty total employees.

Unlike broadcasting, cable is a decidedly blue-collar industry. Most jobs fall into the technical or office/clerical category. Technical jobs generally require training in fields like electronics and engineering. Cable recruits its technicians from trade schools, community college electronics programs, and vocational-technical (vo-tech) schools. Job applicants are also sought from related communication fields, including the telephone and data processing industries.

Because of its history and tradition as a service business (and not "show business"), cable salaries have typically lagged behind those in broadcasting. General managers of local cable systems, for example, earn average salaries of less than one-half those of their broadcasting counterparts.

One reason for lower salaries and wages is the relative lack of union membership in the cable business. Whereas most telephone employees and many broadcast technicians belong to unions, less than 10 percent of the cable work force is unionized. However, as cable operations become more profitable, they are also becoming more like broadcast stations. As we have seen, many cable systems are actively seeking advertising dollars.

Some are programming their own entertainment and information shows. The cable industry has locked horns with the home video business, in direct competition for the entertainment dollar of the TV viewer. These events will increase cable salaries as cable competes with other media for talent in programming, advertising sales, marketing, audience research, and other areas.

ALTERNATIVES TO CABLE

Cable TV is by far the primary means of delivering nonbroadcast signals to American households, but it is by no means the only source. Three other technologies are in use. What each shares with cable is a desire to lure viewers for movies, sports, and other attractive program services.

TVRO: The Home Satellite Dish

As documented earlier, cable will probably never reach all households with TV in the United States. Rural areas in particular lack the population density that would make cable economically feasible. As a result, many people invest in a satellite dish to receive additional TV signals directly from orbiting satellites. The home satellite dish is known as a *television receive-only* earth station, or *TVRO* for short.

TVROs proliferated in the early 1980s, particularly in suburban and rural regions. By 1985 more than a million dishes were in place, and it was predicted that as many as 5 million or 6 million would dot the landscape by the mid-1990s. But the major cable and pay cable services soon realized they were losing valuable subscriber revenues when dish owners received their signals free of charge off the satellite.

In 1985 the services began scrambling their satellite signals, requiring dish owners to purchase a costly descrambler and to pay monthly charges for their signals, just like cable subscribers. Scrambling was encouraged under the terms of the Cable Communications Policy Act, which took effect in that year: for the first time cable operators were assured by Congress that people who "stole" cable service could be prosecuted.

As a result, the bottom fell out of the TVRO business. In the early 1990s just over 2 million dishes were on American lawns, a figure not expected to increase substantially in the near term.

SMATV: Private Cable

Perhaps you have noticed a satellite dish on the grounds of an apartment complex or atop a highrise residential building in your community. This technology is known as *satellite master-antenna television*, or *SMATV*. SMATV is also known as "private cable." Like TVRO, SMATV rose

In many rural areas it is not feasible to deliver cable because of the low population density. Consequently, satellite receiving dishes, called TVROs (Television Receive Only) are becoming common sights.

rapidly in the early 1980s when residents of apartment complexes (known as multiple-dwelling units, or MDUs) wanted cable service but were impatient in waiting for the wire to reach them. Instead, they turned to SMATV operators. SMATV companies provided the dish, downlinked the additional TV services off the satellite, and delivered service to homes in the complex by their own set of cables. In communities where cable was already available, SMATV operators were able to "skim" potential subscribers from the local cable systems by charging lower monthly rates. The lower rates reflected their much lower expenses (smaller headend, no studios, simpler wiring, and so on). This took place particularly in many of the new "upscale" condo communities proliferating in the mid-1980s near many large cities.

However, by the end of the 1980s SMATV growth slowed. One reason was a slew of lawsuits filed by the cable industry, which argued that private cable violated their rights to provide TV service in their franchise areas. Like TVRO, SMATV suffered from scrambling. Moreover, cable was becoming increasingly available to many communities that had originally relied on the satellite dish. Cable systems offered incentives to attract SMATV subscribers, including more channels and even lower rates. As a result, by the early 1990s fewer than 500,000 homes—less than 1 percent of TV homes—were linked to SMATV services.

In early 1992 SMATV got a boost that may yet help its growth. The courts ruled that private cable systems could compete with local cable in their communities. Look for more community-owned satellite dishes to dot the grounds of MDUs in the mid-1990s.

MMDS: SMATV for the Inner City

As we have seen, because of construction costs, utility problems, and franchise disagreements, cable service has lagged in the inner cities. In major urban areas satellite dishes are impractical. There is simply too little space. Yet people in major urban areas also want cable services, especially movies. A potential solution was the development of the *multichannel, multipoint distribution system*, or *MMDS*. MMDS makes use of short-range microwave transmissions to beam a small number of channels of video programming from a central transmitter location, such as the top of a tall office building. Receiving households use a small microwave antenna to pick up the signals and a special decoder called a *downconverter* to turn them into TV channels.

Thus far the MMDS service has been more promise than reality. Like SMATV, fewer than 500,000 households have MMDS service. The future of MMDS may be brighter as the service moves from movies and sports to private video channels (those used for business communication as discussed in Chapter 7), data services (like stock tickers and airline schedules), or perhaps high-definition TV (discussed in Chapter 20).

TVRO, SMATV, and MMDS have achieved only limited success in bringing cable-type services, especially movies, to the homescreen. But a major competitor to cable has emerged in recent years. That foe is the home video business.

HOME VIDEO TODAY

For nearly one-half a century Peter Goldmark was one of the true visionaries in telecommunications. From his laboratory at CBS had come high-fidelity sound recording, the long-playing phonograph record, and some of the basic research that resulted in color TV. Regarding home video, Goldmark made this prediction in 1976 (cited in *Videography*, July 1986, p. 61):

I doubt that packaged video programs for home entertainment would be economically justifiable. It's speculative in so many ways, and basically a question of cost. Will people pay . . . for a first-rate entertainment program they know they may view only once or twice?

In fairness, Goldmark with talking about buying movies for home viewing. Nobody had yet conceived the idea of video rentals. However, the statement does reveal the speedy germination of a new industry that would have immediate and long-term impact on American TV. In less than fifteen years the home video market exploded onto the media scene. Today broadcast and cable TV are just part of a total "home video environment." The TV set is now a video monitor, or video display terminal (VDT). Wedded to many home screens today is a range of attachments and accessories (which computer people call *peripherals*) that have helped transform the way we watch TV. This revolution began with the introduction of the videocassette recorder.

The VCR: An Overview

The spectacular proliferation of the home videocassette recorder (VCR) is graphically illustrated in Figure 6-4. The VCR revolution began with the introduction of the Betamax VCR by Sony in 1975. Crude by today's standards, the table-model machine could record up to one hour of video. However, the machine touched off a fiery court battle between Sony and Universal Pictures. Shockingly (at least to the movie studios and their major clients, the broadcast networks, for whom they produced the majority of programs seen on prime-time TV), Sony was promoting the machine's ability to tape broadcasts off the air! "Piracy!" claimed the studios. After a much-publicized legal battle, which came to be known as the "Betamax case," in 1984 the U.S. Supreme Court ruled that home taping did not violate copyright law. Not that they needed further encouragement, but this ruling essentially gave Americans a green light to tape TV shows.

In 1978 there were 175,000 VCRs in use in the United States. By 1982 nearly 5 million units were in use, representing about 9 percent of TV homes. In 1985 some 26 million homes had VCRs, about one-third of all households. By 1988 the figure had doubled: 52 million VCRs were in use, representing just under 60 percent of homes. Today 75 million households own a VCR, representing eight in ten homes in the United States.

At first all home VCRs were table models, designed mainly to record TV shows off the air for later playback (a phenomenon known as *time-shifting*, the term reportedly coined by Sony executive Akio Morita). However, in a few short years technological development had reduced the size of the VCR and had made home color TV cameras practical realities. A second growth industry was created: home video moviemaking. Weddings, confirmations, bar mitzvahs, and other cultural rites are captured by camcorders (portable combination camera and VCR units), which have replaced 8-millimeter film as the medium of record. The proliferation of home camera equipment has been almost as spectacular as the rise of the VCR itself. In 1985 about one-half of a million homes had portable video equipment. Today there are over 10 million camcorder units in use, about one in every nine homes!

All of this home taping, off the air and in the backyard, has created still another growth industry: the sale of blank videotape. In 1985 the sale of blank tapes accounted for $233 million dollars in consumer expenditures. Today blank tape sales produce over $500 million in annual revenues. Of course, much of this revenue does not benefit the

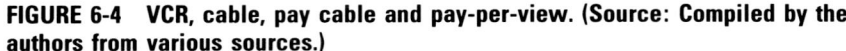
FIGURE 6-4 VCR, cable, pay cable and pay-per-view. (Source: Compiled by the authors from various sources.)

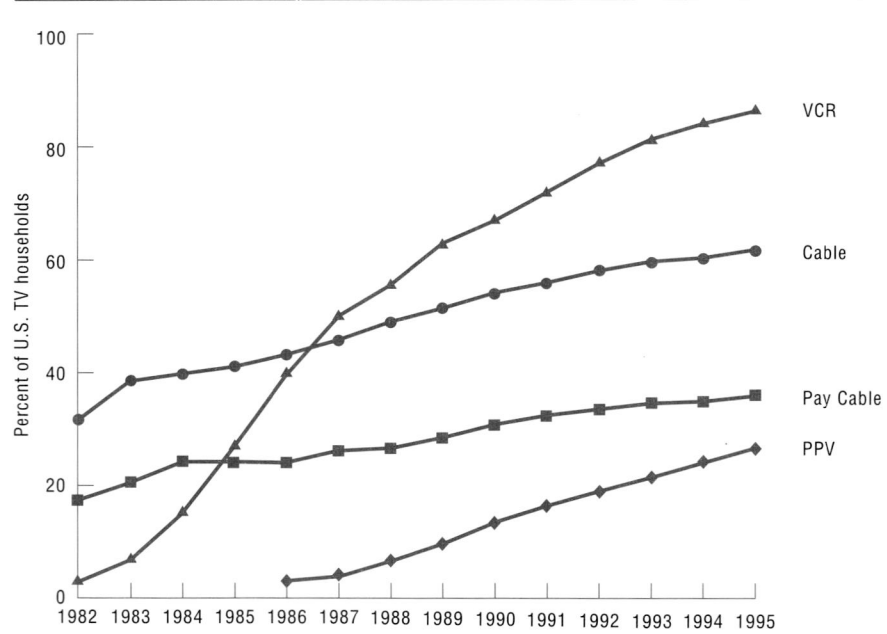

U.S. economy, as the leading tape producers are companies located in Japan, Germany, and Mexico.

The Video Store

A major reason people buy VCRs is to screen movies. In scarcely ten years a new commercial establishment—the video store—has become an indispensable feature of our shopping districts and malls.

The pioneer of this portion of the home video business was Andres Blay. In 1977 Blay convinced 20th Century Fox to lease to him videotape versions of a few films, including *Patton*, *M*A*S*H*, and *The French Connection*. His assumption that a few consumers would be interested in watching films at home was a gross miscalculation. Thousands of people were interested! Within months he was cranking out more than 20,000 copies per month (at a retail cost of $50). Home movie viewing had taken off.

Other entrepreneurs soon entered the picture. Some reasoned that it would be better to offer movies for home viewing on a rental basis than to require consumers to buy copies of the films. A leading entrepreneur in this phase of the business was Arthur Morowitz. The owner of a chain of movie theaters in New York (in Times Square and elsewhere), Morowitz reasoned that adults might want to rent X-rated (and other) films to watch in the privacy of their own homes. In the late 1970s he opened a chain of Video Shacks in the New York area, the prototypes for today's video retail establishments.

By 1984 about 20,000 video shops were operating. By 1986 there were over 35,000. Today more than 50,000 retail video establishments dot the landscape. Tapes are offered for both sale and rent at convenience stores, record shops, camera stores, and many large grocery and discount stores.

The leading chain of video stores today is Blockbuster Video. The rise of Blockbuster has been nothing short of phenomenal. In 1985 there were seventeen Blockbuster stores; by 1988 the chain had grown to about 600. As of this writing there are over 1600 Blockbuster outlets, with annual revenues over $700 million.

Home Video and Hollywood

When the VCR was introduced, Jack Valenti, the president of the Motion Picture Association of America, called the fledgling home video industry "parasitical" and claimed that home movie view-

A Blockbuster Video rental superstore. About $11 billion was spent on tape rentals in 1991.

ing would turn Hollywood into "an entertainment desert." It didn't happen.

Home video has created an additional revenue stream for Hollywood. In 1986 watching videotapes surpassed theater-going as America's favorite way to view movies. Today the video rental business accounts for over $4 billion dollars in revenue for the major film studios, about twice the volume of box office receipts.

In the motion picture business the major studios license their films to a distribution company. The distribution company contracts with the movie houses to play the film. The distributor and the exhibitor divide the revenues with the movie company based on box office receipts. The most popular films obviously earn the most money. Since ticket "take" is the key, box office receipts are followed religiously in the trade press (*Variety* and *The Hollywood Reporter*).

The system works somewhat differently in home video. The motion picture producer sells the VCR rights to a movie to a distribution company for a flat fee (perhaps $10 per copy). The distributor sells the tape to the video retailer at a markup (let's say $30). The retail video operators can then use the tape however they choose: they can rent it (for $1 to $3 per night) or they can sell it. Most new titles are first rented and then sold at a discounted rate as used copies. Many used tapes retail in the $10 to $20 range.

Despite the positive impact of the VCR on the motion picture business, there are some storm signals on the horizon. For one thing, the way VCR distribution operates, most of the financial risk is shouldered by the video retailers. They pay up-front for the movies at a premium price. They must decide how many copies to stock of each tape ordered. Then, as the appeal of a film diminishes over time after most potential viewers have seen it, they "eat" the inventory.

In addition, the motion picture companies have discovered that bypassing the distributor and selling tapes directly to consumers can be a lucrative venture. "Sell-through" tapes have been offered as low as $10 to $20 by the studios, which have sold over 2 million copies or more. Recent hits marketed in this fashion include *Batman, Who Killed Roger Rabbit,* and *Top Gun.* In 1990 the sell-through market reached $3 billion in revenues, nearly one-half the total for video rentals. Many video stores have suffered as a result, with a number of specialty video stores going out of business. Even the powerful Blockbuster chain saw its revenues flatten in the early 1990s.

Because of these and other factors, a new VCR distribution pattern is expected to emerge. *Pay per transaction (PPT)* will enable video retailers to pay a license fee of $8 to $12 (instead of $30 or more) for a title and to pay the studios a share of the rental receipts. Essentially the procedure will match traditional movie theater distribution. The retailer would be able to rent the movie at the peak of its popularity, without having to absorb it as inventory. However, the plan has met with some resistance from both the distributors and dealers. First, standardized financial and computational equipment will be required to keep accurate records of the rental receipts for each title. In the film business it's a simple matter to tally up the box office receipts for each film playing at a given theater. Moreover, the studios are somewhat reluctant since the initial risk will shift from the video retailer back to them to provide the popular titles that the public will want to see.

Home Video and Broadcasting

Clearly the VCR has had enormous impact on the movie business. What has it done to traditional broadcast TV? The answer is somewhat less clear.

The evidence of declining network shares indicated earlier supports the notion that, like cable, the VCR has diminished the network audiences. It is estimated that VCR playback has cost the networks from 2 to 6 percent of the audience, depending on the time of day and the season. As much as 40 percent of all VCR use is on Saturday and Sunday nights, traditional network strongholds. To date prime time in the summer months has been most affected: 10 percent or more of the prime-time audience is tuning out network reruns and playing tapes instead. In addition to prime time, a second area of impact is the late afternoon, when increasing numbers of children are watching videotapes instead of their traditional choice of cartoons from broadcast stations.

Figure 6-5 illustrates how the VCR has come to be used in the "typical American household." Most recording on the VCR (42 percent) is done during prime time. However, weekday and weekend daytime programming accounts for almost a third of all taping, no doubt when people are out of the house at work or play. About one-half of all

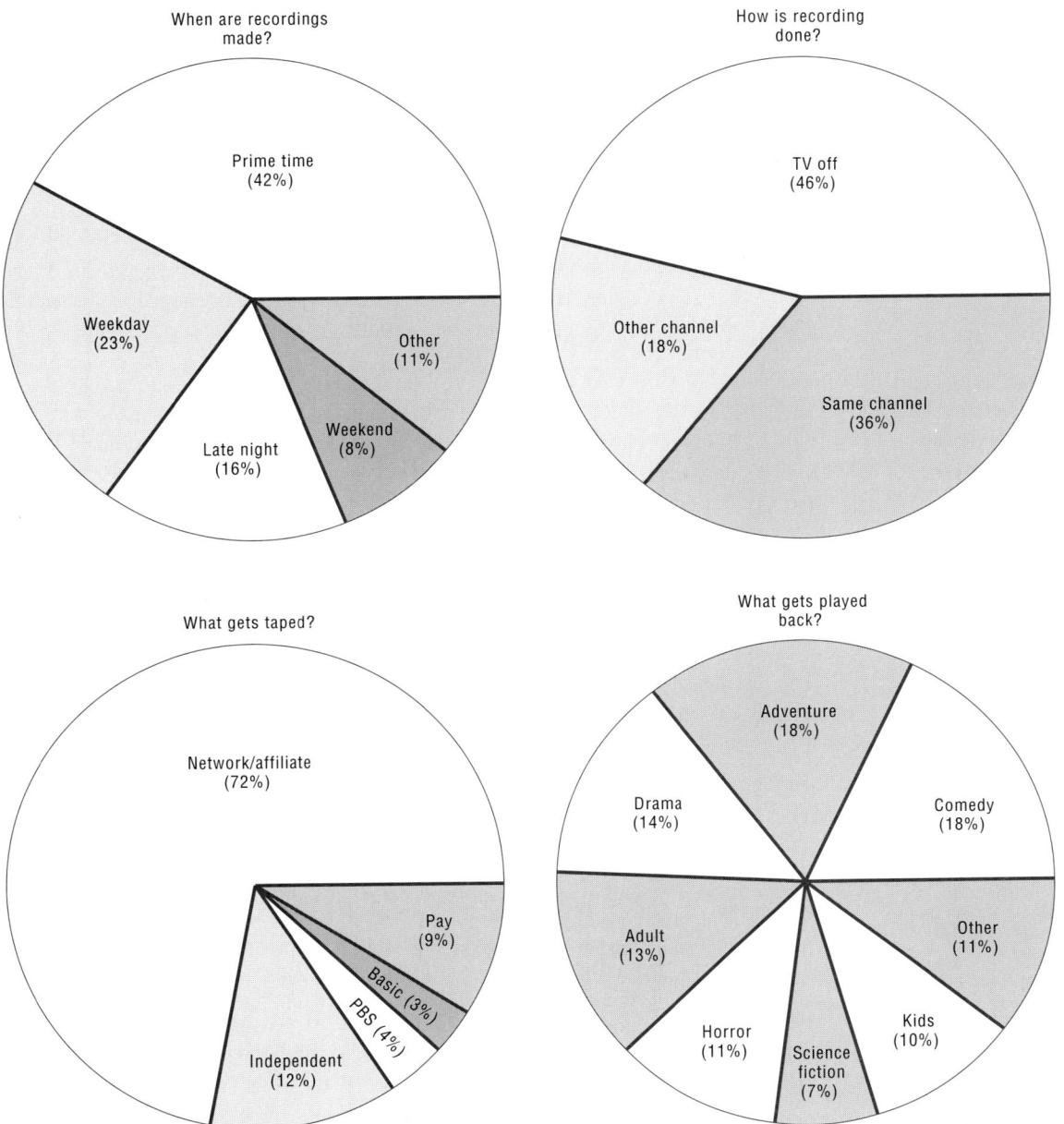

FIGURE 6-5 VCR activity in the home. (Source: *Electronic Media,* August 17, 1987, p. 32.)

home taping occurs with the set off, further indication of the use of the machine to tape things while we're away from our TV sets. Just over a third of all taping occurs with the set tuned to the same channel that we are currently watching. This indicates that many viewers are building tape libraries of their favorite shows.

Interestingly, only one tape in five is made of a show on a different channel from the one being watched. This may be due to the programming procedures required to perform this task, which seem very complex to many home viewers, especially those who grew up before the arrival of VCRs.

What do we tape? Almost three-fourths of all home taping is of network or affiliate programming. Daytime serials (soap operas) have picked up as much as 10 percent in bonus audience: people who set VCR timers to record while they are at work or elsewhere during the day. Other program types to receive a viewing bonus include network sports and top-rated entertainment shows like "Murphy Brown" and "Designing Women." Independent stations account for just over 10 percent of taping, followed by pay cable (10 percent), public TV (4 percent), and basic cable (3 percent).

Zapping One area of great speculation has been the effect of VCR use on TV commercials. VCR technology and TV remote controls allow viewers to avoid commercials. *Zapping* means using the pause feature of the VCR to bypass recording a commercial in the first place. TV networks, stations, and advertisers are concerned that zapping threatens the foundation of the "free" broadcasting system. If people don't watch the commercials, they claim, why should advertisers pay huge sums for thirty- and sixty-second spots?

To date research on the effect of zapping has been inconclusive. Current estimates suggest that under 5 percent of ads are effectively zapped and that most of the zapping is done by those people who are least likely to be sought by the advertisers: teenagers and adult males. However, the advertising agencies are keeping close tabs on zapping.

Zipping A cousin of zapping is the *zipping* phenomenon, the process of using a remote control to speed through commercials recorded when taping shows off the air. Ingenious methods to counter zipping include longer segments with the product name superimposed to make sure the viewer sees it (even at high speed) and positioning the product in virtually every frame of the spot!

With impact ranging from bonus audiences to lost viewers, from bonus advertising to zapping, VCRs overall appear to have been a mixed blessing for the broadcasting business.

Home Video and Pay Cable

Without doubt, the telecommunications industry most affected by the rise of the VCR has been cable TV—specifically pay cable. Survey after survey reveals that the prime reason people buy a VCR is to watch movies—precisely the reason people heretofore subscribed to HBO, Showtime, Cinemax, the Movie Channel, and other pay

---

ECONOMICS

BLIMEY! VIDEO PIRACY ON THE RISE!

Pay TV and home video have more in common than programming. Each is being increasingly victimized by video pirates.

In the cable business it is estimated that 20 percent of all HPs receive cable services illegally. Cable thieves have devised all manner of means to receive HBO and other pay services free of charge. Strange antenna arrays and illegal descramblers and decoders (cryptically called "black boxes") are sold by electronic "blackbeards." The cable business has partially contributed to this piracy. Lax work by field crews and office personnel have put free Showtime, Disney, and other services into thousands of homes that haven't requested them. Not surprisingly, few households have called the cable companies to report this fact.

On the home video front, video pirates make illegal copies of theatrical motion pictures, either during their first run at movie houses or later, by copying from the video release. These "bootleg" copies are then sold to program-hungry video enthusiasts, even some unscrupulous owners of video stores.

Video piracy has generated a vigorous response from the industry. Cable systems have begun full-scale investigations on theft-of-service grounds, sometimes offering "amnesty" to illegal viewers before commencing prosecution. The Video Software Dealers Association (VSDA) maintains an 800 number "Piracy Hotline," whereby video dealers call in information about illegal taping.

Although illegal receipt of cable service and copying of movies both violate the law, many consumers continue to "rip off" cable and home video. To these people, stealing TV is seen as a "victimless crime."

channels. As a result, the advent of the retail video establishment directly eroded the subscriber base and therefore the revenue stream for pay cable.

Today more than twice as many American homes have VCRs as have pay cable. More than 80 percent of pay homes also own a VCR, making it likely the consumer has seen a movie before its much ballyhooed "premiere" on a pay service.

Not surprisingly, pay cable has entered an era of retrenchment. Profit margins are down (as much as one-half, according to a recent report), layoffs at HBO and other channels have been on the increase, and the pay services have begun massive promotion and incentive plans to woo cable systems and subscribers to their services.

THE FUTURE OF HOME VIDEO

All indications point to continued growth for the home video industry. The TV is becoming more like a movie screen: more than one-half a million large-screen sets were sold each year in the late 1980s, until a recession slowed sales in the early 1990s.

Table 6-7 charts the revenues for various program sources, including movies, cable, pay cable, home video, and PPV for 1987 and offers projections for 1996. By the late 1980s almost two-thirds of consumer spending for entertainment stayed at home: cable TV and videotape rentals. By the middle of the 1990s that total is expected to rise to over 70 percent of all consumer expenditures for entertainment programming. Add PPV and pay cable, and nearly nine of every ten entertainment dollars will be consumed in the home TV environment.

It is no wonder that in a recent survey that asked people whether they would rather see a movie at home or in a theater more than two-thirds preferred to stay at home. Will the viewer of the future be an electronic agoraphobic who is confined to the home media room? Only time will tell.

TABLE 6-7 Consumer Spending for Entertainment, 1988 and 1997

Medium	1988 Spending (billions) ($)	Percentage of Total (%)	1997 Spending* (billions) ($)	Percentage of Total (%)
Pay per view	0.21	1.0	4.13	7.0
Cable TV	7.89	32.0	21.76	39.0
Home video	7.90	32.0	16.06	29.0
Movies	4.46	18.0	8.13	15.0
Pay cable	4.30	17.0	5.39	10.0

* Projected.
SOURCE: Paul Kagan Associates, *The Kagan Media Index,* April 14, 1989, p. 1.

SUMMARY

Cable TV has experienced tremendous growth since the 1970s. Many cable systems now offer more than fifty channels. Cable programming consists of local broadcast stations, cable networks, superstations, premium channels, and specialty services such as PPV.

Cable subscribers fit into several categories. Basic subscribers pay for those services offered in the lowest tier of cable pricing only. Pay subscribers take both the basic channels and one pay service. Multipay subscribers take basic plus more than one pay service; these are the customers who are most desirable to cable companies.

Deregulation has changed the operating environment of the cable industry. Thanks to the Cable Communications Act of 1984, cable systems are basically free to set their own rates and do not have to follow most of the programming regulations imposed on traditional broadcasting. However, concern over rising cable rates may lead to reregulation of the industry.

Alternatives to cable are TVROs, which are popular in rural areas where cable is unlikely to penetrate; SMATV, which operates in high-rise apartment buildings; and MMDS, a type of "wireless cable" popular in urban areas.

The home video industry came of age in the 1980s. The video industry and Hollywood are now in a position where it is to their advantage to work together. The popularity of prerecorded videocassettes has hurt the viewership of traditional TV and the premium movie channels carried on cable.

SUGGESTIONS FOR FURTHER READING

Baldwin, T. F., & McVoy, D. Stevens. (1988). *Cable communications* (2nd ed.). Englewood Cliffs, N.J.: Prentice-Hall.

Chaffee, C. D. (1988). *The rewiring of America: The fiber optics revolution.* New York: Academic Press.

Ferris, C. D., Hoyd, F. W., & Casey, T. J. (1985). *Cable television law.* New York: Matthew Bender.

Garay, Ronald. (1988). *Cable television: A reference guide to information.* Westport, Conn.: Greenwood Press.

Heeter, C., & Greenberg, B. S. (1988). *Cableviewing.* Norwood, N.J.: Ablex.

Jones, K., Baldwin, T. F., & Block, M. P. (1986). *Cable advertising: New ways to new business.* Englewood Cliffs, N.J.: Prentice-Hall.

Levy, M. (1989). *The VCR age.* New York: Sage Publications.

Roman, J. W. (1984). *Cablemania: The cable television sourcebook.* Englewood Cliffs, N.J.: Prentice-Hall.

Yates, Robert K. (1990). *Fiber optics and CATV business/strategy.* Norwood, Mass.: Artech House.

CHAPTER 7: CORPORATE AND ORGANIZATIONAL VIDEO TODAY

The scene on the TV set shows a woman on a street corner buying drugs. Next, she's in a bar, drinking. On the soundtrack there is a woman's voice describing how she battled drug and alcohol addiction. The scene shifts to show the woman having trouble on the job. The narration tells of how she went from being just a casual drug user to a full-fledged addict.

A segment on the evening news? "60 Minutes"? "20/20"? No, this is a video produced by the Seattle-based Boeing Company to inform its employees of its new policies on drugs and alcohol. The woman is not an actor; she is a current employee of the company. The narrator is another employee who describes himself as a recovering alcoholic. The sixteen-minute video, produced at a cost of $40,000, is part of a campaign at Boeing to introduce employees to its new Employee Assistance Program designed to help those workers with a drug problem.

Boeing's use of video for corporate communications is not unique. Thousands of corporations and institutions are using TV for training, employee communications, and various other tasks. Although you might not have seen the preceding example, a sizable number of college students have some familiarity with this area because they have worked at fast-food restaurants. Not surprisingly, because of the high turnover in the fast-food industry, TV is used extensively to train new employees. Perhaps you or some of your friends have seen videos entitled "Servicing the Salad Bar," "Running the Cash Register," or "French Fries." In any case, video is a vital part of many organizations and this chapter examines the applications of TV in the corporate setting.

DEFINING THE FIELD

Given the wide range of companies and organizations using TV as a communications tool, coming up with an all-encompassing definition of—or even a name for—corporate video is nearly impossible. Labels such as nonbroadcast TV, private TV, industrial TV, institutional video, business TV, corporate video, and organizational TV have all achieved some degree of popularity. For simplicity's sake, we shall adopt the name *corporate video (CV)*. Keep in mind, however, that this label covers video used in other organizations that are not, strictly speaking, corporations.

As far as a definition goes, CV is any type of TV program that serves the particular needs of the producing organization, is aimed at a specific audience, and is typically distributed through channels that are not usually available to the general public. These programs generally deal with training, internal communications, public relations, and marketing. Although this definition may help a little, perhaps the best way to get a feel for the area is to examine how CV started and how it's used.

HISTORY OF CORPORATE TV

Corporate TV has been influenced by advances in video technology and changes in the environment of American business. Back in the early 1950s many large companies had audiovisual departments that specialized in making films, slides, or audio tapes that helped train new employees. Television was impractical for the needs of business because production was very costly and there was no practical means of storing programs. Corporate communication was accomplished primarily through face-to-face communication, by telephone, or in printed memos and brochures. In the mid-1950s the invention of videotape solved the problem of storing programs, but early videotape machines were expensive, cumbersome, and hard to use. Videotape was 2 inches wide and had to be edited by physically cutting the tape and pasting it back together.

The 1960s brought advances in technology that made videotape more accessible to many large and medium-sized companies. The Ampex Company introduced 1-inch-wide black and white, reel-to-reel videotape that was cheaper and more manageable than its 2-inch predecessor. New machines that used the 1-inch tape were produced initially for the home market. They did not catch on in this market because they had several critical drawbacks.

First, machines made by different companies were incompatible. Tapes recorded on a machine made by one company could not be played back

Corporate video in the workplace. Company executives watch a videotaped training series about management.

on machines manufactured by another. Second, the reel-to-reel format made the tapes awkward to work with. They could snag or get twisted and had an annoying tendency to unravel at the wrong time. Third, the quality was poor. Productions were restricted to black and white and the tape had a peculiar grainy look.

Despite these initial drawbacks, the new 1-inch machines attracted private companies and government organizations because they were affordable and the quality was good enough to do many of the training tasks that the companies needed to accomplish. Thus a new profession was born. The glamour and excitement usually associated with traditional TV did not carry over into the name given this new endeavor: *industrial television*, a name that conjures up images of drill presses and assembly lines. Despite its drabness, industrial TV was a name that would stick until the next decade.

In the late 1960s and early 1970s the technological breakthrough that enabled industrial TV to blossom was the invention of the U-Matic, $\frac{3}{4}$-inch color videocassette. As before, the $\frac{3}{4}$-inch cassette was originally designed for the home market but did poorly. Cassettes were too bulky and expensive for home use. Corporations, however, found the new, smaller tape format ideal for their training and communication needs. The quality of these cassettes was much better than earlier tapes, and their standardized format allowed them to be interchangeable. What's more, cassettes could be easily duplicated and distributed. Companies that had operations in many parts of the country could make one program on videocassette, copy it, and send it to all their branches. Several firms started video networks with playback units in key locations in all of their plants and offices. Some independent production companies produced programs designed for corporations on videocassettes. Finally, videocassettes were reusable. They could be erased and used again, thus cutting production costs.

Moreover, in the early 1970s many companies began introducing computers to perform data processing or related services. These companies were faced with the immense task of training their personnel on the new computer systems. Literally thousands of employees had to be retrained quickly and efficiently to use the new computer technology. Some big firms, such as AT&T, Xerox, Hewlett-Packard, and Texas Instruments turned to CV for a solution. Companies began to invest in numerous TV production centers, which in turn produced numerous "how-to" cassettes. The value of TV as a training medium became apparent and the boom was on.

From 1973 to 1977 the size of the private TV sector doubled. About 700 business or nonprofit organizations produced 46,000 programs comprising 15,000 viewing hours. Corporate video grew even more during the last half of the decade thanks to the development of portable, low-cost cameras and editing systems. First used by TV stations for electronic news gathering, these new, easy-to-operate systems allowed CV to escape the studio and to go almost anywhere.

Along with the technological revolutions, the growth of CV was aided by changes in the business environment during the 1970s. First, business itself became more complicated thanks to new advances in computer and communication devices. As a result, employees required frequent retraining and continuing education to do their jobs effectively. Corporate video could do this task inexpensively and efficiently. Second, because of mergers, buyouts, and takeovers, corporations became more complex and segmented. Communication and coordination were major tasks. Corporate video proved itself effective in this area as well. Finally, business travel became increasingly expensive. An improved domestic satellite communications network allowed many companies to use TV transmissions of sales presentations instead of sending employees on the road.

There was also a social reason for CV's growth. The generation that entered the work force in the 1970s was a generation of people who were raised with TV. They grew up with it, got their news and entertainment from it, and considered it a necessity of everyday life. For them, it was natural to use video in the workplace. Taken together, all of these factors resulted in a CV industry that by 1980 was ten times its 1973 size and generating $1.5 billion annually.

The 1980s saw continued growth, most of it again fueled by advances in technology. The $\frac{1}{2}$-inch VCR became a huge hit among consumers, penetrating 50 percent of American homes by 1987. Once again, corporate users took advantage of this new format and employees could check out videocassettes of training and orientation sessions for home viewing the same way that they might take home a book or brochure. Half-inch cassettes were such a convenient communications medium that "video memos" to employees replaced corporate executives' traditional memos. Low-cost $\frac{1}{2}$-inch videocameras and editing gear prompted many companies to produce in-house company newscasts or video newsletters.

The recession of the early 1990s and the huge debt load acquired by many corporations during the mergers and acquisitions of the 1980s caused business to monitor spending carefully. As a result, CV's growth was curtailed and some companies downsized their operations. Nonetheless, in 1992 CV was a $7 billion industry with about 9000 organizations producing about 60,000 hours of programming. To put these figures in perspective, the three major broadcast TV networks produce about 18,000 hours annually. More recent developments include the introduction of interactive videotapes and videodiscs to encourage self-paced learning, more widespread use of teleconferencing, and the utilization of local cable TV as another channel of corporate communication. Although it may not be expanding as rapidly as before, the future of CV still looks optimistic.

USERS

Who uses CV? It ranges from the one-person operation at the local health department to the dozens of people employed in the video department at a big insurance agency. Table 7-1 lists the major categories of users. Service industries are the biggest users of CV. Of those in the service category, most are utility or insurance companies. For both of these communication-oriented industries, informing employees and customers is an important task.

Small and medium-sized companies are now the dominant CV users. This is a change from the 1970s when big companies (those with more than 25,000 employees) were the places where CV operations were most likely to be found. To illustrate, in 1973 a survey of corporate TV users found that 41 percent had 25,000 or more employees and 19 percent had 5000 or fewer. In 1985 those percentages had virtually reversed themselves. Organizations with fewer than 5000 people accounted for 40 percent of corporate TV

TABLE 7-1 Type of Organization Using Corporate Video

Classification	Percentage (%)
Service	56
Manufacturing	35
Medical/educational	4
Government	4
Other	2

SOURCE: Judith Brush and Douglas Brush, *Private Television Communications,* Cold Spring, N.Y.: HI Press, 1986, p. 10.

use while the large companies accounted for only 24 percent. Additional characteristics of the typical company using CV include:

1 has a highly diversified work force consisting of scientists, technicians, managers, and clerical and unskilled personnel;
2 is geographically dispersed with plants and division headquarters located across the country or abroad; and
3 has a management which believes that communication helps achieve organizational goals.

APPPLICATIONS

The functions of CV are many, but, at the risk of oversimplification, five categories seem to predominate: training and orientation, internal communication between employees and management, education, external public relations, and sales and marketing. We'll examine these areas in more detail and provide examples of each.

Training and Orientation

One of the first and still one of the biggest uses of corporate TV is training and orientation. A recent survey disclosed that video was used for skills training and employee orientation by about three-quarters of responding companies. Perhaps no activity lends itself better to video than employee training. Companies use video to teach basic operational skills and to train employees in sales, management, and interpersonal dynamics. The approaches and formats of such programs are varied: demonstrations, lectures, role playing, skits, dramatizations. Most presentations take the viewer through the process step by step and incorporate drills or practice sessions that help viewers evaluate their understanding of a subject. Some tapes are designed to be used with a human teacher or with a workbook or in a laboratory situation.

Many examples illustrate the widespread use of video training. The Adolf Coors Company, brewer of Coors beer, produces training tapes for all brewery departments. One such presentation is a ten-minute tape designed for loading dock workers on how to couple properly a tractor to a trailer. Another Coors production teaches office personnel how to operate an office machine called a jet printer. Many hospitals use video to educate and inform patients before they undergo medical procedures. The Metro-Dade Police Department in Miami, Florida, prompted by the influx of Cuban and Haitian refugees into the south Florida area, produced a series of programs on the unique social and historical perspectives of these groups and the different values and perceptions of police inherent in the Cuban and Haitian cultures. Police who viewed the training tapes got much-needed insight concerning their work with these groups. The Hilton hotel chain recently spent more than one-half a million dollars putting together twenty-three programs that train new and current employees. Hilton's competitor, Sheraton, uses video not only for training but also as part of an aptitude test for new employees. Applicants are shown a videotape that dramatizes several hotel situations and then take a written test on how they would respond to the situations. Results help identify those potential employees who have a knack for working in the industry. Georgia-Pacific produces a series of do-it-yourself training videos that are available to building supply retailers to be used as a sales incentive. A customer interested in vinyl siding is offered a video on how to install it as part of the deal. Supermarket companies are also heavy users of training video. The Los Angeles–based Ralph's Grocery Company produces more than 100 training videos a year with topics ranging from how to operate the cash register to how to identify various types of vegetables.

Interactive video is a technique now commonly used in education and training. In a typical system a videodisc is linked with a computer and informa-

Training is one of the biggest uses of corporate video. Here an employee works with an interactive videodisc teaching system.

tion is displayed on a TV monitor. At predetermined points the video message stops and the viewer is asked to make a choice, answer a question, or input some information. The viewer's response determines what is seen next on the screen. If the right answer is given, the lesson proceeds. A wrong answer will prompt a review. The potential of interactive video for education and training is enormous. For example, an English professor developed an interactive program that allowed students to use computer-drawn characters to block out scenes from *Hamlet*. Then, using a videodisc on which various productions of *Hamlet* were recorded, the students were able to compare their blocking with that done by several distinguished professional directors.

Most new employees are somewhat anxious when starting a job and televised orientations and introductions to a company or organization can reduce tension and provide a strong and uniform welcome to newcomers without consuming much of management's time. These programs can be motivational ("Welcome to the company. You've made a smart decision. You'll accomplish much.") or informational ("Here are the specifics of the company's pension plan."). Most videos are practical, dealing with such common problems as where to park, where and how to collect a paycheck, and even where to find the best places to eat lunch in the neighborhood. To illustrate, the International Paper Company produced a series of tapes that took new employees on a video tour of the company's mills and paper plants. A second set of tapes served as an orientation for new managers and contained a general overview of the company, its locations, organization, products, and marketing techniques. The Mutual Benefit Life Insurance Company showed its new salespeople a ten-minute tape called "Career Path" about the stages of a life insurance agent's career.

Internal Communication

Employee communications is the next most common application of CV. More than 70 percent of firms responding to a 1989 survey reported this use, which is likely to grow further in the future. Organizations use TV to communicate information on such topics as employee benefits, layoffs, safety, controversial issues affecting the company, economics, and productivity. The programs used in this area are highly varied. Television newsletters, video memos, documentaries, teleconferences (discussed in more detail later), and feature programs have all been used to get the company's message across to employees. In addition, these presentations carry news of fellow employees and can provide a real boost to organizational spirit and morale.

Probably the most popular corporate TV program in terms of audience appeal is the news and information show. Quite a few of these have adopted formats used by commercial TV and resemble corporate versions of "Hour Magazine" or "PM Magazine." At the SmithKline Beckman Corporation employees can watch the company news program while eating in the cafeteria, waiting in the lobby, or riding the elevator. The show includes features on intramural company basketball games, employee photography awards, and new company products. Pacific Telephone's corporate newscast is designed to boost employee morale. Every two weeks a fifteen-minute story highlights one of the workers on the company assembly lines. With more than 60,000 employees across the country, Georgia-Pacific found

PROFILE

MONTY PYTHON AND CORPORATE VIDEO

John Cleese, the tall, gangly member of Monty Python, has also made a name for himself in corporate video (CV). Cleese's company, Video Arts, specializes in producing training programs that deal with sales, management, and finance. Video Arts has been remarkably successful, with revenues of $10 million in 1985. Their programs are used regularly by more than 50,000 organizations in twenty-six countries and have been dubbed into Russian, Finnish, Cantonese, and Mandarin. (The Russian reaction to John Cleese's brand of humor would be interesting to examine.) In all of the programs Cleese injects his own zaniness. In one program dealing with the right and wrong way to sell, Cleese is listening to a boring training tape while driving to his next appointment. Cleese is angry about having to listen and keeps talking back to the tape. Finally, the tape pauses, in hurt silence, and talks to him directly, "So, if you know it all then, turn me off. Go ahead. Turn me off." Cleese softens and lets the tape go on about the importance of doing research and remembering key facts about clients. Cleese is unimpressed. "Kid stuff," he says.

Next, Cleese, the salesperson, walks into an office and greets his customer. The dialogue goes like this:

CLEESE: Nice to see you, Mr. Jennings.
CUSTOMER: Jenkins.
CLEESE: How's Mrs. Jenkins?
CUSTOMER: I haven't the faintest idea what's happened to her since our marriage broke up.
CLEESE: Oh dear, I'm sorry about that.
CUSTOMER: You were sorry the last two times I told you as well.

Needless to say, Cleese doesn't make the sale.

Monty Python's John Cleese (seated) is a well-respected actor, writer, director, and producer in the world of corporate video. In this heavenly scene taken from his video, "The Unorganized Manager," he offers tips to executives about how to manage their time.

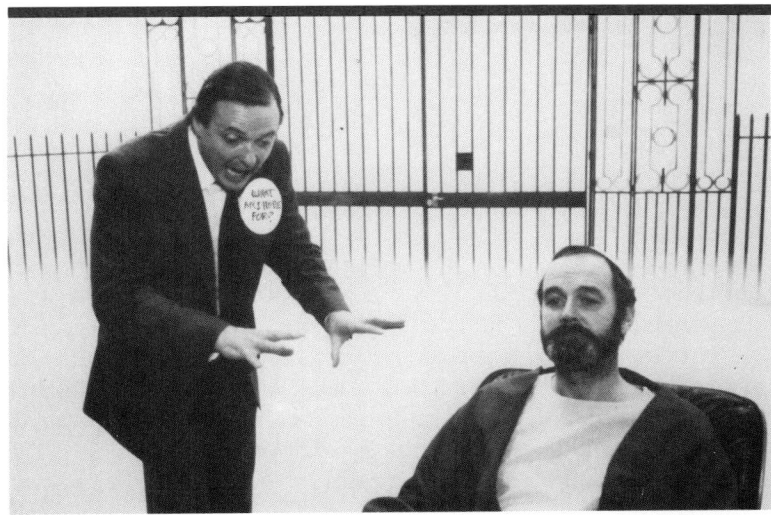

CHAPTER 7: CORPORATE AND ORGANIZATIONAL VIDEO TODAY

that it was difficult for everybody to feel part of the same team. Consequently, the company produced "Kaleidoscope," an in-house video news magazine. The bimonthly program focuses on the company's products, people, and programs. Federal Express beams a daily five-minute company newscast, "FedEx Overnight," to 1100 different sites.

Video is often used to deal with controversial issues within a company. When the Pennsylvania Power and Light Company decided that it could save money by renovating rather than replacing its trucks, many line workers were displaced and argued for more modern vehicles. The company was able to defuse this potentially volatile situation by producing a tape on the renovated trucks. Representative line workers were shown the renovation process and the work-testing of the trucks. Users of the trucks were interviewed, and their answers highlighted safety. The tape not only convinced the line workers that the renovated vehicles were safe, but it also portrayed the company as cost conscious. In another example, when the Adolph Coors Company became the target of a "60 Minutes" segment on unfair treatment of employees, the company produced a tape response to the charges and presented its own side of the story. People who saw the Coors tape now had more information to use to make up their minds.

The use of video to inform managers of events and activities throughout a big company is increasing. By 1990 more than 200 companies had leased satellite time to present their message to managers at locations across the country. For example, when the chairman of the Ford Motor Company retired, a special program on the change of management was beamed to fifty-four Ford plants and offices. In many companies the video memo is replacing the traditional paper version. If top management has made a policy change or some other major decision, a tape of an executive explaining the decision is sent to managers at other levels in the company, who play it before discussing the change at staff meetings.

Video is also used to explain employee benefits programs. Modern organizations are faced with government regulations requiring that they share with employees information about pensions, savings plans, benefits, and health arrangements. For example, one program done in 1979 for

Corporate video helps internal company communication. Here employees watch one of the weekly editions of SmithKline Beecham TV News, produced by the company's Employees Communications Department.

United Technologies Corporation featured comedian Jonathan Winters in a series of humorous sketches describing the various aspects of the company's benefit plan. More than a quarter of a million employees have seen this tape.

Education and Instruction

Elementary schools, high schools, and colleges make use of a wide variety of video applications. These can range from taped lectures to special programs produced by students for the general audience. Many colleges and universities now offer credit for telecourses carried over cable TV. A national cable service, Mind Extension University, for example, conveys educational programs to 15 million subscribers. Ball State University in Indiana transmits classes to high schools and junior colleges throughout the state. The State University of New York at Albany produced a series on the various ethnic cultures in America that was distributed nationally. On a more modest level, elementary students in the Southern Cayuga school system in upstate New York produced their own series of videos for use in their classes.

Video is also used in continuing education. A number of hospitals and clinics can be linked together via microwave to witness the latest surgical procedures or for consultation among a group of medical specialists. Colleges and universities also transmit courses in professional development via microwaves to work sites so that participants can learn without leaving the job. These courses typically have a talk-back system so that viewers at the remote locations can ask questions and participate in discussions.

In addition, organizations such as the Children's Television Workshop produce general education programs for such outlets as the Public Broadcasting Service (PBS) and The Discovery Channel. "Sesame Street," for example, has been on TV for more than twenty years and has helped several generations of children prepare for school.

Public Relations

About one-half of the companies using CV employ it for external public relations: communicating the company's message to audiences outside the organization, such as shareholders, civic and government leaders, and the general public. Programming for public relations includes news about the organization and its products or services and information about projects designed to strengthen the company's image. The common formats used for public relations video are video news releases (VNRs), shareholder meetings, and community relations.

VNRs Organizations that have sent printed news releases to newspapers and radio stations for decades are now relying on CV departments to turn out the TV equivalent. The *video news release (VNR)* is a videotaped presentation of company news that is sent to local TV stations for possible later broadcast. Its format is that of regular news clips: a correspondent on the scene tells the story while location footage illustrates it.

Video news releases are popular for at least two reasons. First, there has been an increase in TV news programming at both the local and national level, which has created a need for more news content. Video news releases help fill this additional time. Second, since VNRs are supplied to stations by independent production agencies, the cost to the station is minimal. In fact, many stations add local material to the VNR to make it more relevant in their market.

For example, Pratt and Whitney covered its own news conference announcing the largest single commercial aircraft sale in aviation history. Within an hour Pratt and Whitney beamed by satellite edited excerpts from the conference to TV stations around the United States. Television stations in seven major aerospace markets carried the release on their evening news within three hours of the conference. R. Dakin and Company produced a news story on their Chatter animal toys, which use a voice synthesizer. The VNR traced the history of voice synthesis, using the Dakin product as an example of its application. The story aired in several big markets during the Christmas season. When bowling was selected as a 1988 Olympic event, Brunswick covered the announcement and made a VNR available to sportscasters. Parker Brothers, the game manufacturer, covered the national Monopoly championship, and many TV stations used their VNR as a program closing feature, giving the company some favorable publicity. A 1991 estimate by Medialink, a major distributor of VNRs, suggests that more than 5000 VNRs are circulated to TV stations each year. About three-quarters of all the broadcast newsrooms in the country use VNRs.

When StarKist Seafood Company announced its new line of Dolphin-friendly tuna, it released a VNR featuring actress–activist Morgan Fairchild that emphasized the animal rights angle of the story. The VNR was seen by about 81 million TV news viewers. San Diego's Sea World released a VNR about the birth of a killer whale (which also showed Sea World in a positive light) that aired on CNN. The Hershey Company arranged a special stunt to spice up a VNR about the introduction of Hershey's Kisses with Almonds. A 500-pound replica of a Hershey's Kiss covered in sequins and gold foil was dropped from the same building in Times Square, New York City, where the New Year's Eve Big Apple is dropped. The unusual visuals of the event helped the video to air on news broadcasts throughout the country.

The use of VNRs, however, is not without controversy. Some news directors view VNRs as nothing more than free advertising or free publicity and are hesitant to use them. They feel that using VNRs means abdicating control over the content of the newscast to an outsider. Others insist that the video should be identified as supplied by an outside organization. Still others may use some of the VNR and edit out other sections that seem too commercial or public-relations oriented. In any case, it is likely that VNRs will continue to crop up in TV newscasts in some form or another.

Shareholder meetings Most of the people who own stock in a company don't attend the annual meeting of shareholders. Video gives companies a chance to reach more of their stockholders in order to make this yearly ritual meaningful. About one-third of all companies use video for this purpose. The Emhart Corporation, for example, taped the entire two-and-one-half-hour meeting and edited it to a twenty-five-minute summary, which was offered to all shareholders. More than 2000 people in the United States and in eight foreign countries screened the tape. As a follow-up the company produced a video version of its annual report, which was broadcast to shareholders and the general public over cable TV. McDonald's used a satellite hookup to broadcast live its 1985 shareholders' meeting from its Illinois headquarters. Stockholders in Washington, D.C., and Los Angeles were able to see the meeting. The W. R. Grace Company, with branches in forty-five states and forty-one foreign countries, distributed edited versions of its annual meeting on videocassette to interested stockholders.

Community relations Organizations are located in towns or cities of various sizes and CV is being used increasingly to improve relations with members of the local community. Many companies rely on taped programs to help them make a contribution to their communities and to live up to their responsibilities as good corporate citizens. The Arkansas Power and Light Company donated its video services to create videotapes for the state's towns and cities that highlighted each community's unique resources, location, and character. The tapes were used to attract industry and to further economic development. The Hewlett-Packard Company produced a fund-raising videotape for Stanford University and also volunteered their tape distribution network as a means of reaching Stanford alumni spread across the country. The Minneapolis *Star and Tribune* provided elementary schools in its market area with a videotaped tour of the paper's facilities. Many companies make copies of their training tapes available to local vocational and technical schools.

In the public relations area, Dow Plastics released a video aimed at high school students that emphasized how plastic products could be recycled. The program was a "Jeopardy" takeoff in which contestants got showered with garbage if they gave a wrong answer. As can be seen, CV functions in a number of different ways to foster better community relations.

Sales and Marketing

Slightly more than one-half of the corporations in a late 1980s survey reported using video for sales and marketing. There are three general areas where video is commonly employed: new product introductions, sales presentations, and point-of-purchase displays.

New product introductions Introducing a new product to distributors and dealers scattered across the country was a difficult task before the advent of video. For example, Pepsico has several hundred bottling plants nationwide that needed to be informed of the introduction of Nutrasweet in Pepsi's diet colas. Pepsi informed bottlers of the change and kicked off a sales-

---

APPLICATIONS

HITTING THE VIDEO JACKPOT

If you are lucky enough to hit the jackpot on a $1 million slot machine in Las Vegas or Atlantic City, don't be surprised if you also become an instant TV star. Several casinos keep a video production cart near the casino floor. When someone wins a big slot machine jackpot, the TV gear is rushed to the scene and a video is made of the winner scooping up silver dollars or being awarded a big check. The video is quickly edited and then shipped off to local TV stations as a video news release. Casino owners agree that this sort of coverage is the best possible publicity they can get.

incentive contest with an elaborate video starring Martin Sheen that spoofed every war movie ever made. Dressed as a World War II Air Force commander, Sheen gave a pep talk to his men before they flew off to the "cola wars." To reinforce the point of the video, the company mailed bomber jackets to all of their distributors.

Sales presentations The development of the combination videocassette player–monitor has made it easier for sales persons to demonstrate their products directly to customers. This is particularly useful if the product is too large to be carried around and demonstrated or if the service is too abstract to be concisely explained. The Wastemate Corporation produced a seven-minute video program demonstrating the company's waste disposal equipment and sent the tape to dealers who used it during sales calls. The San Diego Cruise Industry Consortium produced a videotape that promoted the advantages of San Diego as a port of call. Armed with the tape, consortium officials went to the major cruise ship operators and used the program to persuade them to make San Diego their main west coast port. It was so successful that a second videotape, aimed at travel agents, was produced.

Point-of-purchase displays Video also works as a point-of-purchase sales tool. Programs are shown to consumers right in the place of business next to the product being sold. Video can sell products and services at a wide range of locations and never get tired. People who shop at large store chains, such as Bloomingdale's and Macy's, are likely to notice video programs in many departments, demonstrating the latest fashion trends or new electronic gadgets. When designer Norma Kamali opened a retail store she replaced the traditional mannequins in the front window with a bank of TV monitors that showed continuous videos of her fashion line. Many supermarkets have installed video monitors at key points in the store to show point-of-purchase sales messages. One series, "A Minute in the Kitchen with Mary" (whose programs actually average ninety seconds in length), contains recipes for fruits and vegetables. Supermarkets that have played these tapes in the produce section report that sales of some fruits and vegetables portrayed in the programs actually doubled.

ORGANIZATION OF CORPORATE VIDEO

The first distinction that needs to be made about the organization of CV is the difference between in-house and outside production sources. An *in-house production facility* is located within the company's corporate structure and ultimately reports to top management. An *outside* (or *out-of-house*) *facility* is an independent organization that is hired temporarily to produce one or more programs. An outside production source is not part of the company hierarchy.

Much of corporate TV represents a blending of these two sources. For example, in the mid-1980s about 31 percent of all corporate TV programs were produced by in-house facilities; only 4 percent were produced by outside sources. The majority, about 65 percent, were produced in-house with some assistance from outside sources. Apparently many of today's organizations have the personnel, equipment, and facilities necessary to produce routine programs but need extra people and equipment for bigger projects. Even some firms with fully equipped production centers will occasionally use outside services for special productions. When it comes to actors, hosts, and narrators, for example, most CV facilities will hire them from outside agencies. And to make things even more complicated, some in-house production divisions also work on a contract basis for other organizations or associations. The Georgia Pacific Company, for example, has a well-equipped studio that it uses for in-house production. The company also produces programs for other firms in the area, making it an outside supplier from those firms' perspective. Perhaps the best way to sum up the situation is to reiterate that most CV combines both in-house and out-of-house production.

In the past few years, in an effort to save money, many organizations have downsized to the point where their production staff is made up of only a few producers. The rest of the work is contracted to external production houses or freelancers. What are some of the advantages and disadvantages of in-house versus outside production? Here is a short summary. In-house production has the following advantages:

1 It's close (usually in the same building) and easier for employees to get to.

2. Production staffs generally know the company well and don't have to spend time familiarizing themselves with the organization's special needs.
3. Productions are usually cheaper once the initial cost of the facility is paid for.
4. Information contained in some tapes might be private and necessarily kept in-house.

Using an outside production company has the following big advantages:

1. It produces a slicker-looking final product, since outside suppliers tend to have better equipment and a more experienced staff, both of which help make a more professional-looking program.
2. It has fresh perspectives and insights on a company problem.
3. It has the capacity to supply special services or special effects.
4. It has the ability to meet tight deadlines.

In-house CV organizations are found in various departments, usually as a result of historical accident. Unlike the accounting and customer service departments, which generally appear in predictable places on most company organization charts, video centers show up in many areas: sales, administration, advertising, personnel, and even maintenance. Most recently the trend has been to centralize the video production function in the corporate communications department, but this is by no means universal since many video operations remain housed where they started, in the personnel or training department. Table 7-2 shows the administrative position of video operations in several types of organizations. As is apparent, the manufacturing and service organizations more frequently include video as a corporate communication function, whereas medical, educational, government, and other organizations are more likely to have video housed in the training department.

Most CV departments tend to be close to the top of the managerial hierarchy. A 1986 survey found that 95 percent of the department heads

TABLE 7-2 Administration of Video Operation

Position of Video Operation	Manufacturing (%)	Service (%)	Other (%)
Corporate communication	44	49	27
Personnel and training	13	27	32
Administration/financial	9	5	18
Marketing/advertising/sales	24	14	5
Line	6	4	0
Other	5	2	18

SOURCE: Judith Brush and Douglas Brush, *Private Television Communications,* Cold Spring, N.Y.: HI Press, 1986, p. 127. Based on a sample of 245 companies.

APPLICATIONS

VIDEO LAW

One of the new places where corporate video (CV) is becoming more common is the courtroom. Some of the legal applications include:

- Depositions—A deposition is a procedure in which a witness is questioned under oath before a court reporter. Video depositions are used when the witness can't be on hand for the trial. Some lawyers consider them more effective than written depositions.
- Accident sites—Instead of using charts and diagrams of accident locations some legal firms now reenact a traffic accident using stunt performers, special autos, and sets. Computer animation and video graphics are sometimes used to show judges and juries aspects of accidents that can't readily be observed or described.
- "Day in the Life"—This video is usually used in personal injury or settlement cases to show how a victim's life has changed as a result of injuries that he or she sustained.
- Courtroom rehearsals—Before going to trial, lawyers' presentations are taped and shown to several test juries who then deliberate about the case. The juries' discussions are also taped and analyzed to see what are the strong and weak points of the case.

had no more than five reporting levels between them and the company's top officer. In fact, CV departments are increasingly becoming directly responsible to top management.

Video departments in most companies tend to be small, as seen in Table 7-3. In the mid-1980s the median size of the video staff was three, a manager–producer, a writer, and a technician–engineer. About 25 percent of the respondents to a 1986 survey reported staffs of six or more, and only a fraction of companies had more than twenty people in their video operation.

The CV department stays busy, with most departments producing between twenty and forty programs a year, about one program every two weeks. Most programs tend to be short, averaging about seventeen or eighteen minutes. Programs running longer than one-half a hour are rarely produced. The median cost per program was about $7400.

VIDEO NETWORKS

Although the word "network" might conjure up images of ABC, CBS, and NBC, CV has networks of its own. They are not so elaborate or extensive as those of commercial TV, but their function is the same: distributing programs to an audience of employees or clients scattered across a wide area. Video networks of one form or another are in place at virtually all of the organizations that use CV. In 1985 almost 60 percent of companies had more than fifty remote locations where their programs were viewed. Some large companies, such as Prudential, Westinghouse, IBM, have networks numbering in the thousands. About one-third of all corporate video users distribute the programs internationally. Videotapes are by far the most used distribution medium, with nine out of ten companies producing ½-inch VHS cassettes. Approximately 10 percent of video users send their programs over closed-circuit in-house TV and another 6 percent or so use satellite or microwave systems.

The simplest CV network is a *one-way network*. Videotapes are produced and duplicated at a central facility and then shipped to remote locations equipped with playback units where the tapes are played for the intended audience. If tape duplicating proves too expensive, one copy is made and the videocassette is physically sent from one place to another (this is known as a "bicycle" network).

TABLE 7-3   Size of Video Production Staffs

Number of Employees	Percentage (%)
1–2	35
3–5	39
6–10	15
11–15	5
16–20	3
More than 20	2

SOURCE: Judith Brush and Douglas Brush, *Private Television Communications,* Cold Spring, N.Y.: HI Press, 1986, p. 53. Based on a sample of 241 companies.

A *real-time network* is a little more complicated and expensive. With this arrangement, a central office facility is connected to remote receiving locations via communications satellites (or land-based microwave signals). Picture and sound from the origination center are displayed simultaneously in all of the remote locations, just as the traditional commercial TV networks operate. If one or more of the remote locations is equipped with audio facilities and has a link to the satellite, a two-way network is possible and the different locations may converse directly with one another. (This arrangement is called a videoconference; see below.)

Some examples of companies using video networks include: K-Mart Corporation, the second-largest retail chain in the United States, uses a private satellite network to improve communication between the chain's headquarters in Troy, Michigan, and its more than 2000 stores nationwide. The system has one-way capability and is used for executive messages, sales meetings, training, and introduction of new merchandise. The signals are scrambled so that the company can distribute confidential data. In addition, K-Mart also duplicates and distributes videotapes to all its stores. Federal Express recently started a similar network with about 500 sites around the country. Wang Laboratories has a one-way video and a two-way audio satellite network with about seventy receiving sites in North America, ten in Europe, and three on islands in the Pacific. The company uses the network for training, technical updates, and customer service.

These corporate networks are beginning to resemble their commercial counterparts in at least one way—many of them now accept advertising.

For example, IBM sponsors shows on the real-time video network owned by the Computerland Company. Computerland retailers around the country can learn about IBM hardware and software products that they might want to stock.

VIDEOCONFERENCING

For purposes of this chapter, a *videoconference* (or *teleconference*) is a meeting or other event transmitted via satellite to a number of sites of which one or more are equipped with a communication channel that allows it to send audio and perhaps video messages back to the originating location. Thus a videoconference is two-way communication. This distinguishes it from the one-way, real-time network discussed earlier. At most organizations production of videoconferences is the responsibility of the corporate TV division.

The videoconference gained popularity during the late 1970s and early 1980s as a way to save on travel expenses. As airline fares increased, many companies found that it was cost-effective to conduct business over video transmission links, thus saving time and money. For example, video-conferencing saved executives at the Celanese Corporation ninety-seven business trips and $28,000 in air fares during the first five months of 1981. Honeywell saved $150,000 a month in travel costs by implementing a videoconference setup in that same year. IBM, the Ford Motor Company, and Aetna Life and Casualty have also turned to videoconferences.

Employees participating in a teleconference. Teleconferencing saves companies a great deal of money by avoiding costly travel expenses.

The idea became even more popular as communication satellite capacity increased and it actually became less expensive to send TV pictures on a 45,000-mile round trip into space than it was to send them cross-country through lines owned by the phone company. Simultaneously, a newly invented device called a *codec* (*co*de/*dec*ode) used digital video processing and made it less expensive to transmit video and data from one point to another. Consequently, videoconferencing was enjoying a boom period as the 1990s opened. It was estimated that at the beginning of 1992 there were more than 12,000 satellite receiving sites at corporations and other organizations. In addition, several private communications companies were established to serve videoconference needs. Big hotel chains, like Hilton and Holiday Inn, have their own private networks that they lease to

APPLICATIONS

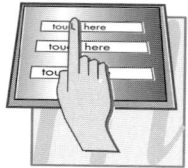

THE VIDEO JOB INTERVIEW

Teleconferencing may alter one integral part of college life—the job interview. In the past students enrolled at west coast colleges generally had a hard time getting jobs with east coast firms. Many eastern firms couldn't afford to send interviewers across the country and most western students couldn't afford a trip east. The video conference may provide a solution. MBA candidates at the University of Southern California (USC) tested such a system when they "met" with recruiters from AT&T in New York via a two-way video connection. About thirteen students went through the interview format of questions and answers in an eight-hour session. The students who participated said they liked the experience and actually felt less pressure than they did in a normal face-to-face situation. The recruiters also seemed pleased, although they wished they had had more control over the interview setting. Plans are in the works at USC to use teleconferencing for interviewing on a regular basis.

others for video meetings. About sixty organizations had their own private satellite networks. Industry statistics indicate that about 36 percent of CV users had also used videoconferences or were about to. The most common type of conference consists of picture and sound sent from a central source via satellite to various receiving sites where the viewers can send back questions using phone lines or satellite links. About 80 percent of video meetings use this one-way video, two-way audio arrangement; the remainder use two-way audio and video.

The actual production involved in staging a teleconference can vary from the simple to the highly complex. At the simple end of the spectrum employees are taken to a hotel conference room to view on a big screen a corporate executive in another hotel room who is making an important announcement. Parked outside the originating and receiving sites are trucks equipped with satellite dishes that handle the uplinks and downlinks. A telephone connection with the sending site permits questions and answers.

The more complicated arrangement can best be seen from the following examples. In 1985 the Compaq Computer Corporation, with dealers across the United States, Canada, and Europe, staged an international teleconference to announce the introduction of two new computer models. Compaq's video department called on a local production company and a private satellite firm for assistance. The origination point was a hotel in Houston. Five uplink trucks outside the hotel beamed the signal to an AT&T communication satellite, which sent the signal to receiving sites in New York, Dallas, Los Angeles, Chicago, Washington, D.C., and Toronto. At the same time, the signal was sent down to an earth station near Houston where it was uplinked to a Western Union satellite from which it was received by another earth station in Andover, Maine. From there it went to an Intelsat satellite to cross the Atlantic and was received in France and England and then sent by land lines to Paris and Munich. About 3000 people saw the ninety-minute extravaganza. The production cost of more than a quarter of a million dollars was cheaper than the travel costs of bringing all of the participants to Texas.

PRODUCING PROGRAMS: CORPORATE VIDEO VERSUS BROADCAST TELEVISION

One thing that CV and traditional broadcasting have in common is top-notch technical quality. People have been watching TV for about forty years and have come to expect a certain professionalism and slickness in the shows they see. If a CV production doesn't measure up to the usual TV standards (let's say the picture is out of focus and the camera is jiggly), the audience will most likely reject it as amateurish and not worth watching. Consequently, producers in CV departments strive to emulate the technical standards employees see on their home TVs.

One area where corporate TV and broadcast TV differ is in purpose. Traditional TV has two general content areas: news and entertainment. As we have seen, corporate TV encompasses

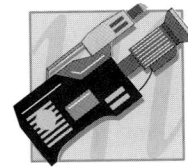

PRODUCTION

THE HAZARDS OF CORPORATE VIDEO

A video production company was shooting a training video at a medical center that housed state prisoners. One scene called for an inmate to enter through a door and walk to a reception area. An actor, dressed in the appropriate clothing, was recruited to play the part of the inmate.

The rehearsal went smoothly. Upon hearing the cue "Action," the actor opened the door and walked briskly to a registration desk. When the scene was shot for real, however, something went wrong. The director yelled "Action" but nothing happened. The director yelled "Action" even louder. No door opening. No actor.

Perplexed, the director opened the door and looked into the hallway from which the actor was to have entered just in time to see a corrections officer haul the protesting actor around the corner. The officer thought the actor was a real inmate who had escaped and was taking him back to a cell.

The taping was delayed until a lot of explaining was done.

training, public relations, employee communications, and marketing/sales. In addition to being informative and entertaining, CV also has specific goals to accomplish: instruction, motivation, persuasion, orientation, and demonstration. A second difference is that CV is usually designed for a relatively small and select audience: company employees, the sales staff, maintenance division, orthopedic surgeons, and so on. Unlike "Cheers" and other traditional TV series, which generally aim for everybody or mostly everybody, CV programs have a very select target audience, all of whom share a special interest in the subject matter. As we shall see, these two differences sometimes pose special challenges for CV producers.

Yet another area where corporate TV differs from traditional TV is time. Broadcast TV programs must be scheduled into precise time slots. Corporate video has no such limitations. There is no necessity for a company newscast to run exactly thirty minutes. Video news releases can be as short as a minute or two. The message can be as long as is appropriate. This gives corporate TV a great deal more flexibility than is typically found in the broadcast sector.

When it comes to the actual process of producing programs, CV and broadcast TV share many similarities. Although the budget, crew, cast, equipment, and sets may be less elaborate in the typical corporate production, the same general process in followed in both arenas. The following is an abbreviated description of the production process in corporate TV. (This discussion should be compared with Chapters 10 and 11, which detail how conventional TV shows are put together.)

The four phases of corporate TV program production are preproduction, production, postproduction, and evaluation.

Preproduction

Major strategic planning decisions are made during preproduction. The CV department or independent production facility consults with top management or other representatives from the organization concerning the objectives of the program, the intended audience, the format of the show, the suggested length, and technical considerations. Communication theory, learning theory, and prior research are examined for guidance and suggestions.

Next, a script is prepared, usually by a three-part process. First comes a *treatment*, a two- to six-page blueprint of what the final program might look like. Following the inevitable revisions, the treatment is turned into a *script*, a document that describes the sounds, words, and pictures that the audience will see and hear. A script alone, however, can't really capture the visual dimension of the program. Consequently, most writers also prepare a *storyboard*, a series of detailed sketches of the major scenes in the program accompanied by key portions of dialogue or narration. The storyboard allows both client and production staff to get a feel for how the finished product will look.

Production

Once the preproduction stage is completed, actual production begins. Although usually not as long as the preproduction phase, the production phase requires more people and is more complicated. During this stage the technical crew, the director, and the performers get together and actually tape the program. There are typically three kinds of CV productions: (1) those done in a studio, (2) those done on location, and (3) those combining the studio and location production.

Studio productions are done in facilities created for the making of TV shows. Programs are usually done in the studio when it's necessary to control the environment: lighting, background, audio, props. Moreover, studio productions usually have a quick turnaround time, making them the logical choice when a deadline is near. And the studio is probably the most appropriate choice when top executives must tape their contributions in the shortest possible time. On-location shooting is called for when natural surroundings are required for realism, accuracy, and audience comprehension. Location shooting might also reduce production costs; sometimes it's cheaper to shoot on location than it is to recreate an office or factory set in the studio. Finally, location shooting offers a variety of visual settings that are not found in a studio.

Postproduction

The biggest part of the postproduction phase is editing. Once the production footage has been shot and previewed, a tape editor selects shots and sequences that go best together. A *rough cut* gives the client a general idea of what the finished program will look like and also gives the director an indication of how the program is paced. After some fine tuning—some scenes might be shortened, others rearranged—the *final cut* is prepared. This is the final version that contains all of the special audio and visual effects, music, graphics, and titles. After the final cut is approved duplicate tapes are made and shipped to the various viewing locations.

Evaluation

The evaluation phase is one that is frequently overlooked by some in CV but is nonetheless important. Most CV producers want to know if their program actually worked. Did it have the intended effect on the audience? There are two basic categories of evaluation: formative and summative. *Formative evaluation* takes place during the preproduction and production stages and includes evaluations of the concept, approach, storyboard, script, and rough cut. *Summative evaluation* is done after the program has been completed and shown to its intended audience. Some of the more common techniques used by organizations to evaluate their programs are audience questionnaires, informal comments, random interviews with audience members, focus groups, pretests and posttests of viewer knowledge and attitudes, and direct measures of performance after watching the program. No matter what technique is chosen, evaluation is important for two distinct reasons: (1) it gives feedback to program producers to allow them to improve their performances in the future and (2) it allows top management to assess the effectiveness of the program against other means of communication.

Producing good CV programs requires a great deal of imagination. Many times it is a heady challenge to make the subject matter interesting. For example, to capture and keep audience interest, preparing a program called "Meclomen: A Unique Compound" for a pharmaceutical sales force or a training tape on how to install a conveyor belt might tax a writer's powers of imagination and creativity. Most CV departments are up to the challenge and some highly creative, imaginative, and effective programs are produced.

To illustrate, when Bell Telephone was forced to divest, C&P Telephone chose a novel way to disseminate information about the breakup to its employees, shareholders, and the general public. It would have been very easy to produce the traditional "talking head" program with a company spokesperson mouthing a lot of dry statistics. Instead, the video department decided to do the whole thing in the style of a 1940s Humphrey Bogart detective movie. Shot in black and white, the twenty-five-minute program had a mysterious blonde hiring private eye Sam Slade to track down the impact that the phone company's divestiture would have on her phone bill and her stock. Slade had to get the information from local stoolies and actors who took the parts of Peter Lorre and Sydney Greenstreet. Entitled "The Big Bypass," the tape was such a hit it was aired on cable TV channels and copies were sent to related businesses.

The Red Lobster restaurant chain trains waiters and waitresses using a tape entitled "Guest Service That Pays." Realizing that a lot of its waiters and waitresses are in their late teens and early twenties and might be bored with a traditional instructional approach, the restaurant produced its tape MTV style. The program runs twenty minutes and is introduced by a video DJ. The audio track is all rock and roll and divided into sections so that employees can stop and review the information when needed. The notion that business video is dull and predictable is obviously wrong.

CAREERS IN CORPORATE VIDEO

Working in CV has several advantages over the traditional broadcast job market. First, the pay seems to be better. A 1985 salary survey found that the median annual salary for managers, producers, and technical staff in CV was several thousand dollars higher than that of their counterparts in broadcast TV. Second, jobs in CV tend to be more secure than those in the sometimes volatile world of over-the-air TV. Private TV employees are generally not exposed to the pressure of ratings or declining sales of commercial time. Finally, there is more creative freedom in CV. In commercial TV, pressures abound and creativity is sometimes sacrificed to raise ratings or satisfy outside consultants. In contrast, CV departments are usually left alone to produce the best shows they can. On the downside, the jobs in CV may not be so glamorous or exciting as those in a typical Hollywood production. Making programs such as "Roofing Inspection Procedure" or "Double-Entry Cost Accounting" is probably not so exciting as putting together an episode of "Northern Exposure." This does not mean that CV is boring. In fact, it probably requires just as much imagination and creativity to work in CV as it does in broadcast TV. If, however, you hunger for the bright lights and tinsel of Hollywood, corporate TV might not be your best choice.

Getting an entry-level job in CV is similar to getting a position in broadcast TV. The marketplace in CV is slightly better than in traditional TV, but it is still highly competitive. The traditional method of preparing a résumé and going on interviews is the norm in the private TV sector as well. The trade press provides information about the major companies in the field and carries job announcements. Before starting a job search in corporate TV, read the following periodicals: *Videography*, *Video Manager*, *Educational and Industrial Television*, and *Millimeter*.

What job skills are companies looking for? First and foremost, employers in the business sector want employees to have superior communication skills. This is understandable since the main job of CV is communication. Good writing skills are essential, as are good speaking and listening abilities. At a job interview most potential employers will give the applicant a chance to demonstrate his or her writing and verbal skills. Many will also note how the candidate pays attention and comprehends what is being said during the interview. A basic knowledge of equipment and production is helpful, but most companies would probably not want someone who was primarily a lover of hardware. A broad-based background that enables a person to think conceptually and to creatively visualize the finished product will probably be more attractive to an employer than rote knowledge of what button to push or what knob to turn. Many students have achieved success by putting together a dual major that combines communication skills with a specialization in a substantive field such as medicine, social services, or agriculture, which makes them an asset to a company within that field.

Consider two other helpful hints. An internship at a CV department while still in college will be an advantage. Employers like to know that prospective employees have some practical experience in the field. Second, develop a professional portfolio that you can bring with you to the job interview. If you are interested in writing, bring along a sample

RESEARCH

THE IMPORTANCE OF THE TARGET AUDIENCE IN CORPORATE VIDEO

No matter how slick the message, no matter how professional the production, no matter how clever the script, a CV program will fail if the audience is not considered. To illustrate, a chief executive at a large petroleum company decided to calm employees by issuing a video report of himself describing the state of the company during a time of falling oil prices. The company's video staff had just acquired a new mobile production van and decided to tape the message in a nearby park. The chief executive, well dressed and well groomed, sat on a bench and gave his message amid a sylvan setting of trees and birds. After looking at the completed tape, everybody involved agreed that it had gone well. The executive looked relaxed and read his lines smoothly; the park looked wonderful on tape. Pleased with itself, the video unit ordered copies made and sent videocassettes to all work sites.

Unfortunately, the company didn't take into account that the tape would be viewed by workers at oil refineries and offshore drilling rigs. When these workers filed in to see the tape, they were hot, tired, dirty, and greasy. After a few minutes of watching their cool chief executive in an expensive suit and expensive haircut, most got up and left long before the tape was over. What had gone wrong? The workers felt that their physical labors were providing some country relaxation for an overpaid bureaucrat. Instead of enlightening and informing the employees, the tape simply made them angry. The company had neglected to consider the needs and situation of its audience when it taped the report. Consequently, the plan backfired.

of scripts that you have written. If some of your scripts have been produced, even better—bring along a tape of them. Those interested in production slots should edit together a highlight tape of the programs they have worked on. Some industrious types have even put their résumés on video, providing prospective employers with a concrete demonstration of how well they can use video to communicate. Finally, students can join the International Television Association (ITVA), the leading professional organization in this area. ITVA, which publishes a newsletter and runs a placement service, is a good place to make contacts.

SUMMARY

Technological advances in CV can be attributed to the following developments: videotape, 1-inch black and white tape, $\frac{3}{4}$-inch color tape, and portable, low-cost camera and editing systems. Aside from changes in video, changes were also occurring in the business world and in our social environment. The combination of these factors led to the CV of today.

A typical company that uses CV has a diversified work force, is geographically dispersed, and has organizational goals.

Corporate video can be used for training and orientation, internal communication between employees and management, external public relations, education, and sales and marketing. In the category of public relations, video is useful for news releases, PSAs, shareholder meetings, and community relations.

Video networks and videoconferencing are being used by many organizations.

Corporate video is similar to broadcast in technical quality as well as production techniques. They differ in their purpose and their timing.

To determine if a program is successful, producers use formative and summative evaluation.

SUGGESTIONS FOR FURTHER READING

Brush, J. M., & Brush, D. P. (1986). *Private television communications: The new directions.* Cold Spring, N.Y.: HI Press.

Brush, J. M., & Brush, D. P. (1988). *Update '88.* LaGrangeville, N.Y.: HI Press.

Budd, J. (1983). *Corporate video in focus.* Englewood Cliffs, N.J.: Prentice-Hall.

Bunyan, J. (1987). *Why video works.* White Plains, N.Y.: Knowledge Industry Publications.

Carlberg, S. (1991). *Corporate video survival.* White Plains, N.Y.: Knowledge Industry Publications.

Cartright, S. (1986). *Training with video.* White Plains, N.Y.: Knowledge Industry Press.

Degen, C. (1985). *Understanding and using video.* New York: Longman.

Gayeski, D. M. (1991). *Corporate and instructional video.* Englewood Cliffs, NJ: Prentice-Hall.

Hausman, C. (1991). *Institutional video.* Belmont, Calif.: Wadsworth.

Marlow, E. (1989). *Managing corporate media.* White Plains, N.Y.: Knowledge Industry Publications.

Matrazzo, D. (1980). *The corporate scriptwriting book.* Philadelphia: Media Concepts Press.

Sambul, N. (Ed.). (1982). *The handbook of private television.* New York: McGraw-Hill.

Stokes, J. T. (1988). *The business of nonbroadcast television.* White Plains, N.Y.: Knowledge Industry Publications.

Whittaker, R. (1988). *Video field production.* Mountain View, Calif.: Mayfield Publishing.

CHAPTER 8: THE INTERNATIONAL SCENE

Let's take a quick trip around the world and turn on some TV sets. First stop, Australia. You click on the TV, scan Australia's nine TV networks, and feel right at home. "The Simpsons" is popular on Australian TV. Next a long trip to Japan where you discover that the Tokyo Broadcasting System airs "The CBS Evening News with Dan Rather." Across the sea to Hong Kong where "L.A. Law" attracts a large audience on one of the two commercial stations. Another long trip to the Slovak Republic where there are only three TV channels and one of them is showing the cartoons "Duck Soup" and "Chip 'n' Dale." On to France, which has eight privately owned and two state-run TV channels where you find "Knots Landing" is a hot show. A stop in Holland reveals that "Married . . . with Children" ranks in the top-25 shows on Dutch TV. Great Britain is next where the most popular show on the commercial-free BBC is "Eastenders," a long-running prime-time soap opera. On the commercially supported ITV network a popular show is the U.S.-made "Baywatch." Back to North America for a stop in Canada where there are both English and French TV networks. In Canada "America's Funniest Home Videos" garners large ratings and competes with such local favorites as "Hockey Night in Canada."

What can we learn from this quick trip? Obviously, U.S. TV shows are popular all over the globe. But on a more general level it suggests that studying broadcasting as it exists solely in the United States might be too narrow a focus. Radio and TV signals and programs do not stop at national borders. Other countries have developed systems different from ours. It might be valuable to examine electronic media systems in a world context since such a perspective might give us a broader view of our own system. Furthermore, American production companies make a great deal of income from programs that are sold to foreign systems. Total revenue in 1991 was more than $2 billion.

As will be noted later, in Europe and many parts of the globe commercial broadcasting has expanded and U.S.-made programs have become a staple on many systems. Although the initial surge of demand for U.S. programming has subsided and a number of countries have enacted quotas that limit the amount of foreign-made programming on their systems, income from international sales of programs will continue to be an important consideration. In fact, many U.S. production companies now routinely test the ideas for new series to see if they have international appeal.

Moreover, many U.S. companies are entering into coproduction agreements with foreign firms. ABC and German broadcaster ZDF are working together on a movie package called "True Stories." Showtime is coproducing a comedy series, "Funny Business" with the BBC. Hanna-Barbera, a U.S. company that produces cartoons, is working with a Canadian company to produce "Young Robin Hood." Public broadcasting station WGBH in conjunction with a British company, Central Television, is producing a documentary on the evolution of the Russian dissident movement. Finally, foreign ownership and involvement in U.S. media companies has grown significantly over the past five years. CBS Records is now Sony Records, the Japanese conglomerate Matsushita has bought MCA, the parent company of Universal, and the Fox Broadcasting Company is controlled by the Australian-based News Corporation, to name just a few examples. For all these reasons it's important to know about the global context within which the U.S. media operate.

To begin, we need to define some terms. There are two basic ways to study world broadcasting. First, there's *comparative electronic media systems*, the analysis of media systems in two or more countries. Thus if we took the U.S. system and compared it to the electronic media operations of Canada and/or Great Britain, we would be practicing comparative analysis. Next, there's the study of *international electronic media*, the analysis of radio and video services that cross national boundaries. Programs can be deliberately aimed at other countries (as is the case with the Voice of America or Radio Moscow) or they can simply spill over from one country to its neighbor (as happens between the United States and Canada). This chapter first looks at international systems and concludes with an examination of comparative analysis.

INTERNATIONAL ELECTRONIC MEDIA SYSTEMS: A HISTORICAL PERSPECTIVE

Earliest attempts at wireless broadcasting were designed to see how far a signal would go. Chapter 2 notes that one of Marconi's striking achievements was the transmission of a signal from England to Newfoundland. Thus it was only natural for other inventors to try to send their signals even farther. One of the things they quickly realized was that the frequency of the radio signal was directly related to the distance the signal would travel (see Chapter 12). For example, using reasonable power, medium-wave signals (300–3000 kilohertz) travel a few hundred miles during the day (farther at night). This makes them appropriate for domestic radio services but not suitable for international purposes. On the other hand, shortwave signals (3000–30,000 kilohertz) travel much farther. Using sky waves, they are reflected by the ionosphere and with the right combination of power and atmospheric conditions they can travel thousands of miles. Thus it is not uncommon for a listener in North America to be able to pick up signals from Europe and Africa. Consequently, the earliest international efforts were generally confined to this part of the spectrum. (Of course, this increased distance has its price: shortwave signals are subject to fading and interference.)

The first radio service designed to be heard by overseas listeners was started by Holland in 1927. The program was directed at Dutch citizens in Holland's colonial empire. The purpose of this service was to keep these people in touch with what was happening back in the homeland. Germany followed suit in 1929 and Great Britain began its Empire Service in 1932.

These "colonial service" stations, however, began to attract a secondary audience of listeners who were native citizens of the country receiving the signals. This paved the way for propaganda broadcasts from one nation to another for the purpose of political persuasion. In 1929 the Soviet Union All-Union Radio, with announcers speaking in both German and French, was broadcasting appeals to the working classes of Germany and France to unite under the banner of socialism. The late 1930s brought tension and war clouds to Europe. Nazi Germany's mobilization under Hitler included an active propaganda program over shortwave. By 1939 approximately twenty-five nations were involved in international political broadcasting, with Germany, Italy, France, the Soviet Union, and Great Britain the most active. Note that the United States was not a leader in early shortwave broadcasting. This period was the golden age of radio (see Chapter 2) and the commercial networks gave Americans all the news and entertainment they wanted. Further, the memory of World War I was still fresh, and a mood of isolationism—of not getting involved in Europe's problems—permeated the nation. Consequently, Americans paid scant attention to shortwave. World events, however, soon changed everything.

In December 1941 the United States was swept into war against Germany and Japan. Almost immediately the U.S. government instituted a program of shortwave broadcasts to blunt the impact of German propaganda. The new U.S. service, called the Voice of America (VOA), made its debut February 24, 1942. VOA was quickly placed under the jurisdiction of the Office of War Information and spent the duration of the war encouraging America's allies and announcing to the rest of the world that the United States was determined to win the war.

World War II also saw a sharp increase in clandestine radio services. A *clandestine station* is an unauthorized station that broadcasts political programs, usually in the name of exile or opposition groups. The United States, Britain, and Germany all operated such stations. For example, Radio 1212 claimed to be a German station but was actually operated by the U.S. Army's Psychological Warfare Branch. The station broadcast to Germany news reports about air raids and battles that were kept secret by the Nazi high command. The real purpose of the station was to cause confusion and dissension in military ranks. On one occasion it misdirected a convoy of German trucks behind Allied lines where they were captured. Most of these wartime clandestine stations vanished when the war ended.

Yet another type of international broadcasting also began in this era—commercially supported shortwave stations. Radio Luxembourg started sending music, news, and ads to other European countries as early as 1933 and continued to broadcast until the war erupted. Radio Monte Carlo signed on in 1943 (it derived some money from ads but was primarily supported by Italy and Germany), and even the United States got involved in broadcasting commercially sponsored shortwave radio to Latin America and Europe. The war, of course, curtailed the U.S. effort, which would resurface a few years later (see below).

Although the shooting war stopped in 1945, the propaganda war between East and West continued. The Soviet Union established international stations in Poland, Hungary, and Bulgaria that supported international communism. Not to be outdone, the VOA stepped up its propaganda efforts. In 1952 the VOA became part of the newly formed United States Information Agency (USIA). One of the stated goals of the VOA was "to multiply and intensify psychological deterrents to communism." The BBC's External Service also entered into the propaganda fray. Broadcasts in Russian began in 1946 and the BBC was urged by many politicians to stay tough on communism. Budget cuts and aging equipment, however, plagued the service for many years.

Other countries also became involved. International operations started in Japan and West Germany in the 1950s. China increased its shortwave programming from next to nothing in 1949 to about 700 hours weekly by 1960. Fidel Castro's Radio Havana, an AM station, signed on in 1961. The Soviet Union debuted a second international service in 1964, Radio Peace and Progress. For its part the United States answered with several new systems. Armed Forces Radio Service (AFRS) started in 1942 to serve military forces around the world, had about 300 stations by the late 1960s, and established a large foreign audience. Radio in the American Sector (RIAS), broadcasting primarily to listeners in East Germany and East Berlin, began operation in 1946 and ten years later was one of the larger systems operating in Western Europe. Radio Free Europe (RFE) and Radio Liberty, founded in the early 1950s, were given the task of encouraging dis-

The Voice of America reaches more than 120 million listeners every week. VOA will have to redraw its coverage map thanks to the recent political changes in Europe.

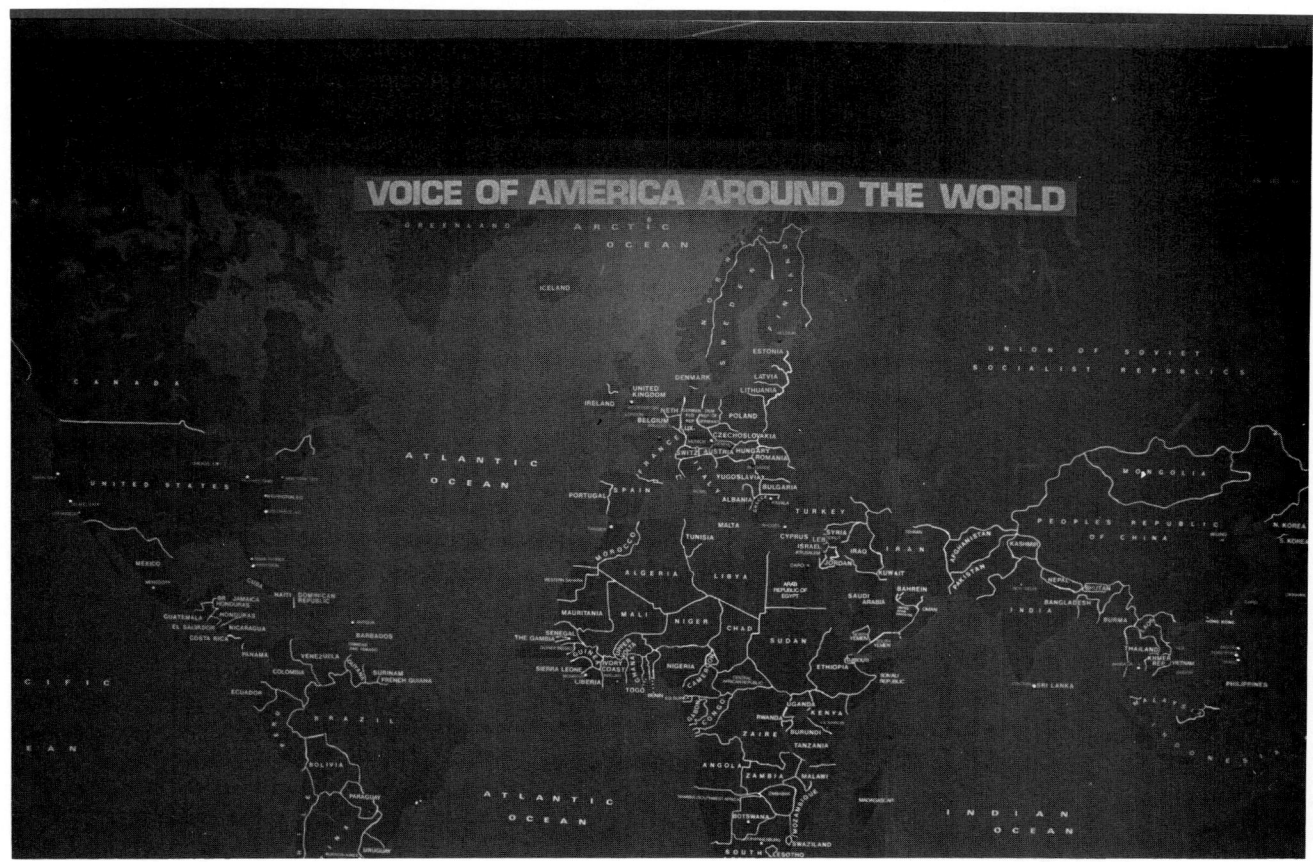

sent in Communist-controlled countries. Both stations were secretly financed by the Central Intelligence Agency until 1973, when Congress approved a new independent governing board.

International broadcasting since 1960 is characterized by five major trends. First, international broadcasters are growing in number. Although the total is hard to pin down, about 150 countries have external shortwave stations. Many former colonies and developing nations, commonly referred to as third world countries, began their own international services. Some developing nations started broadcasting for political reasons; others entered it for prestige. Whatever the impetus, by the early 1990s India, Egypt, Nigeria, and North Korea were among the top-10 international broadcasters in terms of hours of external broadcasts.

Second, the thaw in the Cold War at the beginning of the 1990s has diminished the role of propaganda in international broadcasting. Much of the content has shifted toward public affairs, news, and cultural programming. Nonetheless, the political dimensions of international broadcasting are still important, as illustrated by the events in Russia described in Chapter 1. In addition, the United States continues to operate Radio Martí and TV Martí, two broadcast services whose prime purpose is to offer Cubans an alternative interpretation of national and international events.

Third, there was increasing interest in the United States in private (nongovernmental) international broadcasting. In the early 1980s WRNO in New Orleans became the first licensed commercially supported station. Others soon started in Saipan (a U.S. possession in the Pacific), Alaska, Pennsylvania, Texas, California, Utah, and Indiana. Bangor, Maine, is the site of a high-powered station that is the international voice of the *Christian Science Monitor*. Some of these stations were supported by advertising; others broadcast religious programming and were supported by donations. All in all, private international broadcasting appears to be a growing area.

Fourth, TV became an international medium during the 1970s and 1980s. Programs originally produced in more developed countries were shipped overseas where they were shown on national TV systems. By 1973 imported programs, many of them from the United States, dominated foreign systems. By the early 1990s there was more variation by region in the proportion of imported programs, but imports still accounted for about one-third of the world's total TV air time. Latin America was the area most dominated by foreign shows. This situation was partly responsible for the controversy over the "new world information order" (see below). In addition, the Gulf War coverage on CNN firmly established it as a global news service.

Finally, new technologies have become important factors in international electronic media systems. As of early 1992 there were two global satellite systems in operation, INTELSAT and INTERSPUTNIK, but plans called for a merger. The INTERSPUTNIK system is operated by Russia and serves Eastern Europe, Cuba, and Asia. INTELSAT is a consortium of 121 nations and 170 territories linked together by sixteen satellites. Russia joined INTELSAT in 1991, prompting plans to join the two systems. Currently INTELSAT handles about two-thirds of

---

ECONOMICS

LEARNING THE LANGUAGE OF CAPITALISM

The BBC has come to the aid of budding business operators in the new Commonwealth of Independent States (CIS). When much of the CIS belonged to the Soviet Union, there was little need to understand the terms used to describe doing business in a free marketplace. Now, however, many entrepreneurs in the CIS are searching for a common vocabulary with business persons in the West. In that connection the BBC has supplied Moscow Radio with a series of programs that define and translate many of the expressions of basic capitalism.

"Discount," "profit margin," "product launch," "collateral," and the like are concepts that are still murky to the average CIS businessperson. The BBC series, called "Trading Words," will explain in detail these concepts and other nuances of doing business with the West. Working in conjunction with Moscow University, the series will start with simple words and phrases and then work up to those that are more subtle, such as "Let's do lunch" and "the bottom line."

the world's telephone and data transmissions and virtually all live TV transmissions beamed to Western countries. There are also several regional systems in existence that serve several nations, such as ARABSAT, which covers about twenty Middle Eastern countries. Shortwave broadcasters, plagued by technical problems of interference, turned to satellite broadcasting as a means of increasing and supplementing their coverage.

International TV has grown rapidly because of satellite technology. The U.S. Information Agency (USIA) started a satellite newsfeed to national systems in North America, Europe, and Asia. Other countries are investigating the use of direct broadcasting by satellite (DBS) (see Chapter 13). DBS satellites beam radio or TV signals direct to homes via an umbrella-sized rooftop receiving dish. Japan started a DBS system in the late 1980s. British Sky Broadcasting, partly owned by Rupert Murdoch, supplies a service to DBS-equipped households all over Europe. Both France and Germany have DBS systems that are designed to serve viewers in several countries. The regulatory problems raised by these systems are obviously great.

VCR penetration has increased not only in the United States but also abroad. As of 1992 about 70 percent of TV-equipped households in Australia also had VCRs; Canada had about 66 percent penetration as did the Netherlands. Mexico had about 40 percent. Countries in which local programming is limited or heavily censored, such as the Middle East, also have a surprisingly high percentage of VCR-owning families. The recent surge in VCR popularity has several implications for international broadcasting. First, since most of the prerecorded cassettes that are available consist of U.S.-produced movies, developing countries who wish to maintain their own culture are faced with another challenge. Second, viewer support for nationally owned TV systems might dwindle as audiences spend more time viewing imported cassettes. Last, a thriving video piracy business that illegally duplicates and distributes tapes is costing the legitimate motion picture and TV industry millions of dollars a year. Some companies might be forced out of business if international piracy remains high.

It seems safe to conclude that the increased popularity of these technical advances will force many nations to rethink and perhaps reshape their rationale and procedures for reaching audiences in other countries. The next ten to twenty years may be a transition period for international radio and TV.

INTERNATIONAL RADIO BROADCASTERS

Table 8-1 displays the major international radio broadcasters as of 1991. This section briefly ex-

BACKGROUND briefing

CABLE AND VCR PENETRATION AROUND THE WORLD

There are wide variations among countries in the degree of VCR and cable penetration. In general, however, it appears that the VCR is rapidly becoming a global appliance while cable is lagging behind in most countries. Below are figures from 1991.

Country	Percentage of Households with VCR	Percentage of Households with Cable
Australia	72	0
Brazil	45	0
Canada	66	82
Germany	30	33
Great Britain	60	7
Japan	68	18
Poland	5	0
Saudi Arabia	60	0
Spain	40	0
Switzerland	58	75
United States	72	60

TABLE 8-1 Major International Broadcasters

Country	Hours Broadcast per Week (includes all languages)
United States	2401
Russia*	1951
China	1537
Germany	1287
United Kingdom	797
Egypt	593
North Korea	535
India	472

* Includes Radio Moscow, Radio Station Peace and Progress (shut down in 1991), and regional stations.
SOURCE: *BBC Annual Report and Handbook, 1991*, London: BBC.

amines the organization of each of the top-5 international broadcasters.

United States

In terms of program hours per week, the United States is the most prolific international broadcaster. The most powerful U.S. system is the Voice of America (VOA), now in its fifth decade of operation. The VOA has been at the center of several recent political controversies. During the administration of President Jimmy Carter the mission of the VOA was to educate and inform the rest of the world about the United States. The VOA's propaganda role was deemphasized and nearly eliminated; it was not used as an instrument of U.S. foreign policy. The Reagan administration reverted back to the original mission of the VOA. The service resumed its role as champion of American policies. The administration also spent more than $1 billion to modernize VOA's studios and transmitters.

The United States also operates two other services that are targeted to Eastern European countries: Radio Free Europe (RFE) and Radio Liberty (RL). After the CIA was forced to sever its ties with these stations in the early 1970s they have been operated by the Board of International Broadcasting, a nonprofit corporation, and supported by Congress. Together, the two services broadcast about 1100 hours of programming per week in twenty-one languages using forty-six transmitters located in Germany, Portugal, and Spain.

The collapse of communism in Russia and in Eastern Europe caused the United States to reexamine the mission of the VOA and RFE/RL. Everyone agreed that the VOA must continue its operations. The Gulf War and the failed coup in Russia both demonstrated the need for the United States to have the potential to broadcast anywhere in the world. The future of RFE and RL is a little less clear. For the near future, the services are expected to stay on the air and to continue providing a Western version of the news. In addition, both will report stories that Russian and Eastern European news agencies do not have facilities or personnel to cover. In the long run, in order to reflect changing world politics, the three radio services will probably be reorganized or consolidated. In 1991 President Bush formed a task force to make recommendations for streamlining the three organizations. On the technical front, in Europe the VOA and RFE/RL are expected to move away from shortwave transmission to regular AM and FM broadcasts. In another decade VOA plans to start direct radio broadcasts from satellites and more extensive TV services.

The VOA broadcasts news, editorials, features, and music in forty-four languages through a satellite-fed system of more than 100 transmitters and thirty-two domestic studios, most of them in Washington, D.C. Its major domestic transmitting facility occupies about 6000 acres in Greenville, North Carolina, and other domestic transmitters are in California, Ohio, and Florida. The VOA also maintains relay stations in many foreign countries.

The VOA estimates that about 120 million people worldwide are regular listeners, with about 60 million of these in eastern Europe and Russia. In the last two years the VOA has provided more "how-to" programming for the emerging democracies in Eastern Europe. Recent programs have focused on explaining how the free-market system operates, how laws are made, and how the judicial branch of the government functions.

Another VOA agency is Radio Martí, which went on the air in 1984 as a special service directed toward Cuba. It remained, however, outside the normal VOA chain of command, with its own presidential advisory board. In 1990 it broadcast about 122 hours per week. Radio Martí's transmitter is located in the Florida Keys, and since most Cubans have conventional rather than shortwave radio receivers, it operates at 1180 kilohertz in the standard AM band. In the early 1990s Radio Martí was joined by TV Martí.

One other U.S.-operated system figures in international broadcasting—AFRTS (Armed Forces Radio and Television Service), which has its impact because of spillover. Its programs, as its name suggests, are aimed primarily at military personnel in foreign countries but many people in the host country listen and watch as well. AFRTS has about 100 stations in fifteen countries, many with low power but some are more formidable. The facilities in Germany include a 150-kilowatt AM transmitter and twelve FM and eight TV stations. AFRTS radio stations play popular American music and carry U.S. news and sporting events. The TV stations carry U.S. domestic programs (minus commercials) and in some countries provide significant competition for local TV broadcasts.

Radio Moscow

As of 1991 Radio Moscow (RM) was still one of the major international radio broadcasters. The recent dissolution of the Soviet Union and the formation of the Commonwealth of Independent States, however, have made its future somewhat uncertain. As of early 1992 RM was still broadcasting but at a reduced scale. Several foreign language services had been dropped, hours of operation had been reduced, and equipment and frequencies were shared with other broadcasters. Long-range plans called for RM to remain on the air (although it might be renamed Radio Russia) with limited services in four languages.

At the same time, republics of the former Soviet Union were increasing their efforts in international broadcasting. Using frequencies and facilities once controlled by RM, such services as Radio Kiev in Ukraine, Radio Belorus in Minsk, and Radio Latvia in Riga have begun broadcasts in Russian, English, and other languages.

RM's programming had also changed as of 1992. Newscasts were scaled back and more music was played. Occasionally, the fluid state of affairs in Russia gave rise to some curious juxtapositions of programming. A program honoring the famous Russian composer Prokofiev was followed by a newscast whose lead item reported that only a two-day supply of bread was left in Moscow stores. Moreover, many of the recent newscasts were devoted to explaining the workings of the new Commonwealth to the rest of the world. The English-language service had a large number of light entertainment and feature programs as well, including "Joe Adamov's Mail Bag," "Audio Book Club," and "Culture and the Arts."

The large landmass of Russia enables RM to locate its transmitters on native soil and still be in a position to reach most of the world. RM broadcasts on a large number of frequencies to reach its various target audiences; broadcasts to North America, for example, are carried on fifteen different places on the shortwave dial. This effort,

---

PROGRAMMING

J-WAVE

Listening to American rock music on the radio in Tokyo is no big deal. Many stations play U.S. hits. What is unusual is turning on your radio in Tokyo and hearing "Hi. This is Charlie Burger . . . the Burgerman . . . on WBLS in the Big Apple. I'm on the KDD Sound Call Request Line, live in New York City." A New York City DJ on in Tokyo? Yes . . . thanks to Tokyo Radio Station J-Wave. Every Saturday night Charlie Burger appears as part of the KDD Sound Call—Dancing Freak 001. KDD is a Japanese telephone company and Charlie urges his listeners to call New York City and to make a request.

When a listener calls from Tokyo, the listener is actually connected with two Japanese women in a Manhattan office who write down the requested songs. A list of songs is then faxed back to Tokyo where an English script (with phonetic spelling of Japanese names) is written for Charlie Burger to read. The script is then faxed back to New York where Charlie studies it. Charlie then calls Tokyo, where a technician records him reading the script. Charlie's comments are then spliced into the program along with the requested songs. Finally, the program airs on Saturday night in Tokyo and the process starts all over again.

PROGRAMMING

BROADCASTING IN THE COMMONWEALTH OF INDEPENDENT STATES

Describing the broadcasting system in the Soviet Union was relatively simple. The government ran both TV and radio; censorship was common and programming was basically lackluster. Now that the Soviet Union has given way to the Commonwealth of Independent States (CIS), its broadcasting system has gone through a period of upheaval. In fact, the pace has been so hectic that it's difficult to predict what shape the new broadcasting system will take. Consider some of the following recent changes in CIS radio and TV.

- Gostelradio, the branch of the government that ran Soviet broadcasting, has been replaced by a new governing structure, the Russian State Radio and TV Co. Plans call for this organization to be privatized within the next few years. The other Commonwealth republics are expected to put forth their plans for regional broadcasting.
- Commercial broadcasting is now allowed in the CIS. An American company with local partners has announced plans to start two commercial TV stations, one in St. Petersburg and one in Kiev. Programming, dubbed into Russian, would consist of some U.S. shows, including "Geraldo," "CBS Evening News," and "The Joan Rivers Show," along with some European-produced shows.
- Moscow TV recently aired such U.S. series as "Love Boat," "Little House on the Prairie," "Dallas," and "Beverly Hills, 90210."
- MTV-Europe is carried 24-hours a day on St. Petersburg cable.
- Commercials for products such as Benetton, L.A. Gear, Wrangler, Coca Cola, Pepsi, and Renault are now common on TV.
- The Disney Company is offering Russian TV a ten-year deal in which Disney would supply five hours of programming per week.
- A Wyoming firm is planning to start a commercial radio station in Moscow. Disc jockeys will be recruited in Russia and then flown to Wyoming for training.
- Rox Radio, an FM station that is a Russian-Norwegian co-venture, beams rock music into Moscow. Another popular music station in the city is Maximum FM, a joint venture among Harris, the American radio manufacturing firm, Westwood One, and an English-language Moscow newspaper.

Obviously, things will never be the same in the new Soviet Union.

---

however, will probably decrease in the future as the service is cut back. In addition, RM has recently started renting its transmitters to Deutsche Welle (the German International Service—see below) in an attempt to generate some needed revenue. RM has done little audience research so that precise figures on the size and makeup of its listeners are hard to determine. Surveys done by Western researchers suggest that RM has fewer listeners than the VOA or BBC. A second service, Radio Station Peace and Progress, begun in the mid-1960s during the Soviet Union's ideological break with China, was shut down in May 1991.

The BBC

The BBC's World Service came into existence in 1948 when Parliament decided to merge the old Empire service with the European operations started during World War II. The BBC service is different from the VOA and RM in that the British system is independent of government ownership. (It does, however, work closely with the British Foreign Office so that official government policies are represented accurately to the outside world.) This official separation helps explain the BBC's reputation for accurate and impartial newscasts, a fact demonstrated during the aborted Russian coup described in Chapter 1.

Programming on the BBC World Service is diverse and imaginative. Along with its highly respected news programs, there are many entertainment programs, including rock music, serious drama, sports, comedy, and features. For example, 1990 saw the radio broadcast of Chekhov's *Three Sisters* and Shakespeare's *Macbeth* along with *The Mystery of Edwin Drood*. The BBC has

PROGRAMMING

ALL LATIN, ALL THE TIME?

Those of you who think Latin is dead, *cogita iterum*. The language of Cicero and Caesar is alive and well—on Finnish radio. Every week the government-run Finnish Broadcasting Company presents "Nuntii Latini," a five-minute digest of world news in Latin. Somewhat surprisingly, the newscast draws a small but loyal following (one of its big fans is the Pope). Not to be outdone, Austria launched a TV newscast entirely in Latin. Who knows what's next? Movies? Rock lyrics in Latin? *Ad astra*.

---

also introduced a series of international phone-in shows with guests ranging from Prime Minister Margaret Thatcher to Paul McCartney.

The BBC conducts a wide range of audience research. Data suggest that about 120 million adults regularly listen to the World Service in English and thirty-six other languages, including about 14 million in Russia. The BBC has forty-three transmitters in England plus others in such places as West Berlin, Singapore, Ascension Island (in the mid-Atlantic), and Masirah Island (in the Persian Gulf). In addition, the BBC also leases time over VOA and Canadian transmitters in North America. As a result, few areas of the world are outside BBC's range. Finally, the BBC also maintains an extensive monitoring service of international broadcasting. Both radio and TV programs from other countries are taped, translated, analyzed, and sometimes rebroadcast.

Radio Beijing (Peking)

The external service of the People's Republic of China got its start during the Cold War period of the late 1940s. It basically broadcast propaganda programs until the early 1970s when a thaw in U.S.–China relations caused a mellowing in its attitude toward the United States and an increase in nonpolitical content. In the 1990s Radio Beijing was suffering from outmoded and unreliable equipment. Nonetheless, the station has more than doubled its programming hours since 1960 and now transmits more than 1400 hours weekly in about forty foreign languages. Relay stations are located in Canada, Spain, France, French Guyana, Mali, and Switzerland. Reception in the United States is difficult and subject to interference.

During the June 1989 violence in Beijing, Radio Beijing broadcast a strong condemnation of the brutality used by the government in putting down demonstrations in Tianenmen Square. Shortly thereafter, however, the government cracked down on the station, which then began to broadcast the official government version of the events in the square.

About 80 percent of Radio Beijing's programs are devoted to cultural information about China, news, analysis, and commentary. The other 20 percent are mainly music. In addition, Radio Beijing produces a half-hour program called "The China Connection," which is distributed directly to local U.S. and Canadian stations that subscribe to it. Little is known about the audience research efforts of Radio Beijing but surveys done by other international broadcasters suggest that it ranks behind the other major services in listenership. Its highest audience levels are in neighboring countries such as Thailand and India.

Deutsche Welle (German Wave)

The international radio voice of Germany first signed on in 1953, with broadcasts only in German. Operations quickly expanded and by the 1960s it was sending out programs in twenty-six languages. The service continued to grow during the 1970s and 1980s, and by 1992 Deutsche Welle (DW), which incorporated the former East German international radio service when the two Germanies merged, was broadcasting more than 1200 hours per week in thirty-four languages.

Deutsche Welle's (DW's) annual budget comes from the federal government and is funded at roughly the same level as the VOA or BBC. It has two domestic transmitters and six others in Africa and Asia. In 1992 it started leasing transmitters from RM to improve DW's coverage of Asia. Its programs focus on news, music, and German culture, with an occasional feature on the economic and social life within the country. DW has an extensive listenership, ranking just behind the VOA and BBC. It draws particularly well in East Africa but has trouble getting its signals into Asia.

Unofficial International Services: Clandestines and Pirates

As mentioned previously, a clandestine station is an unauthorized station that beams propaganda at a country for political reasons. In contrast, a *pirate* station, while also unauthorized, does not usually devote itself to political messages. Because of their very nature, it's hard to get authoritative data on clandestines. One study published in the mid-1980s found that about ninety-four clandestines were then in operation, broadcasting to about twenty countries, with Iran and Cuba the most popular targets. For example, during 1980, when the Iran–Iraq war broke out, the Iraqis operated the clandestine Ahrav Voice of Al-Qadisyah, designed to reach dissident Arabs in Iran. The Iranians also sponsored a clandestine that broadcast to Iraq—The Voice of the Iraqi Islamic Revolution. Clandestine stations also cropped up during the 1991 Gulf War. After the Iraqi invasion a station in Kuwait broadcast information to resistance fighters and the U.S. started clandestine stations that broadcast into Iraq.

Pirate stations generally program entertainment material. They typically operate outside a country (usually on a ship anchored beyond territorial limits) but can operate from within, depending on governmental leniency. Pirates became big news in 1958 when a number of them set up operations in the North Sea (between Britain and Scandinavia) and beamed rock music to the teenagers of Europe, whose musical tastes had been ignored by established national systems. The most famous of these early pirates was Radio Caroline (named after President John Kennedy's daughter). Supported by the sale of advertising, these stations had a literally stormy existence. North Sea gales would frequently threaten to sink them and, in better weather, government ships would harass them. By 1967 the BBC and other national systems had started their own popular music programs, which weakened the pirates' appeal. Nonetheless, the pirate stations were credited with popularizing British rock music during the early 1960s. Pirate broadcasting is not limited to radio. Television Nordzee operated for a time off the Dutch coast, beaming programs to Europe, and in 1988 a pirate station located on a fishing boat off the Florida Keys was broadcasting video to Cuba.

Even the United States, with its 11,000 radio stations and liberal commercial policies, has had its share of pirates. "WBUZ-FM" on Long Island was shut down twice by FCC authorities. In 1987 a pirate station broadcasting from a ship in Long Island Sound ("Radio Newyork International") was also forced to cease operation. Three years later the FCC shut down pirate radio stations in Michigan, Texas, and California and fined their operators $1000 each.

INTERNATIONAL VIDEO

Most of global video traffic consists of (1) a videotape or film that is shipped from one country to another and (2) TV signals designed to cross international borders.

Looking first at the tape and film area, the trade imbalance that characterizes much of the U.S. economy is completely reversed in TV. The United States imports only 2 percent of its TV shows (most from Britain) while exporting a great deal. West European countries import about 30 percent of their programs with about one-half of this amount supplied by the United States. African countries receive 40 percent of their programs from abroad and, once again, America is the leading source. The situation is the same in the Middle East. American-made movies on videocassette are also popular the world over. In the future, American-made programs might not be so prevalent because of laws that impose quotas on foreign programming and because many domestic TV systems can now produce professional and slick programs that rival those from U.S. firms.

Direct broadcasting across borders is on the rise thanks to communication satellites. CNN now broadcasts a service to more than 25,000 European hotel rooms and to numerous cable systems in Britain, Europe, Africa, and Asia. During the 1991 Gulf War it was estimated that about one-half a billion people all over the globe tuned in to CNN's coverage of the initial days of hostilities.

Other organizations are also trying their hands at international TV. In mid-1991 the BBC launched a new World Service Television (WST) channel. Distributed by satellite to subscribers with receiving dishes, the channel was picked up

PROGRAMMING

SOAPS IN EUROPE

The soap opera, a longtime fixture of American TV, has migrated to Europe. As of the early 1990s American network soaps were pulling down respectable numbers in many European countries. Part of the reason for their popularity is their ability to draw loyal audiences. As in the United States, once hooked, a soap viewer tends to remain faithful to a show for years. A second reason is economic—soaps don't cost much to produce.

Whatever the causes, a U.S. fan of daytime TV would feel quite at home in Europe. "The Bold and the Beautiful" was a smash hit in Italy. "Santa Barbara" aired at 7:00 P.M. in France and was among the top-20 shows in 1991. "The Young and the Restless" was popular in Greece. In Sweden the prime-time U.S. soap "Dallas" was still doing well.

Can soaps from other countries compete with the U.S. version? Not as yet. "La Dama de Rosa" from Venezuela did well on Spanish TV and in Germany but U.S. fare still dominated. The Europeans, however, had high hopes for "Riviera," a soap jointly produced by companies in the United States, France, Germany, and Italy. "Riviera" costs $40 million for 260 episodes and centers on the trials and tribulations of a rich family living in the Riviera region in France (where else?). The series was carried in six European countries but early ratings were not encouraging.

---

by about 1 million homes in twenty-three European countries and plans to go global by 1993. Like CNN, WST devotes much of its programming day to news coverage but will also carry some general-interest entertainment programs. An all-news channel, similar to CNN, was also under consideration by the Japanese state-owned broadcasting firm, NHK.

Russia's TV network has joined with the Florida-based World One Inc. to announce plans for the Global TV network. The design calls for a satellite-based network to deliver both Russian and U.S. programming to cable systems worldwide. The network would be supported by advertising and the programs would be dubbed into four languages.

Back in the United States, the Fox Broadcasting Company announced in late 1991 its plans to start a commercially supported global network. The Rupert Murdoch–owned company is hoping to form a worldwide network of affiliates that would simultaneously carry Fox's programming. This would make it possible for a global advertiser, such as Coca Cola or Gillette, to run ads on the same program at the same time all over the world. In a related development, CBS was negotiating with two European companies for the production of a prime-time action adventure series to air first on CBS and then to be sold to broadcasters in other countries.

In addition, some cable networks are well known all over the globe. Along with CNN, MTV has a global presence. It is currently available twenty-four hours a day in twelve European countries and is carried part-time by networks in Russia, Japan, and Australia. MTV Internacional, a Spanish-language version, is seen in five Latin American countries, and MTV Asia was launched in 1991. All told, MTV, in one form or another, reaches about 180 million households in thirty-seven countries.

In Europe Rupert Murdoch is part-owner of British Sky Broadcasting (BSB), a satellite-delivered system that offers seven channels of programming to more than 10 million subscribers in Britain and on the continent. International TV news exchanges, such as VISNEWS in London and UPITN, a joint venture of UPI and British Independent Television News also in London, supply global video news coverage.

The U.S. government is also actively involved in global TV broadcasting. The U.S. Information Agency (USIA) produces live daily broadcasts and sends them to more than 100 cities in seventy-nine countries. Called Worldnet and described as the "first live global satellite TV network," the USIA service delivers programs about U.S. culture, ideas, and foreign policy. Worldnet is picked up around the world by local cable systems, master antennas at hotels, and privately owned satellite dishes, and is even rebroadcast in a few areas by over-the-air stations. The USIA also makes use of videocassettes to deliver its messages. Feature films and TV series, as well as educational and how-to tapes are available free at about a half of all USIA posts throughout the world. Demand for the tapes has been steady. In short, international TV is one area that will see more growth in coming years.

COMPARATIVE ELECTRONIC MEDIA SYSTEMS

Let's now turn our attention to electronic media systems as they exist within other countries, recognizing, of course, that certain international factors, such as geography, global politics, and culture, have an influence on national systems. The study of other systems permits us to see how other countries deal with basic problems such as freedom of expression, regulation, and access. The purpose of this section is not to prove that one system is necessarily better than any other but to give us a broader perspective from which to evaluate the U.S. system.

To accomplish this purpose, we shall analyze several foreign systems along four key dimensions: structure, economics, regulation, and programming. These are not the only dimensions that could be examined but they appear to be among the most seminal. Moreover, it's not possible to analyze systems in every country. For our purposes we have chosen only four. This should not suggest that these countries are more important than others; they simply offer better examples to bring out comparisons.

The four systems that we shall analyze are those of Great Britain, the People's Republic of China, Canada, and Kenya. The British system has been praised for its excellence; it also represents a good example of a pluralistic broadcasting system in a developed country. China provides an example of a government-controlled monopoly on broadcasting. An examination of the Canadian system will illuminate the problems of maintaining cultural autonomy in the face of broadcast spillover from the United States. In addition, Canada, as an immediate neighbor of the United States, is a good candidate for study. Kenya demonstrates the special role of broadcasting in a developing nation. It also provides us with an opportunity to discuss the "new world information order." Finally, after examining these four systems we present a classification system that will summarize the main differences in broadcasting systems.

The United Kingdom: System in Transition

Broadcasting in the United Kingdom and the United States developed at about the same time but underwent a markedly different evolution. Like many countries in Europe, Britain decided early that broadcasting should be a monopoly public service. Coupling this thought with the British tradition of strong independent public institutions, the government chartered the British Broadcasting Corporation (BBC) in 1927. The programming on this new service was essentially "highbrow," characterized by news, analysis, education shows, and classical music. Over the years, however, the BBC has diversified its radio programming by adding other, more popular content. In the early 1970s there were four BBC radio networks in operation, one devoted entirely to rock music.

The BBC got into TV broadcasting in 1936 but, as in other countries, development was short-circuited by World War II. Television service resumed in 1946 and grew rapidly in popularity. The BBC enjoyed monopoly status until 1954 when Parliament created a competitive commercially sponsored TV system. Further competition came about in the early 1970s when local independent radio stations were authorized. In 1992 the BBC consisted of two national and several regional TV networks, four national radio networks, and several regional and local services. The rival commercial system had two national TV networks, approximately fifty local independent radio stations, and TV and radio news services.

VCR penetration is high, more than 65 percent. Cable TV is not growing so rapidly as in the United States, primarily because of a generally weak economy in Great Britain and because of an environmental protection act that makes it necessary for all cable to be buried rather than hung from poles. As a result, cable penetration was less than 20 percent in 1991 and cable companies, many of them owned by U.S. firms, put their plans on hold.

British broadcasting was deregulated and partially restructured in 1990. A new governing body was created for commercial TV; the way was cleared for more independent radio stations; and a new method, based on sealed bids, was used in 1991 to award commercial programming franchises to production companies. As a result of this bidding, several established and well-known British companies lost their franchises. The government also announced plans to start a fifth network, which would be commercially sponsored. In addition, British Sky Broadcasting began to provide competition to the BBC and the commercial system, using a direct broadcast satellite system. If that weren't enough, the charter of the BBC comes up for renewal in 1996, which means that the renewal legislation must be introduced by the end of 1994. Early plans call for some significant changes in the way the BBC is organized and financed. All in all, the beginning of the 1990s has been a hectic time for British broadcasting.

Structure As portrayed in Figure 8-1, broadcasting in the United Kingdom is organized into a *duopoly*. There are two major broadcasting entities: the noncommercial BBC and the commercial sector. The BBC is a public corporation that is ultimately responsible to Parliament but whose day-to-day operations are generally free from interference. Operating under a royal charter, the BBC oversees the noncommercial national radio and TV system, the regional radio and TV services, and the BBC World Service. An American parallel to the BBC is difficult to find since most government-chartered organizations tend to be politically involved from time to time. Perhaps the Red Cross, chartered by Congress but run by its own board of trustees, comes the closest.

The commercial sector is regulated by the Independent Television Commission (ITC) and the Radio Authority. These organizations, like the BBC, are public corporations but their main function is to administer and regulate the British commercially supported broadcasting system. ITC has charge of Independent Television (ITV), which consists of a network of sixteen regional stations (ITV is also called Channel 3), and a second commercial network, Channel 4. (The third commercial network—Channel 5—is scheduled to begin in 1993). In addition, ITC oversees all cable systems and all satellite systems. The Radio Authority is in charge of all independent radio stations.

It's important to note another difference between the commercial networks and the BBC. The BBC produces most of the programming that is broadcast over its transmitters. In contrast, although it operates its own transmitters, ITV contracts with and supervises private production companies that provide all of its programming. (This distinction is discussed in more detail below).

Economic support Since the BBC is prohibited from making money by selling commercial time, it must turn to other sources for income. The system used by the BBC is a license fee, or receiver fee, that all set owners in the United Kingdom must pay to support broadcasting. This license fee system is fairly common in Western Europe but is unfamiliar to those in the United States who are accustomed to "free" radio and TV supported by the sale of advertising. (A variation of the

FIGURE 8-1 Organization of British broadcasting.

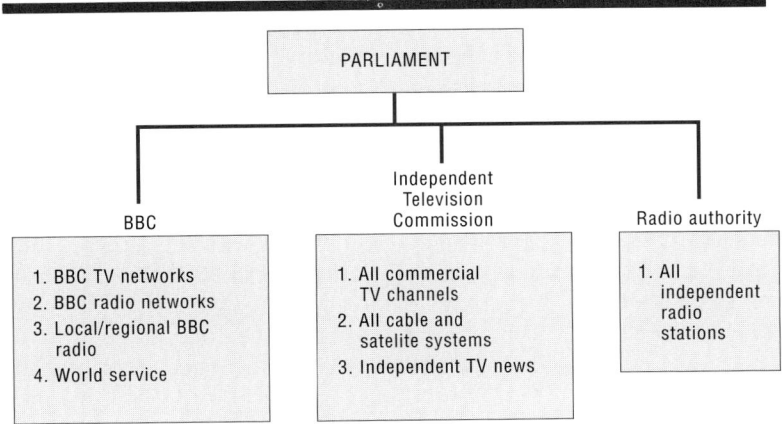

license fee system has been suggested from time to time as a means of supporting the public TV system in the United States, but so far this plan has not been adopted.) As of 1992 the license fee, set by Parliament and tied to the inflation index, was about $140 per year per household for a color TV set. (License fees support both radio and TV broadcasting but are imposed only on TV receivers.) The post office used to gather this fee but in the early 1990s the BBC assumed responsibility for its collection. People who avoid paying it are subject to a heavy fine. The BBC also receives a small amount of income from selling program guides, books, records, and other items. About 60 percent of total license fee revenue goes to TV, about 25 percent to radio, with the remainder going for reserve and miscellaneous expenses.

The problem with the license fee is that the cost of programming is rising faster than the inflation index. As a result, the BBC argues that even though the license fee has increased, the BBC is losing ground. Consequently, in late 1991 the BBC announced that it was planning to eliminate 4000 jobs (out of a total work force of about 30,000) and to cut its budget by about $130 million. The dreary financial condition of the BBC will probably be a focus of any new plans to restructure it in 1996 when its current charter expires. Early plans being discussed by members of Parliament include allowing the BBC to accept limited corporate sponsorship (much like the system of underwriting on public TV in the United States), selling some of BBC's radio networks to private industry, and contracting out much of its programming to independent producers.

The commercial TV networks, on the other hand, receive no part of license fee revenue. Their income is derived from rental fees paid to them by local radio and TV franchises that use the ITC's transmitters and in turn make their money by selling advertising on their stations. The ITC has recently liberalized its rules concerning the placement of advertising. Advertisers may now sponsor programs and have more flexibility in where to place their ads within the show. Despite the more liberal rules, the ITC also has regulations governing the amount of time given over to commercials and the commercial copy itself. In 1991 advertising revenue on ITV amounted to about $4 billion, compared to the BBC's total license fee income of about $2 billion. (This difference is not unusual. In other countries having similar duopolies advertising revenue generally exceeds license income.) When profits exceed a certain level, the government is empowered to levy a special tax on commercial broadcasters.

Law and regulation Government exercises a light hand with the BBC. Although technically the government has the final say, it has rarely exercised that power. The government, however, can exert significant pressure as it did in the late 1980s and early 1990s when it prohibited the broadcasting of terrorist voices in connection with the violence in Northern Ireland. In addition, the Police and Criminal Evidence Act of 1984 offered in principle, at least, some protection to journalistic materials—reporters' notes, photographs, news footage, and so on. In practice, however, requests from the police for such material have generally been approved by the courts.

The BBC's charter obliges it to cover activities of Parliament, forbids editorializing, and, as we have seen, prohibits commercials. In addition, the BBC has imposed regulations on itself. Its board of governors has enacted guidelines that state the BBC will treat controversial issues fairly, operate in good taste, and neither encourage nor incite crime. The Broadcasting Complaints Commission was set up in 1981 to consider charges of unfair treatment or invasion of privacy by both BBC and ITV programs. In 1990 it acted on only twenty complaints, partially or fully upholding eleven of them.

The ITC finds itself in an administrative as well as regulatory role. It operates within a framework set by legislation and is responsible for supervising programs in the public interest. The ITC selects the companies that will produce programs, supervises and controls program schedules, and regulates advertising. It issues lengthy program guidelines to the production companies that cover such things as taste, portrayals of violence, charitable appeals, and accuracy in news broadcasts. Moreover, in accordance with the Broadcasting Act passed by Parliament in 1990, the ITC constructs and enforces a code that deals with advertising. One provision of the code, in contrast to the situation in the United States, prohibits advertising in programs intended for children. Moreover, no ads may be shown for any tobacco products.

In addition, both the BBC and the commercial networks allot time to political parties during election campaigns according to the number of votes garnered by the party in preceding elections. Again unlike the United States, political candidates are not allowed to buy time.

The Broadcasting Act of 1990 established the Broadcasting Standards Council that has jurisdiction over obscene and violent content in programs on both the BBC and commercial stations.

Programming Keep in mind that the BBC produces most of its own programming (about 95 percent in 1990), whereas the ITV gets its programming through contracts with independent production companies. Nonetheless, they both strive for the same programming objectives: information, education, and entertainment.

The BBC, however, has a somewhat stodgy reputation when it comes to programming (the BBC's nickname is "Auntie Beeb"), whereas the ITV has a reputation for more "mass appeal" programs. These images are not necessarily accurate. The BBC, after all, produced "Monty Python," and in 1991 "Auntie Beeb" broadcast the broad sitcoms "Birds of a Feather" and "Open All Hours," and aired the American imports "Cop Rock" and "Dallas." During the same period it also ran the opera *Boris Godunov* and "Joseph Campbell and the Power of Myth." Meanwhile, the ITV, while presenting such popular appeal shows such as "The Benny Hill Show," also broadcast several "prestige" dramas. The ITV also relied on some American-produced shows, including "Baywatch," "The Cosby Show," and the "Father Dowling Mysteries." It would appear that both services offer a variety of programs.

The production process at ITV deserves special mention. Programs are produced by independent companies, which are awarded franchises of up to eight years with the ITV for various dayparts and regions. In 1991 the ITC awarded new programming franchises for Channel 3. More than forty companies, including such U.S. giants as Time Warner, HBO, NBC, and Disney, bid for the sixteen available franchises. When the process was over, four established British produc-

One of the most popular shows on the BBC during the early 1990s was the sitcom "Open All Hours."

tion companies, including Thames TV, Britain's biggest production company, lost their franchises to newcomers. Of the American bidders, only the Disney company was successful. Disney and three British companies were part of a consortium that was given the early morning "breakfast-TV" franchise. Those companies that lost their franchises or that were unsuccessful in securing a franchise have turned their sights to supplying programs for Channel 3 or to gaining a franchise for the new Channel 5 in 1993.

All the BBC's TV production used to be developed by the BBC itself. In 1990 the British government, in an attempt to cut BBC costs, passed legislation that required 25 percent of BBC TV shows to be produced by independent producers by 1993. Production still done by the BBC tends to be localized in London. As of 1990 about three-fourths of all TV shows and 50 percent of all network radio production originated in the British capital. The BBC does maintain, however, a number of regional production centers scattered about the country and in Scotland, Wales, and

Northern Ireland. Production teams tend to be centered around content areas: light entertainment, drama, children's shows, current affairs, and so on.

People's Republic of China

Broadcasting developed in China around 1940. Its development was hampered by World War II and internal fighting among China's various political factions. The first station, built with scavenged parts and utilizing a Russian transmitter, was immediately used by China's Communist party as a political tool. China's leader, Mao Zedong, decreed that the station should be used to keep in touch with areas of China that were controlled by the party and with guerrillas and armed forces in the other provinces.

After the Communists assumed power in 1949, radio became a major propaganda tool. From the beginning the goal of Chinese broadcasting was to shape the development of Communist society. The government built new transmitting facilities and, to make up for the lack of receiving sets, installed loudspeakers for group listening in the rural sections of the country. In 1950 the Central People's Broadcasting Station (CPBS) opened in Beijing (in China, the word "broadcasting" refers to radio and not TV). China's international shortwave service, Radio Beijing, started about the same time. Local stations relayed the programming of the CPBS into the provinces. By 1960 more than 120 radio stations were providing the country with news, entertainment, public affairs, educational, and political broadcasts.

The 1960s and 1970s saw Chinese broadcasting influenced by the cultural revolution proclaimed by Mao Zedong. The programs during this period were propaganda-oriented, strident, and mostly political. Entertainment that was carried had to be in keeping with the goals of the revolution. After Mao's death in 1976 controls over the radio system were relaxed and service was eventually expanded to include six national radio channels. In the 1980s China opened its doors to Western culture and the CPBS even began to play American rock music (called *yaogunyue*, translated literally as "shake-roll" music). At the same time, regulation of news was eased. All of that changed in 1989 with the demonstrations in Tianenmen Square. After a few days of tolerance the government cracked down on both the TV and radio media and instituted strict censorship (see below).

Television started in China in 1958 with one station in Beijing. Lack of hardware and technology, however, hampered the development of TV. In addition, during the cultural revolution

The international scope of modern telecommunications enables many stars to be global figures, as this Chinese billboard featuring Willie Nelson demonstrates. Loosely translated, the message is "Enter into ecstasy with music."

TV development was stopped entirely. Consequently, by 1970 there were only thirty stations in the entire country. Since 1976, however, TV has been China's fastest growing medium. In 1991 there were 400 transmitting stations linked by satellite and ground relay facilities. China was the world's third largest producer of TV sets and about 75 percent of the population now has access to TV. Programs were mainly locally produced but some Western shows and even some commercials were cropping up.

Similarly to radio, Chinese TV was placed under strict censorship after the events at Tianenmen Square. In response, many Western media companies refused to deal with China, hampering its development of TV programming. By 1992, however, China was actively engaged in coproduction agreements with several countries. Nonetheless, the government was still concerned about the potential liberalizing effects of TV. Although entertainment programs were seldom censored, the government kept a close eye on domestic news programs and carefully monitored any news footage it purchased from external sources such as CNN, World Television News, and Visnews.

Another recent development was the increase in VCR ownership and the growing market for videocassettes. Many cassettes entered the country from Hong Kong and were copied many times over by video distributors in China. The government was concerned about the effects of home video since it was much harder to control than traditional broadcasting.

Structure Figure 8-2 illustrates the general structure of broadcasting in China. The Ministry of Radio, Film and Television controls Radio Beijing, CPBS, and the TV service, China Central Television (CCTV). Positions in this Ministry generally go to Communist party members, which means that the party effectively controls the system.

China is a huge country with a large landmass. In addition, its people speak a variety of languages and dialects. To accommodate this fact, China has developed both a national and a regional service. CPBS provides two national channels of programming along with two channels beamed at Taiwan along with special radio services in languages other than Standard Chinese, including Cantonese, Amoy, Hakka, Tibetan, and Mongolian. In addition, there are many provincial and local radio stations serving the various regions of the country. Local FM stations, targeted at a single village or farming community, are the fastest growing sector of Chinese radio.

China Central Television (CCTV) is the center of the Chinese TV network. It provides three programming services. The first (called Channel 2) is the primary service and is sent to all of the country's 400 TV stations. The second service (Channel 8) is seen only in the Beijing area. The third service is an educational channel offering telecourses in economics, language, literature, engineering, and so on. This is a popular service, as evidenced by the more than one-half million Chinese who officially enroll in these classes and receive academic credit.

Cable TV came to China in the mid-1980s. As of 1990, however, only 5 percent of the country was wired. Some urban cable systems were quite elaborate, serving as many as 100,000 subscribers.

Financing Since Chinese broadcasting is operated by the government, its financing is simple to describe. Money for radio and TV is simply allo-

FIGURE 8-2 Organization of Chinese broadcasting.

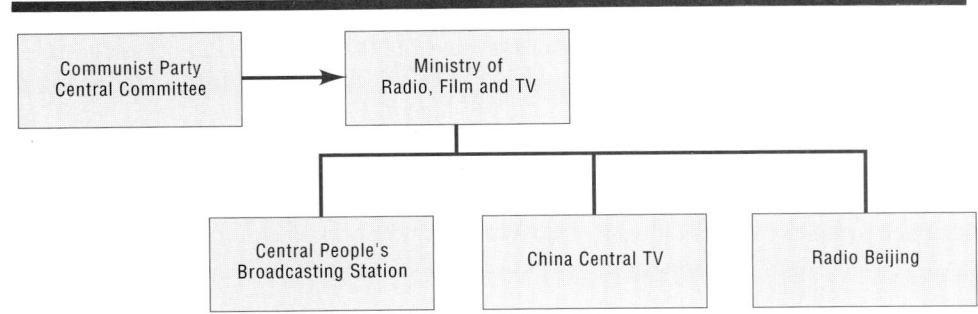

cated from the national budget. Chinese TV does accept commercials and the revenue generated by these ads supplements the budget of CCTV. About 70 percent of the money comes from the government and the remaining 30 percent comes from advertising.

Regulation China exemplifies the Marxist–Leninist philosophy of total integration between the government and the media. This is illustrated by the fact that ultimate control over the CPBS and Chinese TV is exercised by the Central Committee of the Communist party. According to the philosophy expressed by Mao Zedong, the functions of the media are threefold: (1) to publicize party decisions, (2) to educate the people, and (3) to form a link between the party and the people.

Censorship is a given on Chinese TV and radio. The most common type of censorship is self-censorship since journalists and media managers know from training and experience what stories and programs might upset the government. In addition, the party monitors the content of radio and TV stations and, if it wishes, can intervene.

This intervention was never more vivid than in the events that surrounded the events in Beijing's Tianenmen Square in mid-1989. As the pro-democracy movement gained momentum in China during the early part of 1989 a breakdown in the authority of the Communist party's Politburo allowed a brief liberalization of the media, particularly TV. As students demonstrated for democracy in the square, TV viewers saw scenes that were unprecedented on Chinese TV. Frustrated by years of tight government controls, TV journalists dropped the usual news reports about the official doings of the government and showed instead scenes of students on a hunger strike. Emboldened by a lack of government response to this coverage, further TV reports focused on the growing popular support for the students. Four CCTV correspondents reported live from the scene, their cameras showing prodemocracy banners, one of which read "The Ministry of Radio and Television supports the student hunger-strikers." Television also caught the scene in the Great Hall of the People when student leaders were shown questioning the wisdom of the decisions made by Chinese Premier Li Peng. This may not sound like much to Americans, but it was significant because in China no one ever questions the wisdom of the government's decisions and the media never show a leader in such an embarrassing situation.

Radio stations throughout the rest of the country reported on the demonstrations in Beijing and on local rallies in support of the students. CPBS, meanwhile, began live reports from the scene. Press organizations released statements praising the goals of the students.

By June, however, the hard-liners in the Communist party had reestablished political control. The period of liberalization ended dramatically on June 3 when about 50,000 combat troops riding in tanks and armored personnel carriers fought their way through the crowd and occupied the square. The next day an announcer on Radio Beijing read the following statement: "Remember 3rd June, 1989. . . . Thousands of people, most of them innocent civilians, were killed by fully armed soldiers when they forced their way into the cities. Among the killed are our colleagues at Radio Beijing." The announcer was not heard from again.

For most Americans the most memorable image from those days was the scene of the lone Chinese man facing down a tank. The Chinese people never saw this footage. The government quickly seized control of the media and began telling the "official" story. According to a forty-five-minute documentary that was broadcast over and over again on national TV, the demonstrations in Beijing had been taken over by "ruffians" and "hooligans" who had attacked the soldiers, and the government had no choice but to use force. The documentary used footage that was taken by the military showing the bodies of soldiers and scenes of demonstrators who had taken over an armored personnel carrier, driving the vehicle back and forth while firing its machine gun in the air.

News reports going out of China to the rest of the world were censored. The VOA was intermittently jammed. CCTV was taken over by the military and the internal security forces throughout the month of June. Journalists who had sided with the students were ordered to do hours of grueling political study aimed at making them recant their prodemocracy attitudes and acknowledge the supremacy of the party. In short, the government was back in firm control of the media.

Since 1989 the restrictions on the broadcast media have been relaxed but not by much. CCTV had returned to civilian control but newscasts were carefully monitored. Chinese TV carried several Western entertainment shows but TV news footage from external sources, such as World Television News, was edited for political content. Signals from Hong Kong that spill over into China ("The CBS Evening News" is shown in Hong Kong) were blocked by having powerful Chinese stations on those same channels. The use of satellite dishes has been heavily regulated. Apparently, China is still fearful of the unsettling effect that TV might have on its citizens.

Programming Radio programming in China resembles the content of public radio in the United States. Programs consist of news, humor, storytelling, and music. AM radio is the mainstream service and has the biggest audience. FM radio tends to be localized and relies more on music. The two main CPBS channels provide about forty hours a day of service, two-thirds of which is general entertainment. News, features, and advertising make up the rest of the schedule. Regional stations will also carry some CPBS programming but will originate about twelve hours per day of local programming, mainly music. Western music is played on many stations, and most Chinese are familiar with the latest trends in rock music, thanks to stations in Hong Kong whose signals reach into China.

In the wake of the crackdown after Tianenmen Square Chinese TV was limited to a maximum of 15 percent air time for imported shows. In 1991 this restriction was eased and entertainment series from Japan, Taiwan, England, Hong Kong, and the United States aired on CCTV. China was also in the midst of a coproduction venture with the Tokyo Broadcasting System in Japan. The U.S. series that were most popular were "Hunter," "Remington Steele," and "Mi Laoshu," Mickey Mouse. Budgetary restrictions limit China to purchasing only about 100 foreign-produced TV shows a year. Nighttime TV is generally devoted to news and entertainment shows. Daytime TV generally shows educational or informative programming. As of the early 1990s some popular locally produced TV shows were the national news, consumer news, "Tonight We Meet," sort of the Chinese equivalent of "The Dating Game," and "Expectations," a prime-time soap opera that has garnered a huge following. Commercials are run in a block, sometimes as long as thirty minutes, between programs.

Canada

Broadcasting in our northern neighbor has been shaped by its geography, cultural heritage, bilingualism, and the importance of its proximity to the United States. Almost 90 percent of the Canadian population lives within range of U.S. TV signals. Because of these factors, Canadian broadcasting has to face some unique problems.

The origins of Canadian broadcasting generally paralleled those in the United States. Experimental licensing began in 1919 and three years later private commercial stations began broadcasting. As radio grew so did concern about the ultimate future of broadcasting. On the one hand, the country's British heritage suggested a model such as the BBC. On the other hand, the proximity of the commercial system in the United States suggested another approach. In 1929 a government commission recommended the creation of a national company to provide a public broadcasting service throughout the country. This ultimately led to the creation of the Canadian Broadcasting Corporation (CBC) in 1936, whose main tasks were to provide a national radio service and to regulate Canadian broadcasting. This may sound reminiscent of the BBC, but there was a well-established system of private broadcasters in operation at the time who were allowed to continue operating but under the regulation of the CBC. Thus, from its beginnings, Canadian broadcasting has represented a blend of the U.S. and British approach.

Radio grew quickly, reaching 70 percent of the population by 1937. After World War II the CBC instituted FM broadcasting and started investigating the development of TV. Television became a reality in 1952 and, like radio, both private and CBC stations were established. In 1958, after pressure from the private sector, a body separate from the CBC, known as the Board of Broadcast Governors, was established to regulate broadcasting. This in turn was followed by the Canadian Radio–Television Commission—now called the Canadian Radio, Television and Telecommunications Commission (CRTC)—which has the power over private stations and more limited authority over the CBC. Television networks, in both French and English, also started during the 1960s as did cable TV. In fact, cable grew so rapidly in Canada, it is now one of the most "cabled" countries in the world, with about 80 percent of its homes wired. Much of this growth can be accounted for by the desire of Canadians to receive the full lineup of U.S. cable channels. The Canadians also pioneered satellite communication to cover the vast distances of their country. VCRs are also popular, with about 66 percent penetration.

The 1970s and 1980s saw a conflict arise over the problem of spillover of U.S. TV signals into Canada. Concerned about the loss of advertising revenue and cultural invasion from the United States, Canada enacted legislation that would deny business tax deductions to companies that advertised in U.S. media. The United States responded with a similar measure. A United States and Canadian Trade Agreement, ratified in 1988, finally did away with these tax barriers. In fact, many U.S. TV series are now produced in Canada to save money.

Structure Broadcasting in Canada is officially described as a single system with two components. In practical terms, the Canadian system, like the British system, is a *duopoly*, a mix of the public and private, commercial, and noncommercial. On the public side, as we have mentioned, the CBC is owned by the federal government and is the major broadcasting force in the country. It provides national radio and TV services through separate English and French networks. Unlike the BBC, however, the CBC relies on a number of privately owned stations that serve as affiliates and provide CBC programming to areas not served by CBC-owned stations. In addition, the CBC operates a special service for the sparsely populated north regions, an armed forces network and Radio Canada International, its shortwave international broadcasting operation.

The private sector in Canadian broadcasting is made up of all commercial radio and TV stations that are not affiliated with the CBC. There are no privately owned radio networks in Canada but there are private TV networks as well as several regional satellite and cable networks. Figure 8-3 provides a simplified representation of the Canadian system. Of course, as we have already men-

Public Sector (CBC)
1. English TV Network
2. French TV Network
3. French Radio Network
4. English Radio Network
5. Radio Canada International

Private Sector
1. English TV networks
2. French TV networks
3. Independent radio stations

FIGURE 8-3 Organization of Canadian broadcasting.

tioned, most Canadians can also pick up the major U.S. networks and many independent stations or see American programs that are provided directly to Canadian stations.

Economic Support The financial system that supports Canadian broadcasting also combines elements from the British and American systems. The CBC is currently supported by funds appropriated to it by Parliament and by revenue collected from advertising. (True to its British heritage, from 1923 to 1953 the CBC was also supported by a license fee on receivers.) In 1987, of CBC's total $1.1 billion budget, about 80 percent came from the government and 20 percent came from the sale of advertising. The revenues of private radio and TV broadcasters are generated largely by advertising. In addition, similar to the U.S. system, some cable channels are supported through direct subscriber payments.

As the 1990s began, Canadian broadcasters, both public and private, were suffering financial hardship, thanks to a recession and the fragmentation of the broadcast audience. In late 1990, in an attempt to cut costs, the CBC cut 1100 jobs (about 10 percent of the total workforce) and closed thirteen stations. In that same year private broadcasters reported that cost increases had outstripped advertising revenue by 50 percent over the last five years. As a result, profit margins had dropped precipitously. On the other hand, many TV production companies have stayed prosperous by discovering a new market for their programs in the United States, particularly on U.S. cable networks such as Nickelodeon and the Family Channel.

Law and regulation Throughout the twentieth century Canada has made extensive use of independent commissions and parliamentary committees in shaping its broadcasting policy. In contrast, the United States rarely relies on this technique, preferring to use Congress and the FCC. In many instances the recommendations of these Canadian commissions and committees have been enacted into law.

The current law governing Canadian broadcasting was enacted in 1968 after much commission and committee study. This law established the Canadian Radio-Television and Telecommunications Commission (CRTC), an independent authority charged with regulating and licensing all broadcasters and cable operators. In terms of structure and overall function, the CRTC is somewhat similar to the FCC in the United States. Working with the CRTC is the Department of Communications, which does research and formulates national communications policy.

The one area in which Canadian regulation sharply differs from that of the United States or Great Britain has to do with regulation of programming. The Broadcasting Act of 1968 emphasizes the nationalistic purposes and cultural goals of Canadian policy. It states that the broadcasting system should be owned and controlled by Canadians, should be predominantly Canadian in content and character, and should promote Canadian identity. Why is such language necessary? Because nine out of ten Canadians live within 100 miles of the U.S. border and because the economics of TV program production make it cheaper for Canadian broadcasters to buy a U.S. show than to produce a local one, Canadians are worried that their national identity and cultural values will be overwhelmed by American broadcasting. Consequently, during the 1970s, quotas were imposed on all domestic broadcasters governing the minimum amount of Canadian content required to be broadcast. These rules are too complicated to explain in their entirety but some specific examples will illustrate their scope and intent. For radio, 30 percent of all recorded music must be Canadian in origin (this means either the performer, composer, lyricist, or recording studio is Canadian). In addition, there are rules regarding diversity in radio formats. Canadian radio stations are allowed to play any kind of music but are not allowed to mix formats. For example, if a station has a country format, it has to play at least 70 percent country music. The other 30 percent can be anything the station wants but no less than 70 percent must be country. There are also rules limiting how often a hit record can be played in a given time period. Many Canadian radio stations employ one full-time person just to deal with these federal regulations.

For TV, programs are scored according to a complicated point system with points awarded to a program according to the nationality of the creative team (director, writer, performers, etc.). For CBC stations, Canadian programming must account for 60 percent of all content during the day. For private stations, 60 percent must be Canadian during the day and 50 percent at night. (Keep in mind that most Canadians also have access to U.S. stations so that even with these regulations about 70 percent of all English-language TV available in Canada is from the United States.) In addition, a Broadcast Program Development Fund, supported by a surcharge on pay-TV fees, was created in 1983 to help the CBC "Canadianize" its shows. Moreover, in cable TV priority is given to the carriage of Canadian services on basic channels (the ones viewed without the need of a converter.) Nonetheless, when it comes to popularity, particularly among English-speaking viewers, American TV shows predominate. One survey in 1991 showed that the top two English language TV shows were "America's Funniest Home Videos" and "America's Funniest People." French TV is more Canadian in origin. A similar survey found that only two foreign shows were in the top 15, neither of them American.

Canada is not alone in having legislation to protect its cultural heritage (Australia and several European countries have similar rules) but few countries seem to devote as much time and energy defending national identity. Although much of the justification for quotas is based on an appeal for cultural autonomy, it should be noted that there are economic reasons as well. Broadcasting in Canada is a multi-billion-dollar industry, but it can ill afford to lose substantial advertising and production income to foreign countries.

Programming Since Canada is a multilingual nation, it must provide radio and TV programs in English, French, and other native languages. The CBC administers both an English language all-news radio network and a French language network that links most of the French language stations in Montreal, Quebec, and Ottawa. In addition, the CBC provides a native language service to residents in the far northern provinces. Network radio in Canada generally follows an eclectic approach, emphasizing music, news, and features. Private radio stations generally feature a format that emphasizes popular music and news. The CBC also operates an English and a French TV network. There are also several privately owned TV networks: CTV, an English net; TVA, broadcasting in French; Global Television, another English service; the Atlantic Satellite network, a regional satellite to cable service in Atlantic Canada; and Quatre Saisons, a private French language service started in late 1986.

Looked at in its entirety, Canadian TV tends to be entertainment oriented, with about 75 percent of its programs in this category. News and public affairs account for about 14 percent, with sports programming (particularly hockey) making up the remainder. The private TV system tends to be a bit more entertainment-centered than the CBC. In a typical year CBC programs average about 55 percent entertainment, 25 percent news and public affairs, and 20 percent sports.

Canadian-produced TV drama demonstrates high quality. In the past several years Canadian productions such as "Anne of Avonlea" have captured large segments of the viewing audience. The CBC National News is also popular. Nonetheless, a substantial part of both the CBC's and the private networks' schedules is American made.

Kenya

The Republic of Kenya is located on the east coast of Africa. About the same size as the state of Texas, Kenya has about 20 million people who speak either Swahili or English. The literacy rate averages about 50 percent, higher in the urban areas, particularly around Nairobi, the capital, and lower in the rural areas. Kenya was a British possession until it achieved its independence in 1963. It is a republic, although in recent years Kenya has become a rigidly controlled, one-party nation and government power over the press has increased.

Kenya belongs to that set of nations typically described as the third world. (A third world, or developing, country is one that is changing from an agrarian economy and colonial rule to an industrial economy and sovereignty.) A majority of countries in the world belong to the third world. As in many of these countries, the critical issue for Kenya is development. Indeed, the stress of political, economic, educational, and cultural development has strongly influenced its broadcasting system.

Radio began in Kenya in 1927 when the British East African Broadcasting Company relayed BBC broadcasts to British settlers in the area. In 1930 this company was renamed the Cable and Wireless Company and was given a monopoly on broadcasting for the East African Colony. In 1959, aided by a grant from the British government, this station was replaced by the Kenyan Broadcasting Corporation (KBC). Kenya received its independence in 1963 and the new government dissolved the KBC and turned it over to the Ministry of Information and Broadcasting, which renamed it the Voice of Kenya. Today the Voice of Kenya offers radio programs in Swahili and English as well as in about a dozen local languages.

Television began in 1963 with a station in Nairobi. A second station went on the air in 1970 in the city of Mombassa. Most of the audience then as today tended to live in the urban areas of Kenya. In 1991 it was estimated that about 2 million Kenyans, about 10 percent of the population, are regular viewers of TV. About 90 percent of these regular viewers live in the Nairobi or Mombassa area. The government-run Kenyan Broadcasting Service enjoyed a monopoly in TV until 1990 when the Kenyan Television Network, a commercially sponsored network owned by the late Robert Maxwell, started broadcasting from Nairobi. VCRs are not prevalent in Kenya, with only 20,000 in the whole country.

Structure   In Kenya, as in other third world countries, radio is the primary means of mass communication. Many citizens, particularly in the rural areas, cannot read, limiting the impact of newspapers. As we have mentioned, TV is generally confined to the urban areas of the country. Radio is inexpensive and portable; it can run on batteries and can be understood by all residents.

Consequently, in Kenya, radio broadcasting is the primary means of mass communication.

As currently organized, the Voice of Kenya, controlled by the Ministry of Information and Broadcasting, provides both English and Swahili programming. It also provides an educational service for schools and uses shortwave radio to reach people who live in distant villages. It has no international shortwave service. From its origin the Voice of Kenya was charged with aiding in the task of national development. Its goal has been defined as "educating, informing, and entertaining Kenyans."

The TV system consists of one state-run network and one commercial network. Like radio's, the stated goal of the state-run TV network is to encourage national development. The commercial network puts more emphasis on entertainment.

Economic support The state-run Kenyan radio and TV services are supported by grants from the Kenyan government. Both the TV and radio networks are permitted to accept advertising but the revenue they collect is then turned over to the treasury. In theory, this money (and more) is returned to the stations when the government enacts its annual budget for broadcasting. A weak economy has caused problems for Kenyan broadcasting, and both TV and radio services have recently complained that they are underfunded. The commercial TV service, as its name implies, is supported entirely by revenue raised from selling advertising time.

Law and regulation A survey made in the late 1980s rated the media in Kenya as having a moderate degree of freedom. The government-controlled radio and TV services, however, are more tightly controlled than the print media and censorship is common. News programs are scrutinized by the Ministry of Information and Broadcasting for stories that might reflect unfavorably on the government. In the early 1990s, for example, when human rights protests broke out in Nairobi, the government stations did not report them. Intimidation is also utilized. Reporters who tried to report on police brutality were themselves beaten up. News from external sources is monitored. For example, the government TV channel carries CNN but blacks out any mention of Kenya in CNN stories. Conversely, stories about the activities of the Kenyan president and the Kenyan Parliament are common features on the state-run channel.

Somewhat surprisingly, the state has taken a hands-off approach to the new commercial TV station, KTN. It has carried, without consequence, unflattering reports of the government that are not seen on the state-run channel. A CNN correspondent called KTN the "freest in sub-Saharan Africa." Part of the explanation for the government's reluctance to crack down on the station might be the fact that it reaches only 50,000 or so viewers in Nairobi. Nonetheless, KTN is significant for its uncensored reports.

Programming Radio programs are geared for national development. Typical programs include literacy instruction, farming guidance, health education, and technological training. Television, centered in the urban areas, tends to be more news and entertainment oriented. The most popular shows on Kenyan TV as of 1991 were "Trickery," a sitcom; "Family Affairs," a drama show; "Sing and Shine," a program of gospel music; and sports, particularly soccer and American wrestling. American situation comedies are also popular. Like many of its neighbors, Kenya has rules concerning the amount of foreign material that can be shown. Despite the entertainment emphasis, the goal of national development is still present on the state-run channel. One program in Swahili, loosely translated as "What Has Happened," builds a sense of nationalism by focusing on the achievement of the country and its leaders. Another program, "Rural Development," as the name suggests, focuses on farming activities.

NEW WORLD INFORMATION ORDER

The Kenyan system provides a natural springboard for consideration of what's been called the "new world information order." Countless hours have been spent debating this topic and dozens of books have been written about it. (In fact, some critics would say that it has been overdebated.) In oversimplified terms, the new world information order is a synthesis of several issues that have been discussed for many years in forums spon-

International communication satellites now make it possible for world leaders such as Boris Yeltsin and Mikhail Gorbachev to be interviewed live by American journalists. This is a scene from ABC's "Town Meeting" hosted by Peter Jennings.

sored by the United Nations, particularly the United Nations Educational, Scientific and Cultural Organization (UNESCO). The spokespersons for third world and many Communist countries offer the following argument:

1 The United States and Western European countries dominate TV programming and control the flow of news between developed and developing countries.
2 This domination leads to a form of cultural imperialism reminiscent of colonial times.
3 To remedy this situation, the developing nations want more balanced news coverage, restrictions on the amount of information that comes from the industrialized countries, and a code that would regulate journalists.

Those in the developed nations argue that the flow of information from one country to another should be free and open and journalists should not be regulated by any codes or guidelines. In sum, the debate boils down to whether the media should be government controlled and used to promote national goals or privately owned and free to pursue their own interests. The debate between the "free-flow" and new world camps became so heated that it was one of the reasons the United States withdrew from UNESCO in 1985.

This issue promises to become even more volatile with the introduction of DBS. Most developing countries and some Western nations are afraid that unwanted political messages, advertisements, and foreign values might result from unregulated DBS TV transmissions. Accordingly, a resolution was introduced to the United Nations that any country desirous of using DBS to reach another country must first obtain the prior consent of the target country. The United States, noting that no comparable regulations existed concerning radio, opposed the motion as a violation of the free-flow concept. The United States was outvoted 102 to 1.

Over the past few years the free-flow/new-world debate has become entangled in political rivalry. And, like most political arguments, it's unlikely to be settled quickly. Nonetheless, there are some hopeful signs. The United States has promoted plans to channel millions of dollars into developing countries to help them strengthen and enlarge their mass media. The International Telecommunications Union (see Chapter 17) has established the Independent Commission for Worldwide Telecommunication Development. The idea behind this group is to stimulate Western investment in and trade with the third world and to promote open markets in some areas of mass communication. Perhaps as the gap between the developed and developing countries narrows, so will the differences over information flow.

DIFFERENCES IN ELECTRONIC MEDIA SYSTEMS: A MODEL

As this brief examination has indicated, the arrangement of a country's media depends on the nation's political philosophy, social history, and economic system. To make sense of the variations that exist among countries, several writers have classified electronic media systems according to four categories based on the system's (1) ownership, (2) control, (3) financing, and (4) programming. Figure 8-4 displays this fourfold classification. Let's examine each of these dimensions and see where our example countries fit.

Looking first at ownership, we find that systems can vary along a continuum ranging from government ownership through public corporation ownership to private ownership, with, of course, some combination of methods. The dominant system in the United States, for example, would be found toward the private end of the

OWNERSHIP	
Government	Private
CONTROL	
Strong	Weak
FINANCING	
Government	Advertising
PROGRAMMING	
Ideological	Cultral enlightenment · Mass appeal

FIGURE 8-4 Typology of broadcasting systems.

spectrum, whereas the system in China and Kenya would fall on the government-owned end. The BBC comes closest to the public corporation model, and Canada's system falls somewhere between private ownership and public corporation. If we broadened our scope and examined the radio and TV systems of all countries, most would be government controlled. Private ownership, however, is growing in Western Europe and Latin America.

Control over a broadcasting system can range from strong, as seen in countries like China and Kenya, where government ownership is the model, to relatively weak, as in the United States, where the government licenses stations and generally leaves the day-to-day regulation to a federal commission. The British system, with its autonomous public corporation, also falls toward the weak side of the scale. The Canadian system, given its content restrictions, falls a little more toward the other side. The mode of control has probably changed most drastically in the United States, where there recently has been a noticeable trend toward deregulation (see Chapter 18).

A broadcasting system can be financed in several ways. The state can subsidize it through taxes or annual license fees can be collected or the system can be supported by advertising, or there can be some combination of these. A government-controlled system, such as that of China, generally supports its broadcasting entirely or almost entirely through subsidy. In the third world, where government ownership is also common, most broadcasting systems receive direct state support supplemented with additional revenue from advertising or license fees or both. Privately owned systems, like the U.S., British, and Canadian private networks, depend almost entirely on ad revenue. In fact, the popularity of

A production scene from Saudi Arabian TV.

advertising support is growing even in those countries where it traditionally has not been allowed on the government system.

The programming supplied by these systems can be classified along the following scale: ideological (typified by propaganda, information, and educational programs); cultural enlightenment (characterized by arts, information, education, and light entertainment programs); and mass appeal (marked by entertainment and commercially successful programs). The Kenyan system, with its emphasis on nation building, best exemplifies the ideological end of the spectrum. The programming in China falls somewhere between the ideological and cultural enlightenment divisions. The BBC and CBC fall somewhere between cultural enlightenment and mass appeal. To no one's surprise, the U.S. commercial system finds itself firmly at the mass appeal end of the spectrum. Public TV in the United States, however, would be shaded toward the cultural enlightenment category.

SUMMARY

In the early years of shortwave radio several countries used it for propaganda. The United States, on the other hand, did not broadcast internationally at that time.

In 1941 the United States entered World War II and began to broadcast internationally in an attempt to counter German propaganda. The United States also established the Voice of America. In the 1950s both Communist and anti-Communist messages were being sent over the airwaves.

Five notable conclusions can be made about the past thirty years of broadcasting: the number of international broadcasters is growing; propaganda is still important; there is growing interest in the United States and among private international broadcasters; TV is becoming an international medium; and new technologies have been developed.

The top five international broadcasters are the VOA, Radio Moscow, the BBC, Radio Beijing, and Deutsche Welle. There are also unauthorized stations on the air in the forms of clandestine and pirate stations.

The rest of this chapter compares the radio systems of Great Britain, China, Canada, and Kenya. This is accomplished by looking at the structures, economics, regulations, and programming of these countries. The comparison concludes that a country's system is dependent on political philosophy, social history, and economic structures. To make sense of these data, the analysis is further subdivided on the basis of ownership, control, finance, and programming of stations.

SUGGESTIONS FOR FURTHER READING

Bishop, R. (1989). *Qi Lai: The Chinese communication system.* Ames, Iowa: Iowa State University Press.

Boyd, D., Straubhaar, J., & Lent, J. (1989). *Videocassette recorders in the third world.* New York: Longman.

British Broadcasting Corporation. (1990). *Annual report and handbook.* London: British Broadcasting Corporation.

Browne, D. (1982). *International radio broadcasting.* New York: Praeger.

Canadian Radio Television Commission (1986). *Annual report: 1985–1986.* Ontario, Canada: Canadian Radio Television Commission Information Service.

Chang, W. (1989). *Mass media in China.* Ames, Iowa: Iowa State University Press.

Gertner, R. (1987). *International television and video almanac.* New York: Quigley Publishing Co.

Head, S. (1985). *World broadcasting.* Belmont, Calif.: Wadsworth.

Howell, W. J. (1986). *World broadcasting in the age of the satellite.* Norwood, N.J.: Ablex.

Laurien, A. (1988). *The Voice of America.* Norwood, N.J.: Ablex Publishing.

McPhail, T. (1987). *Electronic colonialism.* New York: Sage.

Merill, J. (1991). *Global journalism.* New York: Longman.

Mickiewicz, E. (1988). *Split signals: TV and politics in the Soviet Union.* New York: Oxford University Press.

Mowlana, H. (1986). *Global information and world communication.* New York: Longman.

Passport to World Band Radio. (1990). London: Internationl Broadcasting Services.

Rosen, P. (1988). *International handbook of broadcasting systems.* Westport, Conn.: Greenwood Press.

Soley, L., & Nichols, J. (1987). *Clandestine radio broadcasting*. New York: Praeger.

Stevenson, R., & Shaw, D. (1984). *Foreign news and the new world information order*. Ames, Iowa: Iowa State University Press.

Wedell, G. (1986). *Making broadcasting useful: The African experience*. Manchester, United Kingdom: Manchester University Press.

World Radio-TV handbook. (1990). New York: Billboard Publications.

PART FOUR

HOW IT'S DONE

CHAPTER 9: RADIO PROGRAMMING

To unlock the secret of radio programming today, it's helpful to turn to the biological sciences. Biologists define symbiosis as "the living together in intimate association or close union of two organisms," especially if mutually beneficial—like silverfish and army ants or coral and sea creatures, for example.

Symbiosis is also an especially good term to use to describe radio programming today. Radio enjoys a close and mutually beneficial relationship with a variety of other "organisms." The popularity of music from feature films (from *Saturday Night Fever* in 1975 to *Dirty Dancing* in 1987 to *Beauty and the Beast* in 1991) illustrates how radio is intertwined today with the movie business. The rise of MTV and its host of imitators (like NBC's "Friday Night Videos" and "Night Tracks" on the TBS Network) point to radio's interrelationship with TV, especially cable. But radio's most symbiotic relationship is with the popular music business, the world of records, tapes, and discs. This chapter on how radio is programmed will focus on that key interrelationship: how did it come about and how does it work today? Since radio is more than an electronic jukebox, we will also examine the dynamics of information programming on radio: the nature of "news radio" and "talk radio."

RADIO REGULATION AND FORMAT DESIGN

For this symbiotic relationship to work, it's necessary for radio stations to have the freedom to choose the programming they want to provide to their communities. Section 326 of the Communications Act, the law that empowered the FCC to govern broadcast operations, states:

Nothing in this Act shall be understood or construed to give the Commission the power of censorship over the radio communications or signals transmitted by any radio station, and no regulation or condition shall be promulgated or fixed by the Commission which shall interfere with the right of free speech by means of radio communication.

In short, the FCC has neither the right nor the power to control radio programming. Radio stations are free to program their air time however they may. In certain specific areas, such as political advertising, obscenity, and indecency, the FCC has introduced legislation regarding programming. However, the bulk of radio programming—music, news, and information—is largely free of governmental intrusion. In fact, this characteristic is one of the fundamental distinctions between the sound of American radio and that of the rest of the world, which was described in Chapter 8. Basically, American radio is programmed to satisfy listener tastes and not, as in government-owned systems, to serve political or bureaucratic interests.

We call this situation "format freedom." Faced with the task of filling twenty or more hours per day, 365 days a year, radio programmers are "on their own." Their task is simple: to provide attractive programming to meet the informational and/or entertainment needs of an audience. In commercial radio, the audience must be large or important enough to be of interest to advertisers. Public stations must entertain and inform their listeners to an extent that justifies financial support from government agencies, foundations, business underwriters, and the listeners themselves. If the task seems especially formidable, at

Cross media symbiosis was vividly demonstrated by the success of the movie *Ghost*. The movie spawned a hit album which sold more than a million copies and a single—"Unchained Melody"—which became a hit nearly thirty years after its original release.

least the programmers don't have to worry about direct governmental intrusion. (Maybe life is easier for the program director of Radio Libya.)

A MATRIX OF RADIO PROGRAMMING

Types of radio programming today are mapped in Figure 9-1.

Across the top of the matrix of radio programming are the sources of radio programming. *Local programming* is original programming produced by the radio station in its station or from locations in its immediate service area. *Prerecorded* or *syndicated programming* is programming obtained by the station from a commercial supplier, advertiser, or program producer from outside the station. The most common sources of programming of this type are records, tapes, and compact discs. Prerecorded programs may also be received by stations through telephone lines, by microwave relay, or by satellite. Stations that belong to a network, such as Mutual, CBS, or National Public Radio, are permanently interconnected, usually by telephone lines or satellite transponders. Unlike syndication, *network programming* is regularly scheduled, that is, with few exceptions, network programs run the same time each day at every station on the network.

Top-to-bottom in Figure 9-1 are the two main types of radio programming. Most plentiful in radio today is *music programming,* from opera to country, from "beautiful music" to progressive jazz. *News/talk* covers the broad spectrum from news and traffic reports to sexual advice, from stand-up comedy to stock tips.

Now, let's examine the kinds of radio programs that fall into each box. In box 1 is locally produced music programming. Once a staple of radio programming, when many stations employed their own orchestras, original music emanating from studios or area concert halls is heard today on only a few stations (mostly noncommercial). Some rock stations have had success promoting the music of local bands. For example, "America's best rock and roll band" (according to *Rolling Stone*), R.E.M., owes much of its success to widespread airplay in Athens, Georgia, the group's hometown. Some programs heard nationally today began as local productions. American Public Radio's "Prairie Home Companion," hosted by affable Garrison Keillor, started in this fashion. However, locally produced music is becoming increasingly rare and is thus the smallest segment of the matrix.

By far the biggest element of radio programming today is box 2, prerecorded and syndicated music. Nearly nine of ten radio stations rely on some kind of music as the backbone of their schedule, and that music is most likely coming from a record, a CD, a tape recorder, or a satellite transponder. This is the high-intensity world of format radio, presented briefly in Chapter 4 and described in detail later.

Pronounced dead and buried by industry observers just a few years ago, network music programming (box 3) has undergone a renaissance in recent years. Joining the long-running orchestral and opera broadcasts (such as the Texaco-sponsored opera and the New York Philharmonic, Philadelphia Orchestra, and Chicago Symphony broadcasts) have been the live broadcasts of Westwood One, the Los Angeles–based radio network. Rock and roll music has been the network's strong suit, featuring live national broadcasts of concerts by the Rolling Stones, Eric Clapton, Genesis, and Billy Joel, among others.

On the talk radio side, the bulk of the programming is produced locally (box 4). Indeed, as examined in Chapter 4, news and talk stations have the largest staffs in radio today, including reporters, writers, editors, producers, and a host of technicians.

A fast-growing area of radio programming in the 1990s fills boxes 5 and 6. On the syndication side (box 5), popular talk radio personalities include Dr. Ruth Westheimer, Sally Jessy Raphael, and conservative standard-bearer Rush Limbaugh. Box 6 is illustrated by the one-man radio network Paul Harvey, whose unique commentaries are heard on over 200 stations. Mutual's late-night host Larry King is also on more than 200 stations. Former TV newspersons now on network radio include Tom Snyder (whose eve-

FIGURE 9-1 Types of radio programming.

	Source		
Type	Local programming	Prerecorded/syndicated	Network
Music	1	2	3
News/talk	4	5	6

ning program is immodestly called "The Radio Show") and Deborah Norville, who hosts a daily radio talk program from her New York apartment.

MODES OF RADIO PRODUCTION

Just as radio programmers have a full menu of types and sources of programming, they likewise have a range of ways to produce those programs for their audiences. This is just one example of the many decisions that have to be made by radio managers. We call these choices modes of radio production. The various modes are depicted in Figure 9-2.

At the left end of the continuum is *local live production*. When radio stations employ their own announcers or newscasters locally and play records and tapes that they themselves own, they are using this mode of production.

Live-assist production occurs when radio managers use syndicated programming, such as reels of taped prerecorded music and satellite-delivered music services, but retain local announcers and DJs as the backbone of their program schedule. In this case the live air personality assists in the implementation of the syndicated schedule, hence live-assist.

Semiautomation refers to the reliance of the local radio station on the services of the syndicated program producer. The music is typically played on large tape machines. When a break point for a commercial or program announcement is reached, smaller cartridge tape machines are triggered to play by a subaudible cue tone on the master tape. In a semiautomated system the station occasionally inserts live personalities, perhaps in morning entertainment programs or for news and local weather breaks. But the backbone of the programming is the syndicated music schedule.

At the far end of the radio production continuum is *turnkey automation*. This refers to fully automated radio stations that take one of two main forms. Some automated stations consist largely of a satellite dish and a control board. The satellite dish downlinks a radio program service, such as country, rock, or beautiful music. In some cases the service has been so localized that time, weather, and news information is telephoned to the program producer in time for the announcers thousands of miles away to prepare the inserts.

Other turnkey automation systems rely on tape machines interfaced with a computer console. The program director uses a computer program to prepare the logs, which plan all the music, information, and commercial elements, including their order, length, and frequency. Once the manager has approved the logs, the same or another computer at the station controls the program schedule, selecting the music commercials, and cartridge tapes with news and weather from bins of prerecorded tapes, discs, and other recording media.

New developments in radio automation in the 1990s rely on the enhanced audio processing capabilities of today's computers. It is now technically possible to have a music radio station without a single record, tape, or compact disc. All voice and music elements can be stored in digital computer format and played on demand from the desktop computer of the general manager or program director.

The task facing radio program managers is formidable. They must decide whether to emphasize talk or music or strive to operate a full-service station. Having made that decision, they must determine where the programming should come from. Will it all emanate locally? Should the station acquire a music library? If not, which program sources should be used? Should the station purchase a satellite dish? For what services? Should it affiliate with a network? If so, which one(s)? And to what degree of commitment? In addition, the programmer must decide on how the programs will be produced for the audience. Will format freedom reign, giving local control to program directors and disc jockeys? Will outside consultants make many program decisions, with the goal to assist local announcers and personalities? Will the station be programmed from afar via satellite, with only occasional local break-ins? Or, will the station be essentially a radio music box, completely automated and controlled by management personnel and their desktop computers?

FIGURE 9-2 Modes of radio production.

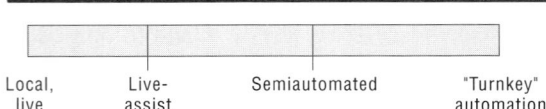

THE "FORMAT HOLE"

Whether placing a new station on the air, acquiring an established station, or reevaluating the programming of an existing station, radio broadcasters have two main choices about how to fill their air time. They may try to outdo the competition in a given service by programming the same format. Most of the nation's largest markets boast at least two of radio's major formats, from news, to contemporary hit radio, to country and rock. The station may decide instead to inventory and analyze radio programming in the market in search of an unfulfilled programming need. Managers may decide to try all-sports, classic rock, progressive jazz, or other formats in areas where these services were previously not heard.

Whatever the choice, the secret to successful radio programming is to carve a unique niche, one that will deliver a large enough audience to attract advertising revenues to that station. In the radio business the phrase used is "find the format hole." The process begins with a competitive market analysis, which examines the existing stations, including their technical properties, ownership, financial performance, their ratings, and, of course, their current formats. For example, let's take a look at radio in 1991 in the nation's largest radio market, New York City (Figure 9-3).

By 1991 there were nearly fifty stations available to listeners in New York, competing for the listening attention of about 15 million people. Simple arithmetic would indicate that if listening were distributed equally, each station would attract about 300,000 listeners. In actuality, the ten top stations attracted at least three times that number in the average week, with the top-ranked station (WCBS-FM, an oldies station) garnering about five times the average for a weekly circulation topping 1.4 million people.

A check of formats reveals almost as many program formats as stations available. In the spring of 1991 the New York listener could choose from about fifteen different formats. There were several adult contemporary stations, six news/talk operations (two all-news, one all-

FIGURE 9-3 Local radio in New York City, 1991. (Source: *Duncan's American Radio*, Spring, 1991. Used with permission.)

Radio Stations		Owners	Format
AM Stations			
WABC,	770,	Cap Cities/ABC	Talk
WADO,	1280,	Tichenor	Hispanic
WCBS,	880,	CBS	News
WEVD,	1050,		Big Band/Ethnic
WFAN,	660,	Infinity	Sports/Talk
WINS,	1010,	Westinghouse	News
WKDM,	1380,	United	Hispanic
WLIB,	1190,	Inner City	Black News/Talk
WMCA,	570,	Saleva	Religion
WNEW,	1130,	Westwood One	Nostalgia/MOR
WOR,	710,	Buckley	Talk
WPAT,	930,	Park	EZ
WQXR,	1560,	New York Times	Classical
WSKQ,	620,	SBS	Hispanic
WWDJ,	970,		Religion
WWRL,	1600,		Gospel
WZRC,	1480,		AOR
FM Stations			
WBLS,	107.1,	Inner City	Black
WCBS,	101.1,	CBS	Oldies
WHTZ,	100.3,	Malrite	CHR
WFME,	94.7,	Family	Religion
WHUD,	100.7,		Btfl/Ez
WLTW,	106.7,	Viacom	Soft AC
WNCN,	104.3,	GAF	Classical
WNEW,	102.7,	Westinghouse	AOR
WNSR,	105.1,	Bonneville	AC
WPAT,	93.1,	Park	Btfl/EZ
WPLJ,	95.5,	Cap Cities/ABC	CHR
WQCD,	101.9,	Tribune Co.	Jazz
WQHT,	97.1,	Emmis	CHR/Dance
WQXR,	96.3,	New York Times	Classical
WRKS,	98.7,	Summit	Black
WSKQ,	97.9,	SBS	Hispanic
WXRK,	92.3,	Infinity	Classic AOR
WYNY,	103.5,	Westwood one	Country

Average Quarter-Hour Persons/Share of Top 25 Radio Stations

12+ METRO		1/4/SHARE
1. WCBS-F	(O)	1418/ 5.2
2. WRKS-F	(B)	1373/ 5.1
3. WPAT-AF	(EZ)	1329/ 4.9
4. WLTW-F	(AC)	1301/ 4.8
5. WOR	(T)	1237/ 4.6
6. WINS	(N)	1158/ 4.3
7. WNSR-F	(AC)	1133/ 4.2
8. WQHT-F	(CHR)	1093/ 4.0
9. WBLS-F	(B)	1046/ 3.8
10. WHTZ-F	(CHR)	972/ 3.6
11. WCBS	(N)	936/ 3.4
12. WXRK-F	(CL AOR)	908/ 3.3
13. WYNY-F	(C)	902/ 3.3
14. WNEW-F	(AOR)	850/ 3.1
15. WFAN	(T)	793/ 2.9
16. WABC	(T)	723/ 2.7
17. WQCD-F	(J/NA)	662/ 2.4
18. WPLJ-F	(CHR)	602/ 2.2
19. WNEW	(BB)	578/ 2.1
20. WSKQ-F	(SP)	507/ 1.9
21. WADO	(SP)	467/ 1.8
22. WQXR-AF	(CL)	445/ 1.7
23. WLIB	(B/T)	443/ 1.6
24. WNCN-F	(CL)	366/ 1.3
25. WSKQ	(SP)	346/ 1.3

PROGRAMMING

RADIO'S DIRTY TRICKS

Radio program directors go to great lengths to get an edge on competing stations. Some of their efforts approach the boundary between fair competition and outright unethical (and possibly illegal) behavior. A recent article in a trade publication listed some "dirty tricks" employed by PDs:

- Finding a job in another city for a highly rated DJ at a rival station.
- Sending congratulatory telegrams to air personalities at a rival station when their ratings have declined.
- Searching the trash at a rival station for copies of their confidential programming and management memos.
- If the trash is empty, paying the graveyard announcer at a rival station to "borrow" confidential programming information from the day shift.
- Crashing a rival station's concerts and promotional events, handing out bogus tickets and fake backstage passes, busing in residents of local homeless shelters, and generally wreaking havoc.

---

sports), five black (urban contemporary) stations, and at least four stations programming beautiful music, Spanish, and various kinds of album and classic rock. In supersophisticated New York, at least one station was full-time country (WYNY). With all of this, where's the format hole?

One could decide to go "head to head" in one of the more popular formats, like adult contemporary or beautiful music. This would require a large bank balance, since the station would have to spend a small fortune to attract listeners with expensive advertising campaigns and special promotions, such as commercial-free weekends and merchandise giveaways. For this reason the major formats tend to be programmed by big, "deep-pocket" radio station owners like Malrite, Viacom, Tribune, and Bonneville.

The other option would be to try to devise something new for the market. This was the strategy behind Emmis Broadcasting's WFAN, a full-time sports channel. Emmis strategy apparently worked. In 1988 Emmis purchased the five NBC owned-and-operated stations for $122 million. Three years later they sold all-sports WFAN (formerly WNBC) to Infinity Broadcasting for $70 million, believed to be the highest price ever paid for a stand-alone AM station.

The competitive programming environment is not limited to the nation's largest markets. The search for the format hole extends to the nation's smaller cities and towns. For example, examine the situation in Wichita, Kansas (Figure 9-4). With a population approaching half a million,

Radio Stations	Owners	Format
KFDI, 1070,	Great Empire	Country
KFH, 1330,	Midcontinent	Oldies
KNSS, 1240,		Talk
KQAM, 1410,		Oldies
KBUZ-F, 99.1		Black
KFDI-F, 101.3,	Great Empire	Country
KICT-F, 95.1,	Lakoduk	AOR
KKRD-F, 107.3,		CHR
KOEZ-F, 92.3,		EZ
KEYN-F, 103.7,		Oldies
KRBB-F, 97.9,		AC
KRZZ-F, 96.3,		Classic AOR
KYQQ-F, 106.5,		CHR
KXLK-F, 105.3,	Midcontinent	AC
KZSN-F, 102.1,	Southern Skies	Country

Average Quarter-Hour Persons/Share of Top-10 Radio Stations

12+ METRO	1/4/SHARE
1. KFDI-F	102/ 15.5
2. KRBB-F	60/ 9.1
3. KICT-F	59/ 9.0
4. KFDI	52/ 7.9
5. KZSN-F	48/ 7.3
KEYN-F	48/ 7.3
7. KKRD-F	44/ 6.7
8. KDEZ-F	32/ 4.9
9. KRZZ-F	30/ 4.6
10. KNSS	22/ 3.3

FIGURE 9-4 Radio programming environment—Wichita, Kansas, 1991. (Source: *Duncan's American Radio*, Spring, 1991. Used with permission.)

Wichita boasts almost twenty stations competing for a share of the listening audience.

As might be expected, country music leads the pack, with three stations in the format (including the leading station, KFDI-FM). In addition, in Wichita there are three oldies stations (two on AM, one on FM), two adult contemporary stations, an album rock station, a classic rock station, a black station, and one full-time talk station.

Where's the format hole? Perhaps there's room for something new in Wichita, like classical or all-news. What would you do?

Of all the options facing radio management regarding programming, clearly the primary one is selecting the format. Today more than eight of ten radio stations overall and nine of ten on FM choose some form of music as the backbone of their programming.

For this reason we now turn to the crucial interrelationship between radio and popular music, the symbiosis between the radio and recording businesses.

RADIO AND THE MUSIC BUSINESS

Since its beginnings in the early part of this century, the radio business has been intertwined with the musical field. The leading figures in classical and popular music, from Enrico Caruso to Eddie Cantor, were frequent performers on radio. Many, like Kate Smith, Rudy Vallee, and Bing Crosby had their own shows. Although for many years radio stations were prohibited from playing recorded music (and the quality of the recordings was usually unfit for broadcasting anyway), radio appearances contributed immeasurably to record sales.

Unquestionably, however, the dynamic interplay between radio and the music business that exists today began in the late 1940s, in a small bar in Omaha. One of the patrons, Todd Storz, noticed how night after night people would amble up to the jukebox and select the same tunes. The jukebox had room for more than forty records, but he noticed that only a few (less than half a dozen) kept getting played. Why couldn't radio do the same thing? With help from dad, Storz bought KOHW, implemented his "top-40" idea in major markets, and became highly successful. Another pioneer, Gordon McLendon, further refined the top-40 format and introduced a heavy emphasis on station promotion. The rest, as they say, is rock 'n' roll history.

Fueled by the success of top-40 radio, since more airplay led to greater record sales, the recording business grew by more than 20 percent per year for twenty-five years! In the late 1970s the business peaked: the BeeGees' soundtrack album for *Saturday Night Fever* and "Rumors," an album by "supergroup" Fleetwood Mac, sold more than 20 million copies each.

IMPACT

"DON'T BELIEVE THE HYPE"

So said rap group Public Enemy a few years ago. It's getting harder and harder to avoid the hype, even if you don't believe it. Record companies are going to unprecedented promotional extremes to bring their new music to the attention of radio programmers, distributors, and consumers. Here are a few recent examples:

- Metallica booked Madison Square Garden in New York City, not for a concert, but to play its new record to its fans before releasing the record.
- Record stores opened up at midnight to begin selling two new Guns n' Roses albums.
- Prince held private, invitation-only concerts for media and radio executives to promote his "Diamonds and Pearls" record.
- Jermaine Jackson released a song critical of her brother ("Word to the Badd"), which mysteriously appeared at radio stations shortly before Michael's new album ("Dangerous") mysteriously began receiving airplay in advance of its official release date.
- Singer Richard Marx performed half-hour concerts on the same day in Baltimore, New York, Cleveland, Chicago, and Los Angeles, the so-called "Rush-in, Rush-out, Rush Street Tour."

Like the radio business, the recording industry suffered hard times in the late 1970s and early 1980s, with sales dipping nearly 10 percent by 1983. But by the late 1980s the magic was back. MTV had arrived and introduced another example of cross-media symbiosis. Top-40 radio stations were reluctant to feature a lot of new releases by artists without a track record. MTV, however, was willing to take a chance. Radio stations waited until the reaction to possible new additions to their playlists appeared on MTV. If the new releases were well received, the stations added them to their lists. In effect, MTV was a testing ground for some radio stations. The record industry benefited from the additional exposure on TV and from the subsequent radio airplay.

Like other sectors of the economy, record sales suffered from the recession of the early 1990s. However, by 1992 a turnaround was seen, fueled by new releases by Michael Jackson, Guns n' Roses, and Bruce Springsteen.

The interplay between radio and records becomes immediately apparent through an examination of Figure 9-5. As the figure illustrates, there is almost a direct overlap between the popularity of the major radio formats and the sales of popular recordings, an obvious example of the symbiosis mentioned at the beginning of the chapter.

All told, the various rock formats, including adult contemporary, contemporary hit radio, and album rock, comprise just over 40 percent of the radio formats listened to today. This coincides with record sales, of which 36 percent are in the rock vein. Country music stations account for about one in every ten listening hours; country records account for about one of every nine records sold. Similar parallels between format listening and record sales are noted in black/soul music (11 percent of listening; 13 percent of sales), beautiful music (9 percent of listening; 14 percent of sales), classical and jazz (each with about 2 percent of listening and 4 percent of sales).

Further corroboration of the close relationship between radio listening and record sales has been provided by industry research, such as that conducted by CBS and Time Warner. In a recent study Time Warner found that more than one half of all record sales and nearly two-thirds of industry revenue were produced by people who listened to at least ten hours per week of radio.

On the surface, then, it might appear that programming a music station is a simple matter: contact area record stores for the lists of the most popular records and play them in approximately the same percentage on one's radio sta-

FIGURE 9-5 Radio formats and record purchases. (Sources: *Duncan's American Radio,* Spring, 1991; *Music Business Handbook,* 1985.)

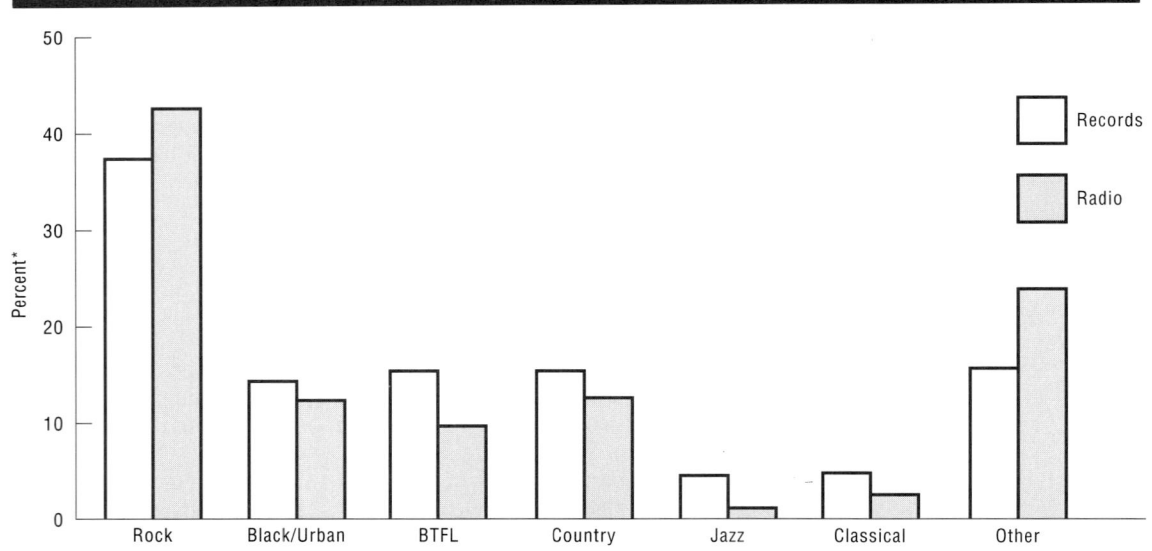

*Records: Percent of sales for each format (1984)
Radio: Percent of listenership to each format (1991)

tion. But, of course, the task is much more complex. Obviously, the people who buy country records are different from those who buy rock and roll. And some country fans of Hank Williams, Jr., wouldn't be caught dead with a Garth Brooks album. There are other problems. Repeatedly playing the same songs causes listener fatigue. Did you ever hear a tune so often you became sick of it? Radio programmers call this "burnout." And unlike records, radio stations need commercials to survive economically. Listeners don't like to have their music interrupted, so how often do you stop the music? And for how long?

We thus enter the sophisticated and semiscientific world of radio music formatting. The process of radio programming is equal parts instinct and intelligence, research and random thought. Its standard equipment includes a personal computer and a Ouija board!

Radio Music Formatting

The process of developing a radio music format is detailed in Figure 9-6. Of course, just as no two radio stations are exactly alike, all stations do not follow the identical order of steps illustrated in the

FIGURE 9-6 Developing the music format.

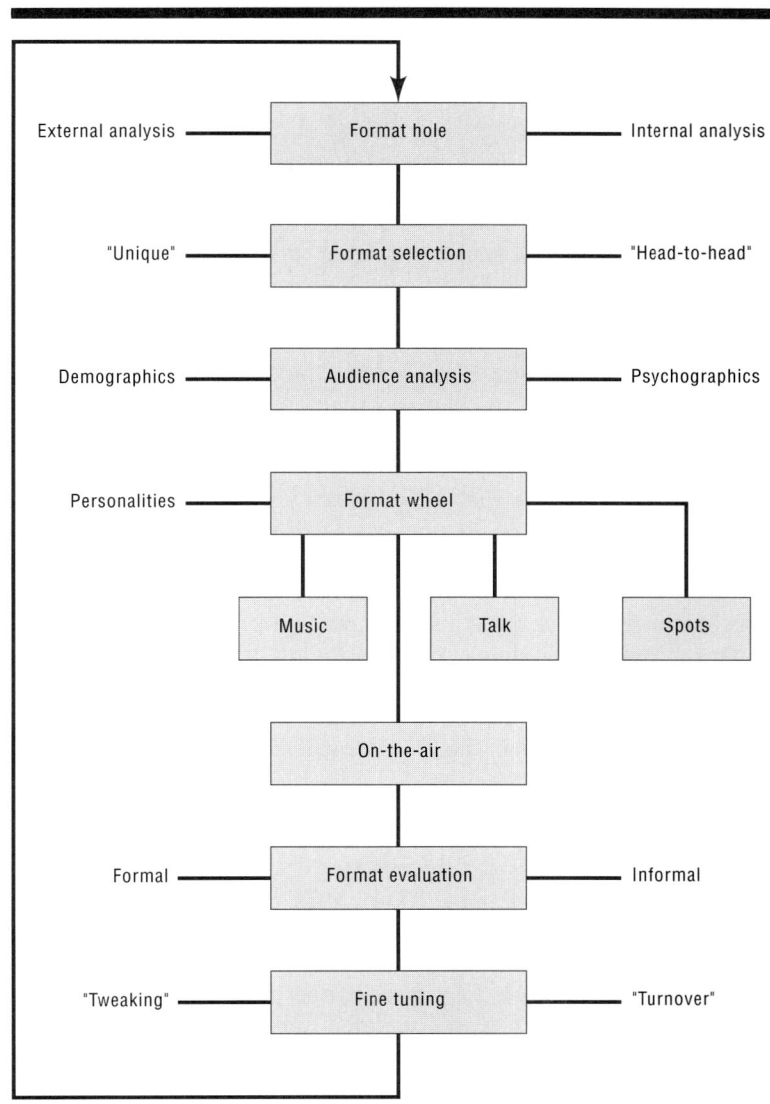

diagram. However, at some point and to some degree, every station addresses the questions raised in each box.

As we have seen, the process begins by finding the format hole. This process is dependent on two sets of factors: internal and external.

Internal factors affecting the analysis include the ownership of the station, its dial location, power, technical facilities, and management philosophy. For example, it would not make much sense for a station at 99.5 on the dial to play beautiful music if a competitor is already playing this format at 101 FM. It would not be wise for a 5000-watt AM station at 1570 on the dial to compete with a full-service news/talk format against a 50,000-watt powerful station in that format at 720 on the dial. Some station groups specialize in one format (such as Great Empire, with eight stations, all country); others have a range of station types (like the Gannett Corporation with sixteen stations in eight formats).

External factors to consider in finding the format hole have been mentioned in our discussion of New York City and Wichita. Paramount among these is the number of stations, if any, programming the same or a similar format. If there are competitors, are they strong? Or do they have program weaknesses (bad musical selection, poor announcers, weak promotions, and so on) that make them vulnerable to competitive attack?

Other external factors include geography and population characteristics (country/western in Wichita, laid-back "new age" in Los Angeles, nostalgia for the aging population of West Palm Beach, for instance).

Recall that the search for the format hole yields one of two outcomes: the station selects a new or different format that is unavailable in the market or it decides to compete in the same format with one or more existing stations. The former choice makes the station unique; the latter we refer to as a head-to-head competitor.

Audience Analysis

The goal of radio programming is to attract and maintain an audience. So it makes sense that one of the primary steps in creating a music format is audience identification. In format radio, every station must ascertain its *target audience*, the primary group of people sought by the station's programming. As we traced briefly in Chapter 5, the target audience is usually defined by its principal demographics, specifically age and sex. Figure 9-7 tracks the format preferences for some important demographic groups.

Format preferences: Women As Figure 9-7 indicates, current popular hits are what most women seek in their radio listening. Among younger women (ages 18 to 24), top 40 is the most preferred format, with its emphasis on current, up-tempo hits. Younger women also have an ear for new modern or "progressive" music, likely to be played on an album-rock station. But they avoid radio stations playing classic rock, the tunes of "supergroups" of the 1960s and 1970s, from the Rolling Stones to Aerosmith.

Older women (if ages 25 to 34 can be considered old) prefer a softer edge to their radio stations. Among this age group, the preferred format is soft hits, a version of adult contemporary that emphasizes light, bouncy tunes, oldies, and lots of love songs. Notice how, as the demographic gets older, interest wanes in heavy metal, new or progressive music, and percussive urban sounds.

Format preferences: Men As might be expected, format preferences for men differ from those for women, even in the same age groups. Younger men (those who are 18 to 24 years old) like to "rock out," with mainstream rock (AOR), classic rock, heavy metal, and new music leading the pack. This group is less interested in urban music, top 40, and (perish the thought) soft hits.

But as men age, their format preferences mellow. Their tastes for the classics improve, and classic rock surpasses mainstream as the most preferred radio format. Their taste for heavy metal declines by almost two-thirds; their interest in soft hits triples. But neither group, ages 18 to 24 or 25 to 34, thinks much of top-40.

Of course, there are many individual exceptions to the demographic "rules" regarding sex and age. Many grandparents still worship the Grateful Dead. The success of the Suzuki method and other approaches to the study of classical music has made this type of music popular to countless preteens and teenagers. There are thousands of women who prefer album rock to adult contemporary, just as there are perhaps millions of men who hate "speed metal." This is why radio stations cannot rely on demographic

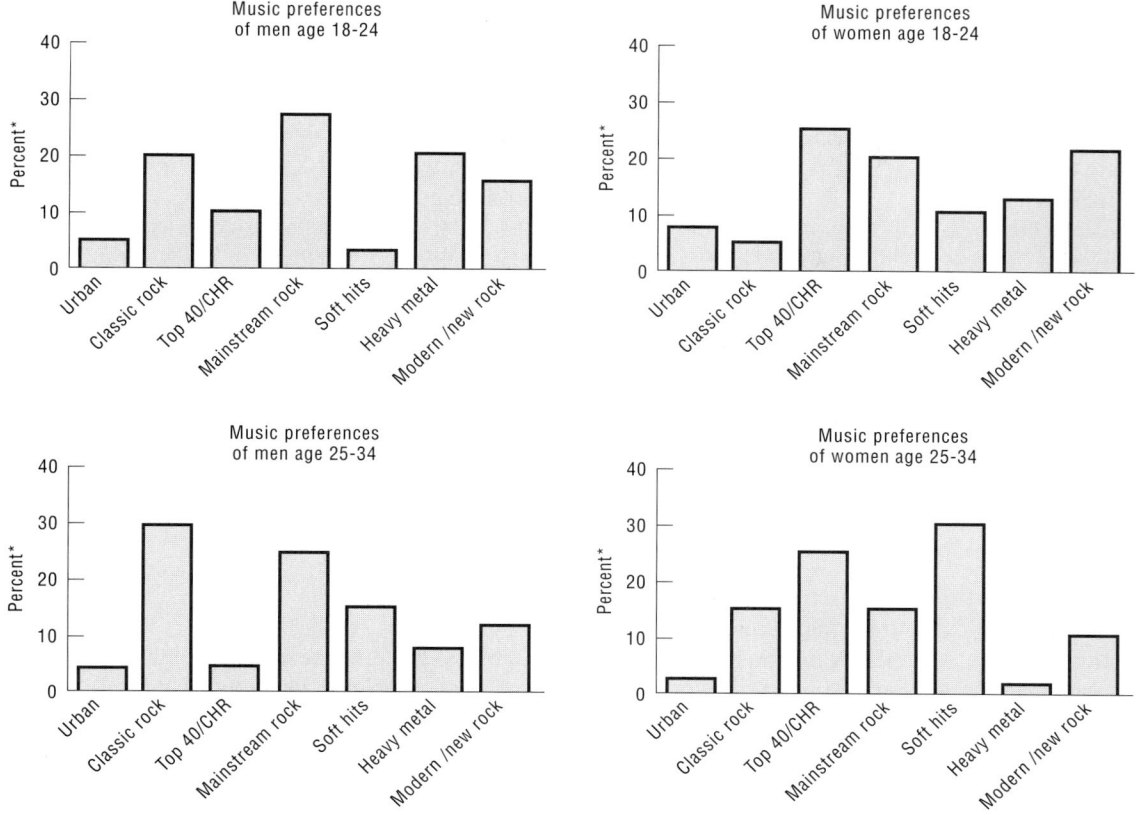

FIGURE 9-7 Radio format preference graphs. (Courtesy Burkhart/Douglas and Associates.)

information alone in ascertaining the musical tastes of their target audiences.

Listener psychographics With about twenty distinct formats competing for a share of the radio audience, and with more than one station and as many as five or six programming the same format in many cities, stations have developed more detailed methods of identifying their audiences. The current rage in radio research is called listener psychographics (see Chapter 15), also known as lifestyle or qualitative research. *Psychographic research* is an attempt to understand radio listeners according to their attitudes, values, beliefs, leisure pursuits, political interests, and other factors. For radio programmers today the age and sex of the audience are insufficient data: they need to know how their listeners view the world (and how their selected radio stations fit into that world).

A number of interesting psychographic studies of the radio audience have been conducted in recent years. Perhaps the best known of these was the "Radio W.A.R.S." series. Conducted in the mid-1980s for the National Association of Broadcasters by the research firm of Reymer and Gersin, these studies are still pertinent today despite some format evolution. Radio W.A.R.S. examined the psychological dimensions of radio listening. In other words, what are the listeners to the major formats *really* like? Let's take a look at some of the results.

Contemporary Hit Radio The research revealed that listeners to top-40 stations tend to fall into two main psychographic groups. One group listens primarily to hear new music with an uptempo beat and a lot of urban rhythms (the kind of music heard on boom boxes). Reymer and Gersin called these listeners "new music trendies" and "get-me-up-rockers." Another subset of the format seemed attracted to CHR because it returned them to the format's heyday in the late 1950s and early 1960s. For them, listening to CHR was motivated by the desire to be put into a romantic, nostalgic mood. Songs about young love, funny DJs, and lots of oldies are what they want. Images of "Wolfman Jack" playing "Teen Angel" on a late Saturday night in Southern California come to mind.

As the 1990s unfold, this research seems quite predictive. Today, two forms of CHR are becoming increasingly popular. The "urbanized" approach has led to the rap, Latin, and dance-influenced CHR, known in the trade as "churban" (CHR + urban contemporary). The romantic, nostalgia-fed CHR that treasures older standards has led to the rise of the pure oldies format (as exemplified by New York's number one station, WCBS-FM).

Album-Oriented Rock Album-oriented rock (AOR) listeners were found to be radio's most "socially motivated" listeners, considering themselves music experts. Three main subgroups were discovered. About 22 percent of AOR listeners were labeled "uninvolved disloyals," listening solely for music, with disdain for DJs, contests, and most of all, commercials. One of five listeners to the format were found to be "mindless loyalists," plugging into radio rock to tune out the world. Another 20 percent of AOR fans were "plugged-in smarts," looking for sophisticated and unpredictable music.

Again, trends in radio programming today reveal the predictive nature of psychographic research. In major markets like New York and Los Angeles, plus some smaller markets rife with "music trendies (such as Boulder, Colorado, and Seattle, Washington)," some AOR outlets emphasize alterative and modern rock, such as Nirvana, Jesus Jones, and The Red Hot Chili Peppers. Other stations target the "mindless loyalists" with a steady fare of predictable metal (the Scorpions, Van Halen, Cinderella, Metallica, and so on).

Beautiful Music Listeners to beautiful music were found to be people seeking companionship, relaxation, and escape from talk, talk, and more talk. For them, the purpose of a radio station was to

ECONOMICS

RADIO FORMAT "POWER RATIOS"

One intriguing radio statistic that has emerged in recent years is the power ratio. A power ratio is obtained by dividing the revenue earned by radio stations in a given format by their share of audience. A ratio of 1 would mean that a format's earnings are in direct proportion to its audience size. The following table presents the power ratios of the major formats:

Format	Power Ratio
Country	1.49
News/talk	1.43
AC	1.41
AOR	1.13
Oldies	1.12
CHR	1.07
Easy listening	.85
Urban contemporary	.75

SOURCE: *Radio and Records,* March 31, 1989, p. 1.

Some formats earn substantially more in revenue than they achieve in audience size. "Overachieving" formats include country, news/talk, and AC. Other formats receive less revenue than their size of audience would predict. In this group fall easy listening and urban contemporary.

These results are provocative. Are advertisers racists, spending less on radio stations with large audiences simply because they consist largely of African-Americans? Do country stations earn more because their programming is somehow "safer" than that on AOR or Urban Contemporary? Like many statistical tests, power ratios raise more questions than they answer.

help them achieve a more cheerful state, away from the stresses of everyday life.

Today a range of easy listening formats is available, depending on the psychographic profile of its audiences. For older, established, and upscale listeners, instrumental and orchestral sounds are preferred (like WPAT in the New York market). Younger audiences seeking escape and relaxation are apt to seek a beautiful music station with a more adult contemporary/oldies sound (such as WPCH in Atlanta). Those influenced by "new age" concepts, from the power of crystals to the joys of vegetarian cooking, might seek a beautiful music station with a more modern touch, like KTWV in Los Angeles.

Of course, the "Radio W.A.R.S." studies were just one set of psychographic evaluations. The sample was national in scope, thus it is entirely possible (and likely) that listener motivations are different in Keokuk, Iowa, and Honolulu, Hawaii. For this reason many stations conduct their own psychographic listener studies. We shall examine those modes of research in Chapter 15.

The Hot Clock

The next step in radio music programming involves the implementation of the schedule: the planning and execution of the station's sound. Most radio programmers today employ a version of a useful chart known as the hot clock, format wheel, or sound hour. Figures 9-8 to 9-10 are sample hot clocks for various formats.

Radio Dayparts The format wheel looks like the face of a clock, with each element of the station's on-air sound—music, commercials, news, sports, promotions, and so on—scheduled at its precise interval in the programming hour. The hot clock performs two main functions. First, it enables programmers to get a visual image of an otherwise invisible concept, their "sound." Second, it enables programmers to compare their program proposals with the competition. In this way programmers make sure that at the time they air music, the other station plays its commercials; that the news does not air on both stations at the same time; that it is unlikely that the same record will air simultaneously on both stations; and so on.

Normally programmers use a different clock for each important scheduling period. Figure 9-11

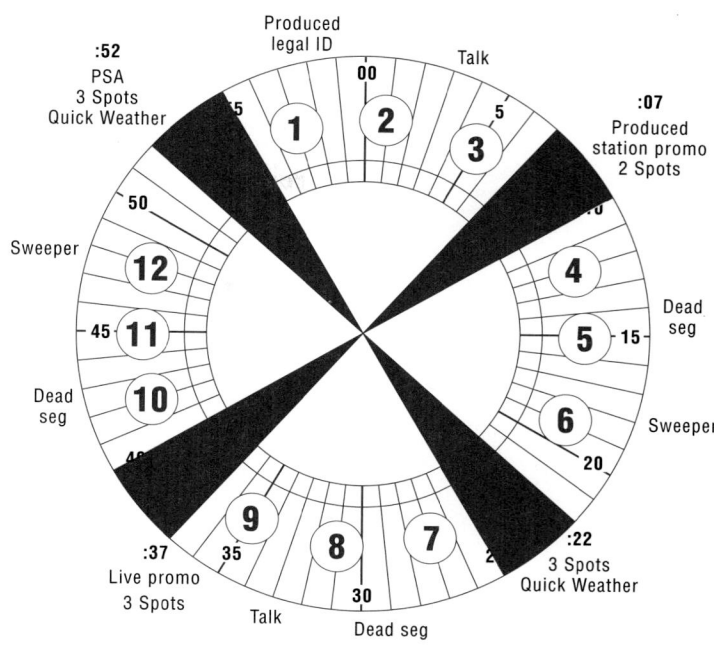

FIGURE 9-8 Hot clock for a country station.

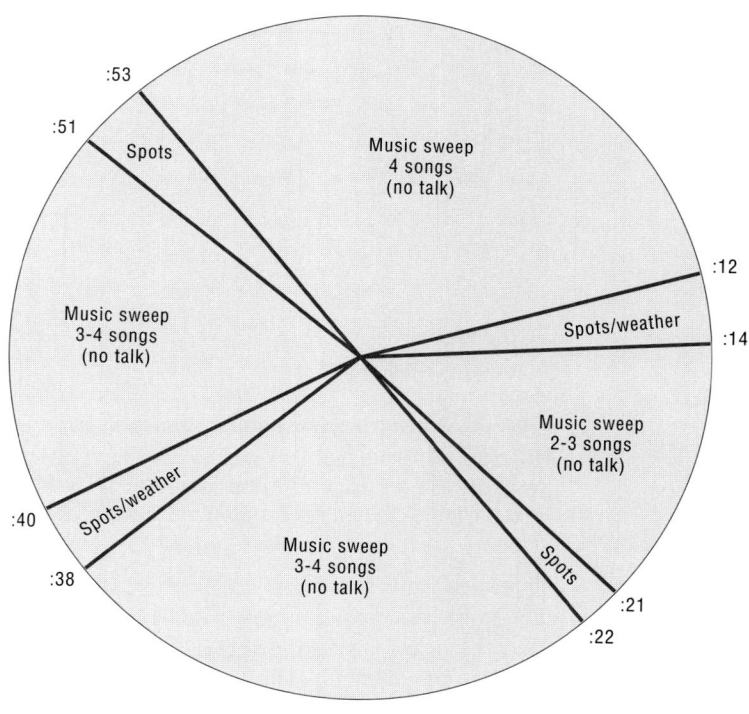

FIGURE 9-9 Hot clock for an AOR station.

211

```
*********************************************************************
*                                                                    *
*                              KBDA                                  *
*                                                                    *
*                       CLASSIC ROCK 'N ROLL                         *
*                                                                    *
*  Friday      05/15/92                                     3 PM     *
*                                                                    *
*********************************************************************
  00:00                   CLASSIC ROCK 9 AT 9 INTRO
............................................................................
1517-15        FLEETWOOD MAC            GO YOUR OWN WAY       D  3:38/F  '77
               RUMOURS
1548-06        VAN MORRISON             BROWN EYED GIRL       F  2:56/F  '67
               BLOWIN' YOUR MIND
............................................................................
  06:44                         SPEED BREAK
............................................................................
2072-03        PINK FLOYD               TIME                  D  7:05/F  '73
               DARK SIDE OF THE MOON    BACKSELL!
1511-01        MOODY BLUES              NIGHTS IN WHITE SATIN F  7:10/C  '67
               DAYS OF FUTURE PAST
............................................................................
  21:09                :20 BACKSELL/SPOTS/ID POSITIONER
............................................................................
2025-01        DOOBIE BROTHERS          CHINA GROVE           D  3:14/C  '73
               THE CAPTAIN AND ME
1510-11        DOORS                    TOUCH ME              F  3:15/C  '69
               THE SOFT PARADE
............................................................................
  30:38                      BOTOM OF HOUR PROMO
............................................................................
1531-20        GOLDEN EARRING           RADAR LOVE            D  6:24/C  '74
               MOONTAN                  BACKSELL!
1504-17        GUESS WHO                AMERICAN WOMAN        F  5:03/F  '70
               AMERICAN WOMAN
............................................................................
  42:35                          :35 BREAK
............................................................................
1508-16        STEVE MILLER             THE JOKER             D  3:40/F  '74
               THE JOKER
1515-06        JANIS JOPLIN             PIECE OF MY HEART     F  4:19/F  '68
               CHEAP THRILLS
............................................................................
  52:34                         SPEED BREAK
............................................................................
1559-10        BOSTON                   ROCK N ROLL BAND      D  2:59/C  '76
               BOSTON
1530-14        WHO                      MY GENERATION         F  3:15/C  '66
               THE WHO SINGS MY GENER..
............................................................................
  58:58                          :50 BREAK
............................................................................
1526-08        KANSAS                   CARRY ON WAYWARD SON  D  5:22/F  '76
               LEFTOVERTURE
1563-14        MOUNTAIN                 MISSISSIPPI QUEEN     F  2:29/C  '70
               MOUNTAIN CLIMBING!
Total Time for Hour is 68:19
```

FIGURE 9-10 Although hot clocks help to visualize the sound of a radio station, most programmers, in practice, use a computerized printout for their day-to-day scheduling. Here is an example of such a printout for a classic rock station. (Courtesy: Burkhart/Douglas and Associates. Software: RCS Selector.)

illustrates how radio use varies throughout the day from Monday to Friday and on weekends. Find the peaks on each chart, and you have discovered the medium's key *dayparts* (important time periods).

Morning Drive Let's examine the Monday to Friday graph first. The highest point on the graph is radio's most important time period: Monday to Friday in the early morning. For convenience, programmers usually identify the boundaries of this time period as 6 A.M. to 10 A.M. Since most listeners are preparing to commute or commuting to work and school, this is known as *morning drive time*. In most radio markets this is radio "prime time." This is where radio managers in the major markets commit their greatest program resources. The highest paid radio personalities toil in this time period, frequently earning salaries in the mid- to high six-figure range. Stations go to great expense and effort to have the top-rated morning show. Recently, the competitive battle has taken outrageous turns, leading to the rise of "shock jocks" and "blue radio" (see box).

Evening Drive The next "hump" in the Monday to Friday graph is seen the late afternoon. This is radio's second most important time slot: 3 P.M. to 7 P.M., *evening drive time.*

The audience for evening drive radio is only about two-thirds the size of that for morning drive. This is because not everyone who commutes to work goes directly home. Also, the attractiveness of late afternoon TV (Oprah Winfrey and Phil Donahue, for example) and early TV newscasts captures much of the audience at this time.

Daytime There is a nice plateau in the weekday chart where radio listening holds steady at about

IMPACT

MADNESS IN THE MORNING

Increasing competition for audiences in the lucrative early morning time slot and relaxed regulations at the FCC have led to the emergence of "shock jocks." Located mainly at album-oriented rock and adult contemporary stations in major markets, these men and women of mayhem tested the limits of taste in pursuit of radio entertainment. In Philadelphia one morning DJ played "Happy Birthday" in honor of the survivors of Hiroshima. In New York another invited listeners to call in for an opportunity to sleep with his in-studio guest. In San Antonio a morning jock team followed a news report on the brutal torture of a hitchhiker with sound effects of electric cattle prods mixed with human screams.

One of the "new breed," Chicago's Steve Dahl, shared his ideas for morning programming in *Radio Only,* a leading trade magazine. His "rules" for a successful show included "don't prepare," "speak your mouth, even if it's foul," "find humor in tragedies," and "criticize management on the air."

Steve Dahl of WLUP, Chicago.

PROFILE

"SUPERFLY JOCK"

Believe it or not, one of the highest rated morning men in Dallas and a leading afternoon drive personality in Chicago are the same person. Disc jockey Tom Joyner logs nearly 8000 miles a week, commuting between Dallas's KKDA-FM (K-104), where he's on from 5:30 A.M. to 9 A.M., and WGCI in Chicago, where he holds down the 2 to 6 P.M. shift. Both stations air urban contemporary music.

The commuting is made possible by a special $30,000 airline ticket, which guarantees Joyner a round-trip seat on midday flights between the cities for five years. To defeat jet lag, he consumes high-protein vegetarian lunches and downs plenty of replenishing liquids. He gets weekends off. And, as a further incentive, Joyner has inked long-term contracts worth over a million dollars with each station. That should keep him flying.

His major problem? Whom to pull for when the Dallas Cowboys play the Chicago Bears. Joyner says he roots for the "Cowbears."

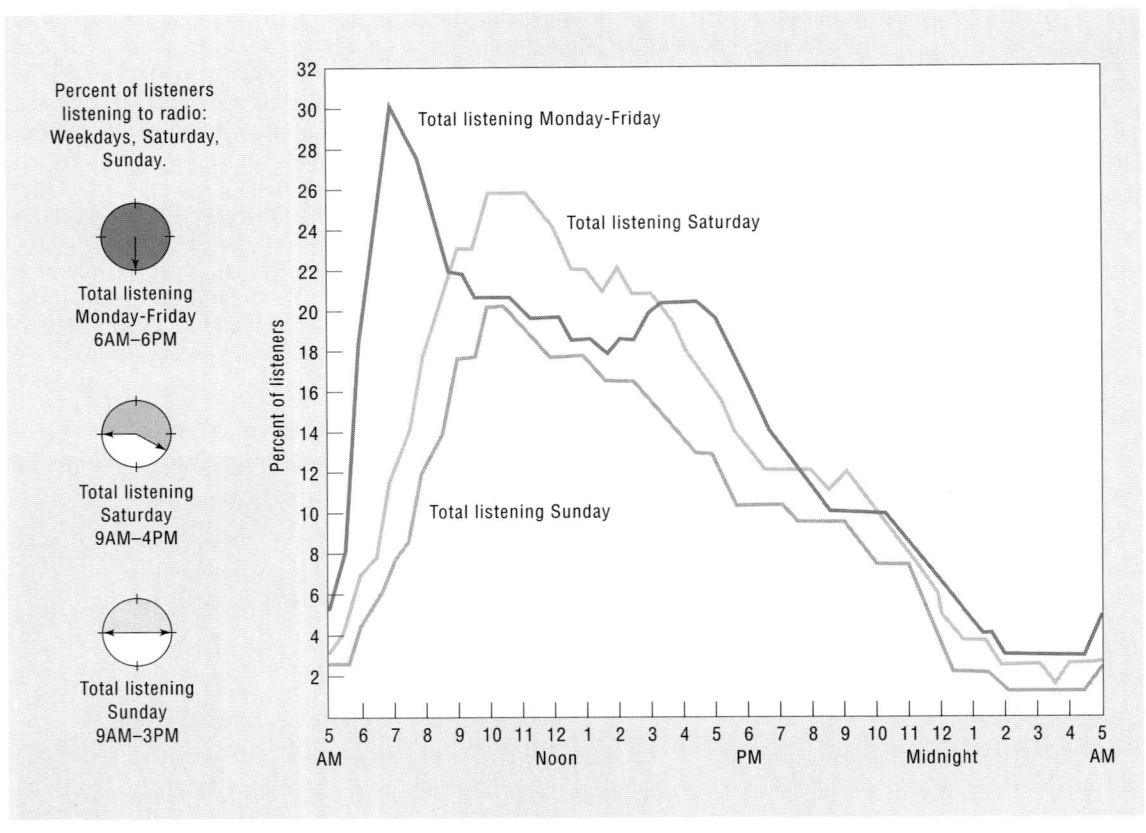

FIGURE 9-11 Radio listening throughout the day. (Courtesy of Arbitron.)

one in five people. This is the fast-growing *daytime* time period, from 10:00 A.M. to 3:00 P.M. The popularity of portable radios in the home and workplace makes this one of radio's most important time periods.

The noon hour is especially critical to radio programmers in daytime. This is because of the popularity of listening to radio on lunch breaks. Many music stations air special programs designed for their target audiences at this time. AC stations may feature an "Oldies Cafe"; CHR stations may feature a "Danceteria"; and an album-rock station may run a "Metallic Lunchbox."

Evening and Late Night Note how radio listening slides in the early evening (as America turns on its TV sets) and continues to plummet through the night until dawn. Thus two less important Monday to Friday dayparts for programmers are *evening* (7:00 P.M. to midnight) and *overnight,* or *graveyard,* from midnight to 6:00 A.M.

Most stations revert to format in the evening and late-night hours. That is, they play long, extended versions of the preferred programming of their most dedicated listeners. Since it is harder to sell commercials in these dayparts (due to the comparatively small audience), the loyal listener is rewarded with longer musical programs, specials, and other extended programming.

Weekend Radio Looking at the weekend charts immediately reveals the most important time period to programmers and advertisers: Saturday in the late morning and early afternoon. This is when we are most likely to turn the radio on, as we clean the house, wash the car (or the dog), and head to the park or beach. There is a modest peak on Saturday evening, when radios are turned on at family or neighborhood gatherings and parties.

Note that at no time on Sunday does radio listening exceed levels achieved on Monday to Friday or on Saturday. The best time on Sunday seems to be in the late morning, around 10:00 A.M. But with many listeners at church or sleeping late, this time slot is usually not a prime programming period.

At minimum, stations will use a hot clock for their Monday through Friday schedule in the morning drive-time period, the midday time slot, the afternoon drive slot, and for the evening schedule. Stations may also employ a weekend clock for the daytime slots comprising late morning through the early afternoon.

Filling the Clock; Radio Programming Terminology

There are three main types of information depicted in the hot clock: commercial and promo-

The computer is becoming a necessary tool in radio programming. Radio Computing Services Tracker system stores station programming logs on digital cassette tape.

tional matter, music, and news/talk segments. The number and location of the commercial positions normally are set first by the general manager in consultation with the sales and program managers. Their job is to decide how many commercials will run and in which parts of the hour. While some beautiful music stations run as few as eight or nine commercial minutes per hour, some rock stations run as many as fifteen or sixteen, up to twenty or more in peak seasons such as the Christmas rush. The FCC has no strict limit on the number of commercial minutes per hour, but many stations hold the line at eighteen. The decision on number and placement of commercials is crucial. Although commercials pay for everything else on the station, too many spots—a situation called advertising *clutter*—may cause listeners to tune out or, worse, to tune elsewhere. In addition to scheduling commercial matter, stations will also use the format wheel to schedule their promotional announcements, including contests and giveaways. The commercial and promotional segments of the hot clock are normally known as *spot sets*. Somewhat confusingly, they may also be called *stop sets* since the music stops during these breaks.

The musical segments of the hot clocks are typically broken down into two or three subcategories, such as current hits (given the most airplay); recurrents, recent hits that are still popular; and gold, for golden oldies. Sometimes programmers use key words to denote the various song categories, such as "power cuts" or "prime cuts" for the most popular current songs, "stash" or "closet classics" for obscure oldies, and "image cuts" for songs that seem to match the format perfectly, such as Beatles' love songs for adult contemporary. Some programmers use color-coding schemes. Red typically denotes power cuts, green may identify a new song in heavy rotation, and gold is used to signify an oldies set.

The area of overlap on the format wheel, where one program element ends and another begins, is known as a *segue,* pronounced "segway." The purest segue is one in which the musical segments blend from one song into another, without DJ interruption. But there are other options. The transition may be made by naming the artists and saying something about the previous song, a procedure known as a "liner." The station identification may be made ("ID"). The announcer might give the time and temperature ("T&T"). She might promote an upcoming song or feature ("teaser") or highlight a list of program and promotional activities ("billboard"). A promotional announcement ("promo") might be made. Or the DJ might cover the time to the next musical set ("sweep") with comedy, ad libs, or listener call-ins ("fill"). With this glossary in mind, see if you can determine the program "sound" of the stations depicted in Figures 9-7 to 9-10.

Once the hot clock has been set, the format is in motion and the station is on the air! But the programming process is far from over.

Format Evaluation

Music programming is a particularly dynamic task. Audience tastes are constantly changing. A new pop star is always appearing on the charts. Listeners tire quickly of some songs. Others remain in our ears and minds seemingly forever. So, how does the music programmer select the songs for the wheel? When does a song get pulled? This is the difficult process of format evaluation, the next step in radio programming.

One good way to start is by keeping track of record sales at local and regional retail outlets. Many record stores provide this information to the radio stations in their area (next time you buy a record, see if it is inventoried at the time of purchase).

Another way to keep track of musical tastes is by checking with trade publications, commonly known as *tip sheets.* Leading stations in the major formats provide these publications with a weekly report of the songs they are featuring, known as a *playlist.* In return they get "intelligence" about the popular songs and artists around the nation and in their geographical region. The major record labels use these report cards to make sure their artists are receiving airplay.

Influential charts include those in *Billboard* and *Radio and Records,* the two leading weekly trade publications. Other tip sheets appear in *Cashbox, Bobby Poe's Music Survey, Hits, Breneman Review,* and *The Gavin Report.*

Most stations also keep track of what their listeners are telling them about their format. *Call-ins,* telephone calls to the station, are logged to determine how listeners feel about the songs, artists, and personalities on the station. Stations use lists of contest entrants and telephone directories to conduct *call-outs.* Short (five- to ten-second) selections of the music, known as *hooks,* are played over the phone and listeners are typically asked to rate the song as one they like a little, are unsure about, or like a lot.

Stations assemble groups of their listeners in large rooms and conduct *auditorium tests.* In this forum up to 200 or 300 songs can be "hook-tested." Or stations may select a small group of listeners (from three to fifteen) and conduct in-depth interviews about their musical preferences; this is a *focus group* study.

If the task of evaluating the format sounds complex, that's because it is. For this reason many stations hire outside experts to help select their music, conduct their audience research, train their DJs, organize their promotions, and perform similar programming tasks. Such services are provided by program and research consultants.

There are essentially two main types of radio consultants. *Specialized consultants* provide expertise in one particular area, such as research, music selection, promotion, or financial management. *Full-service consultants* provide "soup-to-nuts" services, from how to decorate the radio station to how to deal with crank phone callers. A recent survey revealed almost fifty different radio consultants operating in the programming field. Some industry leaders include Atlanta-based Burkhart-Douglas; The Research Group out of Bellevue, Washington; Sklar Communications, founded by Rick Sklar, one of the original masterminds of contemporary hit radio in the 1950s at WABC in New York; and Paragon Research in Denver.

Fine-Tuning the Format

The final phase of radio format evolution is fine-tuning. Using data based on listener reaction, as indicated by audience ratings, station research, phone calls, and other means, the program manager makes changes in the schedule. The changes can range from minor to drastic. Minor changes involve substitutions in the musical mix, reformatting the various time periods, moving personalities around throughout the day, and so on. Major adjustments include replacing air personalities (most typically in morning drive), developing new promotional campaigns, including occasional call-letter changes, and firing music directors (a relatively common occurrence). The most drastic change is to abandon the format altogether and to try a new type of music, targeted to a different audience demographic. In recent years "format turnover" has been increasing at a spectacular rate. In the early 1990s some estimates indicated that as many as 20 percent of radio stations change formats in a given year.

Despite the apparent complexity of the process of radio music formatting, it remains popular and rewarding work. Major market program and music directors can look forward to high incomes, great visibility in the high-gloss world of popular music, and excellent "perks," like backstage passes to concerts and limo rides with the stars. Of course, some of these perks approach the limits of legality, which brings us to the problem of payola.

Paid to Play: Payola

In the late 1950s the DJ became "king" of the radio business. Big-name DJs in major markets, like Howard Miller in Chicago and Murray the K in New York, had become well known to thousands of loyal listeners. And, like kings, they exerted considerable influence in the record business. Simply put, if they liked a song and played it over the air, the song had a good chance of success.

This fact did not go unnoticed by the record industry, which showered DJs and music directors with gifts and gratuities, known as *payola*. Lucky "jocks" received cars, golf clubs, vacation trips—you name it. But the gravy train apparently ground to a halt in 1960, when, following a series of highly publicized hearings, Congress enacted legislation making the acceptance of payola a criminal offense, punishable by a $10,000 fine. Since then most stations have had their DJs and PDs (program directors) sign affadavits guaranteeing compliance with the payola statutes.

However, payola never really vanished. In the ensuing thirty years attempts to influence radio programming personnel by record companies and artist management took on new, sometimes "subterranean" forms. In 1988 two record promoters were indicted for making payments exceeding $250,000 to PDs at nine stations. It was also alleged that at least three of the PDs were provided free cocaine in return for their endorsement of the promoters' artists.

In 1992 a former music director of an Atlanta radio station was arrested for alledgedly accepting money and merchandise in return for playing records.

A related question of potential conflict of interest for radio personalities is the issue of plugola. *Plugola* is the practice of providing product endorsements over the air ("plugs") in return for cash or free merchandise. The listener doesn't know that the DJ is being paid to do these plugs. In the world of sports, product endorsements have become a regular and accepted practice. Many tennis, golf, and basketball personalities have become associated with a line of athletic shoes or brand of soft drink. However, the same practice in the broadcast business presents an ethical dilemma. For example, we might not feel comfortable if Dan Rather wore a visor with a recognizable product logo on the "CBS Evening News." How would you feel if a DJ celebrated one line of clothing or brand of stereo equipment over another, especially if he had a financial interest in the company he was plugging? There are few legal restrictions on plugola. It usually comes down to a matter of management or personal ethics.

NEWS AND TALK FORMATTING

On the surface it might appear that programming a radio station without music is a simpler task than formatting rock and roll or country. The format strategies of talk radio, however, are just as complex as those in the various music formats. How much news? How much telephone interview? What types of personalities? Sports? Which and how much? And, like the music formats, how many commercials? When?

The first consideration for a station planning a spoken-word service is to determine the type and amount of talk. There are two extremes in the format. At one is the all-news operation, providing summaries and spot news reports around the clock. Leading the way in this format are such classic all-news operations as WINS in New York, WBBM in Chicago, KFWB in Los Angeles, KYW in Philadelphia, and KIRO in Seattle. At the other extreme are all-talk stations, which lean heavily on the concept of the telephone call-in as the basis for their programming schedule. On this list are such stalwarts as New York's WOR and WABC, KOA in Denver, and WWWE in Cleveland.

Just like their musical counterparts, most news/talk radio stations use a format wheel to schedule their programming. Examine the clocks for an all-news station and a news/talk station in Figures 9-12 and 9-13.

All-News

First, let's look at the all-news wheel. There are three basic elements in the sound hour: news segments, feature segments, and commercial

FIGURE 9-12 Hot clock for an all-news station. (Courtesy CBS Radio.)

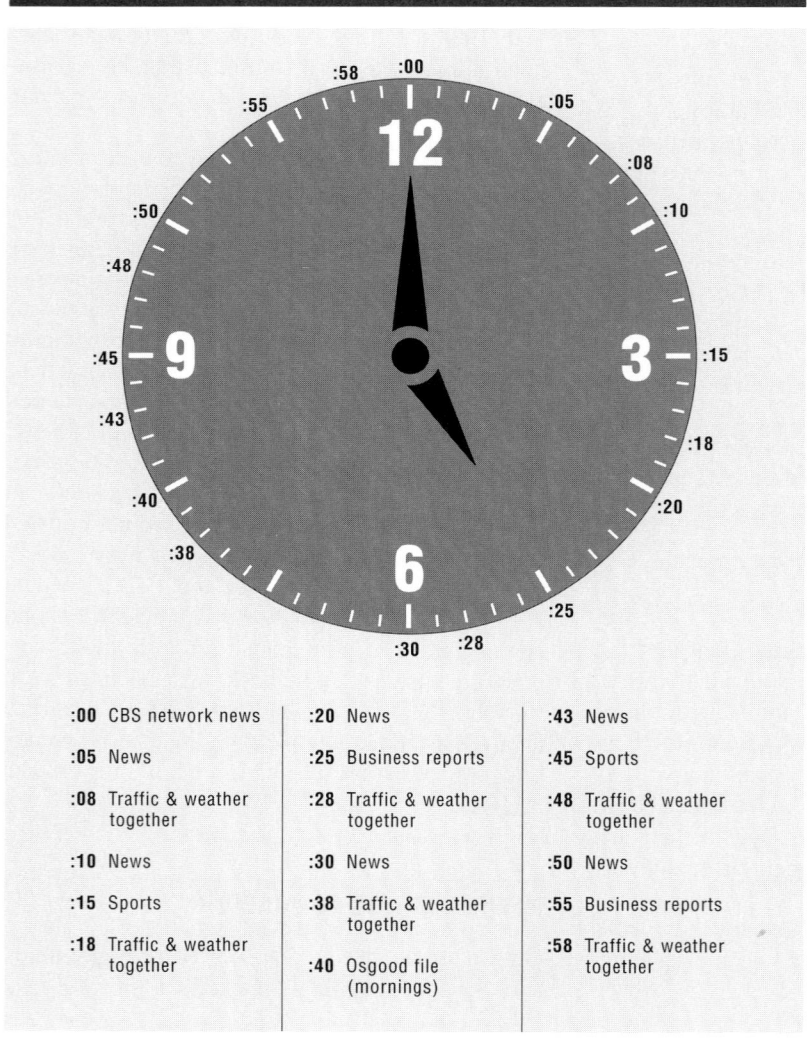

:00 CBS network news
:05 News
:08 Traffic & weather together
:10 News
:15 Sports
:18 Traffic & weather together

:20 News
:25 Business reports
:28 Traffic & weather together
:30 News
:38 Traffic & weather together
:40 Osgood file (mornings)

:43 News
:45 Sports
:48 Traffic & weather together
:50 News
:55 Business reports
:58 Traffic & weather together

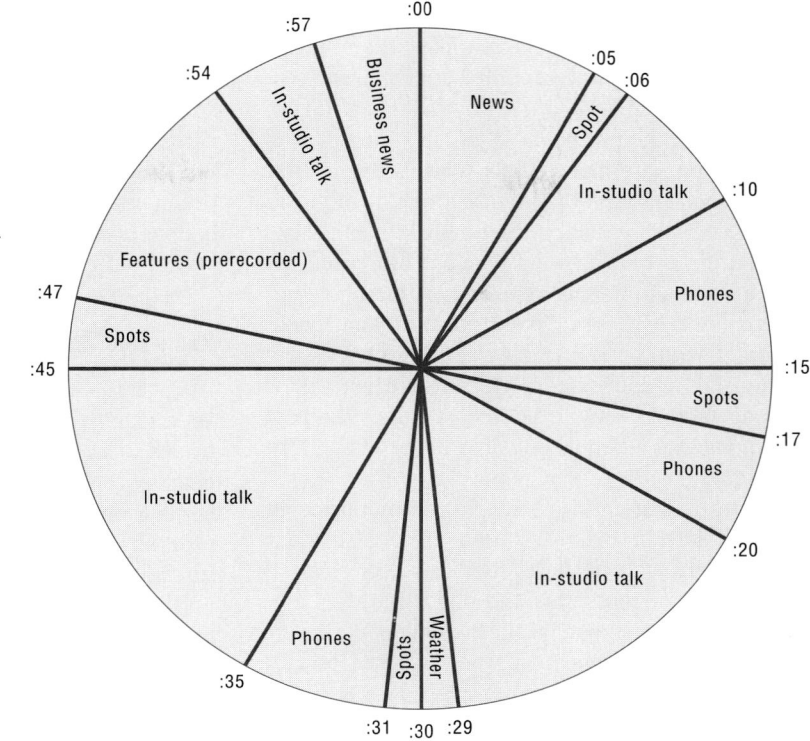

FIGURE 9-13 Format wheel for an all-talk station.

matter. Typically news stations provide network news at or near the top of the hour. This provides the audience broad national and international coverage, which the station can interpret or "localize" for its audience in other news segments. Some all-news stations break for network news reports at other times, typically at the thirty-minute mark (the "bottom of the hour").

Credible announcers are a critical ingredient to the success of the all-news operation. Announcers must sound confident and authoritative. For this reason most all-news operations prohibit announcers from delivering commercial "pitches." This is to avoid hurting the newscaster's credibility, which might result from the announcer reporting a natural disaster, for example, and then segueing to a used-car spot.

Feature programming at an all-news station runs the gamut from the expected (weather, sports, and traffic) to the more specialized (political, economic, and health reports). These reports usually are presented in a streamlined fashion—in units of three minutes or less.

By definition, all-news stations need to be bright, brisk, and dependable. News listeners tend to be "no-nonsense" information-seekers. When they tune in to an all-news station, they expect to become well informed quickly. For this reason the format wheel in all-news tends to spin rapidly: program elements are repeated regularly throughout the day. Phrases such as "give us fifteen minutes, we'll give you the world," and "around the globe in twenty minutes" give evidence of this speedy rotation, especially in key dayparts like morning and evening drive. In this regard, the programming pace at an all-news station is similar to that at a top-40 station.

News/Talk

Compared to all-news stations, those emphasizing talk tend to be "laid back." News segments are common, particularly at the top and bottom of the hour, with the remainder of the program hour filled by features, interviews, and telephone call-in segments. As an example, inspect the hot clock for an all-talk station (Figure 9-13).

Note the more leisurely pace of the talk format.

Talk segments of five to seven uninterrupted minutes are commonplace, like the "music sweeps" in adult contemporary and beautiful music. These are presided over by talk show hosts. Unlike the announcers on all-news stations, who tend to be interchangeable (credibility, rather than individuality, is the key for them), talk hosts are distinct personalities. Some boast political beliefs on the left of the political spectrum, others at the far right. Some are openly brusque and combative with listeners (like Pete Franklin, most recently with WFAN in New York). Others are folksy, home-spun, seemingly in love with their legions of loyal listeners (like Atlanta's Ludlow Porch and the legendary Wally Phillips of WGN in Chicago).

Feature elements are more commonplace in the talk format. Many stations have "resident experts" in such fields as medicine, psychiatry, finance, law, economics, and politics.

The prohibition against the reading of commercials by the announcer in the all-news format rarely extends to talk radio. In fact, talk show hosts tend to be expert "pitchmen" (and women), who frequently give broad testimonials for their sponsors' products, sometimes longer than the thirty or sixty seconds the sponsor paid for.

Both news and talk tend to carry play-by-play sports on their schedule. In large markets there is considerable competition to land the broadcast rights to professional and major college sports in those cities. Smaller news and talk operations sign on to carry regional professional teams, as part of a network. Many also carry local college and high school games.

The outlook for news and talk formats remains promising. Radio's great strength is its immediacy: its ability to respond quickly and effectively to crises and emergencies, particularly local ones like crimes and natural disasters. Thus, despite the growing trend of radio away from talk and toward music, and away from AM toward FM, news and talk remain viable format strategies. In fact, demand for news/talk is partly responsible for the rebirth of radio network programming, to which we now turn.

THE REBOUND OF RADIO NETWORKS

At the end of the 1970s network radio was in decline. Advertising sales were flat. If they affiliated with a network at all, stations used the service almost exclusively for short news broadcasts. Many stations "pulled the plug" on longtime network contracts, preferring instead to rely exclusively on local sources for programming.

Against all odds and counter to most industry predictions, by the early 1990s radio networks had rebounded. A major factor was technological—the arrival of satellite transmission.

Radio and the Satellite Dish

At first the technology of the satellite dish was seen mainly as a boon to the TV industry. As we traced in Chapter 6, satellite delivery was at the heart of the growth of cable and pay cable and the home video revolution. Quickly, however, radio programmers and station operators came to realize the implications of the satellite.

Before the arrival of satellite communication, in order to network, radio stations had to lease telephone lines from "Ma Bell," the phone company. These lines were expensive and occasionally had technical problems that affected sound quality. With satellite communication a station could buy a small receiving dish and receive crystal-clear stereo transmissions at nominal cost. Many stations saw this as a means of obtaining current music without the expense of maintaining a record collection. Rural stations, in particular, envisioned an opportunity to obtain a "major-market sound" by importing a satellite network service complete with music, jingles, and personalities. By the mid-1980s satellite radio was in full swing.

Today more than three-fourths of the nation's radio stations own at least one satellite dish. More than a half own more than one, making the "dish farm" part of the radio station landscape. It was estimated in 1992 that up to 20 percent of radio stations, in excess of 2000, were using satellite network services.

Full-Service Satellite Programming

Satellite program services are available in virtually every format, from country to oldies, from adult contemporary to new age. Leading companies providing satellite programming services are Broadcast Programming Incorporate (BPI) and the Satellite Music Network (SMN). Each provides twenty-four-hour-a-day satellite-delivered programming in each of the major radio formats to more than 1000 stations. In 1992 BPI offered ten different formats, SMN offered twenty-four. Most of the stations affiliating with these full-service networks were in the nation's smaller radio markets.

However, the major drawback of going full-time satellite is the loss of a "local" sound. Owners ask, "How can a big-name disc jockey in New York, Los Angeles, or Dallas relate to my audience?" To counter this problem, some stations go "local-live" in the morning. They make sure that the distant announcers make recordings of the local call letters and promotional "hooks" to give the illusion that the DJ is in Poughkeepsie or Peoria, not in Los Angeles. In some cases stations pay travel expenses for the network announcer to attend station promotional events.

Part-Time Satellite Programming

Of course, the bulk of the satellite-fed network services is not full-time. Just as in the "old days"—the 1950s and 1960s—radio stations are turning to their networks for news, features, and special-interest programming. Leading the pack of this type of network service are Capital Cities/ABC, Westwood One, Unistar, and Sheridan.

Before its purchase in 1985 by Cap Cities, ABC was the acknowledged leader in network radio. It had perfected a target-marketing approach to network news and information, tailoring the "sound" of the news and information provided to the format of the affiliated radio station. Depending on their format, stations could select to air the ABC Contemporary, ABC Entertainment, ABC Information, or ABC-FM network service. Under Cap Cities's management, in the late 1980s ABC extended its network services.

Today ABC Radio offers six networks: ABC Contemporary, Direction, FM, Information, Entertainment, and Rock. The network has a deal with *USA Today*, which sends lifestyle, feature, business, and health items to affiliated stations by computer before their morning drive shows. Stations targeting young audiences can also receive "Morning Prep," a service prepared by comedy writers to provide funny, topical bits for raucous morning shows.

Approaching ABC in its number of services, affiliates, and, most important (to its owners, at least), profitability was Los Angeles–based West-

Two of the best-known radio personalities: Larry King of Mutual and Charles Osgood of CBS.

wood One. In 1987 Westwood One purchased NBC's four program services for $50 million. At the time of the acquisition the NBC Radio Network had 370 affiliates. Its young adult–oriented service called The Source had 122 stations in its lineup. Nearly 300 stations ran all or part of NBC's Talknet menu of radio advisors. A smaller number of stations subscribed to NBC's Radio Entertainment network.

The four NBC networks added to Westwood One's formidable roster of services. Two years earlier it had acquired the Mutual Broadcasting System, with over 700 stations and the popular "Larry King Show." Westwood One itself is a highly popular programmer of music and entertainment specials, with recent events like the Who's world tour and concerts by Eric Clapton to its credit. Westwood One has signed an agreement with Gostelradio, the Russian broadcast distributor, to produce rock radio for the new Commonwealth of Independent States.

Another leading radio network is Unistar, which offers two well-known services, "Power" and "Ultimate." Stations subscribing to these services receive music, news, sports, features, all fed from the network's Washington, D.C., headquarters. Unistar has been a leader in providing capsule news for its FM music stations. In addition, Unistar is a leading satellite music programmer, having acquired the Transtar service of nine formats in 1990.

Another major player in network radio is Sheridan, which targets the black audience. In 1991 Sheridan acquired the National Black Network. Today Sheridan offers network services to over 300 stations; most are in the urban contemporary format. Hispanic-oriented Caballero boasts nearly 100 affiliated stations.

Some "old-time" networks persist, including CBS, with its CBS Radio Network news and information heard on over 500 stations. The Associated Press Network still provides news to over 1000 affiliates.

Thus, as the next century approaches, network radio might respond somewhat like Mark Twain, who said, after reading an incorrect newspaper story: "Accounts of my death have been greatly exaggerated."

PUBLIC RADIO PROGRAMMING

As we traced briefly in Chapter 4, noncommercial radio programming is a "mixed bag," depending largely on the type of facility. Recall that there are three main classes of noncommercial facilities: public, university, and community.

Public Radio Stations

Public radio stations are those that meet guidelines for federal grant assistance through the Corporation for Public Broadcasting. CPB-qualified stations must employ at least five full-time employees, must broadcast at least eighteen hours per day, must have annual budgets exceeding $105,000 per year, and must meet minimum criteria for signal strength and "broadcast quality." By the early 1990s over 350 stations met these guidelines, enabling them to obtain the programming of the National Public Radio (NPR) network.

NPR feeds funding and programming to its member stations. It was a leader in satellite delivery, having replaced traditional land lines with dishes in the late 1970s. Some of the most lis-

One of the most popular shows on National Public Radio is "Car Talk," hosted by the knowledgeable Tom and Ray Magliozzi.

tened-to programs in public radio are produced by NPR, including the award-winning news and commentary shows "All Things Considered" and "Morning Edition." Its music programs are also quite popular, ranging from classical ("Morning Pro Musica") to jazz ("Jazz Alive").

The Gulf War of 1991 provided an unexpected boost to NPR stations. Research revealed that audiences for NPR news programs increased over 15 percent during the ten-week conflict. Before long, staffers at NPR news were wearing T-shirts declaring "Air Superiority."

NPR is also highly regarded by its peers. A recent poll of journalists reported in the *Washington Journalism Review* ranked CBS's Charles Osgood "Best Radio Reporter." In second, third, and fourth place, respectively, were NPR's Cokie Roberts, Nina Totenberg, and Linda Wertheimer.

Typically, CPB-qualified stations rely on NPR for about a quarter of their daily schedule. The remainder of the schedule is filled with locally originated music and public affairs programs. The overwhelming majority of CPB-qualified public stations play classical music as the backbone of their schedule. Other forms of music receiving airplay include jazz, opera, folk, and show tunes.

College Radio

Although over 150 NPR-affiliated stations are operated by universities, we do not mean these when we speak of "college radio." Instead, we are referring to about 800 stations licensed to American colleges (and some high schools) that do not meet the CPB criteria. Many are operated as student activities or training centers; most feature "alternative" programming schedules to both NPR and commercial radio. The musical mix at most college stations is eclectic and progressive. For example, examine the program schedule for WUOG, the student station at the University of Georgia (Figure 9-14).

As we have seen, commercial radio formatting is tightly structured and controlled. For this reason college radio is where many new performers receive their initial exposure and airplay. This was the case for bands like Nirvana, Jesus Jones, R.E.M., Talking Heads, the Police, U2, and INXS.

Most college stations are active in news and public affairs programming. Again, the structure and approach of these services is an amalgam. Stations affiliated with journalism and mass communication programs may have news and public affairs schedules that emulate those at commercial operations. Others rely for news programming upon volunteer efforts by their staffers. In such cases schedules tend to be sporadic and content varied, to say the least.

Community Station Programming

If diversity is the word to describe college radio, the term is equally suited to the broad category of community radio stations. Community stations are operated by civic associations, school boards, charitable foundations, and, increasingly, religious organizations. Their programming spans a similar range.

Most community stations use a *block programming* approach. Unlike commercial formats, where the music, talk, and commercial elements are generally consistent throughout the day, in the block scheme programming is divided into two- or three-hour blocks appealing to different audiences at different times. Educational outlets, for example, may begin the day with a two-hour block appealing to elementary-age students. Midday might be used for in-home college-level instruction and the afternoon dedicated to social studies programming for secondary schoolers.

Many community-owned and operated stations, like those in the Pacifica group, employ the block approach (also known as "checkerboarding").

The trend toward diversity in programming at noncommercial stations extends to the religious stations. While virtually all religious operations make live broadcasts of church sermons and activities a vital part of their programming day, beyond that there is great variation. Among Christian stations, for example, at various times one can hear old-time Gospel, solemn hymns, even rock, known as "contemporary Christian." And within the religious radio milieu, there are "networks." For example, the Protestant Radio and Television Center makes a range of programs available to over 700 stations. Its flagship service, The Protestant Hour, has been on the air continuously since the 1930s.

Time	SUNDAY	MONDAY	TUESDAY	WEDNESDAY	THURSDAY	FRIDAY	SATURDAY
6:00 am	Format	Format	Format	Format	Format	Format	Format
6:30							
7:00 am	Sunday Morning Jazz						
7:30							
8:00 am			Newscurrent	Newscurrent	Newscurrent	Newscurrent	Newscurrent
8:30							
9:00 am			Newscurrent	Newscurrent	Newscurrent	Newscurrent	Newscurrent
9:30							
10:00 am	Continental Breakfast						
10:30							
11:00 am							
11:30	MLK Speaks						
NOON	Newscurrent Format	Newscurrent Lunchbox	Newscurrent Lunchbox	Newscurrent Lunchbox	Newscurrent Lunchbox	Newscurrent Lunchbox	Newscurrent Lunchbox
12:30		Format	Format	Format	Format	Format	Format
1:00 pm	Newscurrent	Newscurrent	Newscurrent	Newscurrent	Newscurrent	Newscurrent	Newscurrent
1:30							
2:00 pm							
2:30							
3:00 pm	Folk Scene						
3:30							
4:00 pm	Newscurrent	Newscurrent	Newscurrent	Newscurrent	Newscurrent	Newscurrent	Newscurrent
4:30							
5:00 pm	Newscurrent	Newscurrent	Newscurrent	Newscurrent	Newscurrent	Newscurrent	Newscurrent
5:30	Dirt Roads and Honkey Tonks						
6:00 pm			Sound of the City		Sound of the City		Free For All
6:30							
7:00 pm		Viewpoint		Wit's End			
7:30	Blue Laws	Minority Matters	Sports Talk	The Film Thing	Sports Talk		
8:00 pm				Who Put The Bomp		Block Party	
8:30		Crisis Cabaret	Essential Rythm		Caribbean Rythms		Back Track
9:00 pm							
9:30	A Matter of Jazz						
10:00 pm		Newscurrent	Newscurrent	Newscurrent	Newscurrent	Newscurrent	
10:30		Industry Standards	Power of Soul	Loud Fast Rules	Jazz By Numbers		Industrial Pipeline
11:00 pm							
11:30							
Midnight	Format	Midnight Snack	Midnight Snack	Midnight Snack	Midnight Snack	Midnight Snack	Midnight Snack
12:30		Format	Format	Format	Format	Format all night	Format all night
1:00 am							
1:30							
2-3:00 am							

FIGURE 9-14 WUOG program guide. (Used with permission.)

SUMMARY

Radio programming can derive from local, pre-recorded or syndicated, and network sources. Radio shows can be produced in four ways: local-live, live-assist, semiautomation, and turnkey automation.

Stations strive to make their formats unique. One way of achieving this is to analyze the market of a particular city and to find a format hole.

There are strong ties between radio and the music business: it is evident that airplay means higher record sales. The advent of MTV and CDs has also reinforced the interdependence of these businesses.

When a station attempts to choose a format, internal factors that are considered include ownership, dial location, power, technical facilities, and management philosophy. External factors such as strength of competitors, geography, and demographics also need to be analyzed. The new trend is toward psychographic research, which tells programmers the type of listener each musical category attracts.

Station managers plan their programs on a hot clock. This helps them visualize the sound of a station. A program schedule is divided into dayparts, including morning drive, daytime, evening drive, evening, and overnight.

Technology has brought back network radio. The satellite dish has helped network radio to become a desirable source for news, features, and special-interest programming.

SUGGESTIONS FOR FURTHER READING

Keith, M. (1987). *Radio programming: Consultancy and formatics.* Boston, Mass.: Focal Press.

Keith, M. (1988). *Broadcast voice performance.* Stoneham, Mass.: Focal Press.

Levin, M. (1987). *Talk radio and the American dream.* Lexington, Mass.: Lexington Books.

Lull, J. (1987). *Popular music and communication.* New York: Sage.

MacFarland, D. T. (1990). *Contemporary radio programming strategies.* Hillsdale, N.J.: L. Erlbaum Associates.

Matelski, M. (1989). *Broadcast programming and promotions worktext.* Stoneham, Mass.: Focal Press.

Oringel, R. (1989). *Audio control handbook.* Stoneham, Mass.: Focal Press.

CHAPTER 10: TELEVISION NEWS PROGRAMMING

Television news is a search for reality in a time of images. It is a search for coherence in a time of fragments. Can we agree that nothing on television is real? Nothing. Everything is a picture. Most of the time it's a picture of actors imitating life. Pictures of imitations. Some of the time it's pictures of fantasies. Cartoons. Pictures of pictures. Some of the time it's pictures of real happenings. A selection of pictures. A selection of pictures that are available. (And that don't cost too much.) So what is real is a machine that gives us pictures. And us, looking at it.

TV reporter John Hart.[1]

Perhaps no other topic in this text has generated the controversy that surrounds broadcast news. Is it trivial? Is it superficial? Is it journalism or show business? Does it focus merely on the sensational, ignoring important fundamental issues that aren't "visual" or don't make an effective "sound bite"? Has it hastened the decline of newspaper reading? Has it contributed to our lack of knowledge about geography? Economics? The third world? Has it made the American public more cynical, leading to apathy at the polls and a distrust of government? Or, has it produced the most enlightened and well-informed generation in our history? Before we can speculate on these and other claims made about the "effects" of TV news, we must understand what it is, how it is produced, and how it came about.

THE RISE OF TELEVISION NEWS

The Roper Studies

Each year the Roper Organization conducts surveys asking people to identify (1) their main source of news and (2) the news source that they perceive to be the most credible. The trends in popular opinion on these questions are summarized in Fig. 10-1*a* and 10-1*b*. From 1963, on TV was named most frequently as the primary source of news. Its lead over second-place newspapers increased steadily. The proportion of people naming radio fell steadily since the 1960s, as did the magazine category (which was never named by more than one in ten people to begin with). Few people over the years named "other people" as their primary news source.

Television news also received high marks on credibility. Since 1961 TV was cited as the most believable source of news. By 1991 TV was named as most credible by more than two to one over newspapers. Radio news had taken a downturn; in the early sixties, about one in nine people found radio to be most believable; by the 1990's the proportion was closer to one in fifteen.

Although the wording of the questions has caused some debate about the validity of the Roper findings, they clearly indicate the rise of TV as a news source in our era. How did this happen? When did TV become our eyes on the world? The Roper studies suggest that it happened in the early 1960s—November 22, 1963 to be exact.

The Kennedy Assassination: The Death of Camelot and the Birth of Television News

President John F. Kennedy had been warned by his advisors not to go to Dallas. Like many places in the South, the city was torn by political and racial unrest. But he decided to go. The press perceived that there might be trouble when a president perceived as a liberal traveled to a staunchly conservative city. A large contingent of the media was on hand as the president's motorcade passed the Texas School Book Depository. Shots rang out. The president was mortally wounded.

Within five minutes news of the shooting moved on the United Press wire service. Within ten minutes the three TV networks had interrupted their afternoon lineups of game shows and soap operas. Receiving the news from a young Dallas reporter named Dan Rather, an emotional Walter Cronkite told the nation on CBS that its president had been slain.

Assassinations in other nations might have led to governmental chaos and public violence, but in the United States it turned people to their TV sets. For four days all regular programming was suspended and people sat, seemingly transfixed

---

[1] *RTNDA Communicator*, September 1988, p. 46.

228

PART 4:
HOW IT'S DONE

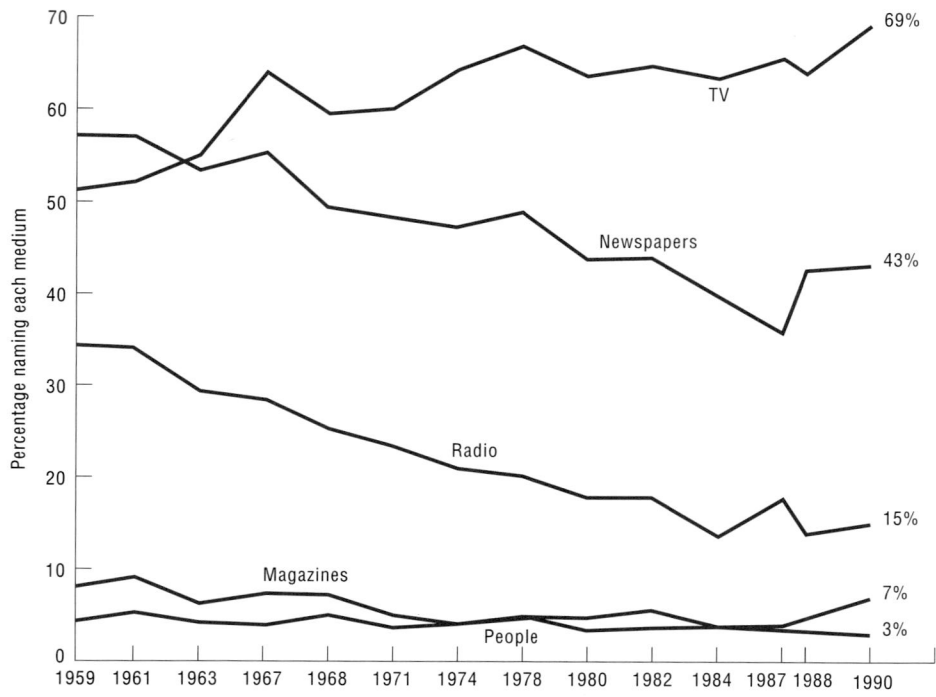

FIGURE 10-1a Sources of news, 1959–1988. Percentage of people naming various media as their source of most news.

FIGURE 10-1b Media credibility, 1959–1988. Percentage of people naming various media as the one they would believe if they got conflicting versions of a news story.

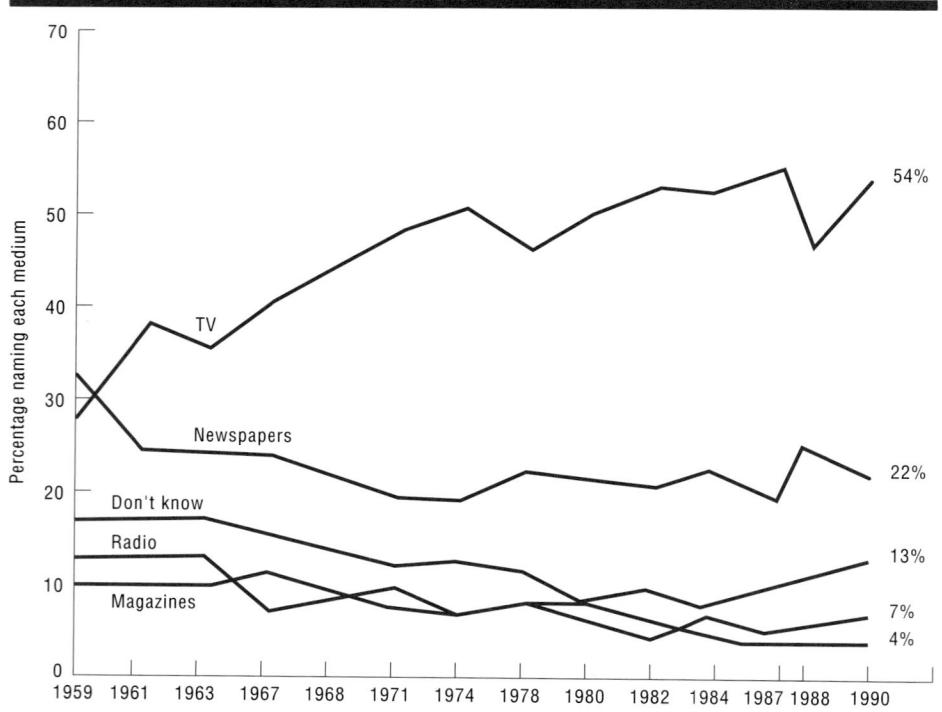

by the story unfolding on TV. The full resources of the TV medium were turned to this one event. Television was there when President Kennedy's widow Jacqueline returned with the coffin to Washington and when the accused assassin, Lee Harvey Oswald, was himself gunned down by a Dallas nightclub owner, Jack Ruby. And TV permitted the nation to attend the funeral, as world leaders came to pay their respects.

More than nine in ten Americans were said to have watched the TV coverage that fateful weekend. Watching, too, were over 500 million people in twenty-three countries, as the coverage was fed to a new device that had recently been launched: the communications satellite.

Network news had just been expanded from fifteen to thirty minutes. It consisted mostly of silent newsfilm and "talking heads." Local news was even more primitive: many stations focused their efforts on sassy "weather girls"; their anchors would read a national story and a dog food commercial with the same officious presentation that they had employed in radio. But those days were over. In one weekend, TV news had seized the public consciousness. Things would never be the same.

The Prehistory of Television News

Although the Kennedy assassination hastened the maturity of TV news, it was by no means the first important story covered by the medium. In December 1941 WCBW, an experimental TV station operated by CBS, reported live for nine hours on the events from Pearl Harbor. Comprehensive coverage of the political conventions began on CBS and NBC in 1948. Famous correspondent Edward R. Murrow had taught his old radio dog new tricks: his influential radio news program "CBS Is There" became "See It Now" in 1951. Murrow and the CBS news team had reported from the world's capitals, the nation's leading cities, even the battlefields in Korea, making that conflict the first "television war."

By the mid-1950s TV's first "news anchors" had emerged. Heading the parade of nightly news reporters were Douglas Edwards on CBS, John Cameron Swayze on NBC, and John Charles Daley on ABC. Spectacular events—perfect for pictures—were occurring. In England a new queen was crowned (Elizabeth II, in 1952.) In July

TV news came of age with its reporting of the assassination and funeral of President John F. Kennedy. At times during the four-day coverage, nine out of ten Americans were watching the events on TV.

1956 the luxury liner *Andrea Doria* sank after colliding with another liner, the *Stockholm*. Douglas Edwards scored a major coup for CBS, providing on-the-spot narration as the liner gurgled out of sight. In 1957 Americans sat spellbound as reporters and scientists tried to explain what a satellite was, and why the public should care. It seems that the Soviet Union had launched *Sputnik*, an "artificial moon," according to one ABC commentator.

Indeed, the "space race" itself was quite an impetus to TV journalism. Beginning with the 1961 launch of the first American into space, Alan B. Shephard, "space shots" were covered by all the networks and were used as a showcase for the new wares being developed by the communications industry, including satellites, graphics, special-effects devices, and videotape recorders.

But for sheer compelling power, no event could match the impact of the Kennedy assassination on TV journalism. No longer was TV news the stepchild of print or its progenitor, radio. As 1963 drew to a close both NBC, with the formidable team of Chet Huntley and David Brinkley, and CBS with the "most trusted man in America," Walter Cronkite, were telecasting thirty minutes of nightly news. By 1965 the national news was in color. This became important as events both at home and abroad began to capture and command the TV news eye.

TELEVISION NEWS IN THE TURBULENT SIXTIES

Today there is a certain nostalgia for the era of the 1960s and early 1970s, a period in history bounded by the Kennedy assassination in 1963 and the Watergate scandal, which ended in 1974 with the resignation of President Richard Nixon. Aging "baby boomers" (people born after World War II, who reached adolescence in this period) tend to recall the sixties as a period of playful experimentation, blue jeans, rock and roll, and "flower power." There certainly were moments of fun, but the decade was marked by violent confrontations, social upheavals, and cultural change. And for the first time in history it all happened in front of the TV cameras. Two major events occurred in this period. Both became forever intertwined with the growth of TV journalism.

Television and Civil Rights

In 1962 Dr. Martin Luther King, Jr., outlined a new strategy:

We are here today to say to the white men that we will no longer let them use their clubs in dark corners. We are going to make them do it in the glaring light of television.[2]

And it worked—TV was there. It was in Little Rock in 1957 to capture the violence following the integration of Central High School. It was in Montgomery, Birmingham, and other southern cities to witness sit-ins at lunch counters and bus stations. It was in Washington to cover the hundreds of thousands who rallied for civil rights in 1963. Television was in Detroit, Watts, and Newark to cover civil disorders. Television presented searing images of white police chiefs turning dogs and fire hoses onto defenseless demonstrators. It revealed the hatred of white supremacists, including the Ku Klux Klan, as throngs of peaceful protesters filed through their towns. Later it demonstrated black anger and frustration, as it turned its cameras on arsonists and looters in the summer riots of 1965 and 1967.

The civil rights movement was one of the key events of the century in this country. And it was perfectly suited for the new news medium. As social scientist Molefi Asante wrote, "Television found this confrontation with guns, whips, and electric cattle prods on one side and love and nonviolence on the other as the classic drama of black and white, good and evil."[3] It had conflict. It featured "big-name" personalities on both sides (Dr. Martin Luther King and Alabama governor George Wallace, for example). It was uniquely American, being played out across the country "on location" from Washington, Birmingham, Detroit, Miami, and countless other cities and towns. And with images of marches and fiery speeches, taunting mobs and soulful singers, raging fires amid clenched fists, it was certainly "visual."

Television in Vietnam

A second major news event of the 1960s took place thousands of miles away. But it, too, harnessed the power of TV—particularly its ability to bring distant events into America's living rooms. The event was the war in Vietnam. Although Vietnam has been called "the first television war," TV had gone to Korea in the early 1950s. This early coverage, however, lacked the immediacy that would characterize Vietnam reporting. The years in which the war was actively fought by Americans (1961–1975) were indeed the "television years." During this period TV news made virtually all its major advances: portable cameras, satellite relay systems, color, videotape replay, and on and on. This was the era when TV's first generation of reporters (Eric Sevareid, Chet Huntley, Charles Collingwood, and others), who had been trained in radio or print journalism, gave way to a new wave of youthful reporters who had grown up in the age of TV. In this group were Ed Bradley, Ted Koppel, Steve Bell, and Don Farmer.

Like the civil rights movement, the war in Vietnam provided a training ground for the new people and techniques coming to TV. As a result, the public was deluged with daily reports, with illustrations of American and enemy dead, with dramatic "point-of-view" shots from cameras mounted on helicopter gun ships and later in the bellies of evacuation aircraft.

---

[2] In David J. Garrow, *Protest at Selma* (New Haven, Conn.; Yale University Press, 1978), p. 111.

[3] "Television and Black Consciousness," *Journal of Broadcasting*, Vol. 26, No. 4, Autumn 1976, p. 138.

Walter Cronkite reporting from Vietnam. The Vietnam war has been called the first television war because of the major role played by TV in covering the conflict.

Regardless of one's political affinity or personal feelings about the Vietnam War, TV coverage undoubtedly had an effect. For many who lived through those years, the war is remembered as a series of indelible TV images: the whir of engines as choppers evacuated wounded soldiers; GIs "torching" a village with their Zippo lighters (reported first by Morley Safer on CBS in 1965); antiwar demonstrators chanting "the whole world is watching" as they clashed with police during the 1968 Democratic National Convention in Chicago; the execution of a prisoner by Saigon police chief Lo An (aired on NBC in 1968); panic-stricken South Vietnamese clutching the landing gear of evacuation aircraft in 1975; flotillas of rafts and fishing boats crammed with refugees as the "boat people" fled their country in the late 1970s; and so on.

Grist for the News Mill

By no means were Vietnam and the civil rights movement the only major stories in the 1960s and early 1970s. Nor were they the only major focus of the TV news camera. The decade was marked by other momentous occasions. There were the assassinations in 1968 of Dr. Martin Luther King and presidential candidate Robert Kennedy. Both were marked by exhaustive coverage, from the announcement of the tragedies to extensive live coverage of the funerals.

In the summer of 1969 Neil Armstrong set foot on the moon, an event witnessed by the largest global TV audience up to that time, an estimated 600 million people on six of the seven continents. NASA had carefully orchestrated the event as a TV program, having mounted a small camera (provided by RCA) on the steps of the landing craft, in front of which astronauts Armstrong and Edwin "Buzz" Aldrin would cavort. The two unfurled and planted an American flag, stiffened by an aluminum rod to give the appearance that it was fluttering in some imaginary lunar breeze. Within minutes President Nixon was on a split-screen to talk to the astronauts by telephone. Two hours later the world audience watched as the astronauts blasted off for the return trip to Earth. The astronauts had provided for this unique TV angle (the first point-of-view shot from outer space) by setting up another RCA camera in the lunar soil.

TELEVISION NEWS IN THE TORTUOUS SEVENTIES

The next phase of TV journalism reflected an era of skepticism and cynicism that many observers believe persists today.

Watergate

On June 18, 1972 a strange news report made its way into the *Washington Post* and later into TV news reports. It seems there was a burglary at the Watergate, a Washington office building and headquarters of the Democratic National Committee. In an address book at the scene was written "Howard Hunt–WH." A link to the White House was found.

The next two years of the Nixon presidency would be filled with revelations about espionage and "dirty tricks," with a web of intrigue leading closer and closer to the president himself. Much of the drama was played out on the nation's TV screens. In the spring of 1973 the Senate authorized a Watergate inquiry, chaired by Senator Sam Ervin of North Carolina and open to TV coverage. By the summer the Watergate hearings had become TV's most popular program, outranking game shows, soap operas, and movies in the daytime and producing staggering numbers for public TV when repeated at night. Watching the hearings became a public obsession.

Shortly after his 1972 landslide victory, President Nixon's approval rating in the Gallup polls had reached 68 percent. By the following August it had slipped to 31 percent. After another year of televised hearings, revelations, damaging transcripts of tapes recorded during Oval Office meetings (one with a glaring gap of almost twenty minutes), and impeachment proceedings, on August 8, 1974 President Nixon resigned. There was no doubt: TV news had helped bring down a president.

In retrospect, the Watergate episode had a permanent and profound effect on both politics and TV. At the national level an adversarial tone developed between the president and the White House press corps. Television journalists and their national viewers became increasingly skeptical of political leaders and their policies.

Television became proud of its role in the Watergate coverage. The mid- and late 1970s were marked by a self-assurance (some would say "pushiness") never seen before. At the national level CBS's "60 Minutes" zoomed to the top of the ratings. Its aggressive techniques of doggedly tracking people and showing up with cameras rolling became known as "ambush journalism."

At the local level, TV stations began their own version of aggressive coverage. Local newscasts expanded, from a half hour in most cities to an hour or more in the early news period (from 5:00 P.M. to 7:00 P.M.). In the mid-1970s portable cameras and videotape recorders enabled local news to adopt some of the techniques of "60 Minutes" and other national programs. A new language of TV journalism emerged. Electronic news gathering and electronic journalism, covered in detail later, were the bywords for a new assertiveness in local TV news. The camera could go anywhere and it did: extensive coverage of local political scandals, murders, and natural and man-made disasters filled America's TV screens as local news managers placed a premium on stories that were "visual." At the same time, there was a curious sort of behavior on the sets of the newscasts.

"Happy Talk"

Throughout its history, TV news had been identified with its anchors and reporters. Newscasters, at both the national networks and local stations, tended to appear as strong, serious, some would say staid, and conservative. At the networks

ECONOMICS

WE DON'T NEED YOUR NETWORK NEWS!

For more than a generation, the backbone of TV news was the half-hour evening newscast on the three major networks. Not any more. In the age of global news at one extreme and local cable news at the other the power of network news has diminished considerably.

Nowhere is this trend more apparent than at network affiliated stations. Many program directors and general managers would like to pull the plug on their network news half-hour and in its place either run their own newscast or a syndicated program. No longer do all three network newscasts vie for the audience against one another at 7:00 P.M.. In some cities network news runs early, at 6:30 P.M., for example. In others, the show is tape-delayed, for a 7:30 P.M. run.

A recent survey of station management (reported in *Mediaweek,* September 1991) may have sounded the death knell for network news. A staggering 44 percent of general managers in the nation's top-50 TV markets said network news was of little or no importance to them. Only one in three felt that keeping the news would be important for them to compete for audiences in their marketplace. Network news was not deemed as significant to the ratings of their local newscast as were other factors, including their competition, cable news, and declining public interest in TV newscasts.

Will we soon see the end of Dan, Tom, and Peter? As they say, stay tuned!

there were Walter Cronkite, Eric Sevareid, John Chancellor, David Brinkley, Nancy Dickerson, and many others. In New York there were Bill Beutel and Jim Jensen; in Chicago, Fahey Flynn; in Los Angeles, Clete Roberts. In every major city there were established figures in news, each with impeccable credentials and an aura of journalistic integrity and self-assurance.

In the late 1970s a new breed of newscaster began to appear. Men were younger, more daring, even "dashing," frequently dressed in the most modern clothing styles, many with facial hair (previously avoided—except for the stately moustache sported by Cronkite).

For the first time women began to appear in the anchor position. Like the new breed of male newscasters, they were young and attractive. On-camera looks, charisma, charm, and sex appeal became at least as important as journalistic training and ability.

Not only did this new breed look different, but they acted differently from their predecessors. They talked to each other on the air. Sometimes they talked about the news reports they had just seen; sometimes they just seemed to be engaging in the kind of gossip and repartee found in most offices. This new approach to TV news was lambasted by the more serious print media, which called it "happy talk." There was concern that TV news was moving away from issues toward personalities, away from information to entertainment, and away from serious news to "sleaze." Research seemed to support these claims. Studies found that stations with the happy talk format featured more sensational and violent content than did traditional newscasts. They also had higher ratings.

TELEVISION NEWS IN THE HIGH-PROFIT, HIGH-TECH EIGHTIES

In the past many station owners and network executives felt that news was a "necessary evil," a costly enterprise that existed mainly to meet FCC requirements for public service or to promote good will in the community. The news budget was only a fraction of the amount spent on programming.

In the 1980s broadcast managers discovered that TV news could be an enormous profit center.

High ratings for early local newscasts meant substantial lead-in audiences for the prime-time or early evening programming that followed. Popular prime-time shows meant large audiences for the late news. And although stations affiliated with the major networks could sell only a few advertising "spots" during primetime, they could sell *all* the commercial positions during their newscasts.

By the late 1980s three-fourths of all TV stations were reporting profits from their news operations. Most of the remaining stations were breaking even; only one in ten stations claimed they were losing money. And the profit margins were astounding. The leading station in large markets, like Phoenix, Atlanta, New York, Houston, and Cincinnati, could make more than a third of the annual profits of the entire station from local news alone. As Professor Vernon Stone, longtime broadcast news observer, concluded, "It takes special effort to lose money on TV news."[4]

News All the Time: Longer Newscasts

Another trend in the 1980s was the advent of longer newscasts. This trend spans all types of news. At the local level the early evening newscast was no longer just thirty minutes; in some larger cities the block ran as long as two hours. By 1989 over 250 stations had at least one hour of early local news (more than a dozen had two hours or more). In addition, many major-market stations had a noon newscast, prime-time news capsules, and, of course, at least a half hour of late-night news.

Longer news was also a feature at the national level. When Iranian militants seized the American embassy in Teheran in 1979, taking fifty-two Americans hostage, ABC decided to follow the event on a nightly basis. The report evolved into the award-winning "Nightline," then and now hosted by Ted Koppel. Not to be outdone, NBC followed suit with a late-night news program ("NBC Overnight"), and CBS experimented with its own "Nightwatch." These programs got competition from an unlikely source: entrepreneur and maverick personality Ted Turner.

---

[4] "Newsrooms Remain Profit Centers," *RTNDA Communicator*, April 1988, p. 52.

CNN and Headline News

In 1980 Atlanta-based cable entrepreneur Ted Turner launched a twenty-four-hour news service, Cable News Network. A second service, originally named CNN II but now called Headline News, followed. Traditionalists, particularly those at the three commercial networks, scoffed and predicted the early demise of the ventures. Thus far Turner has had the last laugh. CNN has become a prime source for national and international news. Its twenty-four-hour schedule and its ability to cover breaking stories live and in their entirety have given it the edge on the networks, which are often reluctant to interrupt soap operas in the daytime and comedies and dramas at night. CNN coverage of events like the assassination attempt on President Reagan in 1981, the space shuttle disaster in 1986, and the stock market plunge in 1987 earned the channel many awards. In 1987 the prestigious Peabody Awards board termed CNN "the channel of record."

The success of CNN was not lost on the traditional networks. One network tried its own service. In 1982 ABC and Westinghouse launched Satellite Newschannel. Turner promptly acquired it and shut it down.

Turner proved that twenty-four hours of reliable, credible news could be produced at reasonable cost. By the late 1980s the annual budget for both CNN and Headline ran at approximately $100 million. By contrast, the budget to produce thirty minutes per day by each of the three networks was running in the $200 million to $300 million range. Partly for this reason the networks trimmed their news operations significantly in the late 1980s, led by a series of highly publicized layoffs. Ted Turner, maverick cable manager, had outdone the networks, at least in the news area.

The New Technology of Television News

Another major trend in TV news in the 1980s was its dependence on new communications technologies. The trend can be summed up in two three-letter abbreviations: ENG and SNG.

Electronic news gathering, known as *ENG*, emerged in the mid-1970s, when portable video cameras and recorders became commercially available. The basic elements of ENG include a lightweight TV camera and a small videotape recorder, which can be hand-held by the cameraperson. Today's ENG equipment includes camera-recorder combinations, known as camcorders.

ENG revolutionized TV news. Prior to the introduction of ENG, TV news relied on cumber-

The CNN news floor. Because of its large size, shape, and technological complexity, the CNN news set is known as "the pit."

News reports can be sent live from almost anywhere these days thanks to flyaway uplink equipment. The apparatus shown here can be transported in a couple of suitcases.

some and costly film equipment. Film stock needed to be processed and physically assembled or edited. It could not be reused. Film cameras required unwieldy and obtrusive lights. With ENG, cameras and recorders could go anywhere. Events could be recorded in full, with natural sound, and the most important parts could be edited together electronically, in a speedy and cost-efficient manner. By the end of the 1970s TV news had scrapped its 16-millimeter film processors and cleared the shelves of its film vaults. The age of ENG had arrived.

At the same time, the dawn of *satellite news gathering (SNG)* had arrived. Satellite news gathering refers to the use of mobile trucks mounted with satellite communications equipment to report local, national, and international events. SNG trucks and vans could transmit the pictures and sounds gathered by ENG up to communications satellites. The satellites sent the signal back down to local stations and networks to use in their news programs. It was thus possible to obtain live news pictures from virtually anywhere in the world.

Although satellite technology had been available to the major TV news organizations since the 1960s, the breakthrough came in the 1980s when the technology became available to local stations. Today many stations own "flyaway uplink" vans, capable of transmitting events live to a satellite and back to earth, to the studio.

The News as "Show Biz"

With increasing profitability inevitably came the "ratings war." News directors began to follow ratings with at least as much concern as their counterparts in the sales and programming departments. At local stations the competition for audiences was most intense during the periods when the leading ratings companies, Nielsen and Arbitron, conducted surveys of local TV viewing (see Chapter 14). This period is known as *sweeps*.

In TV the sweep periods occur four times a year—in February, May, July, and November. In a quest for "number 1" status many local news operations took to programming series of short documentaries, known as *minidocs*, during these times. Many minidocs aired during the sweeps focused on sensational topics, including teenage prostitution, spouse and child abuse, and religious cults. They were accompanied by extensive advertising campaigns on the air and in print, on billboards and buses. In many ways the hoopla surrounding news was not unlike that which accompanied the premiere of a new motion picture. In fact, TV news became the stuff of movies and TV shows—*Broadcast News, Switching Channels*, "Max Headroom," "Murphy Brown"—suggesting that the TV newsperson had become a modern icon.

A number of *news consultants* emerged to provide the expertise required of news managers in this competitive environment. Leading news consultants included such firms as Frank Magid, Reymer and Gersin, and McHugh-Hoffman. Consultants provided advice to stations to use in selecting their talent, designing their sets, using graphics, devising sweeps-weeks promotions, and in other areas. Their research techniques, ranging from focus group tests to viewer surveys, are analyzed in Chapter 15.

IMPACT

NEWS POWER TO THE PEOPLE: THE CAMCORDER REVOLUTION

If TV news reporters developed a conceited, "holier-than-thou" attitude over the years, it may be in part because they wielded considerable power. After all, *they* had the cameras and microphones, not the president, the mayor, the police chief, or "Mr. Average Citizen." But in the 1980s a quiet revolution was going on: TV cameras were making their way into the hands of the public. By 1992 about one in eight U.S. households owned a camcorder, a combination video camera and recorder.

More and more, the TV networks and local stations are finding a place in their newscasts for "home-grown" news reports. One celebrated news event covered by a so-called amateur occurred in 1987. A British doctor attending a medical conference in Moscow was standing in Red Square when a West German teenager, to the consternation of Russian police and politicians, landed his small plane among the throng of startled tourists. The doctor got it all on his camcorder. Within hours, the tape was transmitted by the NBC Moscow bureau to a worldwide audience (at much embarrassment to Russia, which hauled the precocious pilot off to jail).

Perhaps the most famous piece of amateur news video to date was shot by George Holliday late one evening in 1991 in Los Angeles. From his balcony Holliday saw a young man on the ground, apparently under orders from one of the numerous police officers on the scene. When the officers started beating the man, Holliday picked up his newly purchased camcorder and began taping. The next day he sold the tape to his "favorite news station," KTLA, which put it on the air. Within two days the tape had aired nationally on ABC, CBS, and NBC, and internationally on CNN. The taped beating of Rodney King caused an international uproar, eventually leading to an FBI probe, a review of police procedures, the resignation of the chief of police, and a wave of riots following the not-guilty verdict in the trial of four Los Angeles police officers.

A frame from one of America's most watched home videos: the beating of Rodney King.

BACK TO REALITY: TELEVISION NEWS IN THE NINETIES

By the early 1990s some of the gloss had rubbed off TV news, especially at the local level. As costs continued to climb, station owners, many of whom were saddled with debt from the purchase of their stations in the 1980s, began to question the value of their local news commitment.

In 1991 and 1992 shrinking local advertising revenues had a significant impact on TV news. Staff cutbacks became routine, such as fourteen staffers at KIRO in Seattle, twenty-two layoffs at WLNE in Providence, and ten layoffs at WKXT in Knoxville (which dropped its 11:00 P.M. newscast). By 1993 some stations that were third in the ratings were considering dropping their newscasts in favor of syndicated reruns.

PROFILE

CONSULTING RUN AMOK? THE CASE OF CHRISTINE CRAFT

The rise of the consultant to prominence in local news has generated some controversy. Some critics fear that consultants by their nature turn the attention of management from substantive matters—like news—to trivial concerns—like the makeup and hair styles worn by anchorpersons. This particular concern became paramount in a highly publicized incident. A consulting company had been hired by the Metromedia company, which then owned KMBC in Kansas City, Missouri. Station management was interested in viewer perception of the local newscast. Part of the research called for a focus group, to see how men felt about the co-anchor Christine Craft. According to a tape transcript, it was concluded that Craft was "too old, too unattractive" and "not deferential enough to men."

Shortly after the consultant's report was issued, Craft was fired by station management. She sued, and after three tumultuous years of litigation finally reaching the U.S. Supreme Court, she won the case. Although she successfully proved that Metromedia was guilty of fraud (Craft had been assured she had been hired for her journalistic skills and not her looks), she never received a monetary judgment.

In 1985 Metromedia sold its stations to Rupert Murdoch's News Corporation. Most recently, Craft was anchoring for KRBK in Sacramento, California. And today more news consulting firms exist than at any time in the past.

The Rise of Independent News

The news was not all bad at the local level. Many independent stations began offering news at alternative times to the local network affiliates. Independent news at 9:00 P.M. and 10:00 P.M. (instead of at 10:00 P.M. and 11:00 P.M.) became attractive to an increasing number of viewers. Leaders in this trend included the stations owned by Tribune (such as WGN in Chicago and WGNX in Atlanta) and some Fox stations located in major markets. Some affiliates used their news operations to aid independents in their markets. For example, the CBS station in Tampa (WPEC) began producing a newscast for independent WFLX. Similarly, longtime CBS affiliate and local news leader KOIN in Portland began to produce a 10:00 P.M. show for Fox affiliate KPDX. These stations and others agreed to pool resources and, most importantly, to share revenues produced by the new newscasts.

Conventuring with Cable: "Local-Local" News

If cooperation with competing independent stations wasn't shocking enough, by the early 1990s some local TV stations began to "sleep with the enemy"—local cable systems. The idea is to use the resources of local affiliates to feed specially customized newscasts to cable subscribers in a given community. Typically special short newscasts, known as "inserts," are produced for the residents served by a leading cable company. These air only in one cable area; over-the-air viewers stay with the affiliate newscast. The phenomenon is known as "local-local" since the cable subscriber receives an even more localized version of a local newscast.

Toys and Toasters: Low-End Technology

Another trend for TV news in the 1990s was a move from expensive "high-end" equipment, like satellite trucks and elaborate graphics computers, to low-cost equipment that is very similar to what is available on the consumer market. Bulky and expensive $\frac{3}{4}$-inch camcorders were replaced by very small (and much less expensive) 8-mm and $\frac{1}{2}$-inch camcorders.

The fall of communism and the Gulf War helped usher in this new era. Major networks relied on VHS and 8-mm cameras for their video, mostly because they were small (and could be smuggled easily into important locations), and at $1000 to $3000 each they were expendable. News executives were impressed with the quality of the images that these cameras produced. There were additional advantages for news producers. A separate reporter, soundperson, and cameraperson were no longer necessary, ushering in the era of the one-person crew, known in the trade as the "one-man band."

Microprocessors, developed for personal computers, were inside the editing VCR's, enabling video signals to be stored and manipulated like other digital computer data. This made it possible to create elaborate editing and graphic effects (such as show openers, titles, view boxes over anchors' shoulders, and so on) much quicker and cheaper than in the past. One early system created for Commodore's Amiga computer became known as the "video toaster."

Global News

The success of Cable News Network engendered both domestic and international competition. On the domestic scene, Minneapolis-based Hubbard Broadcasting expanded its CONUS satellite news service. Both NBC and CBS organized newsfeed services to provide video to their affiliates, after being embarrassed to see their stations turn to CNN during crises, such as the San Francisco earthquake in 1989 and the outbreak of the Gulf War in 1991.

International media companies began to position themselves to battle CNN for a share of the global market for TV news. In Britain BBC World Service Television began offering eighteen hours per day of coverage for viewers in Europe, with plans to sign affiliates in North America and the Far East. Sky News, a service of News Corporation, was launched as an alternative to CNN in the growing European cable market. In France TF-1 and Canal Plus planned an all-news service in French. Germany's Bertelsmann was considering all-news for its newly unified country, as well as Switzerland and Austria. In Japan NHK was producing an eight-hour global news service, including English-language news and financial reports designed for Western media audiences. At the time of this writing CNN was undaunted. It had expanded its CNN international service into over 100 countries and had developed a separate Spanish-language service targeted to Latin America.

Mini-CNN's: Cable News Channels

A final trend in TV news for the 1990s was the advent of full-service newscasts offered by large cable systems. Unlike "local-local" news provided in tandem with area broadcast affiliates, a number of cable systems either developed their own news service or enlisted the aid of a local newspaper to offer a dedicated local news channel. The leader in this new trend was Cablevision Systems in Long Island, a suburb of New York City. Their news service "News12 Long Island," became an important source of news for Nassau and Suffolk counties, which had been mostly ignored in previous years by affiliates in New York City. In the early 1990s additional all-news cable service became available in cable households in suburban Washington, D.C., Orange County (near Los Angeles), California, and Chicago, Illinois.

THE TELEVISION NEWS TEAM: TELEVISION NEWS COMMAND STRUCTURE

Television news is a collaborative craft. It is not unusual for a large-market station to boast more than 100 people in the news department. And despite highly publicized layoffs, the major networks still maintain hundreds of news personnel. CNN alone employs over 600 people at its Atlanta headquarters. Coordinating the efforts of these small armies is no easy task, especially with the pressures of both the deadline and the bottom line. Thus TV news organizations tend to follow rigid command structures, outlined in Figure 10-2.

For convenience, Figure 10-2 separates TV news production into two phases. Phase I, the left side of the chart, refers to tasks and processes that take place primarily before the broadcast begins. This phase is called preproduction. The right side of the diagram covers the personnel involved on the air, or the production phase.

Phase I: Preproduction

The news director Overall responsibility for the news department falls to the news director. News directors tend to have a solid background in TV news, many as former reporters or anchors. The news director hires news personnel, establishes news and editorial policy, and evaluates the newscast in "postmortem" screenings with the staff of the news department. Today's news directors are increasingly concerned with budgets, making sure the news is produced at a profit. The exciting, high-pressure nature of the job has led

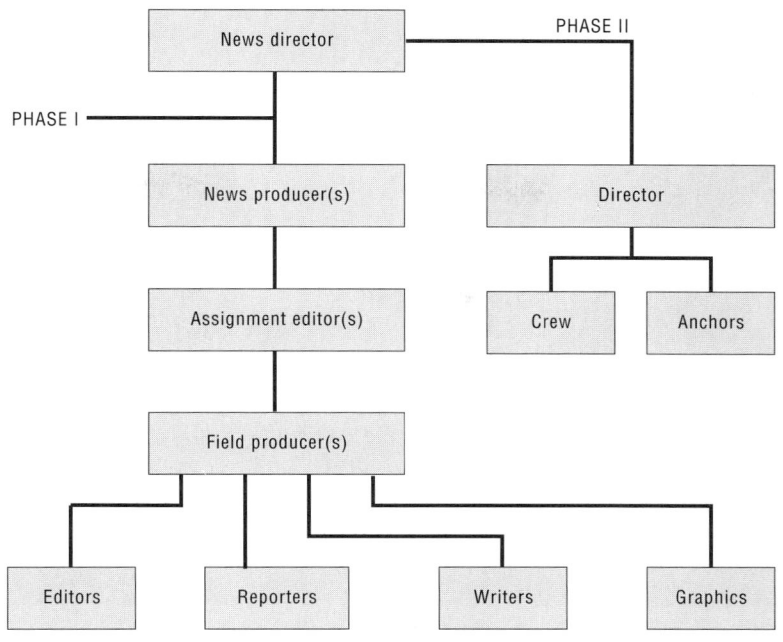

FIGURE 10-2 TV news command structure.

to a typical job tenure in the range of about three years (see box).

News producer If the news director is the "boss" of the news operation, the news producer is the czar of any given newscast, such as the "11:00 O'Clock Report," or "The Noon News." The producers maintain editorial control over the stories that make up their individual TV newscasts. The producers prepare the story lineup, determine the stories' length, and decide how they will be handled. The producer of each news-

PROFILE

LIFE IN THE FAST LANE: THE NEWS DIRECTOR

In a recent poll nine in ten TV news directors reported they liked their jobs because they work best under pressure. They find deadline pressure "exhilarating." But there is a flip side to their exhilaration. More than a half of the male news directors reported they had job-related personal problems. Although most were married at the time of the survey (80 percent), one-half of those polled admitted they were having marital problems. Nearly a third had been divorced at least once. Forty-two percent admitted they had a least one drink daily; a third were smokers. Not surprisingly, the lifestyle of the news director was extracting a high price in health: one in four admitted to having serious job-related health problems, ranging from anxiety and depression to coronary artery disease.

The forty-four women news directors polled matched their male counterparts in some lifestyle characteristics and not in others. Like male news managers, more than nine in ten relished the pressure of the business. About the same proportion of women as men smoked cigarettes. However, unlike the male news directors, most women managers were unmarried. Fewer found the need to drink alcohol on a daily basis. But women news managers reported job-related stress more than their male counterparts (or they are more likely to admit their problems than are the men)! In any case, the job of TV news director clearly takes a sizable toll on one's personal life.

TV news camerapeople on the job. Such a sight is a common one to most newsmakers.

cast will proofread all copy, select the graphics, and be present in the control room during the newscast to make last minute changes.

Most news operations have special segments or "beats," such as consumer affairs, health, arts and entertainment, and, of course, sports and weather. These beats are headed by unit producers, who assume overall responsibility for these segments.

Assignment editor   The main job of the assignment editor is to dispatch reporters and photographers to cover news stories. The job requires great organizational skills. At a local station, as many as five different crews may be out in the field at one time. At a network literally dozens of crews need to be assigned and returned to the studio for editing and other production tasks. To assist in this complex task, assignment editors maintain a *future file*. This is an annotated listing of upcoming news stories, scheduled as many as thirty or sixty days in advance. Traditionally a series of file folders is each labeled with upcoming dates and filled with events for each day; many future files today are stored in computers. Of course, the future file is useless when breaking news occurs: a plane crash or tornado really taxes the skills of the assignment editor.

Field producer   If there is sufficient lead time to prepare for a story, such as a space shuttle launch or a murder trial, it may be assigned to a field producer. The field producer will prepare research and background information, scout locations, and perform other valuable advance functions for the reporter and photographer, who will be busy working on other stories. On breaking stories of great magnitude (an assassination attempt or a skyjacking, for example), the news director may assign a field producer to coordinate the coverage. Field producers generally are found in the large markets.

Reporter   Reporters receive their orders from the assignment desk and begin their research, usually on the telephone. They set up interviews and determine shooting locations. Typically the reporter has been teamed with a photographer. At small stations and some budget-conscious large ones the reporter may be a "one-man band," responsible for shooting her own stories. The reporter proceeds to the scene, obtains interviews and other video, and returns to the studio to supervise the editing, write and announce the voice-over and lead-in copy, and, if part of the newscast, get dressed and made up for the newscast.

Writers It is a sad fact that many of today's reporters may be selected for their on-camera appearance rather than their journalistic skills. For this reason most stations and networks employ newswriters to script the newscast. Writers may prepare the material for reporters and anchors. They frequently prepare the clever lines used to introduce the newscast (*teasers*) or to attract the audience to return after commercial breaks (*bumpers*).

Tape editors Most newscasts consist of separate stories, called *packages*, linked by the anchors and reporters. Each package is a complete story, edited together with its own voice-over and graphic material. The tape editor is the person who puts the package together. Typically the reporter returns from the field with too much material, all of it shot out of sequence. He may have done the introduction to the story six or seven times; there may be dozens of separate pieces to an interview. The reporter (and sometimes the field producers, unit producer, producer, or news director as well) retreats with the editor to a small editing room to make sense of the mayhem of the typical field report. In many small news operations, reporters edit their packages.

Phase II: On the Air

Director As air time approaches, pressure shifts from the news producer to the director. The director is the studio boss. The director is the person who "calls the shots" during the newscast, selecting the camera shots and calling for the packages to roll. The director is assisted by the full range of TV crew personnel. Pushing the buttons that correspond to the many cameras and videotape recorders is the technical director. Selecting the sounds, from microphones to music, is the audio director. In the studio are two or three camera operators. A floor director gives cues to the anchors and reporters, instructing them as to which camera is on and when the commercials are coming. The director is in communication with the people who put words and pictures on the screen, the graphics and videotape operators.

Anchors and reporters rarely read from the printed copy they appear to be holding. Reporting to the director are the people who operate a machine that projects the script using mirrors and reflectors into a space just in front of the cameras: the teleprompter or "prompter," for short. This is where the anchors look to read their copy.

Today the director has the added burden of coordinating live reports from the field, which

Local TV news is a profit center for most stations. In this scene, the floor manager (kneeling) is getting ready to cue the anchors.

may involve pressurized last-minute harangues with satellite companies and panicky reporters. Helping the director with this range of diverse tasks will be one or more assistant directors, who may be assigned to help with time cues, supervise the insertion of video material, or monitor the preparation of graphic material.

As if the job wasn't difficult enough, during the broadcast the director is in constant communication with the news producer to keep the program on time and to allow for last-minute additions and deletions. Most of this activity takes place during commercial breaks, the only respite from the mania in the studio and control room during a broadcast.

Anchors It is important to note how far down we have come in Figure 10-2 before mentioning the people who get most of the attention in TV news: the anchors. Like it or not, the anchors are the celebrities in TV news. Most have the bankrolls and wardrobes to prove it.

Simply put, the anchors represent the newscast to the viewer. The viewing public recognizes the anchors as the news spokespersons for the station. For good or ill, frequently the anchors *are* the station's newscast. For this reason anchor salaries often reflect their celebrity status. It is estimated that Dan Rather of CBS earns $3 million annually, Peter Jennings of ABC, $1.2 million, and NBC's Tom Brokaw, $2 million. At the local level large-market anchors have been known to command upward of $500,000 per year, with the median in the $200,000 range.

What do anchors do to earn these salaries? Their primary responsibility is to read the news in a trustworthy and authoritative fashion. In some cases they actually write their news copy (their networks claim this is the case with Rather, Brokaw, and Jennings). In addition, anchors need to be good interviewers, especially when handling live reports from the field. The venerable Walter Cronkite (and his heir, Mr. Rather) carried the additional duties of executive producer, with editorial control of the "CBS Evening News."

Most analysts (and critics of the movie *Broadcast News*) point out that the network anchors remain solid journalists with excellent credentials. However, this is not always the case at the local level. There is growing evidence that some anchors are selected primarily for their on camera charm and good looks, independent of their writing and reporting abilities. Stories abound in the trade and popular press about anchormen and women who look great but are largely ignorant about the nature and practice of journalism. In a recent popular song Don Henley sang of "the bubble-headed bleached blond (who) comes on at 5:00." Sadly, there may be more than a kernel of truth in that stereotype.

One of the new realities of broadcast news for the 1990s may be the dimming of the star for news anchors. Like automobile workers and college professors, some well-known news anchors have been laid off and scores of others have had to take pay cuts in order to keep their jobs. For example, when Deborah Norville left *"Today,"* NBC replaced her $1 million salary with Katie Couric's estimated $550,000. Pat Harper ($600,000) and Kaity Tong ($750,000) were replaced at WNBC in New York by Dawn Frantangelo ($180,000) and Susan Roesgen ($175,000), respectively, a net savings to GE of nearly $1 million per year.

Industrywide, starting salaries for reporters and anchors have been reduced. At the time of

IMPACT

TO BE YOUNG, GIFTED AND BLOND

One way to save money in TV news may be to replace higher-paid, established news anchors with younger, prettier, and lower-salaried news personalities. There seemed to have been a lot of this going on in newsrooms around the country as the go-go 1980s gave way to the belt-tightening 1990s.

One study conducted by the Gannett Center for Media Studies found that for over three years in the late 1980s, male TV anchors—including Tom Brokaw, Dan Rather, and Peter Jennings—had gotten older and grayer (and presumably richer). On the other hand, women in network TV positions had gotten younger and blonder. On average, women anchors were twenty years younger than their male counterparts, and they earned 23 percent less.

contract renewal, many reporters have been forced to take pay cuts. As one well-known CBS correspondent said, "It's hard to get a job right now. We're not in the best bargaining position." CBS News President Eric Ober added, "This is not the high-spending, can-you-top-this of the 1980s."

THE JOB OUTLOOK IN TELEVISION NEWS

If one were to rely on the evidence from recent national publications, like *Time, Newsweek*, and *People*, the end is in sight for a career in TV news. There has been great publicity about the layoffs affecting the news divisions of the major networks and many affiliates. However, as we have discussed, what occurred is a realignment in TV news. Clearly, network opportunities in news are in decline as new ownership seeks to control escalating costs. At the same time, however, job opportunities may increase as independent TV stations and cable companies enter TV news.

All told, about 25,000 people work in TV news. Although men outnumber women in TV news by about three to one, more than 85 percent of TV news operations reported women on their staffs in a recent survey (compared to about 50 percent in 1972). Women (particularly white women) are slowly making inroads into management in TV news. Nearly 40 percent of producers and about one in three reporters, anchors, and supervisors are female.

The outlook is less promising for ethnic minorities. Blacks and Hispanic Americans comprise about 13 percent of the TV news work force, a similar proportion to that reported in surveys in the 1970s. Gains in employment have been slowest for Hispanics. While the national Hispanic population increased nearly 30 percent in the 1980s, the number of Hispanics working in TV news has dropped in recent years. Management and supervisory positions in TV news appear to be blocked to blacks and Hispanics. About one in twenty news managers is black or Hispanic, but there are almost none in major cities. Generally, blacks and Hispanics in TV news tote cameras: the largest proportion of minorities (about 30 percent each) can be found in field reporting and camera positions.

Salaries have remained flat in recent years, but there have been gains in some areas.[5] Nationwide the average salary for TV anchorpersons is about $25,000; producers make $21,000 and reporters about $20,000. Their supervisors make more money: the typical news director makes $45,000 and executive producers about $34,000. Executive producers, with increasing responsibility for "sweep weeks" series and special reports, have seen their salaries rise in recent years. On the negative side, anchor and reporter salaries have remained the same at most affiliate stations and have dropped at some financially strapped independent stations.

SUMMING UP: THE PUBLIC'S APPETITE FOR TELEVISION NEWS

When we began this chapter, we pointed out that TV news emerged in the 1960s as the public's most common and most trusted source of news. This is borne out by the spectacular growth in TV news viewing observed in recent years (see Figure 10-3).

In the 1960s the average American adult watched about ninety minutes of TV news per week. In the early 1970s the figure grew to just over two hours. By the end of the 1970s "Mr. and Mrs. Average American" viewed three hours of news per week. Today we watch over five hours of news each in a given week.

For TV news, that's the good news. The bad? With growing consumption has come increased skepticism and distrust of the press in general and TV news in particular. Although viewing TV news is at its highest level yet, there is evidence that the American public is losing faith in the institution.

For many years the National Opinion Research Center has conducted surveys tapping the degree of confidence the public has in various American institutions, including the press.

In the early 1970s (the height of the press's triumphant Watergate period) confidence in the press was high. One in four Americans expressed

---

[5] "Broadcast News Salaries Flat to Down," *Broadcasting*, February 3, 1992, p. 22.

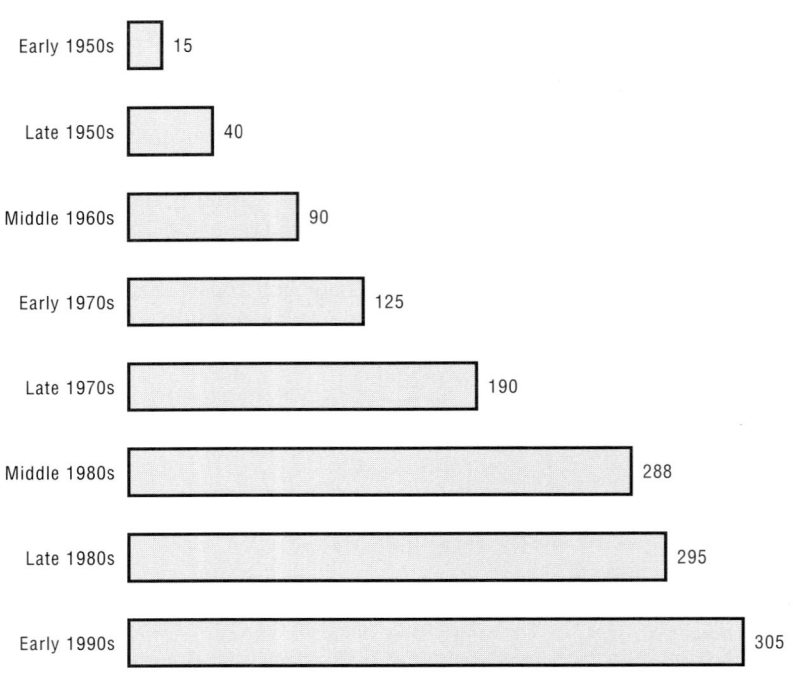

FIGURE 10-3 Trends in the average adult's weekly consumption of TV news, including news, weather, and other cable network news programs. (Source: *TV Dimensions, '88* and *TV Dimensions, '91.* Used with permission.)

great confidence in the press; only 15 percent expressed little or no faith in it. Today the situation has reversed. A quarter of the population has little confidence in the American press; less than one in five place great faith in it.

The medium may not be helping its own cause. With an emphasis on ratings, with stories about satanism, cult worship, prostitution, and aliens replacing investigative reports and public service efforts, today's news has become "tabloid television." Other phrases are less complimentary: they include "sleaze-o-vision" and "trash TV."

The line between news and documentary programs and entertainment shows has blurred. "Infotainment" and "reality-based" programs have become commonplace, like "America's Most Wanted," "A Current Affair," and "Hard Copy." Former news reporters (most notably Geraldo Rivera) have turned up as hosts of their own syndicated programs. Confrontational personalities with little or no news experience (like Rush Limbaugh) conduct programs in which current news issues are highlighted and debated, often in a pseudo-serious manner. The networks, long the home of "serious" news, are taking the plunge. In 1988 NBC aired Geraldo Rivera's special "Satanism in America," which the critics denounced but the public rewarded with the highest ratings in history for a three-hour documentary.

More recently the success of its weekly newsmagazine "48 Hours" led CBS to produce a spin-off called "Street Stories," with a weekly emphasis on crime, drugs, prostitution, and other horrors of contemporary urban life. ABC was peering into the FBI's "Untold Stories." NBC had already dispatched former TV FBI man Robert Stack in search of solutions to "Unsolved Mysteries." There were plans for more reality shows, strong on innuendo, crime, and mayhem, if a bit fast and loose with facts.

Whether or not they cause diminished faith in the press, such shows will continue to proliferate in the 1990s. For one thing, they are very popular with viewers. In one typical week in 1992, for example, six reality shows (including "60 Minutes," "20/20," "Prime Time Live," "48 Hours,"

"Rescue 911," and "Street Stories,") were in the top-20 programs as rated by the Nielsen rating service. And, in view of the costs involved in making movies, dramas, and comedy programs (traced in the next chapter), reality programs were comparatively cheap to produce.

SUMMARY

Television news is the nation's primary source of news and is regarded as credible by a large portion of the public.

Television news reached maturity when President Kennedy was assassinated. The 1960s events such as the civil rights movement and the Vietnam war came to life in people's living rooms.

In the 1970s the Watergate hearings were covered by TV. Soon the trend of aggressive journalism began. Coinciding with changes in reporting styles came advancements in equipment.

The 1980s were marked by money and machines. Television news became a significant profit center for local TV stations and cable. Anchor salaries rose. Electronic news gathering and satellite news came on the scene.

The 1990s have seen significant retrenchment at the TV networks and many large stations. However, there has been continued growth in cable and independent news.

Many people are involved in the preproduction stage of the newscast. They are the news director, news producer, assignment editor, field producer, reporter, writers, and tape editors. Those involved in the production stage are the director and the anchors.

Even though the networks have cut down on employees, local stations and cable companies now serve as alternative places of employment.

With the rise of sensationalized "reality-based" programming, there is concern that the credibility of TV news is being compromised. However, such shows are economical and have proven popular with viewers.

SUGGESTIONS FOR FURTHER READING

Biagi, S. (1987). *NewsTalk II*. Belmont, Calif.: Wadsworth.

Blair, G. (1988). *Almost golden: Jessica Savitch and the selling of TV news*. New York: Simon & Schuster.

Goedkoop, R. (1989). *Inside local television news*. Salem, Wisc.: Sheffield Publishing.

Goldberg, R. (1990). *Anchors: Brokaw, Jennings, Rather and the Evening News*. Secaucus, N.J.: Carol Publishing.

Yoakam, R., & Cremer, C. (1989). *ENG: Television news and the new technology* (2nd ed.). Carbondale, Ill.: Southern Illinois Press.

Yorke, I. (1990). *Basic TV reporting*. Boston, Mass. Focal Press.

CHAPTER 11: TELEVISION ENTERTAINMENT

Most of us use the mass media for relaxation, escape, and diversion—in a word, entertainment. Entertainment does not come cheap in TV and cable. A major TV network might spend more than $20 million a week on nonnews programming. The entertainment programming business is an integral part of modern broadcasting and cable.

This chapter examines how TV programs (particularly those designed to entertain rather than inform) are developed, financed, produced, and distributed. As we shall see, the revolution in telecommunications hardware traced earlier, including cable, VCRs, satellites, and portable video, has had a significant impact on the "software" side, that is, programming. A programming market once dominated by three companies (the major networks: CBS, NBC, and ABC) is now wide open. Today there is unparalleled demand for entertainment programming. To satisfy this demand, programming emanates from a vast array of sources: the networks, Hollywood studios, international broadcasters, sports marketers, even one-person producer–directors.

While programming does indeed come from countless sources, there are three basic ways in which a TV station or cable company can acquire programs: network, syndication, and local origination. First, we provide brief definitions of these terms. Then we examine each major program source in detail.

As you might expect, *network programming* refers to original programming funded by, produced for, and distributed by the major TV networks. Television stations contract with networks to carry network shows. Recent popular network programs include "The Wonder Years" and "Family Matters" on ABC, "Murphy Brown" and "Designing Women" on CBS, "Cheers" and "Seinfeld" on NBC, and "The Simpsons" and "Beverly Hills 90210" on Fox.

Syndication refers to TV programming sold by independent companies to local TV stations and cable services. Syndication companies sell two kinds of shows: off-network (series that have appeared on the networks and are being rerun by local stations or cable systems) and first-run (shows expressly produced for syndication). Some off-net successes in syndication include "The Cosby Show," "Married with Children," and "The Golden Girls." First-run powerhouses include "Wheel of Fortune," "Jeopardy," "Oprah!," and "Phil Donahue."

Local origination refers to programs produced by local TV stations (or cable companies) for viewers in their own communities. The most common form of local origination programs are the local news and talk shows that most TV stations and cable systems carry daily. In the 1950s and 1960s many stations featured cartoon programs for kids hosted by "talent" from the station (typically a cameraman or technician frustrated in an earlier career in vaudeville or theater). If you don't remember these local shows from your youth, your parents probably do.

NETWORK TELEVISION: THE BIG THREE, PLUS ONE

From the beginnings of TV in the late 1940s to the late 1970s, TV programming was dominated by three commercial networks: CBS, NBC, and ABC. For more than a generation these networks were America's great entertainers. Their programs, ranging from Ed Sullivan's "Toast of the Town" to the Super Bowl, attracted millions of viewers. In the late 1980s and early 1990s a new service emerged as a fourth network: Rupert Murdoch's Fox Broadcasting Corporation, or FBC.

Over the past half-century the network TV system has evolved a unique structure for the production, distribution, and carriage of TV programming. Let's examine the dynamics of this system, of which the basis is the affiliation agreement.

The Affiliation Agreement

The backbone of any network is the stations that carry its programs. Stations that receive network programming are known as *affiliates*. At present roughly 200 stations are affiliated with each of the "big three." Fox has about 150 affiliates nationwide. In some small markets with fewer than three stations, the programs of more than one network may be carried, a process known as *dual affiliation*.

The local TV station signs a contract, known as an *affiliation agreement*, with the network. Historically, the network has paid the station a fee for broadcasting network programs. This fee is known as *network compensation*. The fee has ranged from under $500 to more than $10,000 per hour, depending on the station's coverage area, market size, and popularity.

The recent rounds of budget cuts and cost-consciousness at the major networks and the skyrocketing costs of professional sports rights have caused a reconsideration of the affiliation agreement.

Compensation to network affiliates has been cut across the board. ABC and NBC affiliates have been asked by their networks to share the burden of programming costs by accepting less money in compensation. In 1992, CBS announced plans to charge its affiliates for certain programs. However, this controversial move met with resistance from many CBS affiliates and the outcome was unclear as this edition went to press. Fox has adopted a novel plan whereby compensation increases with the ratings achieved by Fox programming. This provides additional incentive for FBC affiliates to carry and promote Fox shows.

Affiliates can elect to carry the network shows, or they may refuse them. Programs that are carried by the station are said to have *cleared*; those that are refused are known as *preemptions*. In recent years many affiliates have exercised their right to refuse network programs, thereby producing some strained network–affiliate relationships.

Even before the restructuring of the affiliation agreement, local stations did not get rich from network compensation. Historically, network compensation has amounted to less than 10 percent of a station's annual revenue. So, why do nearly 800 stations relinquish so much program control to the networks?

For one thing, the network will fill two-thirds of the schedule of the local station with its programs. In the early 1990s it was estimated that each major network delivered more than 5000 hours of programming to its affiliates per year. To the typical station operating 20 hours per day, that represents 250 of 365 days filled by programming. In addition to filling out the program schedule, network affiliation offers other attractions to local station management. Network programming provides attractive lead-in and follow-on audiences for local station programming, particularly local news. And although local stations typically have few commercials to sell in evening prime time, these spots can be sold at rates that are much higher than those charged at other times because of the larger audience.

What does the network get in return? Access to the mass audience. Through its system of local affiliates the network has the potential to pull together a huge simultaneous audience, which it can sell to a national advertiser for a concomitantly large price. The name of the game is the audience. Networks try to attract either a large, undifferentiated audience, as is the case with "Roseanne," or a relatively smaller audience that has the demographic profile that advertisers find attractive, as is the case with "Northern Exposure," with its large 25- to 49-year-old audience. It's a high-stakes game. The network with the most successful programs will make the most money. The one with fewer viewers or the wrong kind of viewers will have bad news for its shareholders, and its suffering affiliates will respond appropriately.

The Network View of the Audience

For many years the primary goal of network TV was to attract the largest possible audiences, and programmers developed interesting theories about how best to achieve this outcome. Television executives operated under the principle of the "lowest common denominator" (LCD). This theory held that to attract the largest possible audience, programming needed to satisfy the tastes of the largest possible audience. It could not be discriminating, highbrow, or elitist. Rather, programming should be broad, even simplistic.

While at NBC master programmer Paul Klein (who later headed the Playboy cable channel) coined the term "least objectionable programming" (LOP). The LOP theory posited that the best shows were those that the fewest people could object to on political, social, economic, or cultural grounds. Under this theory, the ideal show would be perfectly innocuous, steering free of any controversy—"pure video pablum," according to some critics.

LCD and LOP predominated in the late 1950s through the 1960s, when programs like "The Beverly Hillbillies," "My Mother the Car," and

"Gomer Pyle" topped the program ratings. Ironically, this was the time of great social upheaval in the United States, causing critic Michael Arlen to observe that "the mass [was] packing its heads with lime Jell-o as the world burned around them."

More recently the LOP and LCD theories have been revised. Shows targeting a more discriminating, intelligent, and upscale audience (like "L.A. Law" and "Northern Exposure") achieved lengthy network runs. At the same time, more LCD-type programs ("Major Dad," "Full House") remained highly rated. And the surge in popularity of the "reality-based" programs mentioned in the previous chapter ("20/20," "Street Stories," "48 Hours," and so on) has led some observers to describe current network TV as "MOP": most objectionable programming.

The Network Programming Process

Network programming is cyclical: like baseball, it has its own seasons, pennant races, winners, and losers. First, let's examine the ground rules.

Program regulation   The First Amendment to the Constitution specifically mandates freedom of the press. That mandate has been extended by the FCC to broadcasting throughout its sixty-year history (see Part Six). Thus with the exception of matters of national security, political candidates' appearances, indecency, and obscenity, there are generally few federal regulations governing the content of TV programming. However, a number of rules have had a direct impact on modern TV programming practices. These rules were spurred by fears of domination of American TV by the three major networks.

Financial interest and syndication rules   With the exception of news, sports, and a limited number of other programs, the networks do not own the shows that fill their nightly schedules. Fearing monopoly in entertainment production by the big three, in the early 1970s the FCC adopted a set of regulations known as the *financial interest and syndication rules*. In show-biz jargon the regulations are known as "fin-syn." In a nutshell, the rules limit network participation in the ownership of programs produced for them and in subsequent syndication. Rather than paying outright for their shows, the networks pay *license fees* to produc-

"Northern Exposure" is an example of a program targeting a more discriminating (rather than LCD) audience.

tion companies. After the network run, programs can be sold into syndication, but *not* by the networks.

After nearly a decade of legal maneuvering, in 1991 the FCC adopted revised fin-syn rules. While the networks were permitted to own a higher percentage of their programming, the net effect of these new rules was to maintain limited network participation in the ownership of programs in syndication. Essentially, little has changed.

Interestingly, thus far Fox has successfully argued with the FCC that it is not a full-blown network and that it needs relief from fin-syn rules in order to compete against the big three. As we shall see, unlike ABC, NBC, and CBS, many Fox shows are produced, aired, and syndicated by FBC and its subsidiaries.

Prime-time access rule   The FCC implemented the *prime-time access rule (PTAR)* in 1971. The rule restricted the amount of time an affiliate could accept from the network, in effect allowing networks to control no more than three hours of the four-hour prime-time nightly schedule (with some exceptions). Since TV viewing increases through the evening, naturally the networks maintained control of the three hours from 8:00 P.M. to 11:00 P.M. (7:00 P.M. to 10:00 P.M. in the Central

Time Zone). This created a one-hour segment for local affiliated stations to fill by themselves. PTAR further prohibited network affiliates in the top fifty markets from filling this time (known today as "prime access" or simply "access") with off-network syndicated reruns.

Although the intent of the rule was to encourage production by non-Hollywood companies and maybe even stimulate local production, the reality is that most stations turned to syndication in this time period. Evening game shows became particularly popular, especially "Wheel of Fortune."

PTAR had another lasting effect: it enabled independent stations to compete on more equal terrain with network affiliates in the very important hour when people are sampling the TV schedule to make their nightly viewing choices. By programming their best shows here (ironically, a lot of popular old network shows that could not be shown on network stations), many new independent stations were able to siphon viewers from affiliates to themselves.

Like fin-syn, PTAR regulations continue to concern broadcasters. Affiliates are changing their notions of primetime (see box), cognizant of the fact that many Americans today go to bed and wake up earlier than in the past. The growing availability of Fox programming to stations that were formerly independent has made one of the points of PTAR (to enhance competition) moot. Indeed, the networks argue for full repeal of the rule. With profits slipping and new programming services seemingly everywhere, they wonder what became of their dominance (and therefore the need for such rules).

Look for continuing changes to fin-syn and PTAR between this publication and the next edition of this text!

Network seasons Network programming is organized around two seasons. The fall premiere season begins in late September and runs through the end of October. This is when new programs are launched and returning programs begin showing new episodes. The so-called second season runs from mid-January through the end of February. This is when the networks replace low-rated programs with specials and new series.

At one time the two seasons were distinct. The three networks premiered their new shows in one week in late September; the second season almost always occurred in early February. Today, however, there seems to be one continuous season.

Fox has been especially instrumental in this trend. Under former president Barry Diller, Fox staggered the introduction of its new programs, sprinkling premieres throughout the year. A particularly effective strategy included running new episodes of hit shows, like "The Simpsons" and "Beverly Hills 90210," in the summer, when ABC, NBC, and CBS were in reruns.

IMPACT

WHAT TIME IS PRIME TIME?

For the first four decades of TV's existence, time basically stood still. Prime time, TV's most important daypart, ran from 8:00 P.M. to 11:00 P.M. in the eastern and western time zones, and from 7:00 P.M. to 10:00 P.M. in the central and Rocky Mountain states. In 1991 and 1992 prime time began to change. Stations in the western time zone found that their research revealed that viewers were going to bed and waking up earlier. Many were single parents who had to get small children off to day care and school. The failing economy caused some to work different shifts. Others simply couldn't keep their eyes open much past ten o'clock to be around for exciting dramas, local news, and late-night entertainment.

As with other social trends, California led the way in shifting prime time. In February 1992, CBS affiliate KPIX moved its prime-time schedule up an hour. KRON, the NBC affiliate, soon followed suit. Both stations said that the shift was only a short-term experiment, and that they reserved the right to shift back to the traditional 8:00 P.M. to 11:00 P.M. prime time. Like many other decisions in commercial TV, money may be the deciding factor. An earlier experiment by KCRA in Sacramento saw the ratings for its new 10:00 P.M. newscast become 28 percent lower than in its old 11:00 P.M. slot. People indeed seemed to be going to bed earlier: right in the middle of KCRA's newscast!

Despite such "stunting" (an industry term for creative changing of the network schedule, especially during ratings periods), tradition dictates that the bulk of new network programming debuts in the fall and that replacement programs appear in the dead of winter. Late spring and summer programming remains a TV season relatively unchanged over time; these periods are still dominated by reruns.

The network programming process begins early in the calendar year, when program executives begin to map out the coming fall season. The deadline for completing the fall schedule is May, when each network debuts its lineup (typically with much glitz and fanfare) at an annual meeting of network affiliates.

In past years the network season lasted thirty-nine weeks. As many as thirty-two episodes of a new show would be ordered, with the remaining twenty weeks occupied by specials and reruns. Today escalating production costs, competition, and a declining success rate have led the networks to order as few as ten new episodes of a series, with an option to repeat at least two. This is one reason the season has shrunk (or expanded to year round, depending on one's point of view).

Who gets the orders? Production of TV series is dominated by a few large studios, increasingly owned by huge communications conglomerates. These are the major studios, or "the majors," for short. However, some hit shows come from outside these huge conglomerates. A small number of independent producers ("independents," for short) exert significant influence over the network programs we see. Let's start with the majors.

The Major Studios

The major studios in TV production are familiar names, although their ownership may be less familiar. Columbia/Tri-star is one of these, producing such TV stalwarts as "Designing Women" for CBS, "Days of Our Lives" (nearing its thirtieth year on NBC), and "Parker Lewis" for Fox. In 1989 Columbia/Tri-star was bought by SONY for $3.4 billion.

Another Japanese firm that has recently achieved major studio status in Hollywood is Matsushita, which owns MCA-Universal. You may be familiar with Matsushita through its product brand names Panasonic, National, and Sharp. Matsushita bought MCA-Universal in 1991 for $6.1 billion. In addition to acquiring ownership of a movie studio and theme parks in Hollywood and Orlando, Matsushita also gained title to popular TV shows like "Murder She Wrote," "Major Dad," "Coach," and "Quantum Leap." Lieutenant Columbo may be learning Japanese, as Universal TV produces three or four of his mysteries each year for ABC.

Lorimar is a TV studio owned by American communications conglomerate Time-Warner. By 1992 Lorimar had produced fourteen series for network TV, more than any other studio. Included in this list were the venerable "Knots Landing" (completing its thirteenth year in 1992), "Family Matters," "Full House," and "Hanging with Mr. Cooper." The other major Time-Warner studio is Warner Brothers, which has produced such shows as "Murphy Brown," "Room for Two," and "Life Goes On" in recent years.

Another fabled name in Hollywood history, Paramount, has been especially active recently in network TV. In 1991 Paramount lured Brandon Tartikoff from NBC, where as head of programming he had brought to the network hits like "Cosby," "L.A. Law," and "Family Ties." As Chairman, Tartikoff has stepped up Paramount's TV efforts. By 1992 Tartikoff had added "Brooklyn Bridge," and the innovative but expensive "Young Indiana Jones Chronicles," among others.

The TV programming arm of the Mickey Mouse media empire, Disney, is its Hollywood-based Walt Disney Studios and Buena Vista Television divisions. Such shows as "Dinosaurs," "Golden Palace," and "Home Improvement" come from Disney sound stages.

At the time of this writing, MGM (another heralded Hollywood name) faced an uncertain future, following a struggle between its former owner, Italian media magnate Giancarlo Peretti and the studio's financial backers. Executive turnover has been rampant at MGM's Hollywood studios. Some stability has been maintained by a small group of hit shows, including "The Wonder Years," and "In the Heat of the Night." However, long-running programs like "thirtysomething" and "Tour of Duty" are no longer in production, adding to the sense of peril around the studio as this book went to press.

Oprah in action. One of the most successful shows in syndication, "Oprah" is seen in about 99 percent of the TV homes in the United States.

Viacom, which also owns MTV, VH-1, and much of Comedy Central, produces for the networks, cable, and syndication market. Its recent network shows include "Jake and the Fat Man," "Matlock," and the "Perry Mason" telefilms.

The list of majors ends with Fox, which, as we have said, has benefited greatly from its exemption from network regulations, such as fin-syn. Twentieth Century Fox is FBC's studio. Fox has been permitted to produce shows for its own network, which it did with "In Living Color," "The Simpsons," and "Cops." Twentieth TV also produces "LA Law" for NBC, as well as syndicated programming like "A Current Affair" and "Studs."

Independent Producers

The networks also order programs from a small number of independent producers who have had a series of network successes. One of these is Stephen J. Cannell, whose success with "The Rockford Files," "Baretta," and "The A- Team" led to network orders for "Wiseguy," "21 Jump Street," "The Commish," and "Hat Squad."

Carsey-Werner was founded by former ABC programmers Tom Werner and Marcy Carsey. Their first show, "Oh! Madeline!" was forgettable. Their second was "Cosby." Carsey-Werner is now a powerhouse TV independent, responsible for "A Different World," "Roseanne," and Bill Cosby's syndicated show, the new version of "You Bet Your Life."

Stephen Bochco is another influential independent, with a track record on network TV, including "Hill Street Blues" and the development of "L.A. Law." Bochco's recent hits include "Doogie Howser." His most spectacular failure was "Cop Rock," a short-lived series featuring singing detectives on ABC.

Aaron Spelling is a famous independent producer. His two monster hits "Dynasty" and "Hotel" were preceded by "Love Boat" and "Fantasy Island." His current smash hits include "Beverly Hills 90210," and "Melrose Place." Another important independent is Witt/Thomas/Harris (with shows like "Beauty and the Beast," "It's a Living," and, more recently, "Nurses," "Blossom," and "Herman's Head" to its credit).

"Pitching" a Program

Programs get on the air through two primary means. Some are commissioned by the network, whose research and development discovers public interest in a particular program concept. It is

"Tool Time's" Tim Taylor (Tim Allen) tangled up trying to test TVRO. "Home Improvement" was one of the few new shows to become a bona fide hit during the 1992 season.

said that former NBC program executive Brandon Tartikoff ordered "Miami Vice" by telling producer Michael Mann to "give me MTV with cops!" Similarly, after meeting Mr. T at a Hollywood party, Tartikoff is reported to have handed Stephen J. Cannell a piece of paper reading: "'The A-Team,' 'Mission Impossible,' 'The Dirty Dozen,' and 'The Magnificent Seven,' all rolled into one, and Mr. T drives the car!"

More commonly, new program ideas are introduced, or "pitched," to the networks by their producers. The pitch can be based on an idea or *concept*, a short story narrative, known as a *treatment*, or a sample script. It is estimated that the networks are presented with as many as 10,000 new program ideas each year in one form or another. About 500 are chosen for further development.

At this point the lawyers, accountants, and agents get involved. Most commonly, the program (by now known as a "property") is developed under the terms of a *step deal*—an arrangement by which the program is put together in a series of distinct phases. The network will "front" the major portion of the development money for the program, in return for creative control over the show's content. The network gets *right of first refusal*, the contractual right to prohibit the production company from producing the program for another client. The step deal also enables the network to appoint additional writers to develop the concept or "punch up" the script.

About one-half of the optioned ideas will lead to the step deal or fully scripted stage. From this pool the networks will order about thirty or thirty-five *pilots*, or sample episodes, each costing in the range of $750,000 to just under $1 million to produce. By early spring about a dozen of these pilots will result in series orders; ten of these twelve are likely to be canceled before the second season. Sometimes networks will run pilots that didn't make it in the late spring or summer in an effort to recover some of their production costs.

Network program costs While the concept of a show, its location, and its stars are certainly important, for TV executives perhaps most important is its cost. Table 11-1 presents typical production costs for various program types. As might be expected, the most costly programs are the lavish miniseries, with episode costs exceeding $2 million. The most costly miniseries to date was ABC's "War and Remembrance," which cost over $100 million for thirty and a half hours (more than $3 million per hour), or $6 million per episode. Of the regularly scheduled series, crime and adventure dramas are most expensive. These typically feature high-profile stars, location shooting, lavish decor, and lots of costly effects, from car crashes to fantasy scenes.

In recent years cost factors have led to the decline of many "high-concept," expensive programs. Such shows as "Miami Vice," "Moonlighting," "Dynasty," "Dallas," and "Falcon Crest," ceased production, due as much to high production costs as to declining ratings.

TABLE 11-1 Typical Costs for Network Programs

Program Type	Cost per Episode ($ millions)
Major miniseries	2–4
Adventure/mystery	1–3
Movie of the week	1–2
Situation comedy	0.75–1.25
Reality/news magazine	0.35–0.75

PROGRAMMING

HOW BAD IS NETWORK PROGRAMMING?

Many people feel that network programming is bad and getting worse. Usually the networks counter this feeling with evidence of bigger budgets, increased time spent viewing, some widely watched specials, and other arguments. Their own data may be most revealing. Listed below is the renewal rate for network prime-time programs over the past five decades. The renewal rate refers to the percentage of shows premiering in the fall that were "picked up" for a second year.

There is little doubt that today even the networks have less confidence in their programs than they once did. Back in TV's "golden age," nearly one-half of the shows that premiered in the fall returned the next year. Now only about one in four shows lives to see another year. The remainder are killed in the cradle

Prime-Time Television New Show Renewal Rate, 1950–1994

Seasons Beginning with Fall	New Show Renewal Rate (%)
1950–1954	49
1955–1959	43
1960–1964	37
1965–1969	48
1970–1974	37
1975–1979	24
1980–1984	35
1985–1989	31
1990–1994	25

SOURCE: *TV Dimensions '91*, p. 127. Data for 1985 to 1989 are aggregated; data for 1990 to 1994 are authors' projections.

Bucking the trend was Paramount's "The Young Indiana Jones Chronicles," with a typical cost per episode in the $3 million range. Other costly but popular recent programs include "Beverly Hills 90210" ($1.2 million per episode), "Quantum Leap" ($1.3 million per episode), and "L.A. Law" ($1.4 million per episode).

The next most expensive program type are movies made for TV. In industry parlance, these are referred to as movies of the week (MOWs), "made-fors," or more cynically "disease of the week" shows. Typically these shows are contemporary dramas revolving around personal relationships, tragedy, domestic strife, and recent criminal cases. Some made-fors receiving attention recently were "Baby M," about surrogate motherhood, "The Rescue of Jessica McClure," about a child trapped in a well, "Guts and Glory: The Rise and Fall of Oliver North," about the Iran-Contra affair, and "An Early Frost," focusing on an AIDS patient. MOWs are budgeted in the range of $1 to $2 million per hour, but they often exceed this cost.

Fortunately for the networks, the most popular program type is also one of the least costly. Situation comedies cost about $750,000 per episode. Sitcoms can hold the line on cost since most are shot in a studio as opposed to an expensive location. They are recorded in "real time" before an audience, which keeps the need for editing and special effects to a minimum. Today most are shot on videotape, which is cheaper than using film. Most dramas, however, continue to use film.

As situation comedies become more popular, their costs increase, mainly in the area of star salaries. By the early 1990s the most expensive sit-coms included "Roseanne," "Cheers," and "Murphy Brown," at about $1 million per episode. Cheaper to produce (since their stars were lesser known) were "The Fresh Prince of Bel Air" ($550,000 per episode), "Seinfeld" ($575,000 per episode), and "Martin" ($525,000 per episode).

The "bargain basement" of network TV are so-called reality shows, including news documentaries and "infotainment" series (see Chapter 10). For example, CBS's highly rated "60 Minutes" costs about $600,000 per hour and ABC's "20/20" is budgeted in the $500,000 range. "America's Most Wanted" is a bargain for the Fox network, budgeted at only $500,000 per hour. The low cost of reality shows is one reason for their proliferation in recent years.

For example, on ABC, "America's Funniest Home Videos" and "America's Funniest People" have both been popular with audiences and inexpensive to produce (about $400,000 per hour each). "48 Hours" and "Street Stories" on CBS have lured big audiences at less than $500,000 per hour (about one-half the cost of a sit-com).

Deficit finance: Life in the aftermarket The costliness of network programming and the fin-syn rules have led production companies to practice a unique method of financing. Normally, the network will finance from two-thirds to three-fourths of the cost of a program. The production company will borrow the rest and produce the program while operating at a deficit. Table 11-2 lists selected network TV programs and their deficit per episode. Why don't the studios go bankrupt? Since they (not the networks) own the property after its network run, they expect to recover the costs and return a profit in the *aftermarket*. The aftermarket for a show includes several channels of distribution. Many American programs run as series in other countries and some (especially made-fors) are released overseas as theatrical feature films.

There is also the VCR market. And, as we shall examine later, there is the potential goldmine of domestic syndication: the sale of series and programs to individual TV stations and cable networks.

Considerable controversy surrounds this practice of deficit finance. Some stars have claimed that the method allows producers to hold the line

TABLE 11-2 Deficits of Network Series

Show	Cost per Episode ($)	License Fee ($)	Deficit per Episode ($)
"Homefront"	1,260,000	935,000	325,000
"Murder She Wrote"	1,500,000	1,100,000	400,000
"Northern Exposure"	1,150,000	900,000	250,000
"L.A. Law"	1,300,000	1,050,000	250,000
"Married with Children"	675,000	525,000	150,000

SOURCE: *Variety*, August 26, 1991, pp. 48–54; August 31, 1992, p. 30.

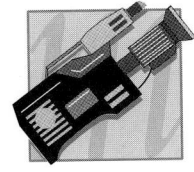

PRODUCTION

WHAT ABOUT "BOB"?—THE PAST AND FUTURE OF "TWIN PEAKS"

In the 1990–91 TV season a sinister and somewhat goofy program briefly took the country by storm. The ABC series, produced by David Lynch, who became famous as the director of such "cult" films as "Eraserhead" and "Blue Velvet," began with the mystery of the murder of a teenage girl and ended with the somewhat obtuse solution to the crime. In between there was a parade of off-beat characters, locations, sets, costumes, and music that was loved by critics, but somehow eluded much of the TV audience. After a few short weeks of solid ratings "Twin Peaks" plummeted. By the end of the season it had been canceled.

The full impact of "Twin Peaks" is being felt long after its cancellation. First, the program proved that unusual locations, characters, and situations might attract a large audience, especially a well-educated, upscale one. In short order, "Northern Exposure," another loopy program set in the Pacific Northwest, became a top-10 hit on another network.

In addition, "Twin Peaks" showed that some forms of TV are not suited to the regular, weekly, season-to-season format. "Imagine if 'Twin Peaks' had settled for just a limited-run," mused ABC Entertainment President Bob Iger, "It would have been hailed as landmark television." The concept of the limited-run "event" series, popular in Great Britain, had come to America.

Finally, "Twin Peaks" showed that TV was ripe for an infusion of new blood and new vision from producers other than Spelling, Cannel, and Bochco. Look for other new talent to appear on the homescreen. In 1992, for example, ABC signed deals with novelist Judith Krantz and film director Oliver Stone to develop limited-run series for the network.

on salaries, while the "losses" mount on paper, if not in reality. The networks have argued that Hollywood has inflated its production costs to artifically cover the deficits. Rising production costs have driven some productions (particularly those on cable and pay cable) to other locations: to Atlanta, Orlando, Toronto, and London. This leads us to the next topic—network programming *not* on CBS, NBC, ABC, and Fox.

Cable Network Programming

We covered the programming of cable and pay cable networks in Chapter 6. However, it is useful to note here how these networks are programmed in comparison to the big three and Fox.

Theatrical motion pictures As we have seen, the bulk of the program schedule (over 80 percent of air time) on the major pay cable services, including HBO/Cinemax and Showtime/The Movie Channel, consists of theatrical motion pictures. Most of these are licensed to the pay service by the "majors," Hollywood's largest studios.

The studios normally make their films (known as "titles" in the trade) available to pay cable a year after their first theatrical release. There is a narrower release "window" for pay per view (PPV) of six months or less. Sometimes studios will delay the pay release of a major box office hit (like *Star Wars, ET* and *JFK*) so that they can continue to reap profits at the box office. Occasionally a theatrical release may be so disappointing that the studio will release the film to cable and home video within weeks of its theatrical debut.

Some distributors make their films available to the pay services simultaneously. However, to offset viewer dissatisfaction with pay services ("they all play the same movies"), there has been a trend toward studios signing *exclusivity deals* with pay services. Such contracts guarantee first cable release to one pay service. HBO has signed exclusivity deals for packages of titles from both Paramount and 20th Century Fox, adding to earlier deals it had made with Columbia and Tri-Star. Showtime has signed exclusivity agreements with other studios, including Weintraub Entertainment and Carolco.

Theatrical motion pictures are also the backbone of a number of the advertiser-supported cable networks, including the USA Network, the "superstations" (WTBS, WPIX, WWOR, WGN, and others), Lifetime, and the Nostalgia Channel. These films are normally sold in *packages*, series of titles made available by a distribution company for sale to cable networks and local TV stations. In fact, it was Ted Turner's acquisition of virtually the entire MGM film library (including *Gone with the Wind*) that enabled his launch of Turner Network Television (TNT). Turner's new Cartoon Channel was made possible by his purchase of the archives of the Hanna-Barbera animation studio.

Cable-produced movies In their quest to keep pay cable attractive to consumers despite competition from video stores and other sources, cable programmers are producing their own movies. Some have been released for theatrical distribution first, to be followed by a pay cable run. Others have been produced for a premiere on the cable network, to be followed by theatrical or home video release.

The increased commitment from cable networks to first-run movies is illustrated in Table 11-3. In 1992 alone cable networks spent over $300 million for over 100 made-for-cable films. While most cable-original pictures receive about the same acclaim (or notoriety) as made-for-TV movies, some outstanding work has emerged. Recent cable-originals include "The Josephine Baker Story" and "Citizen Cohn" on HBO, "Avonlea" for the Disney Channel, and Lifetime's remake of the Hitchcock classic, *Notorious*.

Cable series In their early days cable and pay cable relied almost exclusively on movies, sports, and concerts for their programming. The few cable series that did run tended to be the low-budget, easy-to-film variety (Las Vegas acts, stand-up comedy, show-biz chatter, and so on). One sign of the maturity of the cable business is

TABLE 11-3 Cable-Original Movies

Cable Network	Typical Budget ($ million)	Approximate No. of Films per Year
Showtime	5 million	12
HBO	2.5–5	12
TNT	3–4	18
A&E	2–4	6
Disney	3	10
USA	2–3	15–30

SOURCE: *Cablevision*, December 2, 1991, p. 22.

the emergence of regularly scheduled "high-profile" series from many cable networks.

Many of cable's "home-grown" series imitate network programming. One of the first was "New Leave It to Beaver," produced for WTBS by MCA. For a time WTBS aired its own soap opera, "The Catlins." Talk programs similar to "The Tonight Show" have been commonplace, from "Nashville Now" on the Nashville Network to "The Dick Cavett Show" on CNBC.

But there has been some innovation in original cable series. Two game shows have "broken the mold" and received excellent response. Nickelodeon developed "Double Dare," a participatory quiz program that became a hit with kids and their parents. The tongue-in-cheek "Remote Control" became a mainstay of MTV. Both are now sold in syndication. One of cable's biggest hits in the late 1980s was "It's Garry Shandling's Show," on Showtime. In a manner similar to the "Burns and Allen" program of the early 1950s, the "Shandling" show played with the format of the situation comedy, allowing the host comedian to talk to the audiences and the crew. "Shandling" was the first series produced originally for cable to go to a network (Fox).

In a more serious vein, other cable-bred series include "All Creatures Great and Small," coproduced by the BBC for the Arts and Entertainment Network (A&E), "The Campbells" and "Bordertown" for CBN, "Portrait of America" and "National Geographic Explorer" for WTBS.

Generally, cable services produce their original programs at lower cost than do the commercial TV networks. This cost savings is achieved in various ways. First, smaller independent production companies are often used instead of the costly majors. These companies are frequently allowing them to keep salaries down. To cut overhead expenses that pile up while shooting in a big studio lot, many series are shot on location or in major cities outside the United States, where the dollar is stronger than local currency. Canada and England have become very popular production sites, with Toronto and London "standing in" for Los Angeles and New York.

Although the highest rated programs on cable in many cases are reruns of network shows such as A&E's "Biography" (which aired originally in the early 1950s), "The Dick Van Dyke Show" and "Get Smart" on Nickelodeon, and "Andy Griffith" on WTBS, there is evidence of public acceptance of cable series. The top five programs on the Nashville Network are home-grown, including "Country Kitchen," "This Week in Country Music," "Nashville Now," "Bass Masters," and "Grand Ole Opry Live."

Pediatrician Dr. T. Berry Brazelton educates and reassures parents on "What Every Baby Knows," one of the most popular shows on the Lifetime cable network.

Public Television Networks

Although the dominant networks (in terms of audience size and budget) are the major commercial broadcast and cable firms, public TV remains a vital source of national programming. Today over 350 full-power noncommercial TV stations have the same needs as their commercial counterparts: to fill their schedule with programming attractive to audiences.

The Corporation for Public Broadcasting (CPB) was founded in 1967 to develop noncommercial broadcasting in the United States. Designed to be insulated from government influence through a unique system of advance funding and autonomy from Congress, the CPB is charged with the responsibility of allocating federal monies to produce public programming. In 1970 Congress established the Public Broadcasting System (PBS) as the distribution arm of the CPB.

PBS operates in reverse fashion from the commercial networks. Whereas NBC, CBS, ABC, and Fox funnel money to production companies and pay stations to carry their programs, PBS charges membership dues to its affiliates (over 90 percent of all public TV stations). In return PBS provides programs funded by pooled station funds, the CPB, foundations, individual contributions, and other sources. PBS itself produces no programs. Instead, through its National Program Service (NPS), it serves as a conduit to program producers, usually avoiding the suppliers common to network scheduling.

A key difference between the commercial networks and PBS is that the stations (not the network) decide when to carry national programs. The most popular programs on PBS form the "core schedule," which is designated for same-night carriage. PBS recommends but cannot require stations to air these shows the same night that they are fed by satellite to member stations. Same-night carriage helps PBS promote its shows nationally. Although most PBS affiliates air the core schedule the same day, many PBS shows are taped and delayed for days or weeks. And within limits, which vary from show to show, stations frequently rerun PBS shows. Thus unlike CBS or NBC, a PBS show may or may not air nationally in a given day, and when it finally does run, it may be rerun long before the summer.

In recent years PBS programming has been hurt by cable channels, like Discovery, Arts and Entertainment, and WTBS. Its franchise in international programming (like BBC coproductions) and nature documentaries (like those of Jacques Cousteau) has been broken up. By the early 1990s it was estimated that the national PBS audience had declined more than 10 percent in a five-year period.

In response the public broadcasting network has taken some bold steps. First, in 1990 its board of directors named Jennifer Lawson as the network's first programming czar. Previously, decisions about programming were made by a committee of executives from member stations. Now, like the major networks, program decisions (and the success of the network) are the responsibility of a single individual. Under Lawson's leadership, the enormously successful "Civil War" was broadcast and the network made commitments to more entertainment efforts, including situation comedies and quiz programs.

In addition, PBS implored member stations to stick to the core schedule so that promotion and fund-raising opportunities could be maximized. There is some evidence that affiliates have followed suit, although there has been grumbling. Like their commercial counterparts, some local station managers resent centralized decisions made on their behalf by a large, distant, bureaucratic institution.

Sources of Programming

On average, PBS today distributes about 1500 hours of programming per year to its member stations (about four hours per day).

The source and type of that programming are depicted in Figures 11-1 and 11-2. The bulk of PBS programming consists of news and public affairs (about 38 percent). This includes programs like "The MacNeil/Lehrer Newshour," "Frontline," and "Wall Street Week." Next most common are cultural programs, such as "The American Experience," "Dance in America," and "Live from Lincoln Center." Childrens' programs comprise just over one in ten PBS shows, from "Sesame Street" to "Where in the World Is Carmen Sandiego?" "How-to" shows are nearly as common, from "This Old House" to "The Victory

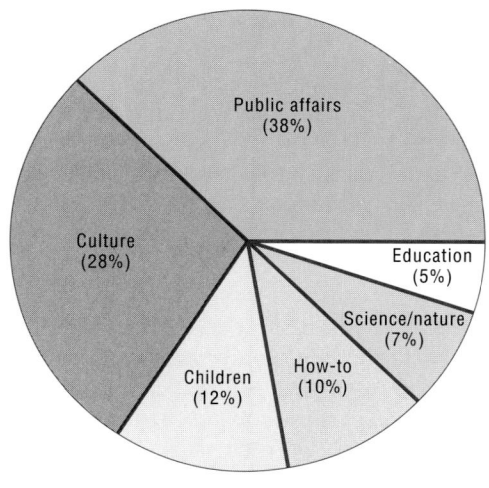

FIGURE 11-1 Types of PBS programming.

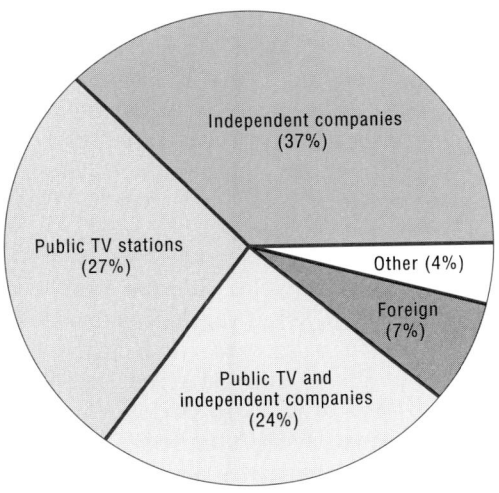

FIGURE 11-2 Producing agencies for public television. (Source: PBS.)

Garden." Filling out the PBS schedule are strict educational shows (like "Eat Well/Be Well," and "College Algebra") and a very small sprinkling of sports (mostly tennis and soccer, although the network did air "Motorweek 91").

PBS strives to avoid the producers that are common to commercial TV. Nearly 40 percent of producers for public TV consider themselves "independent." Just over one in four programs on the network come from one of its member stations, and about 7 percent emanate from foreign producers. The remaining shows are developed and funded by unique consortiums of stations, philanthrophic groups, corporations, foundations; you name it!

The audience Public TV programmers make different assumptions about their audience than their colleagues on the commercial side. In general, research confirms that viewers of public TV are better educated and more selective in their TV viewing habits than the "average" viewer. In addition, they tend to watch less TV than viewers of commercial TV. Thus, rather than rely on LOP and LCD theories, most public TV programmers view their audience as selective and discriminating—not "couch potato" viewers of commercial TV. To attract such viewers, public TV programming must stimulate the mind as well as the eye. Rather than seeking "big numbers" (high ratings), public programmers seek a relatively small number of informed, educated "opinion leaders" in their communities.

Some critics of public TV have suggested that this has created an elitism in production: shows are produced that are biased toward "Washington insiders" or toward viewers who are more liberal and critical in their viewing than the mainstream. In the current atmosphere of federal budget trimming and government accountability this charge and other factors have led to a strong debate about the present and future of CPB and PBS.

THE WORLD OF TV SYNDICATION

After the networks (big three, Fox, cable, and PBS), the largest purveyors of programming are the *syndicators*: the companies that sell programs directly to TV stations and cable services. The syndicators are the "wheeler-dealers" of the TV world. As our discussion unfolds it may be helpful to envision syndicators as direct descendents of the pushcart vendors who populated the streets of America's cities around the turn of the century. Television syndication has been a spectacular growth industry in recent years. The growth of independent TV stations, the arrival of cable, and the introduction of the VCR have created tremendous new demands. Where once syndication meant only two things—movies and network reruns—the syndication universe today ranges from films to game shows, music videos to "how-to" tapes on everything from exercise to hunting water buffalo. Today TV syndication is a $3 billion annual business.

The Syndication Market

There are two primary buyers or markets for syndicated programming. The traditional market for syndication is local TV. Today over 1200 local TV stations obtain syndicated programming to fill their program schedules. On a volume basis, independent TV stations buy more syndicated programs than network affiliates. Naturally this is because they do not have the luxury of having two-thirds of their schedules occupied by network shows.

The second market for syndicated programming is the cable networks. The chief cable buyers of syndication are the advertiser-supported services. As we saw in Chapter 6, the cable superstations are actually independent stations (WTBS in Atlanta, WWOR in New York, WGN in Chicago) available nationally via satellite delivery. Like other independent stations, they rely strongly on syndication, from movies to "Leave It to Beaver."

A new trend in syndication has been the licensing of programming to cable networks other than the superstations. Old network programs now can be viewed on a broad array of cable networks. Recently such network stalwarts as "Cagney and Lacey," "L.A. Law," and "thirtysomething" had been sold to cable for syndication.

The sale of syndicated programming to cable has created problems in the TV business. Local TV stations spend millions to acquire programming. Part of the cost is a guarantee of syndication exclusivity. When a station acquires a program like "Cosby" for its local market, it pays a premium to be the only station with the show. Indeed, stations sometimes buy a program that they don't plan to show precisely to keep a competitor from having it. In the 1980s, however, station managers discovered that the shows they thought were theirs exclusively appeared on channels in their community: on stations in other markets that were carried by the local cable system.

Obviously, this situation infuriated station operators and programmers. Since the spread of cable in the 1970s, TV management has been lobbying for regulatory relief from the Congress and the FCC. Thus far two remedies have been forthcoming.

The compulsory license First, cable systems that import distant, nonnetwork signals from other markets (virtually all cable systems today) have been required to pay license fees under a set of regulations known as the *compulsory license*. The terms of the license were established under the Copyright Act of 1976. The Act created the Copyright Royalty Tribunal, a group of five commissioners empowered to collect and distribute funds under the compulsory license arrangement. To date, millions of dollars have been collected from cable operators and paid to production companies, sports programmers, commercial and public broadcasters, and music licensing groups. However, the FCC has announced plans to scuttle the compulsory license rules. Whether Congress would follow the commission's recommendation and pass legislation eliminating the rules was uncertain as this book went to press.

Syndicated exclusivity The situation has been clouded further by a set of regulations known as *syndicated exclusivity*, or *syndex* in industry parlance. Prior to 1980, FCC rules required cable operators of large systems to black out duplication of programming available to local TV stations when such programming was also being offered on cable services. After an intensive lobbying effort by the broadcast industry, in 1988 the FCC reinstated syndicated exclusivity. Beginning in 1990, cable systems were once again required to black out programming duplicating broadcasts in their marketplace. Immediately after the reimposition of the exclusivity rule, the cable industry brought suit, claiming syndex to be an unlawful infringement on the free speech of cable operators by dictating what shows they could and couldn't carry. As of this writing, the future of Syndex was unclear.

The Syndication Bazaar

In 1963 a small group of TV programmers met with an equally small group of program suppliers

The floor of the annual NATPE programming fair. Millions of dollars of deals for syndicated programs are made during the NATPE convention.

to pool their resources and streamline their efforts. Twenty-five years later over 200 syndicators and 7000 potential buyers met for an annual show-biz ritual known as NATPE, the annual meeting of the National Association of Television Program Executives.

In 1992 NATPE and the Association of Independent Television Stations (INTV) decided to join forces. Today the combined NATPE/INTV Convention is a huge TV programming bazaar, complete with stars, dancing girls, and thousands of extras, determined to sell their TV shows to the throngs of station managers, programmers, and cable operators in attendance. Part Hollywood hype and part consumer trade show, the convention has become the place where new syndicated programs are unveiled and, most of all, plugged. Syndicators lure buyers to screening rooms. Liquor flows. Premiums—like cowboy hats from producers of westerns—abound. And sometimes deals are made.

Types of Syndicated Programming

What does one see at NATPE? Despite the vast array of programming available, it is possible to classify syndicated programming into three main types: motion pictures that have completed their theatrical run, and their home video and pay cable release sold to stations in movie packages; programming originally produced for one of the major TV networks, sold to stations as off-net syndication; and original programming produced expressly for syndication, known as first-run. Each type of syndication is distributed in a unique way.

Movie packages Movies are a mainstay of the program schedule for most independent stations and many advertiser-supported cable networks (including USA, Lifetime, American Movie Classics, and TNT). Even network affiliates need movies for parts of their daytime, late-night, and weekend programming. There are lots of movies out there: it is estimated that since the sound era began in 1927, over half a million films have been produced in the United States alone. Keep that figure in mind when you get the feeling in a video store that you have seen everything.

From the earliest days of TV, movies have been the backbone of syndication. The new TV stations needed something to show; facing declining attendance, the motion picture business needed a new revenue source. It was a match made in heaven (if not Hollywood). Early on, it became unwieldy to sell movies one at a time. Thus the distribution companies (originally sub-

IMPACT

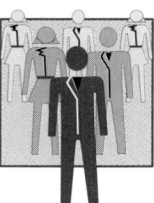

COLORIZATION: PANNING FOR GOLD IN BLACK AND WHITE

In the 1980s the power of computers gave rise to a controversial new technique. Using the digital palette of a complex graphic computer, it became possible to add color to movies and TV programs originally shot in black and white. By the end of the decade dozens of Hollywood movies had been colorized, including *Casablanca, The Maltese Falcon,* and *Yankee Doodle Dandy*. Some greeted the process with ire and outrage. The late director John Huston likened the process to "painting a mustache on the *Mona Lisa*." Martin Scorcese called it "a desecration." Robert Redford agreed, saying, "It's like robbing a grave."

Supporters of colorization argued that it added new attractiveness to "old" movies and TV shows that were otherwise ignored by viewers, especially those under 35-years-old. Ratings for colorized movies supported this view. Colorized versions of *Miracle on 34th Street, It's a Wonderful Life,* and others typically doubled the ratings they had received when shown in black and white.

The real boon to colorization may have been the off-net TV series of the 1950s and early 1960s. Being readied for syndicated rerelease in color were "Hennesey," "I Love Lucy," "Gunsmoke," "Rin Tin Tin," "The Man from U.N.C.L.E.," "Sky King," and the early episodes of "Gilligan's Island." Even "Popeye" was poised for the paintbox, including episodes originally shot in the early 1930s. Clearly, there is more green here than fits in a can of spinach!

---

sidiaries of the major studios) began to package the movies as a collection of titles. Today, with the exception of a very few blockbusters (such as *Rain Man, Terminator-2,* and *Batman*), movies sold in syndication are packaged.

Stations acquire the rights to movies under *license agreements*. The agreements generally run from three to six years and allow the station to show a movie up to six times or more during the period covered. Typically packages of recent box office successes (known as A movies) cost stations more than older and less popular films (known as B movies). In a market like New York City, *Ghost* may license for as much as $500,000, whereas *I Was a Teenage Werewolf* can run for as little as $50. In fact, the primary reason for the invention of the controversial technique of colorization (see box) was to breathe new life in syndication to old black and white movies.

Syndicators have been very creative in packaging movies. Some major distributors include a number of B movies in packages with their blockbusters, forcing stations to take the good with the bad (and sometimes the ugly). Many syndicators, especially smaller companies with a lot of B (and even C and D) movies, put titles together around a consistent theme.

For example, in 1992, Select Media offered a variety of packages, including "The Horror Hall of Fame." Republic Pictures offered "Home of the Cowboys," "The John Wayne Collection," and "All Nite Movies." New World Television's offerings included "The Kid Pix Six Pack."

Recently overexposure has tarnished the attractiveness of movie packages in syndication. As we have seen, by the time a film reaches local TV it has played in the theater and on cable, and it may be on the shelf in the nearby video store. Who wants to see the same film again, this time edited for TV and interrupted by commercials? Because of overexposure, the size of the film libraries at some stations has declined. Independent stations require films, and lots of them. They may maintain as many as 1500 or 2000 feature films in their program inventory. But it is common today for network affiliates to archive 100 or fewer movies. For most stations, independents and networks alike, the search is on for other types of syndicated programming.

Off-net syndication Say what you will about the commercial networks—their imitative schedules, their lowest common denominator programs, and their so-called demise in recent years. The fact

ECONOMICS

SUCCESS IN SYNDICATION

The economics of syndicated programming are simple. The syndication scorecard is kept by noting the number of stations carrying a program, the percentage of the nation's TV households that can see the show, and the average rating it achieves in each of the markets in which the show is aired. Below is a recent syndication report card.

(Nielsen's top-ranked syndicated shows for the week ending Feb. 9, 1992. Numbers represent rating average/stations/percent coverage)

1. "Wheel of Fortune"	15.2/223/99
2. "Jeopardy!"	13.3/215/98
3. "Star Trek"	12.7/244/98
4. "Oprah Winfrey Show"	11.6/225/99
5. "Entertainment Tonight"	9.2/187/96
6. "Current Affair"	9.0/173/94
7. "Wheel of Fortune" (Wknd)	8.8/189/83
8. "Cosby Show"	8.2/208/98
9. "Married...with Children"	8.2/173/94
10. "Donahue"	7.3/227/99
11. "Inside Edition"	7.2/130/89
12. "Sally Jessy Raphael"	6.4/209/97
13. "Hard Copy"	6.1/174/93
14. "WKRP in Cincinnati"	5.8/238/99
15. "American Gladiators"	5.5/177/94

SOURCE: *Broadcasting*, February 24, 1992, p. 56.

The two syndication giants are King World's game shows, "Wheel of Fortune," and "Jeopardy!" The "wheel" turns on 223 stations, covering 99 percent of TV homes in the United States. On average, it delivers 15.2 percent of the potential viewing audience. "Jeopardy!" is carried on 215 stations, which cover 98 percent of TV households. An average of 13.3 percent of those homes watch the show each weekday.

Toward the bottom of this Top-15 list, the data become more revealing. Notice that the rating for "Hard Copy" is 6.1, higher than the syndicated new episodes of "WKRP in Cincinnati." However, "WKRP" has much greater coverage. It is seen on 238 stations and is available to 99 percent of the nation's homes (compared to only 93 percent for "Hard Copy").

CHAPTER 11:
TELEVISION
ENTERTAINMENT

remains that network programs enjoy great popularity, which sustains over time. In the jargon of syndication, network shows have "staying power" and "legs." Old network programs never die: they are sold to stations and cable services as off-net syndication.

Off-net series are packaged for syndication in a manner similar to motion-picture packaging. The station or cable service pays for a certain number of episodes in the series and gets the right to show each title a number of times. Six runs over a period of six years is commonplace. The price per episode for off-net programs varies widely. It depends on the size of the TV market, the popularity of the show in its first run, the number of programs available, and so on.

Normally at least 100 episodes of a network show are needed to launch it in off-net syndication. The ideal number of episodes is 130. Why? Since there are 260 weekdays in a year, an off-net episode can run exactly twice annually when programmed in a time period each Monday to Friday. This process of scheduling a show to run in the same daypart each weekday is known as *stripping*.

You will recall from the section on network programming that in the "old days" (1950s and 1960s) the networks typically ordered thirty or more episodes of a program each year. Thus after three years on the air a network show would begin to appear in reruns. Some of TV's long-running classics had hundreds of episodes for syndication. For example, Viacom International offers 274 episodes of "The Beverly Hillbillies," 271 of "Perry Mason," and a whopping 402 of "Gunsmoke."

Over the years a few off-net series with limited runs have become popular in seemingly endless syndication. These classics include "The Honeymooners" (only thirty-nine original episodes, all in black and white) and the original "Star Trek," whose ten-year mission lasted only three on NBC (seventy-nine episodes).

Today's shorter run of many network programs has created a problem for both syndicators and their buyers. If a station buys a series with a short network run, it can't strip the show without running the same programs over and over (which can cause "burnout" among viewers). However, stations can't run movies and game shows all the time, and off-net series—especially situation comedies—remain very popular. Thus stations can't afford to wait three to five years for the ideal number of shows to accumulate. For this reason programs that have lasted long enough on network air to gain the "magic number" of about 100 episodes have commanded high license fees.

Shows that have recently attained or exceeded the magic number include "The Wonder Years," "Roseanne," "Empty Nest," "Family Matters," "Murphy Brown," and "Designing Women."

To date, the champion of off-net syndication, in cost if not in ratings, has been "The Cosby Show" (see box).

First-run syndication Programs that make their debut in syndication, without a prior network life, are known as first-run syndication. Traditionally first-run syndication has been characterized by cheap, easy-to-produce programs designed to be strip-programmed. Game shows, from "Wheel of Fortune" to "Win, Lose or Draw," have been the most common first-run type. More recently, reality-based programming has emerged, beginning with "PM Magazine," in the 1970s and now including "Entertainment Tonight," "Hard Copy," "Star Search," and "A Current Affair."

Two trends mark first-run syndication in the 1990s. The first is a preponderance of talk shows, for virtually all time periods, from early morning to late at night. By 1993 the "big three" of daytime TV talk, "Oprah!" "Phil Donahue," and "Geraldo," faced foes, including "Jenny Jones," "Regis and Kathy Lee," and "Joan Rivers." With the retirement of Johnny Carson and the rise of Arsenio Hall, late-night talk shows also proliferated. Pretenders to Carson's throne (inherited by Jay Leno at NBC) included Whoopie Goldberg, Byron Allen, Montel Williams, Chevy Chase, and Rush Limbaugh, among others.

The second trend is the advent of high-budget, high-concept programs for first-run. Led by Paramount's success with "Star Trek: The Next Generation," other new adventure dramas are being readied for sale directly to local TV stations. Paramount has introduced a spin-off "Star Trek," called "Deep Space Nine." Cannell has brought out "Street Justice" and "The Renegade." LBS/All American rescued "Baywatch" from the NBC graveyard and brought new episodes about lifeguards at work and play to viewers.

The High Price of Syndication

As we have noted, program costs have escalated appreciably in recent years for both the networks and syndication. In most retail businesses, when costs increase, they are passed on to the consumer. The same situation applies to TV syndication. In this case, it is the local station or cable operator who pays the additional premium. Here's how the bill for syndicated programming is paid (and how the cost is partially recovered at tax time).

Cash Paying cash for syndicated programming is the simplest system. The syndicator will offer a number of titles for a given rate. The station will agree to pay the amount in full at the time of the sale, or the syndicator will accept payments on an installment basis. For example, if the station and syndicator have negotiated a rate of $20,000 per episode for "Golden Girls" and there are 100 available programs, the total cash cost to the station will be $2 million. That may sound like an exorbitant cost. However, keep in mind that the

PROFILE

"DOLLAR BILL" COSBY: VIACOM'S OFF-NET AUCTION

CHAPTER 11:
TELEVISION
ENTERTAINMENT

In the mid-1980s a phenomenon swept broadcasting. After years of declining shares a network TV program attracted audiences approaching one-half the country on a weekly basis! The show, of course, was NBC's "Cosby" series, which not only made a multimillionaire of its star and launched the careers of Lisa Bonet, Malcolm-Jamal Warner, and others, but almost singlehandedly lifted NBC to the status of number one network. After three hugely successful seasons, 125 episodes were available for syndication beginning in the fall of 1988. That's when the fun began.

The show's syndicator, Viacom, offered stations three ways to have a shot at "Cosby." They could indicate their interest by paying a "reserve" price set by Viacom; they could indicate more serious interest by paying the reserve price plus a bonus; or they could present their own financial plan. To guarantee additional income, Viacom withheld one thirty-second spot for its own sale. The auction of "Cosby" became a market-by-market sweepstakes that soon broke all records for syndication.

In New York, WWOR paid almost $44 million for three and a half years of "Cosby" (about $240,000 per week). This was about three times the previous record ("Cheers," at a paltry $80,000 per week).

The New York numbers were the most staggering, but "Cosby" broke the bank throughout the nation. In Los Angeles KCOP shelled out $225,000 per week, nearly twice the previous Los Angeles record ($120,000 for "Webster"). Boston's WCVB paid $102,000 per week; in Detroit WDIV paid over $70,000. Bill brought in $40,000 a week from KIRO in Seattle and $30,000 from WLWT in Cincinnati.

By the fall 1988 debut, as principal owner of the show, Bill Cosby was one of the richest men in the world. Was it worth it to the stations? Early ratings indicated "Cosby" was a syndicated success but did not attract nearly as large an audience as had been anticipated. In the fall 1988 syndicated rankings, "Cosby" was fourth, behind "Wheel of Fortune," "Jeopardy," and "Oprah!" By 1992 "Cosby" had slipped to eighth place.

Maybe the Cosby kids have grown up. Maybe the competition has gotten stronger. Today "Cosby" is largely relegated to the history chapter, although many stations will be saddled with debt due Viacom long into the next century.

---

stations can sell about eight minutes of commercials (sixteen spots) in each program and can usually rerun each episode six times over the length of the contract. Thus it is possible—especially with a popular off-net series—to recover the costs in the second or third run, after which the program makes money for the station.

For a number of reasons cash sales of syndicated programming have declined in recent years. For one thing, the "bottom line" consciousness of the TV industry has made station managers reluctant to take million-dollar gambles on syndicated programs. The record number of mergers and buyouts have left little cash to speculate in program investment. And syndicators have had difficulty extracting payments from stations on installment plans. Many stations have been late in paying their bills; some (especially UHF independent stations with huge inventories) have defaulted altogether.

Barter syndication High prices have combined with low cash flows to create an increasingly common form of syndication finance. Just as the term was used in the days of fur trappers and Indian agents, *barter* refers to the trading of one commodity for another of similar value. In TV the valuable commodity offered by the syndicator is programming; the item of value at the station is its air time. In barter syndication the syndicator provides the program to the station free or at a substantially reduced cost per episode. In return, the station sacrifices some of the advertising slots in the show. The syndicator can integrate its own ads into the show or the syndicator can act like a TV network; calling on major advertisers to place ads in each of the markets in which the program plays.

In barter syndication the key to the syndicator is market clearance. The more markets the program plays in, the larger is the national audience that can be offered to advertisers. Clearing 80 percent of the nation's TV markets is considered good; over 90 percent clearance is excellent.

Advertising revenues from barter syndication have increased dramatically in recent years. In 1988 barter revenues to syndicators represented about $800 million. Today barter syndication is a $1.5 billion business.

The revenue potential for barter has made this the most common form of finance for popular off-net and first-run programs. "Roseanne," "You Bet Your Life," "The Wonder Years," and "Star Trek: The Next Generation," among many others, are delivered to stations on a barter basis.

Amortization: Recovering the costs of syndication
Fortunately for TV and cable operators, acquiring programming has a considerable tax advantage. To stimulate investment, federal law allows companies to deduct from their income taxes the cash value of assets they may have on hand in inventory. For example, an automobile dealer can deduct from the dealership's earnings the value of the unsold cars on the lot; the owner of a shoe store can deduct the value of the unsold shoes on the shelf.

The inventory on hand in TV stations is programming: those shows acquired through syndication but as yet unshown. Stations amortize programming in two ways. Straight-line amortization deducts the program's cost on the stations' tax return in equal payments spread across the length of the syndication contract. For example, if a station acquires a TV show for $10,000 and it can get five runs of the show over five years, it can take a deduction of $2000 per year.

Since programs lose their value as they are rerun (except for "Star Trek," "The Honeymooners," and a few others), stations may choose to recover their costs more quickly. Accelerated amortization allows for this phenomenon. Using the same example, the station could choose to deduct $4000 in the first year, $3000 in the second, and $1000 each for the remaining three years of the contract.

As you can see, amortization helps ease the high cost of obtaining syndicated programming by giving a sizable write off to the buyer. The effect has been to stimulate investment in a new product, in this case, TV shows.

LOCAL TELEVISION PROGRAMMING

The last piece of the programming puzzle is provided by individual TV stations and cable systems through their original, locally produced programs. Once thought to be a disappearing commodity, local programming has flourished because of developments in technology. Although local shows frequently suffer in comparison with the big-budget programs of the networks and the high gloss of syndication, some local programming rivals the production values of the big three and their competitors.

Local Television Stations

Faced with escalating syndication costs, lagging network performances, and the loss of local advertising dollars to barter, TV stations are placing increasing emphasis on their own local programming. As we saw in the previous chapter, the bulk of the local TV budget has gone to news. However, a number of stations have expanded beyond the "news hole" to develop their own talk, entertainment, and dramatic productions.

For example, WDIV in Detroit produced a popular series of thirty-second vignettes called "Stupid Tricks." These vignettes featured people and pets, of whom the best performers were paid by sponsor Taco Bell to attend a taping of "Late Night with David Letterman." WJLA-TV in Washington, D.C. offers "Working Women," a weekly program targeting that growing audience. In Sacramento KCRA offers a half-hour regional magazine program, "The West," which is successfully being syndicated throughout the region. Diversity is the keyword in local TV production. Survey after survey of general managers and program directors suggests that local production of everything from situation comedies to dramas will increase in the 1990s.

Local Cable Programming

Just as the future of TV stations may lie in their ability to develop local programming, the cable industry is making similar forecasts.

IMPACT

WCVB, BOSTON: A LEADER IN LOCAL PRODUCTION

Most of what you have learned in this text would suggest that the odds against WCVB-TV in Boston were stacked against it. First, the station didn't even get on the air until 1972, twenty years after the famous "TV freeze," when most of the nation's leading stations went on the air. In addition, the station was an ABC affiliate. Like most large cities, TV in Boston was dominated by older, established NBC and CBS stations. Added to this was the presence in Boston of WGBH, one of the most influential (and richest) public TV stations.

The last thing one would have expected from WCVB is leadership in the realm of local program production. Yet in the past two decades WCVB has earned a reputation as a leader in this area, perhaps the best local station in America. Here's a short list of the station's accomplishments:

In 1980 WCVB became the first local station to produce a movie-for-TV, "Summer Solstice," with Henry Fonda and Myrna Loy, which aired nationally on ABC. In 1982 its "Chronicle" series became the nation's first daily locally produced newsmagazine show. Its ground-breaking series on health ("Healthbeat") and law ("Miller's Court") were syndicated nationally. "Chronicle" was aired nightly on the Arts and Entertainment cable network. In 1986 its nightly news was the first in the nation to be captioned for hearing-impaired viewers. WCVB has had a long-standing commitment to programming for minorities, women, and children. Its series of sixteen documentaries of the black experience in America can be found in libraries and schools around the country.

Like many stations, WCVB has suffered recently from budget and staff cutbacks. But, at least in the trade press, station management remains committed to the future of local TV production.

As we documented in the previous chapter, cable systems have been especially vigorous in developing news and sports programs as an alternative to local affiliates in their service area. But cable's increasing commitment to local origination extends beyond news and sports.

Actually the new world of cable programming goes back to the medium's growth period of the late 1960s and early 1970s, the era of public access programming.

Cable public access programming The FCC's 1972 cable regulations required systems in the top-100 markets to make a certain number of channels available free for public use on a first-come, first-served basis. Systems were mandated to build studios for this purpose and to make staff available to assist groups and individuals in their programming efforts.

In the ensuing years public access cable became a grand experiment in participatory democracy. Groups from all points on the political spectrum came forward to air their views—often eliciting outrage. Some programs stretched the limits of taste, discussing alternative lifestyles, sexual preferences, and the like.

More in the mainstream, many public access channels flourished as an avenue for new programmers and audiences. Senior citizens and young children alike got their first chance to "make TV."

By 1979 the FCC had rescinded its public access requirement. Although it did allow franchises to negotiate for public channels, the Cable Communications Policy Act of 1985 contained no public access mandate. Yet the concept continues to flourish in some systems around the country. Although few systems still make studio facilities available free of charge, public access remains vital in systems in Michigan, New York, California, and other areas. Channels are made available to a variety of users for an hourly fee, typically ranging from $50 to $500. This is known as leased access.

In addition to conventional TV programming, leased services include security systems and data and text services (stock information, news tickers, and so on).

Community cable channels Unlike cable access, which consists of channels programmed by outside groups, community channels are cable ser-

"Dino & Rocco's Back Alley," a cable access show in Los Angeles known for its controversial, raunchy, and tasteless style. On cable access, no matter how bad you are, you can't get canceled.

vices provided by the cable operators themselves to their subscribers. The most common form of community cable service has been locally programmed pay TV. Indeed, much of cable's growth in the 1970s was due to the availability of "art-house" movies and home-team sports for a monthly fee on local cable. Pioneers in this trend were Madison Square Garden cable in New York (offering home Knicks, Rangers, and other sports events), Prism in Philadelphia (with the 'Sixers and Flyers in its lineup), and the Los Angeles–based Dodgervision.

Community channels have had a tough time in recent years. The sports services have suffered from overexposure: there seem to be hundreds of ball games on at any given time. Movie services have suffered since the arrival of home video outlets and domination of pay TV by the major movie channels (whose exclusivity deals have shut out the smaller channels from obtaining recent box-office hits).

Despite some tough sledding, the outlook for community cable channels remains bright. As documented in Chapter 6, local cable advertising is providing an increasing source of revenue, some of which will be funneled into programming. Once-costly production equipment is being replaced with cheaper, easy to maintain, smaller format gear. This has the potential to enable cable programs to emulate local TV stations in production values, content, and audience appeal.

PROGRAMMING STRATEGIES

Now that we understand the various types and sources of TV programming today, we close this chapter by providing a taste of the techniques of TV programming. Space precludes a full discussion of how TV programmers determine their schedules. This process fills entire books and comprises full semester courses at many institutions. The books listed at the end of the chapter provide a good beginning point to an understanding of TV programming in practice. Now, on to the "taste test."

There is an old saying in broadcasting: "People don't watch stations, they watch programs." This means we select the station to watch based on programming rather than selecting the program because of the station that transmits it. It is also a reminder of the importance of programming to the success of TV.

The first step in programming is to define the potential audience. Cartoons shown while children are in school or football broadcast when men are at work may attract small audiences, but not the numbers possible if the cartoons are run after school or the football game is shown on Sundays when most men aren't working. Ratings services provide programmers with information about what groups are viewing when.

Programmers must also have some idea about what groups prefer the shows that are available to be used. "Murder She Wrote" generally attracts women and older men as its main group of viewers. To have optimum viewership it is necessary to schedule "Murder She Wrote" at a time when these people will be in front of their TV sets. Information about what groups are attracted to a given show is available from the ratings services, from the station's national sales representative, and, though not as objective a source, from the supplier of the program.

Obviously, programming must entail more than merely placing a program that appeals to a particular audience at a time when that audience may view it. There are a number of techniques that programmers use to maximize viewership. Once you have a viewer, you want to hang on to her or him.

Audience Flow

Audiences will tend to stay with the TV station they are watching until something they dislike shows up on the screen. In some ways audiences take on inertia-like properties, viewing the same station until forced to change. The proliferation of remote controls has changed this tendency since viewers no longer are required to get out of their chairs to change the channel, but inertia still influences viewing. The movement of audiences from one program to another is called *audience flow*.

The successful programmer will build and hold an audience from show to show. This means putting together programs that generally attract the same audience. It doesn't make much sense to schedule "The Fall Guy," then "The Love Boat," and then "The A-Team." "The Fall Guy" and "The A-Team" attract a primarily male audience, whereas "The Love Boat" attracts a primarily female audience. The viewers gained with "The Fall Guy" will be lost when "The Love Boat" appears. It would be better to put "The A-Team" and "The Fall Guy" next to each other since much of the audience for one program would flow through to the next program.

As a programmer you want to attract and build your audience. Audience flow is a key way to do this. Pick up a *TV Guide* or the TV section of a local newspaper and look carefully at the programming of one station. Note how the programs from about 9:00 A.M. to 3:00 P.M. would seem to attract primarily women. Some stations stay with the woman audience past 3 P.M.; others switch to children's programming. The decision to go for another segment of the audience is a decision about whether to counterprogram.

Counterprogramming

Counterprogramming is a technique wherein the programmer decides to go for a different audience than the competing station is trying to attract. "Roseanne" has been a blockbuster program on Tuesday nights at 9:00 P.M. (8:00 Central). At its peak the program was attracting nearly a half of all viewers who were watching TV. This was great for ABC—but what if you're NBC or CBS? The decision was to counterprogram.

In the 1992–1993 season NBC decided to go after an older audience than that of teen-favorite "Roseanne," by scheduling "Reasonable Doubts." CBS countered with made-for-TV movies.

At the local level you may observe this same counterprogramming strategy. Some stations will attempt to keep and build on a primarily female audience with shows like "Donahue" while the competing station may schedule "Duck Tales" at the same time to attract children.

Challenge Programming

Let's say that you are assigned to program the 3:00 P.M. time slot on your station. There are two other stations with which you compete. One has "Donahue" and the other has "Duck Tales." Ratings research tells you that two groups, young children and women, make up the audience. What do you do? Counterprogramming doesn't seem to be an option.

In this case you might turn to *challenge programming*: you go head-to-head with one of the two stations and program for the same audience they have. You decide to go after "Donahue" and challenge with "Geraldo." Generally, this means that you will be splitting Phil Donahue's audience, but you would also hope to pick up new viewers who are looking for material a bit more on Geraldo Rivera's controversial side.

Hammocking

Congratulations—you are now the person who makes the programming decisions for NBC. You have a new comedy that you want to introduce at midseason. The problem is to get people to view the show. If people sample it two or three times, you are sure they will want to watch it. But, how are you going to be sure that people will change from their usual programs and watch your new show?

The most popular solution is to place the new show between two other shows that are doing well, so you premiere the show on Thursday nights at 9:30—right after "Cheers" and just before "L.A. Law." In this case you place your unknown show between two consistent top-10 programs—the *hammock* position. Audience flow should take care of the rest. People who watched "Cheers" and are planning on watching "L.A. Law" stay tuned to watch the new show.

After you have given the audience three or four weeks to get to know the show, you may move it to its "permanent" (although there is hardly such a thing in TV programming) time slot. With any luck the audience that got used to watching the program between "Cheers" and "L.A. Law" will follow.

Of course, this was precisely the strategy employed by NBC to develop "A Different World" and "Wings." More recently ABC created a huge hit in "Home Improvement" by hammocking it between "Full House" and Roseanne."

The hammock position can also be given to less popular shows that you have commitments to run. You attempt to protect the show by securing a strong lead-in audience. In network TV the hammock position can be a blessing and a curse. Although the program is virtually assured of a fairly large audience if the hammocked show merely holds or loses portions of the audience it may be axed more quickly than if the show were trying to make it on its own.

Programming is the "fun" part of TV. It is what attracts viewers so that the station can sell time to advertisers to make money. A great deal of thought and effort are put into deciding which programs fit where. The reason for this concern in programming is not primarily because the station is concerned that viewers are entertained but because this is how they make their money.

SUMMARY

Television stations can obtain programming through networks, syndication, and local origination. The commercial networks—ABC, CBS, NBC, and Fox provide broad, mass-appeal programming. These networks pay affiliates to carry their programming.

In the past, successful programming was that which was the least objectionable to the public. However, this trend has appeared to change. Shows that target a specific audience, such as "Northern Exposure," are now moving into the forefront.

Networks have to obtain their programming from outside sources; thus they work in conjunction with large studios or independent producers. The networks receive a great number of suggested story lines. This means that there is only a slight chance that any one idea will be accepted.

Movies of the week and made-for-TV movies are the most expensive types of programming. Situation comedies and reality programs, on the other hand, are the cheapest to produce.

As cable matures, it has directly affected the networks. Cable has been producing higher quality shows, and the effects can be seen in the ratings.

Public TV is another form of network programming. It attracts those who are well educated and those who do not watch a great deal of TV.

Syndication has become a popular source of programming. To understand fully the syndication business, one must be aware of NATPE trade shows, types of syndication, prices, types of payment, and amortization.

Local TV saves money by producing local shows; therefore, community and cable stations are trying this technique. Because they will be saving money and providing original programming, these stations are likely to survive in the future.

TV programming strategies include counter-programming, challenge programming, and hammocking.

SUGGESTIONS FOR FURTHER READING

Auletta, K. (1991). *Three blind mice: How the TV networks lost their way.* New York: Random House.

Baldwin, T. F., & McVoy, D. S. (1988). *Cable communication* (2nd ed.). Englewood Cliffs, N.J.: Prentice-Hall.

Block, A. B. (1990). *Outfoxed: Marvin Davis, Barry Diller, Rupert Murdoch and the inside story of America's fourth television network.* New York: St. Martin's.

Blum, R. A., & Lindheim, R. D. (1987). *Primetime: Network television programming.* Boston, Mass.: Focal Press.

Brooks, T., & Marsh, E. (1985). *The complete directory to prime time network TV shows: 1946–present.* New York: Ballantine Books.

Carroll, R., & Davis, D. M. (1993). *Electronic media programming: Strategies and decision-making.* New York: McGraw-Hill.

Christensen, M., & Stauth, C. (1984). *The sweeps: Behind the scenes in network TV.* New York: William Morrow and Company.

Eastman, S. T., Head, S. W., & Klein, L. (1989). *Broadcast/cable programming: Strategies and practices.* Belmont, Calif.: Wadsworth Publishing.

Gitlin, T. (1985). *Inside prime time.* New York: Pantheon.

Verna, T. (1987). *Live TV.* Boston, Mass.: Focal Press.

PART FIVE

HOW IT WORKS

CHAPTER 12: AUDIO TECHNOLOGY

Watching TV and listening to radio are the easiest things in the world to do. Just twist that dial, flip that switch, or punch that button and poof: vivid sounds and picturesque images are yours (unless, of course, you're watching a test pattern). The ease with which we command TV and radio reception hides the incredibly difficult problems and complex technical processes involved in moving sound and pictures from their source to you. This chapter and the next attempt to demystify the magic of radio and TV technology. We describe how the process works and, more important, why it matters. In many ways understanding the technical bases of broadcasting helps you to understand its history, legal status, social and political power, and future. We begin with audio: how radio, records, tapes, and discs get from earshot to headset and from soundstage to boom box. The next chapter describes the process of TV "pictures in the air."

TECHNICAL ASPECTS OF BROADCASTING: SOME BASICS

It's helpful to begin a discussion of technical aspects of radio and TV with some basic notions of how objects behave in nature. Broadcasting is based on common physical properties.

Facsimile Technology

All modes of mass communication are based on the process of facsimile technology. That is, sounds from a speaker and pictures on a TV screen are merely representations, or *facsimiles,* of their original form. We all learn and practice facsimile technology at an early age. Did you ever use a pencil or crayon to trace the outline of your hand on a sheet of blank paper? That's facsimile technology. Having one's face covered by plaster of Paris for a mask or sculpture is facsimile technology; so are having your picture taken and photocopying a friend's lecture notes.

In general, the more faithful the reproduction or facsimile is to the original, the greater is its *fidelity.* High-fidelity audio, or "hi-fi," is a close approximation of the original speech or music it represents. And a videocassette recorder marketed as high fidelity boasts better picture quality than a VCR without hi-fi (known as H-Q, to distinguish video high fidelity from its audio counterpart). Indeed, much of the technical development of radio and TV has been a search for high fidelity: finding better and better ways to make facsimiles of the original sounds or images.

The second point about facsimile technology is that in creating their facsimiles radio and TV are not limited to plaster of Paris, crayon, oils, or even photographic chemicals and film. Instead, unseen elements such as radio waves, beams of light, and digital bits and bytes are utilized in the process. Although you cannot put your hands on a radio wave as you can a photo in your wallet, it is every bit as much *there.* The history of radio and TV is directly linked to the discovery and use of these invisible "materials," from Marconi's radio experiments in the 1890s to the microcomputer technology of today.

In the technical discussion that follows bear in mind that the engineer's goal in radio, TV, and cable is to create the best possible facsimile of our original sound or image, to transport that image without losing too much fidelity (known as signal loss), and to recreate that sound or image as closely as possible to its original form.

Transduction

Another basic concept is *transduction,* the process of changing one form of energy into another. When the telephone operator says "the number is 555-2796" and you write it down on a sheet of notepaper, you have transduced the message. Similarly, when you slip a tape in your boom box and stroll through the mall, you are transducing— although not all those around you will appreciate it.

Why does this matter? Well, getting a sound or picture from a TV studio or concert hall to your home usually involves at least three or four transductions. At each phase loss of fidelity is possible and must be controlled. And at each phase the whole process may break down into *noise*—unwanted interference—rendering the communication impossible. Although breakdowns due to transduction rarely occur, they do happen. Technicians were embarrassed when the sound went out during the 1976 presidential debate between Gerald Ford and Jimmy Carter. Closer to home, a broken cartridge on your turntable or a cassette recorder with dirty heads can make Tchaikovsky sound like the Red Hot Chili Peppers.

Television and radio signals begin as physical energy, commonly referred to as light waves or sound waves. When you hear a bird chirping in a tree, your brain directly perceives the event by processing the sound waves that enter your ears and vibrate your eardrums. You see the bird as your brain processes the reflections of light that have entered the eye and fallen on the retina. This is direct experience: no transduction, no signal loss, true high fidelity. To experience this event on radio, however, the following translations or transductions occur.

Let's deal with the chirping first (the visual part is covered in the next chapter). As we stand there with a microphone attached to a tape recorder, the bird's song is first changed into mechanical energy. Inside a dynamic microphone, the various sound wave pressures cause a small coil to vibrate back and forth in a magnetic field. This sets up in the coil a weak electrical current that reproduces the pattern of the original sound waves. Next, this current is amplified and varies the magnetic field in the *recording head* of the tape recorder—the device that stores a new signal on tape by rearranging the metallic particles on the tape. Thus the bird's song is translated into patterns of magnetic blips on a piece of audiotape. How well the transduction occurs is based on the quality of our facsimile. A big tape recorder with tape 2 inches wide, like those used to record orchestra performances, will produce bigger and better blips than those produced by a $20 cassette recorder that uses tape ⅜ of an inch across.

Now we have a mechanical facsimile of the chirping, a pile of blips on a piece of tape. Next, we need to transduce that signal into electrical energy. In playing back the tape the *playback head* of our recorder converts the signal stored on the tape into a small electrical charge. The amplifier detects the electricity, boosts it, and cleans it up. And as stereo buffs know, the quality of our facsimile is now based on the quality or power of our amplifier. We are now halfway to hearing the chirp on our radio.

Next, we transduce the electrical energy into electromagnetic energy. At the radio station we feed the amplified sound from the tape recording to the transmitter. Here the signal is superimposed, or "piggybacked," onto the radio wave (channel) assigned to that station (a process called *modulation,* examined in detail later). The fidelity of the signal is based on the station's power, its location, its channel, and all sorts of other factors.

At home we turn on the radio and tune to the appropriate channel. Our antenna detects the signal and begins to reverse the transduction process. The energy is transduced first to electrical impulses as it enters our radio's amplifier. Next, it is transduced to electromagnetic energy, to vibrate the diaphragm of the radio's loudspeaker. At long last it is transduced back into its original form, physical energy, as we hear the chirp as vibrations on our tympanic membrane (eardrum). Note the many transductions in this process.

What's the point of all this? Why does the transduction process matter?

Signal and Noise

First, the signal, or message, goes through many translations, with the possibility (even the necessity) of losing some information or adding some unnecessary data at each phase. It's rather like playing the game "telephone," where several people whisper a simple message in turn. By the time the message gets to the last person a lot of the original information has dropped out and some irrelevant information has crept in. In electronic terminology, losing information in the transduction process is known as *signal loss.* The unwanted interference is known as *noise.* Now you can understand a term found in many advertisements for stereos and VCRs: the *signal-to-noise* ratio. In common terms, this is a numerical representation of the amount of "pure" picture or sound information that remains after subtracting unwanted noise acquired during the transduction process. The higher the signal-to-noise ratio, the higher the fidelity; the lower the ratio, the "noisier" the sound or picture.

Analog and Digital Signals

How information is converted from one form of energy to another is an important aspect of transduction. Until the 1980s broadcast transmissions utilized *analog signals.* This means that to change the energy from physical to electrical impulses, an "analogy" to the original sound or image replaced the matter itself. It sounds tricky, but the concept is simple. Think of the grooves in a phonograph record. As the needle travels through the grooves, it vibrates in a pattern similar to the vibrations made by the guitar string or vocal chords it represents. This is not the original vibration, but an *analog* to it. Similarly, when the chemicals in a piece of photographic film change in response to the patterns of light reflected on it, the film is an analog recording of the events in front of the camera. By their nature, analog transmissions and recordings are subject to a great deal of signal loss and noise. They merely represent the original signal, itself a major limitation since the transmissions can never include all of the information present in the original sound or picture. And analog signals tend to decay over time and space. After all, records collect dust, get scratched, and warp in hot cars and on window sills. Photographs blur, tear, and fade.

Most of the excitement in broadcasting, home audio, and video today concerns *digital* recording and transmission: rather than creating an analog to the original sound or picture, the signal itself is utilized in the transduction process. Each element of the audio or video signal is translated into its digital equivalent—that is, a number—by means of a binary code. A binary code is one with only two values such as "on–off," "yes–no," "open–shut," or 0 and 1. Computers use binary codes to process information. In digital recording each sound is a unique sequence of binary numbers: 010, 011, 101, and so on. The picture or music is transduced by being "read," or *sampled,* by a beam of laser light, the same way the bar codes on your groceries are at the supermarket checkout. There are no needles, no friction, no wear, no scratches. More important, there is virtually no opportunity to introduce noise. When the signal is transmitted and played back, what we hear through the speakers and watch on the screen is a virtual duplicate of the original signal with the highest fidelity and excellent signal-to-noise ratio, meaning it is virtually noise-free. This is why Guns N' Roses on compact disc sounds as if they are actually inside your speakers! This is also why executives of the recording industry have opposed digital audio tape (DAT; see below) recorders. Copies of CDs made by DATs are just as good as the originals; neither will have significant signal loss or noise!

Oscillation and the Waveform

Another basic principle to both audio and video signal processing is the concept of oscillation. Recall that we hear sounds and see images as variations, fluctuations, or vibrations detected by our ears and eyes and interpreted by our brain. Remember too that every vibration has a unique signature or "footprint": every sound and image has its own characteristics. How well we see, how acute our hearing, and how well we can re-create these signals as broadcast sounds and pictures depend on our ability to identify, store, and recreate those vibrations. In electronic terms, the vibration of air produced by our mouths, the instruments we play, and objects in our natural environment, as well as the vibration of light that accounts for every color and image our eyes can see, is known as *oscillation*. The footprint or image of an oscillation we use to visualize the presence of the invisible is known as its *waveform*. Figure 12-1 demonstrates the phenomenon of oscillation and the common waveform.

The most common way you and I can visualize oscillation is by dropping a rock into a pail of water. You know that depending on the size of the bucket and the height from which we drop the rock, the result will be a series of circles or waves, radiating outward from the spot where the rock fell, until dissipating some distance from the center. All audio and video signals produce a pattern like this, except that they are invisible to the naked eye. However, while we can't see the patterns with the naked eye, using the appropriate electronic equipment (such as an oscilloscope and a waveform monitor), we can detect them, use them, even create them ourselves.

Frequency and Amplitude

A major way in which we describe a wave is its *frequency,* the number of waves that pass a given point in a given time. Originally measured in cycles per second (cps), frequency is more commonly measured now in *hertz (Hz),* in homage to early radio pioneer Heinrich Hertz, whose exploits are described in Chapter 2. In our example with the rock depicted in Figure 12-1 the frequency is simply the number of waves that pass a given point in a single second. Note also from the bottom of Figure 12-1 that the distance between two corresponding points on a wave is called the *wavelength*. Also note that frequency and wavelength are inversely related. High frequency means short wavelength; low frequency means a long wavelength. In the early days of radio U.S. stations were classified by their wavelengths. Today we identify them by their frequencies.

Figure 12-2 describes the frequencies of human speech and various common musical instruments. Note in Figure 12-2 that the human voice is capable of a range of about 10,000 hertz, from the lowest bass voices (like Richard Sterban of the Oak Ridge Boys), at under 100 hertz, to the highest alto voices (like Mariah Carey's) at a frequency approaching 10,000 hertz. Dynamic musical instruments like the violin oscillate from 250 to about 15,000 hertz; by contrast, the timpani oscillates in a range of only 100–2000 hertz. These ranges become important when we begin to try to record, transmit, and recreate the oscillations as radio and TV signals.

A wave may also be described by its height or depth, from its normal position before the rock is dropped to the crest created by the splash. This is known as its *amplitude*. When choosing a sec-

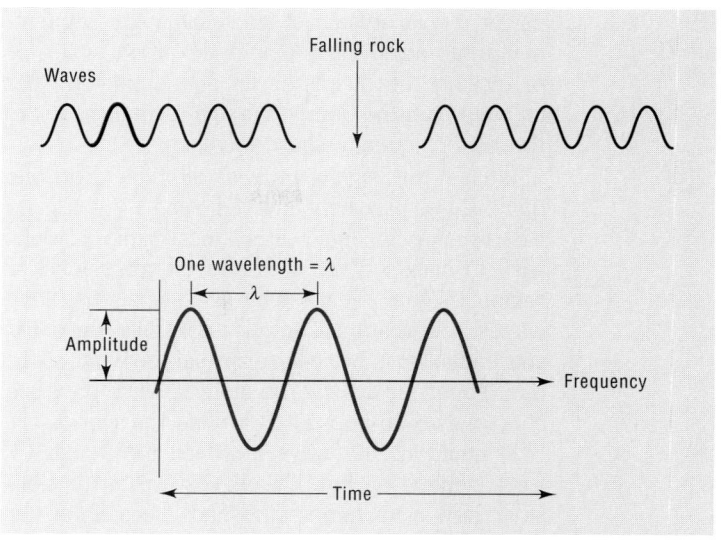

FIGURE 12-1 Principle of oscillation and the common wave form.

tion of beach to swim in, families with small children are likely to select those with slow, low, undulating waves—those of low amplitude or, to use a radio term, the long waves. Surfers will

FIGURE 12-2 Frequency range of the human voice and common musical instruments.

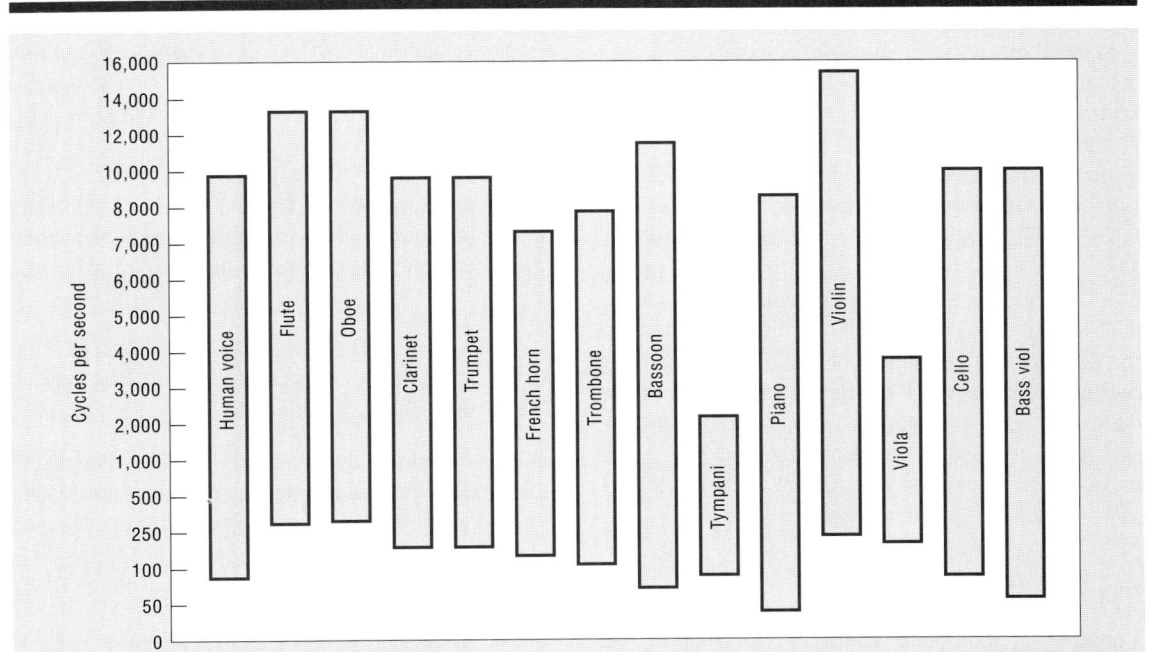

277

select a wild stretch of shoreline with frequent, mammoth-sized (high-amplitude) waves capable of picking the swimmer up and depositing her anywhere on the beach. In radio terms, these are the high-frequency short waves.

What's the significance of all this? For one thing, this is precisely how radio and TV work. As we examine in detail later, local radio stations wish to blanket their area with a strong local or regional signal. That's why AM signals use long waves. International broadcasters like the BBC and Radio Moscow seek to spread the word about their countries around the globe. Hence they use short waves to hopscotch around the world.

In addition, the ways in which radio and TV work depend on how the waves are used. Consider how a swimmer who finds herself too far from shore might get from the water to the beach. She could remain just below the surface, protected in an envelope of water, and try to make it to shore with strong, regular swimming strokes. This would require considerable power and would likely produce a smooth, if unspectacular ride. On the other hand, she might grab a piece of driftwood or a stray surfboard and take her chances at the surface by "riding the wave." She might get to shore quicker but would be much more likely to go off course, become winded, or worse, crash into a pier or jetty.

Well, that's exactly how radio works. AM signals use the "surfboard" method, *amplitude modulation,* where the signal is placed atop the crest of the wave. There's a lot of going off course (fading in and out) and "crashing" (static), but you can hear AM stations over considerable distances. FM stations use *frequency modulation.* The radio signal travels like a torpedo, just under the surface. The oscillations emanate powerfully, in a straight line, and usually hit the right spot, in the form of an excellent noiseless sound in your receiver. However, FM signals usually are inaudible past a certain narrow geographic boundary and are very tough to keep tuned in when the receiver is moving around (a car radio or a jogger's Walkman, for example).

Frequency Response

Consider a final point about oscillation and the waveform. How well we can record and play back music or pictures depends on the range of frequencies that our radio or recorder is capable of receiving or reproducing. This is known as the unit's *frequency response.* A radio set that can be tuned to frequencies below 10,000 and above 1000 only will simply exclude very low and very high sounds. At the same time, a receiver with the best speakers, with a frequency response from 40 to 40,000 hertz, will be able to reproduce virtually all the sounds the human ear can hear. It's the difference between hearing a symphony orchestra on a CD player or through the tiny speaker in a telephone receiver. This is critical since the history of the popular arts is directly linked to the frequency response of the prevailing methods of signal processing.

In the recording industry the early success of banjo-playing minstrel-type performers, who frequently whistled in their acts, was in large part due to their audibility on 78-rpm records of limited, mainly high-frequency capability. Such stars as Al Jolson, Rudy Vallee, and Eddie Cantor fall into this class. Similarly, in retrospect, Elvis Presley's limited tonal range seems to have directly fit the limitations of the cheaply produced 45-rpm record popular in the 1950s, which was meant to be heard on a teenager's much-abused "personal" record player (certainly not on Dad's hi-fi in the den). More recently, is it any surprise that the orchestrations, sound collages, and other experimentations ushered in by the Beatles' *Sergeant Pepper's Lonely Hearts Club Band* and other groups in the late 1960s were aided by the developments of high-fidelity studio recording and FM stereo broadcasting? Today the complexities and acoustic calisthenics of rap performers (like Hammer, Ice-T, and Salt 'n' Peppa) are made possible by the extended range of frequencies available with new audio components, such as digital amplifiers and CD players.

STEPS IN SIGNAL PROCESSING

Having mastered some of the basic media terminology, we turn our attention to a detailed examination of sound operations. The process of getting a sound to your home receiver follows five main steps. These are depicted in Figure 12-3.

Step 1: Signal Generation

This step involves the creation of the necessary oscillations, or detectable vibrations of electrical

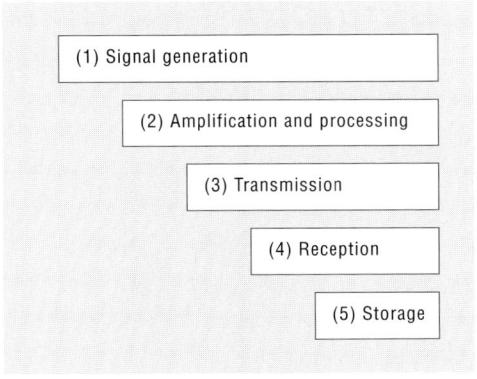

FIGURE 12-3 Steps in signal processing.

energy, which correspond to the frequencies of their original counterparts in nature. In plain language, signal generation involves getting the grooves onto a record, the vibrations into a microphone, and the blips onto a strip of audiotape.

An audio sound may be generated by two main transduction processes. Mechanical generation uses facsimile technology to create an analog of the original signals (by now you should be able to understand that technical-sounding sentence). That is, mechanical means are used to translate sound waves into a physical form, one you can hold in your hand, like a phonograph record or audiocassette. As we have seen, digital generation changes the form of the original wave into its binary or digital equivalent. Let's look at mechanical generation first.

Inside the phonograph record An examination of the grooves in a record under the microscope might look something like Figure 12-4. The sounds produced by musicians in the studio have been converted into cuts on the two sides of the groove (this is probably why songs are known as "cuts" in the popular music business). This process is known as *lateral cut recording*. As the turntable rotates, the stylus (needle) vibrates laterally as it follows the groove, creating the vibrations corresponding to the original words and music.

To ensure the best possible vibrations, the stylus should always be at a direct right angle (straight up and down) to the groove. This explains why the head on the tone arm of a newly designed turntable is bent in a J shape from the edge of the record slightly toward the center of the turntable.

Over the years continuing improvements have been made in the production of records, turntables, needles, and other recording equipment. For example, the frequency response of the gramophone records of the early 1900s ranged from 200 hertz (G below middle C, for you musicians) to around 1500 hertz (second G above middle C). Today a typical high-fidelity LP record ranges over eight octaves, from 40 to 12,000 hertz, which pretty much includes most of the sounds that are produced by voice and musical instruments (see Figure 12-1).

Inside the microphone Another place where speech or music is mechanically recreated to produce electrical signals is inside the microphone. There are three basic types of microphones: dynamic, velocity, and condenser. Each produces the waveforms required for transmission in a different manner. The generating elements of each type are illustrated in Figure 12-5.

In a dynamic microphone a diaphragm is suspended between two electromagnets. In the center of the microphone is a coil of electrical wire, called a voice coil. Sound pressure vibrates the diaphragm, which moves the voice coil up and down between the magnetic poles. The result is an electrical pattern in the mike wire analogous to

Dynamic microphones such as this one are popular in the industry for two reasons: they are durable and unobtrusive.

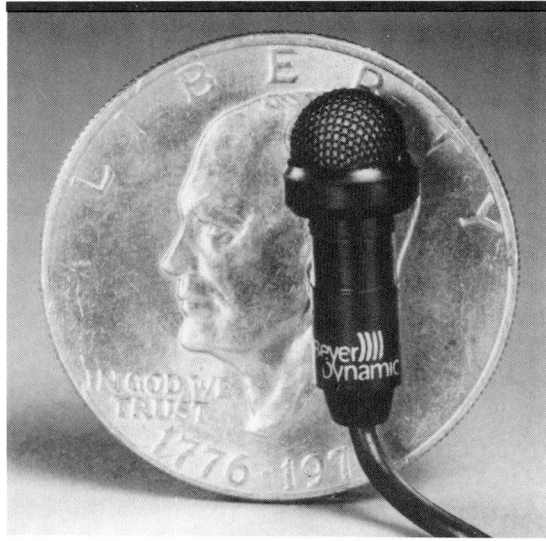

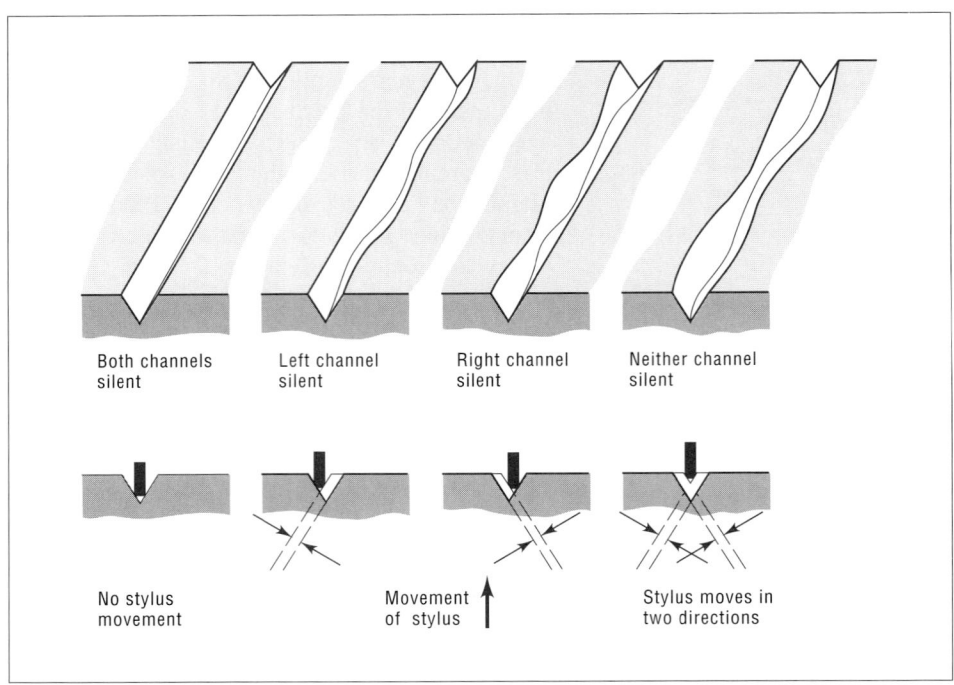

FIGURE 12-4 Grooves in a phonograph record.

the frequency of the entering sound. Thanks to durable and rugged design similar to a sturdy kettle drum and good frequency response with voices and most music, dynamic mikes are frequently utilized in radio and TV productions.

Velocity microphones, also known as ribbon microphones, replace the voice coil with a thin metal ribbon. There is no diaphragm; the oscillations of the ribbon suspended between the electromagnetic poles produce the necessary electric signals. Velocity mikes produce a lush sound, particularly with the human voice, but they are very fragile and highly susceptible to wind damage. Very common in the "golden age" of radio, they are still widely used in recording studios. (You can see one on Jay Leno's desk on "The Tonight Show.")

In place of a diaphragm, condenser microphones use an electrical device known as a capacitor to produce electronic equivalents of sound pressure waves. The capacitor is an electrically charged plate. The pattern of electricity in the plate (its amplitude and frequency) varies in relation to its distance from its stationary backplate. That distance is affected by the pressure of the incoming sound waves. While this might seem complex, the process is quite similar to the workings of your own ear. Without touching the volume knob on a portable stereo, you can vary its loudness simply by moving the headphones closer to or farther from your eardrums.

Condenser mikes feature excellent dynamic range but require external power sources to charge the capacitor and amplify the resulting electrical signal before it enters the standard amplifier. Thus most early condenser mikes were cumbersome and large, limiting their use in broadcast sound applications. Recent developments in microelectronics have led to a special class of condenser mikes in common use in telecommunication today. Electret condenser mikes feature an internal power supply unit, typically indicated by the presence of a small battery compartment inside the body of the mike or attached to its cable connector. Electret condenser mikes can be made extremely small (tie-tack, and clip-on styles) and still produce a crisp sound, especially with speech. Thus they are now the preferred mike for news personnel.

Inside the tape recorder We have now succeeded in transducing sound pressure into electrical os-

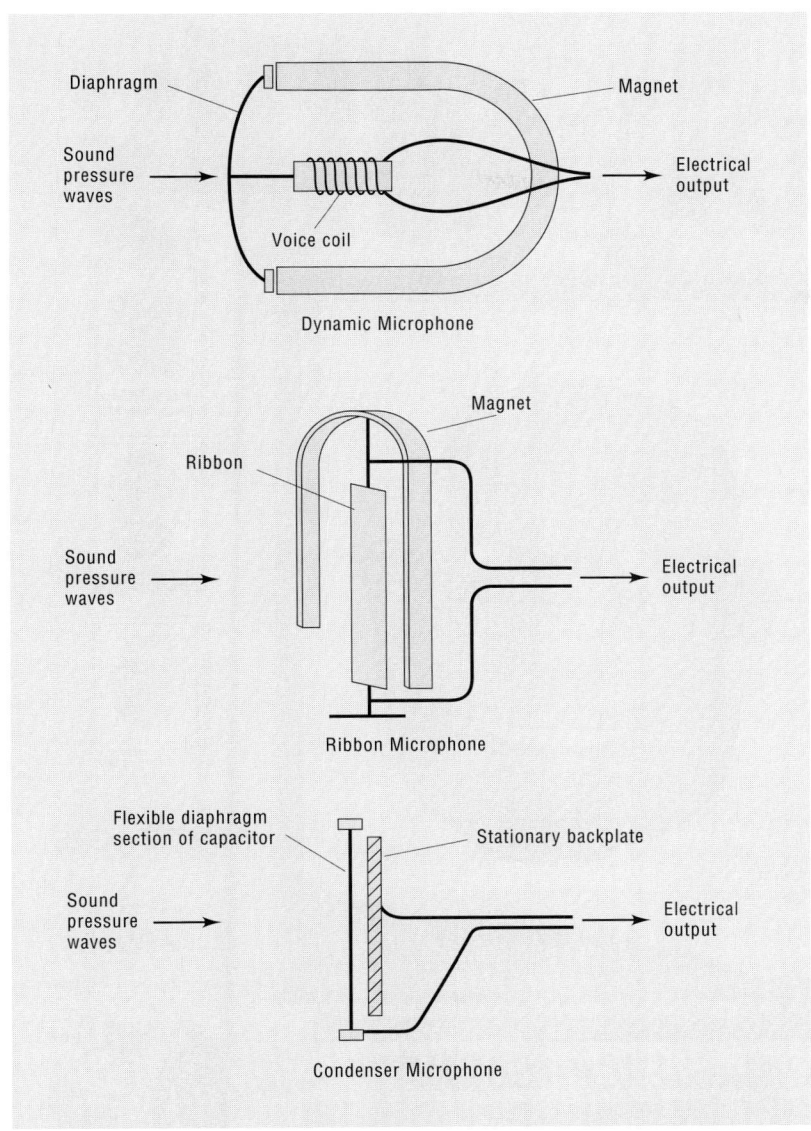

FIGURE 12-5   Types of microphones. (Source: Alan Wurtzel and Stephen Acker, *Television Production,* New York: McGraw-Hill, 1989.)

cillations in the form of grooves on a record and vibrations of a diaphragm or coil in a microphone. The third way we begin audio signal processing is by converting mechanical sounds into those blips on a piece of audiotape we mentioned earlier.

Some people have a hard time understanding the concept of audiotape technology. That is why we call on Wooly Willy for assistance. Wooly Willy appears in Figure 12-6. Some of you may have seen (and played with) Willy or one of his relatives before. The idea is to add a beard, eyebrows, or mustache to Willy (turning him into Dick the Dude or Harry the Hermit, according to the packaging) by manipulating metal filings with the Magic Wand, simply a straight magnet. To move a lot of metal filings you need a strong magnet or a magnet positioned very close to the plastic covering over Willy's face. Placing a few filings (to create a facial mole, perhaps) requires a weak magnetic pull—usually achieved by working the magnet far away from the plastic cover. Believe it or not, in mastering Wooly Willy, you have

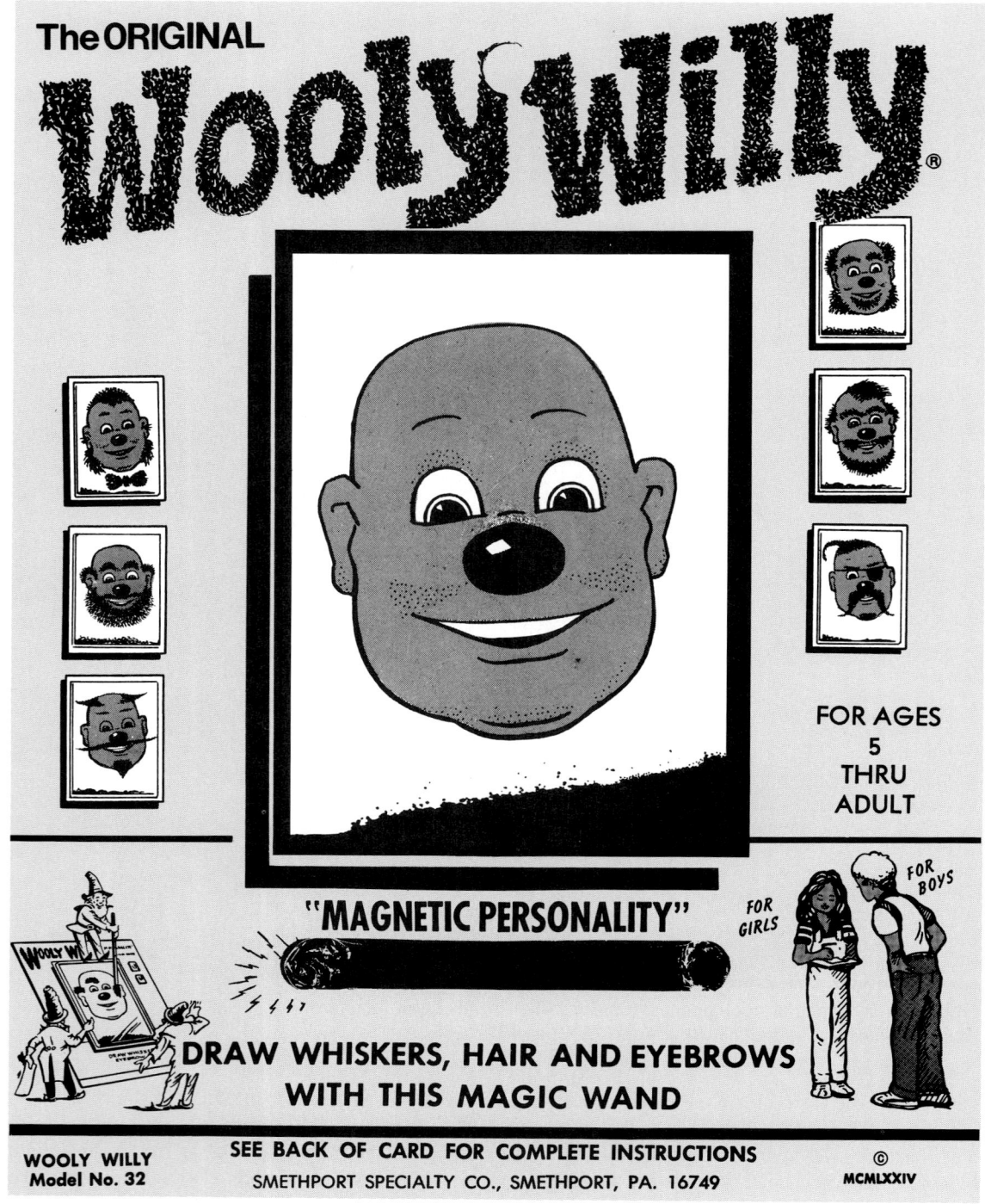

FIGURE 12-6 The magnetic recording principle. (Courtesy Smethport Specialty Co.)

also mastered the technique of audio tape recording!

A piece of audio tape is like the inside of the Wooly Willy package. The tape consists of hundreds of thousands of metal filings suspended inside a plastic covering. As the tape moves from supply reel to takeup reel, the metal filings pass by the Magic Wand, an electromagnet called the tape head. At the center of the head on the tape recorder is a hole, called the head gap. Here's how the process works. Imagine that the microphones described above have generated the electromagnetic energy corresponding to the original sound pressure waves (that is, they worked). It is a simple matter to send that electromagnetic charge through a wire to the tape head. The head is now emitting a signal that is a facsimile of the original sound, only now it is in the form of a magnetic field. A small signal is being emitted throughout the head; it even passes through the gap (just as the Magic Wand can pick up the filings through the plastic cover). As the tape passes the gap its microscopic metal filings are charged: they are shaped into an exact replica of the electrical patterns moving through the head. We have now created an analog signal, just like the grooves on a record or the oscillations in a microphone.

Most sophisticated tape recorders actually contain three different heads, arranged in the pattern seen in Figure 12-7. First, the tape passes the *erase head,* an electromagnet charged with a neutral signal. The erase head returns the metal filings to a noise-free pattern. Next, the tape passes the *recording head,* which stores the new signal in the manner described previously. Finally, the tape passes the *playback head,* which "hears" the recorded signal by reversing the recording process. The playback head sends a neutral signal through the gap. That neutral signal is changed (modulated) by the signal on the tape. In fact, the inside of the playback head is very similar to the inside of a microphone. The electromagnetic patterns on the tape create oscillations in the gap. These are "heard" like the vibrations in the mike coil, as electrical variations between the magnetic poles. Now in the form of electrical energy (waveforms) they can be sent through the wire to the amplification circuits.

Why is it useful to know how tape recording works? A basic understanding of audio technology adds much to our study of the radio industry. First, just as Wooly Willy is never really bald (there are lots of stray metal filings on his face and millions on the bottom of the plastic tray), an audiotape is never really erased or noise-free. The presence of the iron oxide particles on the tape at all times creates an audible hiss, even on the best tapes and best machines. Thus home and studio tape recorders frequently have noise-reduction circuits (such as Dolby) designed to suppress the sound of the hiss. Special tapes have been manufactured, including metal and chromium oxide (CO_2). In addition, since the tape heads are in constant contact with the tapes, there is plenty of opportunity for stray filings, dust, dirt, and so on to come in contact with the gap and cause poor recordings and playbacks. This is why it is critical to clean the heads frequently, both in the radio station and the home hi-fi. Isopropyl alcohol is usually used since, in addition to its cleaning properties, it is electrically neutral.

By varying the width of the audiotape and the location of the head gap, we can create a variety of tape styles and formats. Some of these configurations are seen in Figure 12-8.

Professional audio facilities use tape that is 2 inches wide, capable of recording eight, twelve, sixteen, even twenty-four or thirty-two separate sets of signals on one piece of tape. Consequently, such machines are known as *multitrack recorders.*

A common form of audio tape recorder in radio studios is the open-reel machine. The open-reel recorder employs two reels (the one full of tape is the supply reel; the empty one is the takeup reel). The tape is $\frac{1}{4}$ inch wide and usually can record two tracks in each direction, for a total of four tracks. Thus these are sometimes known as four-track stereo machines.

Radio stations also use audiotape cartridge players, or "carts," for their music, commercials, and station identifications. These machines use a special tape cartridge with only one reel. The tape winds past the heads and back onto itself. This allows for the musical selections to arrive quickly at their original starting point (in radio jargon, they are fast-cued). Cart machines record in one direction only, with a left-channel signal and a right-channel signal for stereo. Thus they are two-track machines.

Cassette tape recorders use a miniaturized reel-to-reel configuration enclosed in a small plastic housing. The tape is only $\frac{3}{8}$ inch wide, and, as

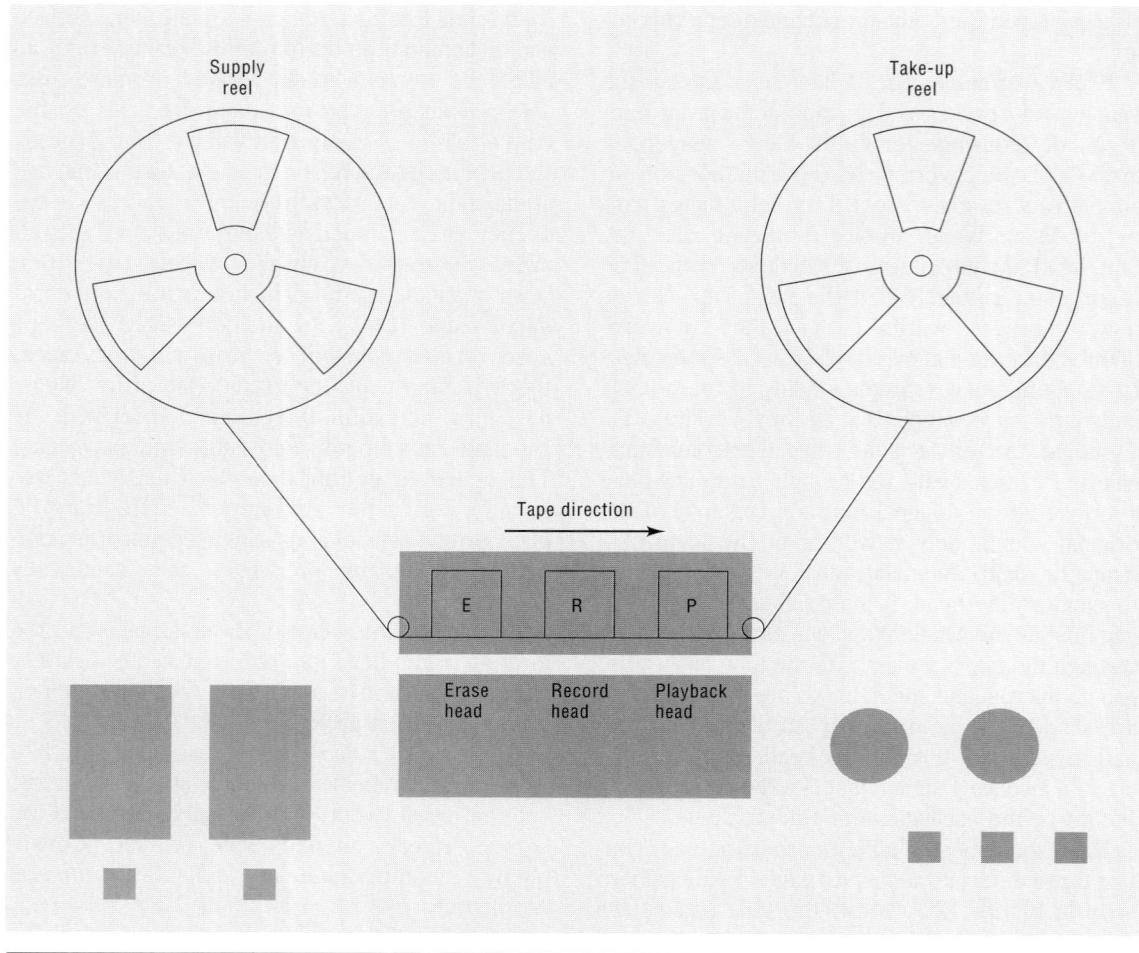

FIGURE 12-7 Tape recording head arrangement.

we all know from our home and car stereos, the tape can be recorded and played back on both sides. Thus there are four tracks of information crunched onto the small tape area. This is why cassette stereo units produce a lot of noise and hiss, and why the better ones allow for two or three types of tape (oxide, metal, standard) and noise reduction options (such as Dolby B and Dolby C).

CD audio As a trip to the nearest record store confirms, there has recently been a revolution in audio signal storage: the emergence of the compact disc (CD). In 1987, for the first time, the sale of CDs actually eclipsed the sale of records to American consumers. In terms used earlier in the chapter, the success of CDs derives from their digital transduction, full-frequency response and dynamic range, and an absence of signal loss or noise. In other words, they sound terrific!

Digital audio was made possible by the development of a different means of signal generation, known as pulse code modulation (PCM). This and other modulation methods are seen in Figure 12-9. At the top of Figure 12-9 is the original waveform: the shape of the original sound we seek to record and reproduce. Let's say it's the sound of a guitar solo. Below the wave form is its shape, transduced into an AM signal. Like a surfer on a wave, its new shape is a series of peaks and valleys, or changes in amplitude. Below that is the same waveform transduced into an FM signal. Now the message is in the form of a series of changes in the number of times it occurs in one second, that is, its frequency. At the bottom is the waveform translated into a digital sig-

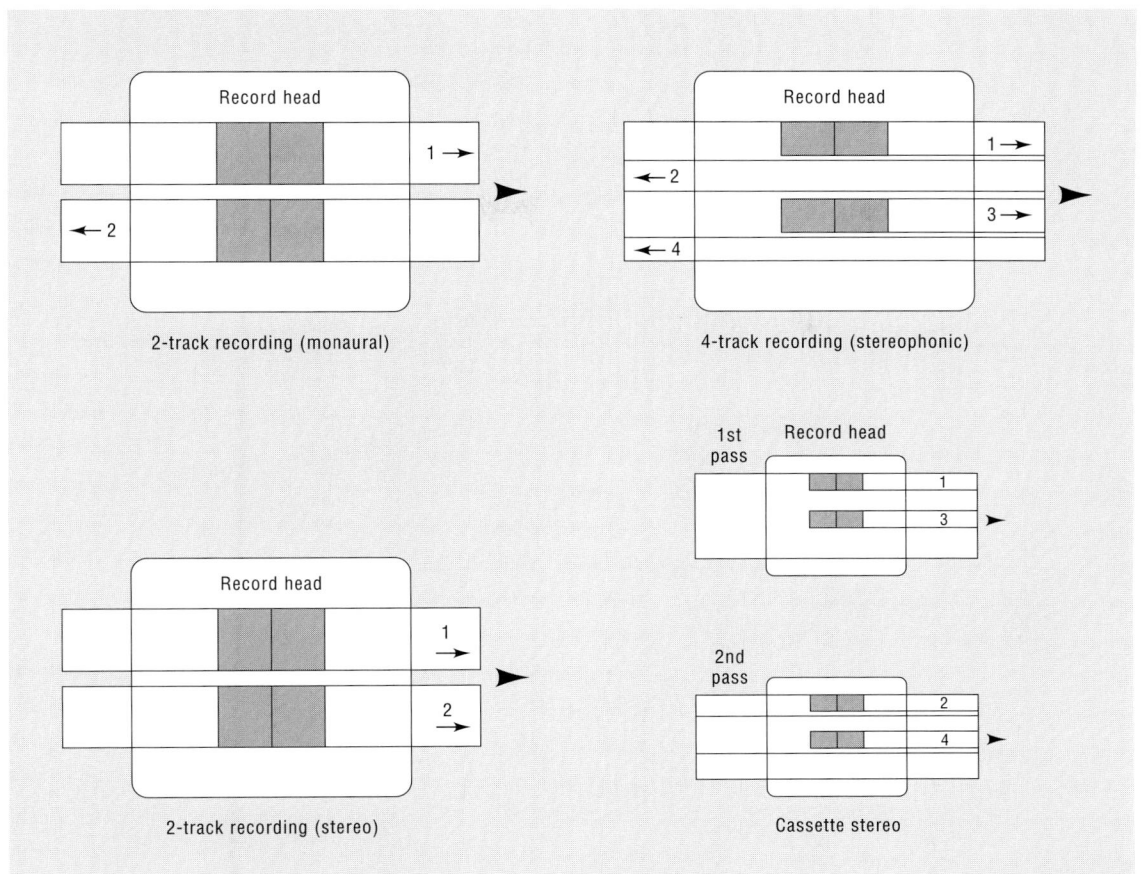

FIGURE 12-8 Head position and tape format configuration.

nal. By sampling the amplitude of the wave at varying intervals (turning a laser beam on and off and measuring the length of the light beam at each interval), a digital version of the wave has been produced. This process is called *pulse code modulation* since the oscillating beam "pulses" and turns the amplitude of the wave into a "code" that can be read by the laser tracking beam.

Foremost, unlike an analog signal, a digital wave is virtually constant—it is the identical shape on recording, on transmission, in the amplifier, and out of the speakers. Second, unlike tapes and standard phonograph records, CDs preserve the original sounds in a sterile, noise-free environment. The mechanics of the CD system are illustrated in Figure 12-10.

The information on a CD is carried beneath the protective acrylic coating in a polycarbonate base. In the base is a series of pits and flats. The pits vary in length in precise correspondence to the amplitude of the waveforms they represent. The flats represent no waveforms: utter silence. As the disc rotates, a laser beam is focused on the disc. Like a mirror, when the beam "sees" a pit, it reflects back a light wave that is a perfect replica of the original sound wave. Now it is a simple matter to transduce that wave into an electrical signal through amplifiers, to antennas, and into speakers. The result is a clean, nearly perfect sound. In technical terms, the frequency response ranges from 20 to 20,000 hertz (remember, the best LP record ranges from 40 to 12,000 hertz), with a signal-to-noise ratio of 90 decibels, the common measure of the intensity of sound, which is pretty close to acoustic perfection.

Unlike records and tapes, there is no friction or wear. Discs are comparatively immune from damage in routine handling and storage. With proper care, they should never warp or scratch,

PART 5:
HOW IT WORKS

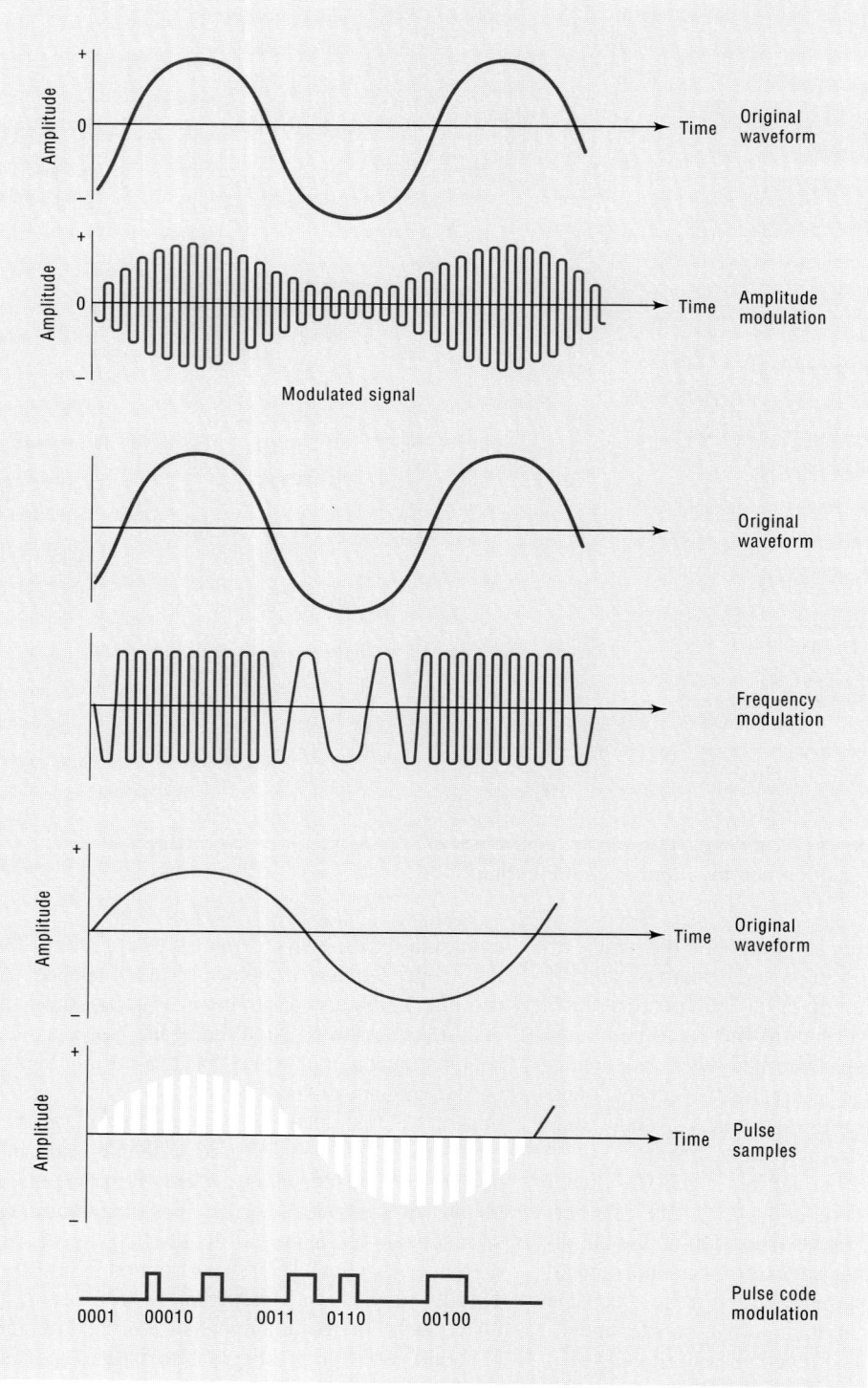

FIGURE 12-9 Modulation methods.

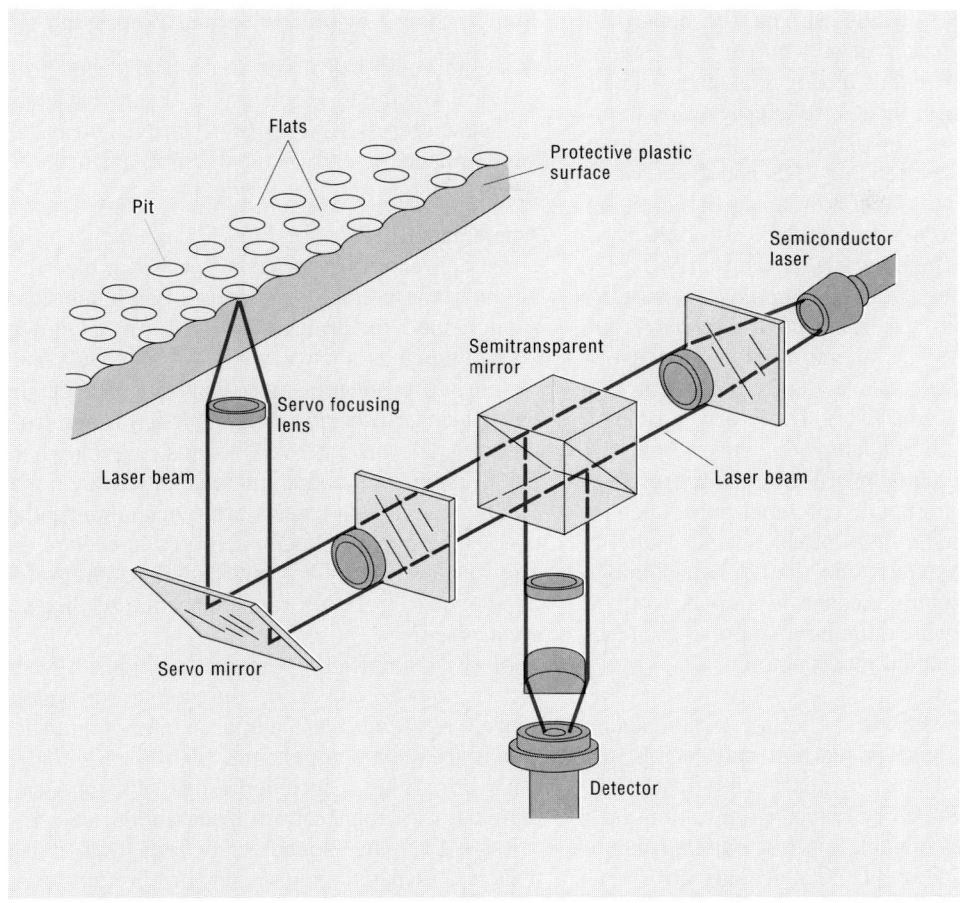

FIGURE 12-10 **Compact disc system.**

unwind, or jam in the player or get accidentally erased. But compared to audio tapes, they have one major drawback: you can't record on them. Enter digital audio tape (DAT) technology.

Digital audio tape recording With the rapid sales of CD players and CDs the logical next step in audio technology is to add digital recording capability to consumer audio components. Indeed, DAT recorders are already on sale in Japan. In 1989 a legal agreement with record manufacturers permitted the marketing of DAT equipment in the United States. As with other new technologies that faced legal roadblocks (cable TV and VCRs come to mind), it is likely that DAT will be a big seller in the 1990s.

For the same reason underlying the growth of CDs—their superior sound quality—digital recorders have been effectively utilized in professional recording situations. Interestingly, digital recording evolved to enhance the audio performance of video and film production, not to stand on its own as a means of audio technology. The first recorders to use PCM technology were developed by Sony in 1977. The early models were coupled with studio video recorders to "sweeten" the sound recordings made in TV and film production and, equally important, to allow for simultaneous electronic editing of sound and picture. By the mid-1980s stand-alone digital multichannel tape recorders were in place in a range of professional audio and video operations.

Like early videotape recorders described in the next chapter, studio digital audio machines are large and cumbersome and require considerable technical and artistic sophistication for effective operation. However, such machines are capable of recording up to thirty-two tracks of material and mixing those sounds without the noise and distortion that occur after only two or three transfers, or "dubs," using analog machines. Their miniaturized consumer equivalent promises to affect home audio in a similar way to the revolution in TV viewing created by VHS and Beta video recorders. In late 1985 formal agreements on technical standards were signed. By 1990 consumer DAT machines were on the market.

DAT machines utilize a compact cassette roughly one-half the size of a standard audiocassette. The digital information is densely packed on the tape, allowing up to three hours of recording time. Remarkably, DATs are capable of a full 48-kilohertz frequency response. And like CDs, the digital signal processing allows for features such as fast-speed searching, quick cueing, track programming, and other special effects. Some observers expect that the DAT machine—with excellent sound encased in a small, convenient package—will be the next component of the expanding home audio environment.

Digital compact cassette Another digital audio tape technology has arrived that may inhibit the growth of DAT. Digital compact cassette (DCC) recorders, introduced by Philips Corporation, use tape that is the identical size of existing audio cassettes. In addition, DCC machines are manufactured that can play both standard audiocassettes as well as the new DCC cassettes. Industry analysts expect this interchangeability to speed DCC over DAT.

SONY's minidisc (MD) As if to steal some of the thunder from a competitor's introduction of DCC, SONY has introduced a compact version of the CD. The minidisc is about 2½ inches in diameter, just about one-quarter the size of a standard CD. Designed mainly for walkabout stereos, the MD eliminates the problem many portable CD players have had: skipping as a result of laser mistracking. MDs can record up to seventy-four minutes of music. They can also read out text (liner notes, song titles, and so on).

As this book went to press, there was a war of sorts going on among proponents of DAT, DCC, and MD. It was unclear which standard of digital audio recording would win out. Regardless, consumer audiophiles can soon expect a new line of products on which they can record and play back their favorite music.

Step 2: Signal Amplification and Processing

Once an audio signal has been transduced from physical reality to electrical or digital facsimile, the next step is to intensify, clarify, and otherwise enhance it into its most usable form.

Amplification An *amplifier* is a device that boosts an electrical signal. Typically in electrical circuitry the voltage or current of an input signal is increased by drawing on an external power source (such as a battery or an AC or DC transformer) to produce a more powerful output signal. Devices to perform this function range from the original vacuum tubes (whose development made radio and sound movies practical in the 1920s), to modern transistors (which emerged in the 1950s to enable manufacturers to reduce the size of radios and TVs), and finally to integrated circuits in the 1970s (which permitted "microelectronics").

Modern amplifiers perform functions beyond increasing the power of the sound (or video) source. An *equalizer* is a special kind of amplifier that is frequency-dependent. This means that it can work within a specified range of frequencies to adjust the amplification. Equalization, referred to as "EQ" by sound engineers and music trendies, enables a sound signal to be fine-tuned for its best tonal quality. For example, equalizers can be used to make bass sections (60–250 hertz) sound more fat, thin, or "boomy." An equalizer can also be used to boost vocal sections out of the "soup" of an orchestrated passage, and even to isolate, diminish, or remove poor-sounding sections or mistakes in vocals or music. Once limited to expensive studios, EQ is now available in home and car stereo systems. In fact, a tiny Walkman may have five EQ faders on it (and cost less than $50, on sale of course).

Amplifiers and associated processing devices also allow sound engineers and home hi-fi enthusiasts to control the distortion or noise inherent in audio. VU meters indicate the changes in amplitude of the sound wave. The meter peaks, or pegs, at the point of highest amplitude. Compressors, limiters, and expanders process the signal to allow for the maximum amplitude (loudness) possible without introducing distortion. It's comparable to the level you can scream before you get hoarse. In practice, compressors are often used to decrease the sibilance (hissing sound) of vocal passages with many *s* and *c* sounds, and limiters are utilized to record or transmit sound with very high but momentary peak periods (like crashing cymbals). Expanders make loud signals softer and soft signals louder to allow for an acceptable mix (recordings of full orchestras and "conceptual" rock bands like Pink Floyd necessitate this technique).

In addition, amplification circuitry allows for electronic special effects to be added. These include reverberation, which is simply the echo effect. Special amplifiers can create all sorts of effects, from echoes to "sing-along" doubling or tripling. They can even create artificial choruses and deep echo chambers. Other devices are available to play tricks on audio signals. Phasers (not the kind used by Kirk and Spock) manipulate frequencies to create the illusion of stereo from mono signals, pitch changers can turn an out-of-tune musician into an accomplished soloist, and tape recorder motors can be manipulated to record sounds backward and to speed up or slow down recordings. In the 1950s and 1960s it was common practice for radio stations to use speeded-up recorders to rerecord popular songs so that they played quicker, allowing for more commercial and promotion time. And there is speculation that the practice continues today. Does it seem to you that top-40 music is faster on the radio than when you play the record on your own machine?

Mixing consoles and control boards The next link in the audio production chain is the audio console, which combines sound sources into a single signal. In radio and TV stations the console is generally referred to as an *audio board* or simply "the board." In recording studios and motion picture sound studios the board is commonly known as the *mixing console*. Regardless of its name, the console is the central nervous system of the audio facility. It is the place where the various sound signals are input, selected, controlled, mixed, combined, and routed for recording or broadcast transmission. Let's examine each of these phases individually.

The first function of the board is to input sound sources. A major-market radio station may have five or six tape recorders, three or four turntables, an equal number of CD players, and perhaps seven or eight microphones spread among several studios but capable of being interconnected.

The audio console at a large multitrack recording studio. The buttons at the top of the console (at left) control the signal processing, including equalization, reverberation, and other special effects. At the right is the patch panel where the many inputs are mixed together.

A recording studio is even more complex. In any event, the central location where each enters the chain of sound processing is the board.

The board usually consists of an even number of dials or sliding bars called *inputs*. Eight, ten, twelve, twenty-four, and thirty-two input boards are common. Some inputs correspond to one and only one sound device. Others use select switches and patch bays to allow for a single input to control as many as four or five different sound signals. In this way it is possible for a thirty-two–channel board to handle fifty or sixty separate sounds in one recording session.

Each input is controlled by a rotating dial, called a *pot* (short for potentiometer) or, more commonly today, a sliding bar called a *fader*. By rotating the pot or sliding the fader, the board operator can control the sound level of each studio input. More elaborate boards allow for equalization and special effects at this stage as well. Boards also allow for each source to be measured or metered and for the outputs of various signals to be amplified similarly.

So, sitting at one location, the audio person can combine and compose the overall sound of a performance, which is called the *mix*. The mix distills all the audio sources. This is sometimes a single (monaural) or, most commonly, a two-channel (stereo) signal. On occasion it's a four-channel (quad) version. The result is preserved on record, audiotape, videotape, or film, or it is broadcast live to viewers and listeners at home.

Having created, amplified, and mixed audio signals, let's examine how the matched and equalized mix is transported over the air in the form of a radio signal.

Step 3: Signal Transmission

The modern world of broadcasting was made possible by the nineteenth-century discovery of the phenomenon of radio waves. As Chapter 2 describes in detail, the radio pioneers found that electrical signals created by human beings could be transported across space so that buttons pushed here could cause a buzz over there. They soon replaced signals and buzzes with a voice at both ends.

The electromagnetic spectrum This magical process was made possible by the discovery and use of the *electromagnetic spectrum,* the electromagnetic radiation present throughout the universe. Figure 12-11 is a chart of the spectrum. A fundamental component of our physical environment, electromagnetic radiation is traveling around and through us at all times. We can see some of it (the narrow band of frequencies corresponding to visible light and color, or the "heat waves" that radiate off a parking lot in the summertime). But most of the spectrum is invisible to the naked eye and must be detected by human-made devices (like radio and TV tuners).

In the past century we have learned how to superimpose, or "piggyback," our own electronic signals on the natural waves in the environment, a process known as *modulation*. This is done by generating a signal that is a replica of the natural wave. This signal, produced by a radio station on its assigned frequency, is called a *carrier wave*. It is "heard" on our radios as the silence that comes just before the station signs on in the morning, or after the national anthem at sign-off. The radio signal is created by varying the carrier wave slightly, in correspondence with the frequencies of the signals we mean to transmit. Our tuner, tuned to the precise middle of the carrier, interprets these oscillations and reproduces them as sounds in the speaker system. While this process may seem hopelessly complex, consider this metaphor. Suppose there is a natural rock formation in the shape of a bridge. Adding a bed of concrete atop the formation, we have propagated a carrier wave. When we ride a car across the bridge, we have superimposed a signal, or modulated the carrier wave.

The radio spectrum Only a small part of the electromagnetic spectrum is utilized for broadcasting and related transmissions. This range spans from long waves of very low frequency to extremely short waves of relatively high frequency. In general, the higher one goes in the spectrum, the more power and sophisticated electronics are needed in the modulation process.

Each new development in electronic media has taken us higher in the spectrum. Radio broadcasting began toward the low end of the spectrum, in the area ranging from 0.3 to 3 megahertz (mega = million), a region known as the medium-waves. Included in this region is the range of 550–1600 kilohertz (kilo = thousand), which is the range of the AM radio dial. In fact, in many countries AM is still referred to as the medium-wave (MW) band.

The high frequencies (which, with satellite and infrared communications, actually aren't so high anymore) range from 3 to 30 megahertz. These waves are utilized for long-range military communications, CB, and ham radio. Since high-frequency waves can be used to transmit signals over greater distances than medium waves, this part of the spectrum has been used for over fifty years by international shortwave stations such as the BBC, Radio Moscow, and the Voice of America. The shortwave band on a radio is sometimes labeled HF, for high frequencies.

The next group of radio waves used for telecommunications applications are the very high frequencies, or the VHF band. VHF ranges from 30 to 300 megahertz. Television stations 2 to 13, FM radio stations, police radios, and airline navigation systems are located in this band.

Above the VHF channels are the ultra high frequencies, or the UHF band, spanning the region from 300 to 3000 megahertz. This part of the spectrum is used for TV stations 14 to 83, police and taxi mobile radios, radar, and weather satellites. In addition, it is UHF radiation that is modulated to cook our food in microwave ovens.

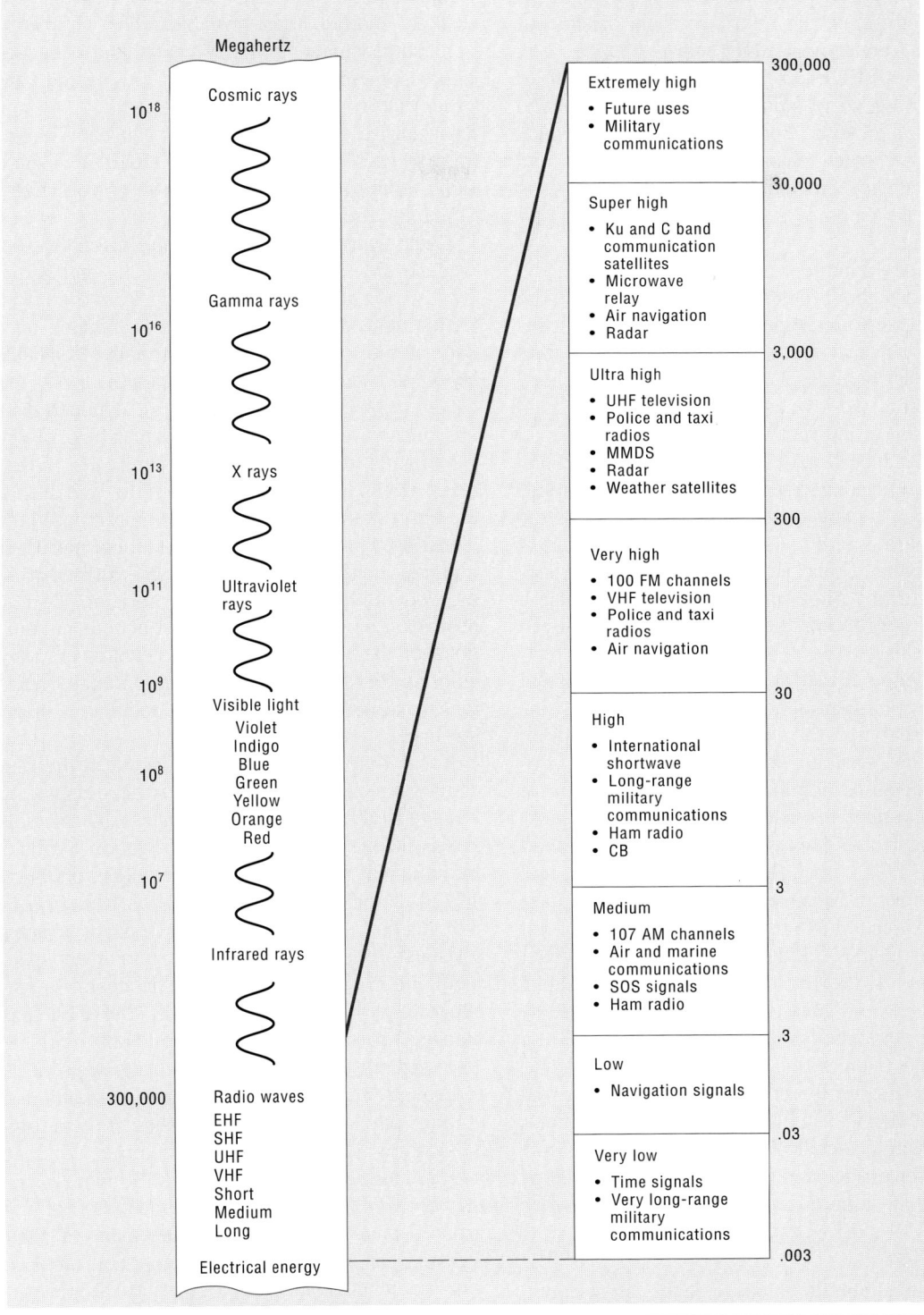

FIGURE 12-11 The electromagnetic spectrum.

Much of recent telecommunications development occurred in the next two radio bands: super high frequencies (SHF) and extremely high frequencies (EHF). SHF spans from 3000 to 30,000 megahertz and EHF from 30,000 to 300,000 megahertz. Commercial satellites, which deliver pay cable channels, superstations, and satellite news, emanate from these bands, as do new developments in military radar and air navigation.

There are a number of reasons that it is important to learn about the radio spectrum. First is the issue of spectrum management. Since the spectrum is a physical entity that crosses national (and interstellar) boundaries, its use must be policed for effective worldwide communication. Nations must meet in international forums to decide on the proper allocation of spectrum space to various services, including radio and TV broadcasting. Sometimes ideological differences lead to deliberate tampering with uses of the spectrum. This is the case when Cuba "jams" our shortwave broadcasts on Radio Martí or when our spy satellites attempt to intercept the transmissions of Libyan fighter pilots.

Knowledge of the radio spectrum is also useful in the development of new technology. Most new devices that use the spectrum require a lot of space. One example is high-definition TV, described in more detail in Chapter 13. To make room for a new service sometimes another service is displaced. This is precisely what happened when the FCC's push for TV's development, and pressure from AM broadcasters, led the commission to "banish" FM radio to the VHF part of the spectrum instead of the medium waves, which would have made the transition to FM cheaper and faster.

At the domestic level the FCC polices the use of the spectrum within our boundaries to make sure that the entire country is within reach of broadcast services and that radio and TV stations operate within technical specifications to prevent undue interference and noise. To do this the commission has adopted certain classifications that govern AM and FM radio.

Radio channel classifications At the latest count more than 12,000 radio stations were on the air in the United States. Yet there are only 107 AM channels and 100 FM channels. How is this unlikely mathematics possible? The answer is spectrum management. By controlling operating hours, power, antenna height and design, and other factors, the FCC squeezes the 12,000 stations into the 207 channels. If you can imagine 12,000 cars competing for 200 parking spaces (kind of like the first day of classes), you have a sense for the task at hand.

Radio transmitters can generate three types of waves: sky waves, ground waves, and direct waves. *Sky waves* radiate upward from the transmitter and either go into space or bounce off a part of the ionosphere (the Kennelly–Heaviside layer) to a distant spot on the Earth, a process called *skipping*. *Ground waves* are conducted by soil and water and follow the curvature of the Earth until they dissipate, or *attenuate*. *Direct waves* travel in a line of sight from the transmitter to the receiver. They can be interrupted or blocked by tall buildings or mountains and their range is limited by the straight line formed from the top of the antenna to the horizon. Certain propagation methods work better in different portions of the electromagnetic spectrum, enabling stations to vary their power and antenna arrays for maximum coverage with minimum interference.

Recall that the AM band is located in the medium-wave portion of the spectrum, the region ranging from 535 to 1605 kilohertz. AM channels are 10 kilohertz wide, with the station's carrier frequency at its midpoint. For example, a station at 770 at the dial actually oscillates from 765 to 775 kilohertz. By studying the unique characteristics of the medium-wave portion of the spectrum and dividing the available AM channels into separate types or classifications, the FCC has managed to make room for over 5000 stations. The medium-wave band is particularly well suited to ground- and sky-wave propagation. Thus AM stations have generally located their transmitters in swampy lowland areas (remember Wolfman Jack in an isolated radio shack in the film *American Grafitti?*). To use the conductivity of the ground wave, AM stations bury part of their transmitters in the ground and may use three or four antennas arranged in a geometrical grid pattern to make sure the signal radiates throughout their coverage area. AM stations also beam a signal upward to make use of the sky wave. This is why some AM stations can be heard over great distances at night, when the ionosphere is more reflective. Figure 12-12 depicts a typical AM station antenna array.

In general, the range of an AM station's ground wave is known as its *primary coverage area;* the limits of an acceptable sky-wave signal is its *secondary coverage area.* For AM stations wet soil, more power, and a lower channel number make for greater coverage. The FCC uses these physical characteristics to manage the AM band.

AM channels and classifications The 107 AM channels are divided into three main types: 60 are clear channels, 41 are regional channels, and the remaining 6 are local channels. Stations are divided into four classifications: class I, class II, class III, and class IV. Here's how the system works.

The *clear channels* are frequencies that have been designated by international agreements for primary use by high-powered stations. Such stations use the ground wave in daytime and the sky wave at night to reach a wide geographic area, often thousands of miles. The United States has priority on forty-five clear channels. Stations on these frequencies include some of our oldest and strongest. They are the ones we tune in late at night for talk shows and out-of-town baseball games. Class I stations, like WABC in New York at 770 kilohertz, WJR in Detroit at 610 kilohertz, and KMOX (1120 kilohertz) in St. Louis, have the exclusive right to the clear channel after sunset. They operate at high power, from 10,000 to 50,000 watts.

Class II AM stations use the clear channels but must operate at reduced power at night to avoid interfering with the class I's. Their wattages range from 50 to 250 kilowatts. Class III stations

FIGURE 12-12 **AM radio transmission.**

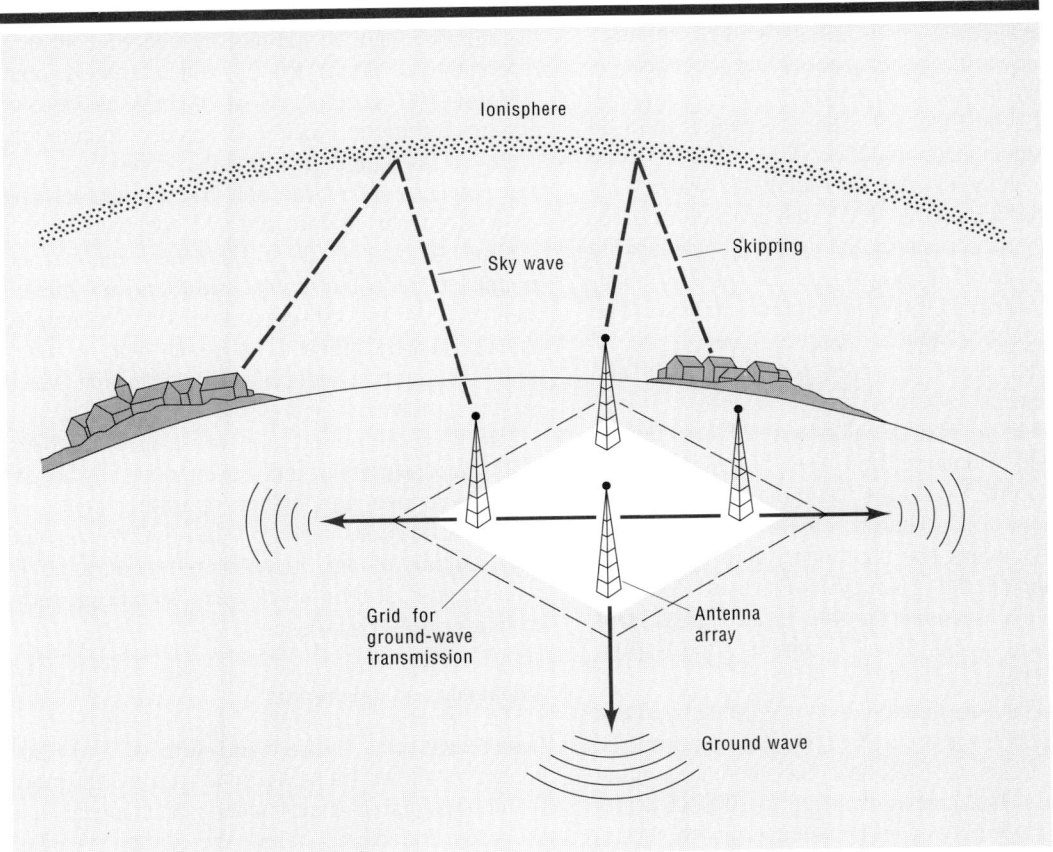

are designated as *regional stations*. Ranging in power from 500 to 5000 watts, class III's must share channels with numerous other stations. To do this they generally use directional antennas and strategic placement to blanket their main population areas. Over 2000 class III stations share the forty-one regional channels.

Class IV stations operate on *local channels*. These stations are mostly at the right or top end of the dial (above 1230 kilohertz). They are limited in power to 1000 watts day and 250 at night.

Together, classes III and IV represent the majority of AM stations in the United States and are the backbone of radio's local service. Because of the sky-wave phenomenon, their location in major cities, their ownership by large corporations, and other factors, the clears are radio's "superstations," providing signals across state boundaries.

FM channels and classifications As we have seen, the advantages of frequency modulation are its high-frequency response and signal-to-noise ratios. But there are disadvantages from a spectrum management point of view: FM requires more bandwidth, higher power, and taller towers to perform its noise-free magic.

In 1945 the FCC set aside the region from 88 to 108 megahertz for FM. This allowed room for 100 channels, each 200 kilohertz wide. This means that when your FM radio is tuned to 97.3, the station is actually radiating a signal oscillating from 97.2 to 97.4 megahertz. Of the 100 channels, 80 were set aside for commercial use. The remaining 20 channels, located from 88 to 92 megahertz, were reserved for educational and noncommercial stations.

To facilitate the development of commercial FM, the FCC divided the United States into three regions. Zone I includes the most densely populated areas of the United States: the Northeast. Zone I-A covers the southern California region. The rest of the country comprises Zone II. Class A FM stations operate in all zones, class B's in Zone I, and class C's only in Zone II (Figure 12-13).

Each class is defined by its *effective radiated power (ERP)*, the amount of power it is permitted to use. Class C FMs are the medium's most powerful. They can transmit up to 100,000 watts (ERP). They may erect transmitters with a maximum *height above average terrain (HAAT)* of 2000 feet. Class B's can generate up to 50,000

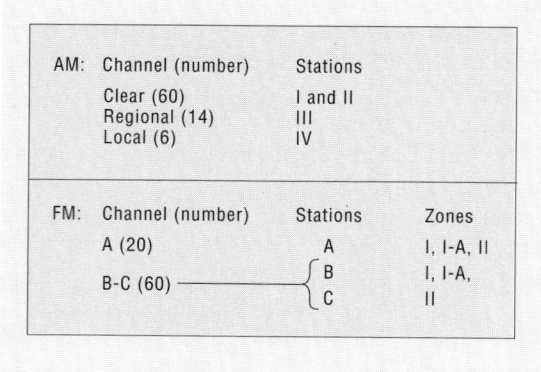

FIGURE 12-13 AM and FM channel and station classification.

watts ERP at 500 feet HAAT, and class A's are authorized to a maximum of 3000 watts ERP at 300 feet HAAT. At maximum power and antenna height, Class C's can cover about 60 miles, Class B's about 30, and class A's up to 15.

In the mid-1980s the FCC greatly increased the number of FM stations by enacting what it referred to as Docket 80–90, which created about 700 new FM stations. Most of these were low-powered stations located in small communities. Many of the applicants for these new licenses were owners of daytime-only AM stations looking for a way to break into the FM spectrum.

Other Docket 80–90 stations were established in smaller cities and towns, and then they moved their towers and directionalized their signals to blanket nearby cities. The result was a much more competitive radio business than that which existed in the past.

The FCC's intention to increase the availability of FM apparently worked. Facing declining advertising revenues and higher operating costs, in 1992 broadcasters called on the FCC to suspend the granting of any more commercial radio licenses, at least until there was a significant turnaround in the economy.

Sidebands and subcarriers

Use of sidebands The bandwidth of FM (200 kilohertz) allows these stations to transmit more than one signal on their channel. Such signals use the area above and below the station's carrier frequency, known as the *sideband*. The most common use of the sideband is to disseminate separate signals for the left and right channel to broadcast in stereo. This is called *multiplexing*. If you have an old FM stereo receiver, it may be labeled an FM mutiplexer.

FM station operators may use additional spectrum space to send multiplex signals, which can be tuned only by specially designed receivers. To do this, stations apply to the FCC for a *subsidiary communications authorization (SCA)*. Such services include the "background" music one hears in malls and elevators, a Talking Book service for the blind, telephone paging, data transmission, and special radio services for doctors, lawyers, and some others. By using subcarriers in this way FM stations have developed an additional revenue source.

Although technically the process is not the same, AM stations can use their carrier waves for other purposes as well. Many AM stations transmit a subaudible tone used in utility load management. A subscriber, such as a business, rents a special receiver. During peak electric use hours the station transmits a tone that turns off the subscriber's appliances. When the peak load period is over, another subaudible tone turns them back on.

Digital Audio Broadcasting

A new means of audio signal transmission threatens to render conventional AM and FM radio stations obsolete. Digital audio broadcasting (DAB) combines the technique of pulse code modulation (PCM), which made CD audio practical and popular, with extremely high-frequency transmission (such as satellites and microwaves). The result is CD quality sound from a home or car radio without the need for records, tapes, or discs. By the early 1990s DAB had been successfully demonstrated in the United States, Canada, and Europe.

Rather than face the possibility of oblivion, AM and FM stations formed a "DAB Task Force," which recommended that the FCC pursue a DAB standard that relied on terrestrial (ground-level) transmissions (from existing radio stations, of course). Other DAB proponents (including long-time broadcast rivals in the cable TV field) would rather see DAB emanate from orbiting communications satellites. In this way a few stations (perhaps one or two in each major format, like rock, country, and so on) would offer digital radio broadcasting on a national basis. From over 12,000 stations American radio might be reduced to as few as two or three dozen.

Step 4: Radio Reception

We have now successfully transduced, modulated, and transmitted a radio signal. The next step in audio signal processing is reception: the process whereby the radio waves are picked up by the radio set and transduced by the speaker into sound waves.

TECHNOLOGY

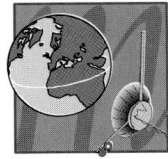

DISC JOCKEYS OF THE FUTURE

Computers and digital audio technology have already profoundly influenced music recording technology. By using a device called a sampler it is now possible to recreate digitally notes played by a famous musician and remix them into a current recording. For example, a note on the trumpet played by Louis Armstrong could be sampled and translated into digital signals and mixed with computerized violin notes as originally played by Itzhak Perlman to create a new composition.

Recently sampling and other digital audio techniques were at the heart of the phenomenally successful "Unforgettable" album by singer Natalie Cole. On the record Ms. Cole was able to perform a seamless duet with her famous father, Nat "King" Cole, who had been dead for nearly thirty years.

The National Association of Broadcasters has demonstrated another use of this technique. The voices of DJs and news reporters were sampled and recreated digitally using a computer program that attempted to capture the unique tonal qualities, pronunciation, and regional dialect of the announcers. The system is yet to be perfected, but the day may arrive when, if your DJ gets sick, he or she will be replaced by a computer that has been trained to mimic his or her voice. . . . Is it real or is it Memorex?

As might be expected, the characteristics of the electromagnetic spectrum and the different modulation methods have led to the development of different types and styles of radio receivers.

AM (MW) band receivers The location of AM in the medium-wave part of the spectrum has several advantages. The effectiveness of the ground wave in AM means that long telescopic antennas are normally not needed and a good signal may be received even when the radio is in motion. This makes AM ideal for car radios. At night the phenomenon of the sky wave allows for listening over vast distances (without the elaborate antennas used by the military, in the high bands, for example, to achieve the same result). AM radios can take almost any form, from microscopic transistor versions to large table-top models. Normally, good reception is only a matter of moving the receiver slightly for best tuning.

However, consider the disadvantage: a check of the spectrum chart shows that the MW band is precariously close to the bandwidth of electrical energy. Thus AM radios are prone to interference and noise, heard mostly as static. In cars, energy produced by engines and electrical systems can often be heard on AM. At home, one can "hear" vacuum cleaners, lights turned on and off, and so forth.

Another limitation of AM reception has to do with its limited frequency response. As you know, AM stations generate signals in a bandwidth only 10,000 hertz wide. Recall that the human ear can hear a bandwidth about twice that. As a result, you may have noticed that AM sounds somewhat thinner or tinnier than FM (with full 20,000-hertz response). AM is best with speech and music of limited orchestration. Hence news/talk and top 40 tend to be popular formats.

AM stereo In the 1980s new developments in transmission and reception devices held promise in improving the AM signal. However, a debate over technical standards and the atmosphere of deregulation at the FCC (see Chapter 17) left first three, then two noncompatible competing systems. This means that some AM stereo receivers needed a switch to allow listeners to tune in properly. Other consumers had to choose one or the other and hope the receiver they chose would

A multiband radio receiver capable of picking up AM, FM, and 13 short wave bands. The radio can also tell you what time it is anywhere in the world.

be the one that the industry adopted. After nearly a decade Motorola's C-Quam system came close to establishing itself as the industry standard. As a result of this kind of confusion, only about one in ten AM stations has converted to stereo broadcasting, despite its somewhat richer sound.

FM receivers The evolution of the FM receiver has followed an interesting path. From the beginning the noise-free dynamic range of FM made it a natural element for the hi-fi enthusiast's home audio system. Thus many FM receivers did not have amplifiers or speakers attached to them; they were separate tuners, which had to be plugged into the hi-fi system. However, when FM boomed in the late 1960s, consumers demanded FM in forms they had become familiar with in AM: in cars, transistor radios, table models, and so on. Thus most radios manufactured after 1970 were capable of both AM and FM reception. As a result, FM moved from an "add-on accessory" to the most important part of the radio receiver.

Since the FM signal requires a line of sight from transmitter to receiver, there are some potential pitfalls in signal reception. First, FM normally requires a long antenna, in the shape of a telescoping rod or a wire. Reception can also be improved by attaching the radio antenna to a TV antenna and in some areas by hooking the receiver to the cable TV system.

Moreover, FM signals tend to be blocked by buildings or moving objects. This situation is commonly experienced in cars. Unlike the AM signal, this type of signal loss is seldom heard as static. Instead, the signal simply disappears and returns, as if it were being turned on and off repeatedly. To solve this problem, many FM receivers have features that lock in the carrier frequency of the transmitting station. Such techniques may be seen on your radio sets as automatic frequency control (AFC) or quartz lock. A car radio typically has a separate switch for local and distance (DX) reception on FM.

Multiband receivers Today virtually all radios offer both AM and FM bands. New radios recently previewed by the National Association of Broadcasters have a continuous-tuning feature that allows a listener to go from an FM station to an AM station without pressing a switch to change bands. In addition, many receivers offer access to a range of other bandwidths that provide various radio services. A radio with the HF or short-wave frequencies provides access to the services of international broadcasters, from Radio Moscow to the British Broadcasting Corporation. Some receivers can monitor channels used by police and fire services. Other radios feature a weather band, capable of tuning the nearest government weather station for use in general aviation. Becoming more popular are radios with "TV sound": these allow listeners to keep up with "soaps" and sports while at the beach or at work.

Digital tuners An exciting and useful feature of many new radios is the digital tuner. Using the same techniques as a digital watch or clock, digital tuners display a station's frequency in real numbers, instead of a line on a dial. The numbers may be presented on a liquid crystal display (LCD) or on a light-emitting diode (LED).

Digital tuners perform some impressive functions. When equipped with a numeric keypad, they enable the listener to program specific frequencies. This can be critical for international shortwave listening, where different stations often occupy the same channels at different times. They enable clock radios and radio–tape recorder combinations to operate with up-to-the-minute accuracy. They can be programmed to tune themselves, either to sample the bandwidth for strong signals (scan) or to find and stop at the nearest strong station (seek).

"Smart Radios"—Radio Broadcast Data System

Digital tuners make possible a new generation of "smart" radio receivers. Using their subcarrier frequencies, in addition to voices and music, radio stations can send other signals to their receivers. On such radio sets, a small display terminal can provide a readout of the station's dial location, call letters, format, even the title and artist of the song being played. Weather forecasts could display a map; news and sports reports could show stock tickers, scoreboards, and so on.

Recently, representatives of the radio and electronics manufacturing industries met to set standards for this new generation of radios. Smart radios will be called radio broadcast data system (RBDS) receivers, and should be on the market by the mid-1990s.

Step 5: Audio Signal Storage Media

We have already discussed some of the most common audio storage devices in the context of signal transduction. We close this chapter with a review of the ways audio signals have been stored for playback or rebroadcast by sound studios, radio stations, and the public.

The golden age of ETs and wire recordings Long before ET entered the language as a lovable fellow from outer space, he was already well known to radio people. In the 1930s and 1940s there was only one means of recording a radio broadcast. That was to make a sound recording of the show. Such recordings were known as electrical transcriptions, or ETs.

An ET was similar to a phonograph record: it was scratchy and unfit for broadcast uses. In addition, ETs were electrically produced. Remember how close AM is to electricity in the band? Thus all the interference and noise were stored forever on the ET. And ETs were often made of glass, not vinyl. This made them extremely fragile, especially in the hands of nervous sound engineers working during a live broadcast. For these reasons, ETs were an unsatisfactory recording mechanism for radio. But for many years they were all radio had.

Wire recordings were another early means of radio recording. The wire recorder was similar in design and look to a tape recorder. Instead of tape, the signal was stored on a length of spun wire. By the 1940s wire recorders were used by some radio news reporters for on-the-scene reports and interviews. Wire recorders also showed promise in the home market. In the 1950s they could be purchased from mail-order catalogues. Although they were more durable than ETs and certainly more portable, wire recordings were also less than satisfactory for professional use. Wire recordings were difficult to

A DAT player. Note the small size of the DAT cassette.

edit and repair. Sound quality was adequate at best. The invention of the tape recorder after World War II solved these problems and relegated the wire recorder to museums.

Magnetic tape recording The development of magnetic tape recording was revolutionary for both radio and TV. Compared to ETs and wire recordings (and in TV, kinescope recordings), tape recordings were suitable in quality for broadcast, were easy to edit, and were small enough to use in the field. Even the earliest tape recorders were small, portable, and of high fidelity. For these reasons the tape recorder became a permanent fixture in sound studios and radio stations. The three most common forms of tape in use today are open reel (reel-to-reel), cassette, and cartridge. Radio stations tend to rely on cartridge and open reel; the cassette recorder is the most common home recording device. High-quality cassette recorders are often used by radio reporters to record meetings and to do interviews in the field.

Phonograph recording The old standby, the phonograph record, has been around since the turn of the century. From that time, the sizes and speeds have been changed, but the basic design has remained consistent. Today, the most common record format is the 33⅓ revolution per minute (rpm), 12-inch high-fidelity recordings. Teens and young adults still buy some 45-rpm 7-inch "donuts," but these are rarely if ever seen in a radio station or sound studio. Of course, with the advent of CDs and DAT, the future of the traditional vinyl disc may be limited.

Compact disc recording As we have seen, the CD has created a minor revolution in sound recording. Today the CD player has become a fixture in a majority of radio stations. In addition, nearly a quarter of the listening public has CD players in their homes, cars, or both.

Digital audio tape Digital tape recorders have long been in professional audio facilities. Now that the legal problems are solved, they will play an increasing role in radio and home audio. By 1995 many radio stations, homes, and cars will be equipped with DAT machines, most likely in the form of a digital compact cassette (DCC).

A FINAL WORD

We have followed the path of the audio signal from original sound wave to storage on record, tape, or disc. The amazing thing is that most of this is taken for granted, and in the case of broadcasting, much is simultaneous. At the same time, someone speaks "out there," we hear it "over here," whether "there" is next door, downtown, or on the other side of the world. "Here" can be at home, from the kitchen to the shower, in the car, on a boat, or at 35,000 feet while on a transcontinental airplane. The magic of radio is that we don't pay much attention to it. But now we'll appreciate the technical wizardry the next time we hook up the Walkman or rev up the boom box.

ECONOMICS

AUDIO PIRACY

To their usual wares of hotdogs, soft-drinks, T-shirts, and leather goods, some street-corner vendors have added prerecorded cassette tapes. Most of the time these tapes look brand new: they're even shrink-wrapped, just like they are at the record store. But beware! Advances in consumer audio have created a new growth industry: the piracy of prerecorded music.

The record industry estimates that between $400 million and $1 billion each year in revenue is lost to illegal tapers. One recent bust of a counterfeit tape ring occurred in a private home in suburban New York. There authorities found home audio equipment alledgedly used to copy over 25,000 bogus tapes, featuring the music of Hammer, Julio Iglesias, Vicki Carr, Ruben Blades, and others. A few days later the police uncovered an additional 32,000 tapes at another home, believed to be the pirates' "distribution center."

SUMMARY

Broadcasting makes use of facsimile technology, reproducing sound and sight in other forms. The better the correspondence between the facsimile and the original, the higher the fidelity.

Transduction involves changing energy from one form to another; it is at the heart of audio and video technology. Transduction can be analog—the transformed energy resembles the original—or digital—the original is transformed into a series of numbers.

Radio waves possess frequency and amplitude. Radio broadcasting makes use of both amplitude and frequency modulation to carry information.

Radio signal processing follows five main steps. First, the signal is generated. Microphones are used to transduce the physical energy of voice and music to electrical energy. The major kinds of microphones are dynamic, velocity, and condenser.

The second step in radio signal processing is amplification. This is accomplished by electrical devices that utilize transistors or integrated circuits. Equalizers also help fine-tune the amplified sound. Sound sources are combined at mixing consoles.

The third step is signal transmission. Radio waves occupy a portion of the electromagnetic spectrum. Radio signals travel by ground, sky, and direct waves. AM radio channels are classified into clear, regional, and local channels. FM stations are classified according to power and antenna height. The wide bandwidth of an FM channel allows for stereo broadcasting and other nonbroadcast services. On the horizon is Digital Audio Broadcasting, or DAB.

The fourth step is reception. There are several types of receivers: AM, AM stereo, FM, short-wave, and multiband. Some are equipped with analog tuners; others use a digital system.

The last step is signal storage. Audio has been recorded by electrical transcription on glass discs, on wire, in vinyl, and on magnetic tape. Compact discs and other digital media appear to be the storage formats of the future.

SUGGESTIONS FOR FURTHER READING

Alkin, G. (1975). *TV sound operations.* New York: Hastings House.

Dordick, H. S. (1986). *Understanding modern telecommunications.* New York: McGraw-Hill.

Evans, C. H. (1979). *Electronic amplifiers.* Albany, N.Y.: Delmar.

Hasling, J. (1980). *Fundamentals of radio broadcasting.* New York: McGraw-Hill.

Mott, R. L. (1990). *Sound effects: Radio, TV and Film.* Boston, Mass.: Focal Press.

Pohlmann, K. C. (1989). *The compact disc: Handbook of theory and use.* Madison, Wisc.: A-R Editions.

Runstein, R. E., & Huber, D. M. (1986). *Modern recording techniques.* Indianapolis, Ind.: Howard W. Sams.

Smithsonian Institution. (1975). *The history of music machines.* New York: Drake.

Sweeney, D. (1986). *Demystifying compact discs: A guide to digital audio.* Blue Ridge Summit, Pa.: Tab Books.

Whitehouse, G. E. (1986). *Understanding the new technologies of the mass media.* Englewood Cliffs, N.J.: Prentice-Hall.

Woram, J. M. (1976). *The recording studio handbook.* Plainview, N.Y.: Sagamore.

CHAPTER 13: VIDEO TECHNOLOGY

We have seen that radio broadcasting emerged in the early twentieth century when inventors, engineers, and hobbyists perfected the use of radio waves to send voice and music over the air. A similar electronic revolution occurred in the middle of the century. Technological marvels in the generation, distribution, storage, and retrieval of video images enabled TV to move rapidly from fantasy to reality. This chapter reviews that technology. As with the previous chapter on audio signal processing, our goal is not to overwhelm you with physics and electronics. Rather, we hope to show that the emergence of TV as a dominant medium of mass communication is directly linked to the advantages and limitations of its technology. In addition, this chapter follows the same organizational structure as Chapter 12, examining video signal generation, amplification and processing, transmission, reception, and storage. But first let's look at the basics.

BASIC PRINCIPLES IN VIDEO PROCESSING

To grasp how TV works, we need first to understand electronic and physical principles that are at the core of film and TV technology.

The Nature of Vision: How the Human Visual System Works

In order to see, our eyes and brains perform some of the same processes of information storage, transduction, and distribution that occur in a TV receiver. The process is pictured in Figure 13-1. What we see as light and color is simply the visible portion of the electromagnetic spectrum, the region at a frequency of 1 quadrillion (10^{15}) hertz. The oscillations of light waves corresponding to each color in the spectrum enter the eye as *photons:* discrete units of light energy, almost like little packets of light. Let's assume we are looking at a tree. The unique pattern of photons that define "green, leafy tree" is focused and directed by the lens onto the retina, where it can be momentarily held for further processing. The millions of receptors in the retina of the eye scan the picture and change it into electrical nerve impulses. These are transmitted by the optic nerve to the brain, where the information is merged with memory and experience; that is, it is transduced into a meaningful visual image. "Aha!" you say, "that green, leafy thing is a 'tree.'" Film and TV are made possible by humankind's ability to understand and utilize this process so that our brains interpret patterns of flickering lights and shadows as symbolic of a tree.

Lenses

The first breakthrough in the development of film and video took place in the Renaissance. Leeuwenhoek's microscope and Galileo's telescope could be used not only to see objects great and small close up, but also to improve visual clarity and to focus, change, distort, or redirect images. No longer was the limit of human experience what we could see with our own eyes. It was expanded to include what we couldn't see, or to see what we could see *differently*. This is the role of the lens on a camera—whether it's on a box camera, a film camera, or a TV news camera. In short, like its counterpart in the human eye, the camera lens collects, clarifies, focuses, and redirects images.

Persistence of Vision and the Phi Phenomenon

Humans are able to perceive TV and motion pictures as moving images because of two characteristics of our perceptual system. The first is somewhat difficult to explain. To make sense of the images we see, the eye must momentarily capture and hold an image on the retina. This gives the sensory neurons in the retina time to process the image and translate it into electrochemical energy. This illusion of continuing to see an object even when it isn't there is known as *persistence of vision,* or visual lag. Film and TV are made possible by our ability to utilize this biological process to create the illusion of continuous motion from still-frame images. As a kid, did you ever play with a riffle book or flip book? To jar your memory, examine the riffle book illustrated in Figure 13-2. When the pages of the book are flipped, the previous picture is still captured by the retina. The mind's eye combines the new picture with the old to create the image of smooth, uninterrupted motion. Keep this process in mind since the principle of persistence of vision

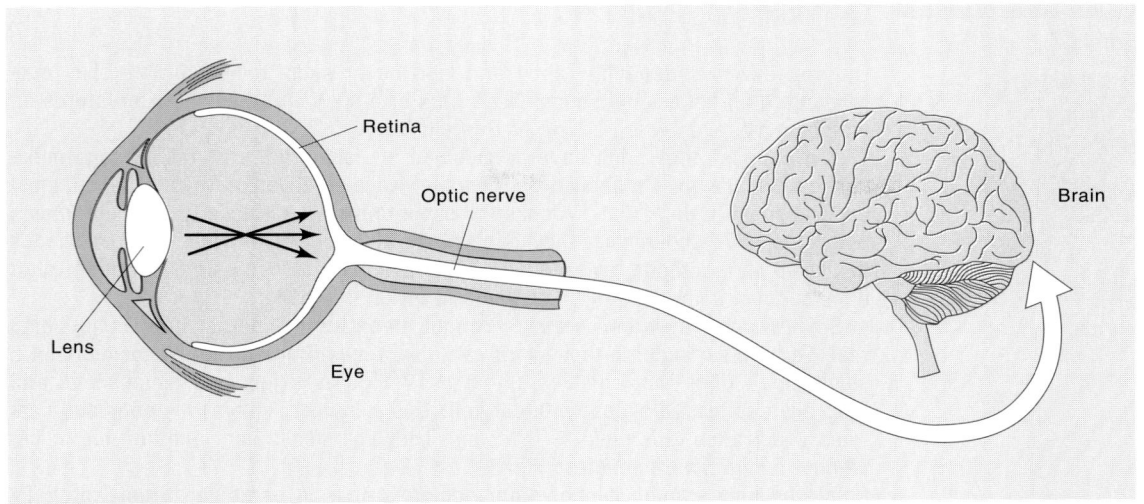

FIGURE 13-1 How the eye works.

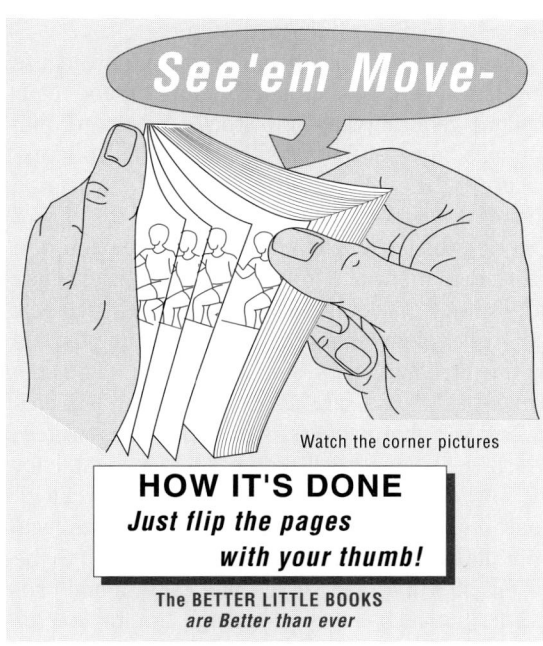

FIGURE 13-2 An ad for a 1940s flip-book, demonstrating how to riffle the corners of the pages with the thumb to create the impression of a figure in motion. Flip-books reappeared in the early 1990s. One popular seller recreated baseball pitcher Nolan Ryan's 5000th strikeout.

illustrated by the riffle book is pivotal to the development of both film and TV technologies.

The second perceptual process is known as the *phi phenomenon*. Suppose a person is sitting in a dark room looking at a small light source. Now suppose the original light source is turned off and another identical light source near the original is turned on at the same time. The brain will perceive the light as moving from the first to the second light source. This tendency to fill in the blanks, as it were, is the phi phenomenon, and it helps explain the perceptual basis behind TV and films: humans also fill in the gaps between rapidly flashing discrete pictures.

All in all, TV and movies are made possible because humans have slow temporal resolution and creative brains. Creatures with better temporal resolution and less creative brains might see TV and film for what they really are: a series of rapidly moving still images.

Facsimile Technology and Transduction

Like radio, TV is made possible through the use of facsimile technology. In radio this involves making a facsimile of a sound: changing sound waves into other forms of energy, like blips on audiotape or grooves on a record. In video we must perform the same trick. Images seen by our eye and interpreted by our brain must be changed into another form, stored on film or tape, sent over the airwaves or through cables, received by

IMPACT

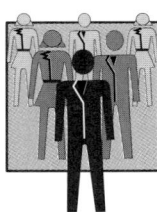

WHY DON'T DOGS AND CATS WATCH TV?

Did you ever wonder why Tabby and Fido never seem to watch TV? They may respond to an occasional sound, like a bark or a mew, but they show no interest in watching TV, even when Lassie or Garfield is on.

Is it because they don't have persistence of vision and, as a result, can't make sense of what's on? Probably not. Of course, we shall never know for sure, but most experts seem to think that TV looks the same to them as it does to us. Experiments with cats have shown that they can tell the differences between certain shapes displayed on a cathode ray tube. This doesn't prove they can perceive motion, but it does suggest some kind of persistence of vision is present.

So, how come they don't watch? Probably because they perceive TV as not being real. To be meaningful in their world, an object must have three dimensions and a smell. Once dogs and cats learn that a TV picture is just two dimensional and basically odorless, they lose interest in it. This also may explain why dogs and cats react every once in a while to TV sound. They sometimes hear another dog or cat without being close enough to smell it.

Interestingly enough, higher-order primates, apes in particular, often watch TV and even develop program preferences. Willie B., a famous ape at the Atlanta Zoo, was said to be a real fan of soap operas and football games.

---

our TV sets, and changed back into images as the glowing light off our TV screens. In the visual arts, from painting and sculpture to TV and film, there are two main ways of creating a facsimile of an image: mechanical and electronic.

Mechanical transduction Mechanical transduction goes back to prehistoric times. The representations of daily activities (such as planting and hunting) found on cave walls were attempts to store and recreate visual experiences common to the culture to which they belonged—a purpose shared with modern TV and film. Throughout history we have discovered and perfected various ways to create and store visual images by mechanical means, that is, through the use of tangible physical commodities to produce images. Until the first half of the last century this was largely limited to the crafts of painting and sculpture. Oils, acrylics, stone, clay, and other elements were fashioned to create facsimiles or representations of people and objects throughout history. How faithful the reproductions were to the original images they represented depended on the individual skill of the artisan as well as on the relative value each particular culture placed on "faithful" reproductions.

In the early 1800s new discoveries in chemistry and optics allowed for more faithful reproductions, independent of the skill of the painter or sculptor. By the 1850s the photography studio (known as a daguerreotype parlor) had become the "video store" of its day, providing low-cost imagery to a public hungry for a visual record of its culture and experience.

The basic element of photography and film is the film itself, which photographers and filmmakers generally call raw stock. A piece of film contains an emulsion sensitive to light. Upon exposure by the camera, the film stock, through a chemical reaction, records the graduations of light entering the lens. When the film is developed, the light graduations are forever captured as changes in gray (for black and white) or hues (for color).

The development of movies was made possible by the introduction by George Eastman of flexible film stock in the late 1800s. Before Eastman's invention, photos were taken on glass plates, which were fine for still pictures but unsuitable for motion. Thanks to Eastman, motion picture stock could now be made using celluloid, which was both flexible and durable. It bent easily to be rolled, shuttled, and stored in a camera and projector, but it was strong enough not to tear or break under normal conditions. With the perfection of celluloid, the technique of the riffle book could be combined with earlier developments in photography and optics. By 1900, moving pictures, or movies, had become reality.

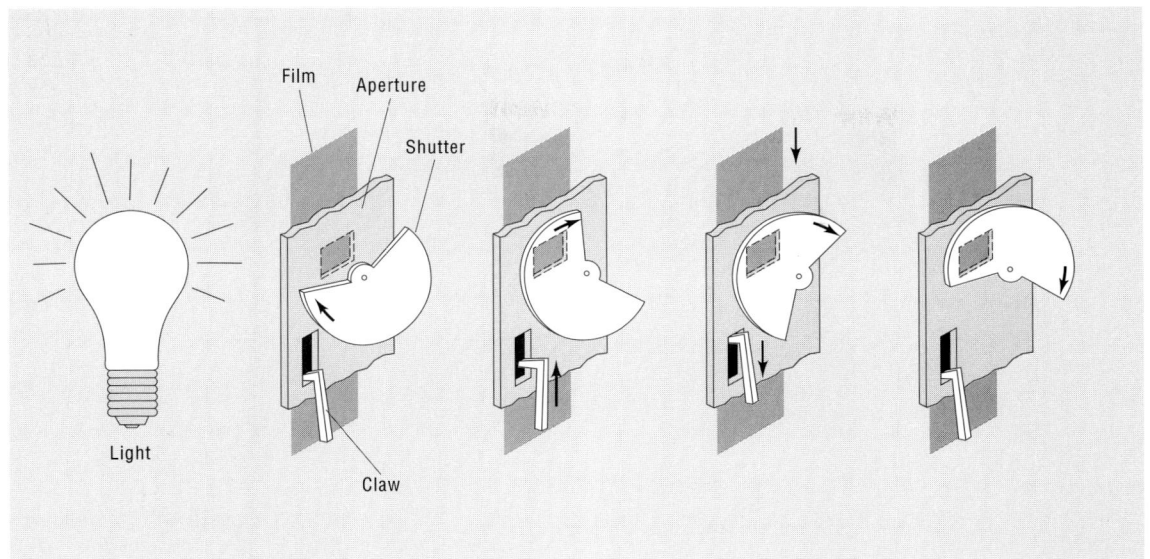

FIGURE 13-3 How a movie projector works.

Figure 13-3 shows how movies make use of mechanical means to transduce visual images. The illusion of film motion is made possible by a mechanism called a *claw,* which grabs each frame on a film by the sprocket holes that run alongside it. The frame is pulled into position, held momentarily, then replaced with the next frame. During the pull-down phase a shutter blocks the projector's bulb, allowing for visual lag, thereby creating the perception of continuous motion. This process occurs twenty-four times per second—like a very high-speed riffle book. We are unaware of the intermittent pattern of picture–black–picture–black unless the projector slows down or breaks and we actually see the flicker. Visible flicker was common in the early days of film, however, when films were projected at sixteen frames per second, too few frames per second to prevent it.

Electronic transduction The technology of live TV is much more complex than the mechanics of motion picture technology. There is no film in the TV camera: the image must be transduced electronically, without a "hard copy" like a photograph or strip of celluloid. Like radio broadcasts, live TV broadcasts must travel over the air; they can't be shipped by U.S. mail in a film canister. Even the means of recording TV signals is more complex: when you hold up a piece of videotape you can't see the images, as you can on film. They are stored in a complex code of video, audio, and synchronization information. Thus, as was shown in Chapter 3, the development of TV occurred later than film or radio and had to await advances in physics, optics, and electronics.

VIDEO SIGNAL GENERATION

Television's ability to transmit images without hard copy (in the form of a strip of film or a photograph) is based on the technology of scanning. The TV camera scans each element of a scene line by line; the picture tube in your TV set retraces the scene. Before we get into the complexities of the process Figure 13-4 presents a simple analogy that will help you understand how it works. Remember playing this game as a child? With pencil or crayon, you traced the outline of your hand, creating a facsimile representation of the hand on the sheet of paper, like the illustration in Figure 13-4*a*. Let's change the game a little bit. Instead of drawing one continuous line, suppose we trace the hand by using a series of five parallel lines, as depicted in Figure 13-4*b*. We move the crayon straight across. We lift it when it encounters the hand and return it to the paper

PART 5:
HOW IT WORKS

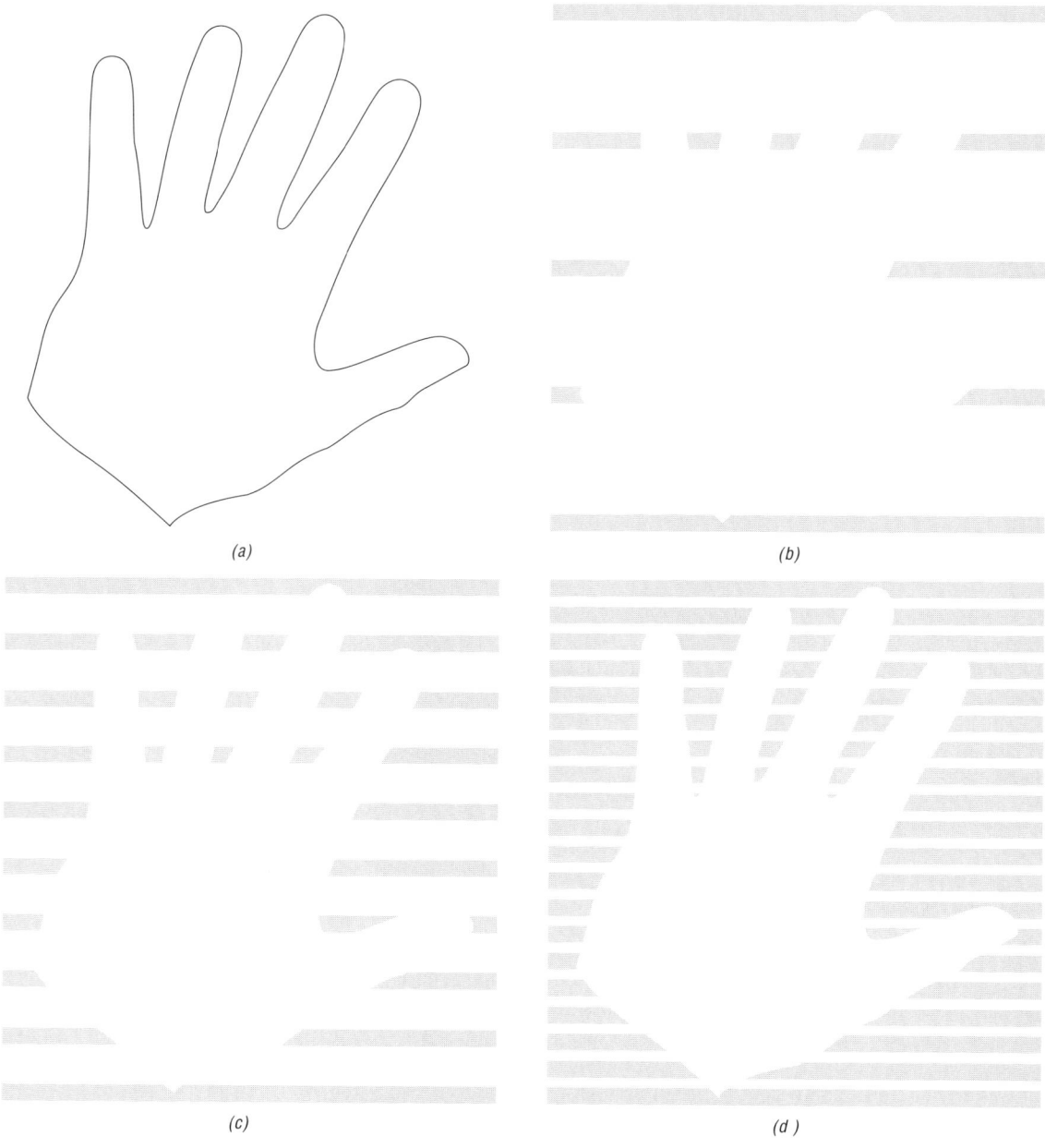

FIGURE 13-4 Examples of scanning.

when it passes by a "hand" area. The result is the rough facsimile of the hand in Figure 13-4b. Now, let's use ten lines instead of five. This tracing will provide a fairly good representation of the hand, as in Figure 13-4c. Just for fun, let's alternate the tracing by doing every odd-numbered line first, then each even-numbered line. After two passes, top-to-bottom, the result is Figure 13-4d. In this exercise we have actually demonstrated the workings of TV technology.

When done invisibly by the TV camera, the tracing process is called *scanning*. The process of alternating lines is known as the *interlace* method. And the process of replicating the scan on the picture tube to produce the image at home is known as *retracing*. The process originates in the

TV camera and is recreated at home on the picture tube.

Inside the Camera: Pickup Tube

The purpose of scanning is to transduce physical light into electronic signals. The process occurs inside the TV camera, as depicted in Figure 13-5.

As we have seen, the first step in creating TV is the collection of an image by a lens. Inside the TV camera the image is captured onto a mirrorlike device called the *target plate* (1). At the base of the camera tube is an *electron gun* (2), which sends a stream of electrons toward the picture on the plate. Force fields created by *deflection magnets* (3) pull the beam up and down across the image, much as our crayon did in Figure 13-4. As the beam scans the picture, an electronic signal (4) is created, each line in the original picture now taking the form of an electronic signal—the common waveform as illustrated in Figure 13-5. To take advantage of the persistence of vision and to reduce flicker, the scanning is alternated. A black line is added between each horizontal line and between each complete scan to allow the image to burn into our retina. This is known as the *blanking pulse*. The signal that exits the camera thus contains two sets of information: picture plus blanking. At the camera control unit (CCU) or in the switcher (see below), a third signal is added. The *synchronization pulse* enables the output of two or more cameras and other video sources to be mixed together and for all the scanning processes to take place at the same time, from camera to receiver. The complete TV picture, picture plus blanking plus sync, is known as the *composite video signal*.

Color television In black and white TV the electron beam responds to the brightness of the picture on the target, its *luminance*. But the color signal is more complex to create in TV. A means had to be perfected to scan for the proper colors in a scene, as well as for the intensity of those colors. In technical terms, each individual color we see is a *hue* or *tint*. The strength of the color (from royal to navy blue, for example) is known as its *saturation*. Figure 13-6 shows how this process developed. The same principle involved in mixing the fingerpaints you played with as a kid is basic to color TV. If you remember that yellow plus blue makes green, you'll probably be able to grasp the basics of color TV technology. Keep in mind, though, that TV is mixing light instead of paint, which makes the process a little different.

The color camera has a special optical system known as a *beam splitter* or *dichroic mirror,* which dissects each image on the lens into its three

FIGURE 13-5 Inside the TV camera.

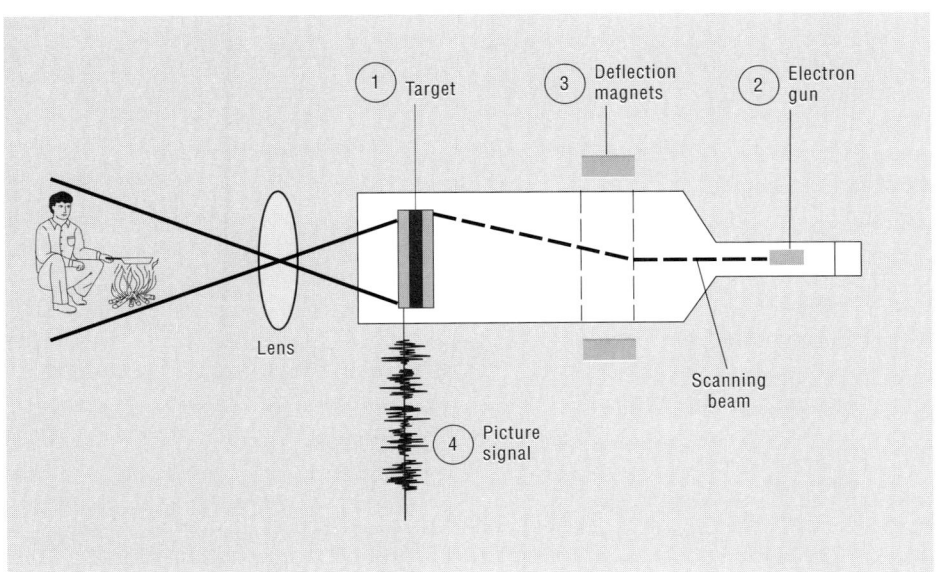

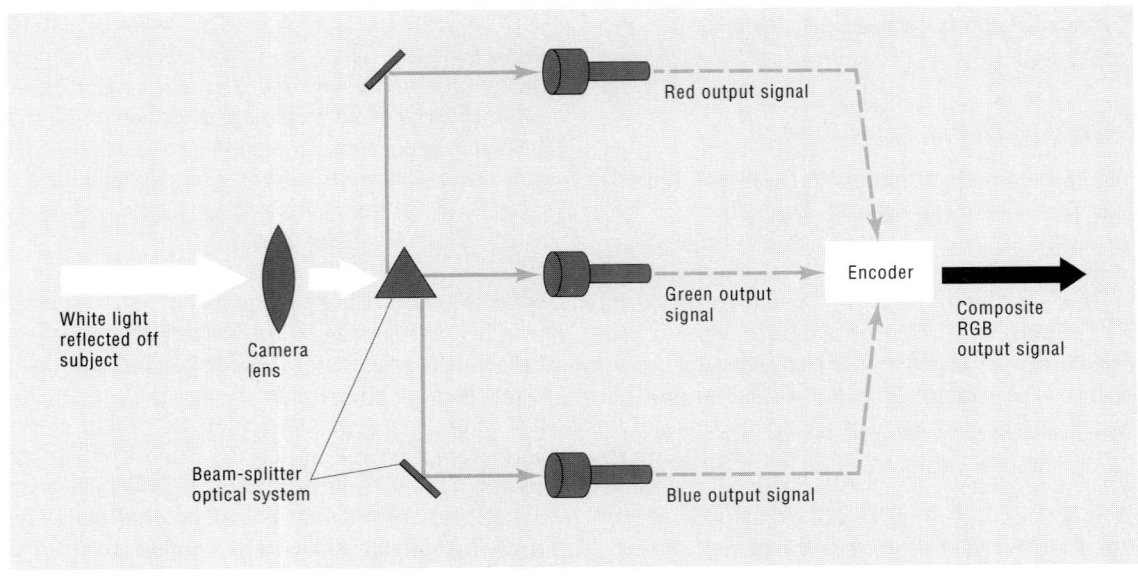

FIGURE 13-6   Inside the color TV camera. (Source: Alan Wurtzel and Stephen Acker, *Television Production,* **New York: McGraw-Hill, 1989.)**

primary colors: red, green, and blue. In sophisticated TV cameras there are three separate tubes: one for each primary color. The electron beam in each color tube scans for the precise tint and saturation of that color and creates an electronic signal corresponding to that color in the scene being transmitted. Once the scene has been broken into its red signal, blue signal, and green signal, brightness (luminance) is added along with a sync pulse, and the signals are recombined into one highly charged picture signal. This occurs in a device called an *encoder.* The original scene has been converted into its red, green, and blue components. Hence it is now known as an *RGB signal.*

Single-gun composite color   A single-gun camera system employs only one pickup tube to scan for chrominance (the color information in a TV picture) and luminance (the brightness information) signals. Most single-gun systems make use of a filter of fine color stripes arrayed in front of the pickup tube. Like a prism, the striping separates or decomposes the scene into its primary colors so that they may be deduced and electronically recreated. Single-gun cameras have the advantage of being simpler in their electronic operations. They typically require less power than studio-grade three-tube cameras and do not have as much difficulty maintaining their picture quality, or *registration,* from production to production. Traditionally, however, composite color cameras have lacked the resolution of three-gun cameras. This limitation has been balanced by the suitability of single-gun cameras to production situations requiring portability rather than overall pictorial excellence (such as industrial video and broadcast news gathering).

Tubeless Cameras: The Camera as Computer

Much of the excitement in TV imaging in recent years revolves around the introduction of the first tubeless, or solid-state, cameras. These cameras replace the pickup tube with a *charge-coupled device (CCD).*

The CCD image sensor consists of three semiconductor chips, which replace the conventional tubes in a standard vacuum tube camera. Each of the red, green, and blue chips can sample over 250,000 picture elements in a manner not unlike the way a printing press uses dots to produce a color newspaper photo. CCD cameras have a number of advantages over traditional vacuum tube designs. First, the chips are much smaller than the tubes they replace, allowing for smaller, lighter cameras. Second, since the CCD is flat, picture distortion is reduced. In addition, CCD cameras are rugged and durable. By the early 1990s these so-called "chip cameras" had become the preferred aquisition medium for much TV news. Inexpensive, small, lightweight, and durable, they proved their worth under fire in 1991: first, in the Persian Gulf War, and later, during the fall of the Soviet Union following the aborted coup to oust Mikhail Gorbachev.

Multichannel Television Sound

For most of the years of the medium's history TV overlooked the opportunity for high-fidelity sound, concentrating mostly on refinements in picture quality. However, the success of enhanced sound in movie theaters, the rise of music video channels on cable, the introduction of stereo sound to home video recorders, and other factors led the TV industry to develop a new means of audio transmission. Today most TV stations in major markets are equipped with *multichannel television sound (MTS)* transmission equipment: stereo sound and a second audio channel multiplexed in the audio portion of the TV signal. Like FM radio, MTS carries separate left and right channel signals and reception requires a special multiplex tuner. In TV the device is known as an *MTS decoder*.

MTS can be used for dual-language TV broadcasting, for example, English on one channel and a foreign language on the other. The subcarrier of the FM audio signal in TV can be used for many other purposes, such as sending cues to ENG news teams in the field.

Adding More Lines

Television engineers have not been content to introduce tubeless cameras and stereo sound. They are now experimenting with the scanning process itself. The idea is to add more lines to the TV screen to improve its clarity and fidelity.

High-definition television (HDTV) utilizes up to 1125 scanning lines, producing an image rivaling 35-millimeter film in quality. It also changes the aspect ratio (the ratio of screen height to screen width) of conventional TV. In standard TV the screen is three units high and four units wide. In HDTV the ratio is 16:9 (sixteen units wide and nine units high), much like the frames of a motion picture. Developed by engineers at Sony and the Japanese Broadcasting System, HDTV has al-

Many companies are now trying to perfect improved definition television receivers. Here Zenith engineers work to develop an HDTV system that would use new spectrum space to carry its signals.

ready been used in the production of TV commercials. In 1987 the first feature length film was produced using HDTV technology. By 1990 the Japanese were broadcasting in HDTV, utilizing high-powered satellites.

The picture produced by the HDTV process is stunning, but a number of problems limit its introduction into the American market. The major problem is that at present HDTV is not compatible with the existing technical standards of U.S. TV. That is, to receive HDTV broadcasts, studios, transmitters, and receiving sets would have to be replaced. In addition, the HDTV bandwidth requires a large amount of spectrum space, which is becoming more and more scarce with the introduction of new distribution methods, like satellites and microwaves, which are discussed later.

The FCC has ruled that American HDTV will have to be compatible with existing U.S. receivers. In response a consortium of corporations established the Advanced Television Test Center (ATTC) in Alexandria, Virginia, to test six different proposed HDTV systems and to evaluate each in light of the FCC's mandate.

By 1992 the ATTC had demonstrated a number of new wide-screen systems. The first, developed by GE/NBC, Philips, and Thomson, was advanced compatible television (ACTV). ACTV offers wider screen TV but uses only 525 scanning lines. While this method ensures its compatibility, critics argue that it is not true high definition. Other HDTV systems being tested for the American market include Narrow MUSE, from NHK (the Japanese Broadcasting System), Zenith and AT&T's DSC-HDTV, and two different systems proposed by a group known as the American Television Alliance (ATVA). Participants in ATVA include General Instruments and the Massachusetts Institute of Technology (MIT).

How soon before we're watching HDTV broadcasts at home? Now that testing is underway the prediction is that HDTV broadcasts could begin as soon as eighteen months after a domestic standard is set, when the first HDTV sets would reach the marketplace. Major networks and large TV stations could be transmitting in HDTV within five years; it might take up to fifteen years for the technology to reach smaller operations. Of course, economics may affect the diffusion of HDTV even more than the problems of technology and standards. Conservative estimates suggest transmitting in HDTV would cost a TV station from $5 million to $15 million. HDTV sets are expected to be priced initially in the range of $2000 to $3000, far beyond the pocketbook of many consumers.

The Importance of Scanning

At this point you may feel saturated by needless technical information. Actually there are excellent reasons why this information is critical to understanding modern telecommunications.

Standardization First is the issue of standards. The number of lines utilized in the scanning process varies throughout the world. The United States uses a 525-line system adopted in 1941 known as *NTSC* (National Television Standards Committee). A complete picture consists of two separate scans or *fields,* each consisting of $262\frac{1}{2}$ horizontal scanning lines. The two fields combine to form a single picture, called the *frame*. In the United States the AC power system we use oscillates at 60 hertz. Thus, our TV system uses a 60-hertz scanning rate. Since two fields are needed to produce one frame, thirty complete pictures are produced each second.

Much of European TV uses a system known as *PAL,* adopted several years after the U.S. system. Based on the DC power available there, European TV uses a shorter scanning rate (50 hertz), with more scanning lines (625). Many Americans are startled to see how clear European TV is; that's due to the fact that it has 100 more lines of resolution.

Beyond better picture quality, the issue of standardization involves a lot of other important matters. Televisions and VCRs produced for one system will not work on the other. The same is true for videotape. As an administrator for the Peabody Awards, one of the authors racks up hours in international telephone calls each year making sure programs are submitted in the proper format.

Innovation The scanning process is also directly involved in many of the current technical innovations in the TV medium. One of these is HDTV. As we have seen, to produce a bigger and higher quality picture, a new scanning rate must be used.

Closed-captioning and teletext use the blanking period between scans (the *vertical blanking interval*) to send additional information with the TV signal. Appropriate decoders provide captions for deaf viewers or specialty text services such as news capsules or stock market reports to specially equipped TV sets. Thus, all parts of the TV signal can be used to send information.

Another innovation made possible by video signal generation is *interactive television*. Interactive TV is a means by which TV viewers can respond to programs they're watching by using a special device, such as a joystick or keypad. Early interactive systems used the blanking interval to enable viewers to play along with game shows, like "Wheel of Fortune." Viewer response was generally underwhelming.

New interactive experiments hold more promise. The idea is to use a region of the audio spectrum to allow TV viewers to engage in all sorts of activities, from polling to playing games, from ordering pizza to selecting from a number of possible TV programs. The viewer uses a small infrared hand-held unit, much like that used by many video game systems. Atop the TV set is a special radio receiver that transmits and receives the various interactive options. Based on prompts on the TV screen, the viewer moves a cursor up and down to order groceries, vote for a candidate, possibly even to choose a program for the VCR to record.

In 1992 the FCC made the first allocation of spectrum space for this purpose, called Interactive Video Data Service (IVDS). TV Answer, a company based in Reston, Virginia, promised a rollout of interactive TV services in short order. It remains to be seen whether people really do want to talk back to their TV sets.

VIDEO AMPLIFICATION AND PROCESSING

As the TV signal travels from the camera to the transmitter several things happen. First, the electrical signal is amplified—increased in electrical intensity—and carried along a wire to a closed-circuit TV set, called a monitor, where it is viewed by the director and other production personnel. In most TV programs the inputs from several cameras and other video sources (tape machines, graphics generator, etc.) are mixed together before they are transmitted. The *switcher,* the device used for mixing, is probably the first thing a visitor to a TV control room notices. The advanced models are impressive-looking devices consisting of several rows of buttons and numerous levers. The switcher is used to put the desired picture on the air. If camera 3 shows what the director wants to see, then pushing the appropriate button on the switcher puts camera 3 on the air. If videotape machine 4 has the desired picture, then pushing another button puts in on the air.

The switcher also lets the director choose the appropriate transition from one video source to another. Simply punching another button generates what's known as a *cut*—an instantaneous switch from one picture to another. By using a fader bar the director can dissolve from one picture to another or fade an image to or from black.

If a special-effects generator is added, a host of other transitions are possible. One picture can wipe out another horizontally, vertically, in a diamond shape, or in many other patterns. In addition, a split screen with two or more persons sharing the screen at the same time is possible, as is *keying,* an effect in which one video signal is electronically cut out or keyed into another. The most common use of this process is *chromakey.* A specific color (usually blue) drops out of one picture and another picture is seen every place where that color appeared in the original picture. Weathercasters, for example, usually perform in front of a blue background, which is replaced by keyed-in weather maps or other graphics. (Performers must be careful not to wear clothing that is the same color as the chromakey blue, or their clothes will appear transparent on screen.)

As might be expected, digital technology has had an impact on video processing. Each TV signal is converted into a series of binary code numbers that can be manipulated and then reconverted back into a TV signal. There are numerous types of digital video effects. They include freeze-framing, shrinking images in size and positioning them anywhere on the screen (as happens when an anchor talks to a field reporter and both pictures are kept on screen), stretching or rotating a video picture, producing a mirror image, and wrapping the picture into a cylindrical shape.

Desktop Video

One of the quiet revolutions in contemporary TV is based in video signal processing. Until the late 1980s, generating video special effects required large, expensive processing equipment. Only a few big-city TV stations and production centers had the capability to produce digital video effects, or DVE.

In the early 1990s the Commodore and Apple computer companies merged the video signal with the personal computer. Today DVE can be produced for less than $5000. The kind of spectacular visuals once reserved for music videos with huge budgets, or for the promotional messages of the major networks, can now be produced by anyone with a camcorder linked to a PC. This revolution is known as "desktop video." As if to underscore the low cost and simplicity of the new DVE machines, Commodore's Amiga setup is known in the industry as the "video toaster."

VIDEO SIGNAL TRANSMISSION

For the fifty-plus years of its history the TV medium has used two primary means of getting the signal from the studio to the home screen: over the air, by using the phenomenon of the electromagnetic spectrum, and via cable. In recent years many new means of getting the signal from point A to point B have emerged. Before we present these, let's examine the traditional means.

Television Broadcasting

As might be expected, the TV signal, with its complex set of information (including picture, sound, color, blanking, and synchronization signals), requires enormous amounts of space in the electromagnetic spectrum. More than any other reason, this explains why there are over 12,000 radio stations but only about 1300 TV stations on air in the United States.

The television channel Each TV station requires a bandwidth of 6 megahertz. This is equivalent to enough space for 30 FM radio stations and 600 AM stations! How this spectrum space is utilized is depicted in Figure 13-7.

Our TV system was perfected in the 1930s and 1940s when amplitude modulation (AM) techniques were considered state of the art. For this reason it was decided to transmit the TV picture information via AM. Two-thirds, or 4 megahertz, of the TV bandwidth is used for picture information. The sound signal is essentially a full-range FM or frequency-modulated signal, oscillating at 25 kilohertz above and below its center (carrier) frequency. The remainder of the video channel is occupied by protective guard bands which keep the various encoded signals from interfering with one another.

TV allocations The complexity of the TV signal and its vast need for space caused the FCC in the 1940s and 1950s to place it higher in the electromagnetic spectrum than had been utilized in prior years.

Channels 2 to 13 are located in the *very-high-frequency (VHF)* portion of the spectrum, the area ranging from 54 to 216 megahertz. Interestingly, a sizable portion of the VHF band is not used for TV purposes. The area between channels 6 and 7 includes the space for all FM radio stations as well as aircraft-control tower communication, amateur or "ham" radio, and business and government applications. This is why sometimes there can be interference between TV channel 6 and public radio stations at the low end

FIGURE 13-7 Anatomy of the TV channel. (Source: Adapted from FCC specifications.)

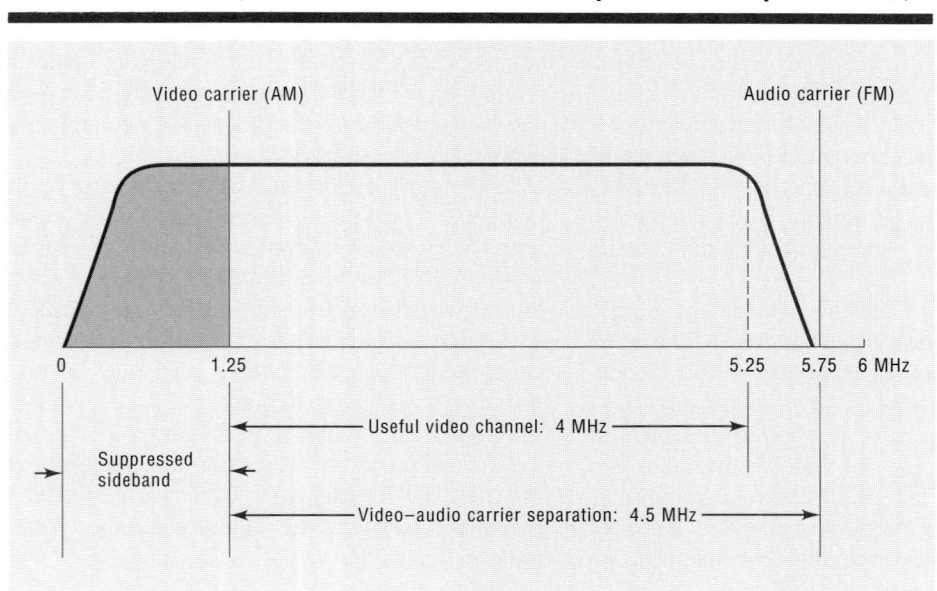

TECHNOLOGY

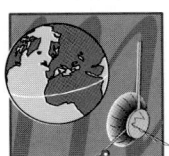

WHATEVER HAPPENED TO CHANNEL 1?

Did it ever puzzle you that the TV dial starts with channel 2 instead of channel 1? How come? Well, it's kind of complicated. Channel 1 did exist for a while, but it was reassigned to other communication services by the FCC. In fact, it was reassigned twice. Here's a simplified version of what happened.

Back in 1938 the FCC assigned nineteen separate channels for TV, each with a bandwidth of 6 megahertz. Channel 1 was assigned the 42–48-megahertz frequency and RCA began experimental broadcasts on it. At the same time, however, FM radio was developing and many experimenters were broadcasting on the frequencies between 42.6 and 43.4 megahertz, the low end of channel 1. By 1940 the FCC gave in to growing enthusiasm for FM and reassigned all of channel 1 to the new service.

But the FCC then renumbered all the remaining TV channels and the new channel 1 was assigned the 50–56-megahertz slot, where channel 2 used to be. An RCA station, WNBT in New York, was one of the first TV stations to transmit on the new channel 1.

About a year later World War II started. The extraordinary strides in communication technology made during the war resulted in tremendous demand for frequencies once thought to be experimental, like the TV band. As the war drew to a close, the FCC had to reexamine totally its spectrum allocations. In 1945 it moved the entire FM radio band up to the 88–106-megahertz position, reduced TV's share of the spectrum to thirteen channels, and ruled that all TV channels must be shared with two-way radio mobile services. As TV became more popular and the number of TV stations grew, interference began to develop between the land mobile services and the TV stations. In 1948 the FCC decreed that the land mobile radio stations had to have their own spectrum space, took channel 1 away from TV, and assigned it exclusively to two-way radio services. This time, however, the FCC did not bother to renumber the remaining channels; thus TV receivers start with channel 2.

---

of the FM dial. It also explains why you can occasionally pick up TV channel 6 by tuning your FM radio all the way to the bottom.

Channels 14 to 83 lie in the *ultra-high-frequency (UHF)* portion of the band, the region between 470 and 890 megahertz. There is another gap, this time between channels 13 and 14, which is reserved for government communications.

You will recall from the radio chapter that the higher one goes in the spectrum, normally the more power is needed, and the more critical becomes the need for sensitive transmitting and receiving equipment. For this reason, inadvertently, two classes of TV stations were created: the "haves"—VHF channels 2 to 13, with favorable dial locations and excellent signal propagation patterns—and the "have-nots"—UHF channels 14 to 83, hard to find on the dial and hard-pressed to generate an adequate signal in their coverage areas. Here's why.

Like FM radio, the TV signal is propagated best by the use of direct waves. You will recall that this means that the best reception is obtained when there is a direct line of sight between the transmitting antenna and the receiver. Thus the coverage of a given TV station is dependent on such factors as its antenna height, frequency, power, and, to a lesser extent, the terrain in its service area.

VHF stations 2 through 6 can achieve excellent coverage transmitting at a maximum of 100 kilowatts. Channels 7 to 13 require power up to a ceiling of 316 kilowatts. However, channels 14 and above need power up to 5000 kilowatts to generate an acceptable signal (the UHF maximum is 10,000 kilowatts). In addition, more sophisticated antenna arrays are required. So not only do viewers often have difficulty locating UHF stations on their dial; once located, tuning and maintaining a clear signal are also problematic. The FCC's freeze on new stations from 1948 to 1952 also hurt UHF's development. The existing stations were all VHFs and had established loyal audiences before UHF even got started.

For these technical and historical reasons VHF stations have tended to travel "first class" in the TV business, while UHF properties have typically been relegated to the "coach" section.

Cable Transmission

Shortly after World War II a new type of cable was developed to support the burgeoning telephone industry. In addition to handling thousands of simultaneous telephone conversations, this cable was soon found to be an excellent conduit for disseminating TV signals. Since then, cable TV has become an increasingly widespread and important means of TV signal propagation. Figure 13-8 illustrates the various elements of this *coaxial cable,* or "coax."

The coax consists of two electronic conductors. At the center is a copper wire shielded by a protective insulator called the *dielectric.* Outside the dielectric is a wire mesh, typically made of copper or aluminum. This conductor is shielded by a durable plastic coating, the outer body of the cable. This design gives the cable some special properties. The shielding keeps the signals within a confined space, reducing the electrical interference common in normal wiring (like that in household appliances). Coaxial cable can also carry its own power source, which allows signals to travel comparatively long distances with little signal loss or noise. In addition, good shielding allows the coax to last many years without contamination by water or other invasive materials.

Cable signals do lose some strength, however, as they pass through the distribution system. As a result, amplifiers must be inserted into the system at certain intervals. Unfortunately, each amplifier introduces some noise and distortion into the transmitted signal. This problem gets worse as amplifiers are "cascaded," or placed in a series along a certain stretch of cable, since the noise and distortion get worse as the signal is passed through each successive amplifier. Consequently, it's necessary to limit the number of cascaded amplifiers to ensure a good signal at the subscriber's set. Current cable systems can cascade about thirty amplifiers, but advances in technology will eventually make larger cascades possible.

Over the years, as the materials used to make coaxial cable and the types of transmission and amplification equipment used by cable systems were refined, the number of TV signals transmittable by cable increased. In the 1950s cable TV systems could carry only three to five TV signals. In the 1960s transistorized components and new cable materials raised cable channel capacity to twelve channels. The cable explosion in the 1970s came about as further technical refinements allowed cable systems to carry as many as sixty TV channels, as well as a range of other services, from FM radio to data, text, and other services.

Figure 13-9 compares the use of the electromagnetic spectrum and coaxial cable in TV delivery and other broadcast services.

Modern cable systems normally have a usable bandwidth ranging from 54 to 450 megahertz. Depending on how the space is used, such systems boast channel capacity from fifty to sixty channels. Whereas broadcast TV has two main bands, VHF and UHF, cable systems have three regions (and therefore classes) of TV distribution.

Note that TV channels 2 to 6 and the FM radio band occupy the same region on the cable as they do over the air. In cable jargon, this is known as the *lowband.* Receiving signals in the lowband requires no special equipment: plugging the cable into the back of the set guarantees reception of these channels. Simply hooking the cable to your stereo will deliver crisp radio reception (if your cable system offers audio services).

As we have seen, there are big gaps in the broadcast spectrum between channels 6 and 7 and between channels 13 and 14 reserved for a range of uses, from aircraft navigation to government mobile communications. Consequently, when cable increased its channel capacity, it could not place all the new channels precisely where they occur in the spectrum. The space assigned to business and government in the electromagnetic spectrum is used by cable to carry channels 14 to 22; channels 7 to 13 follow, in their usual spot. This region is known as the *midband.* Standard TV receivers cannot tune channels 14 to 22

FIGURE 13-8   Coaxial cable.

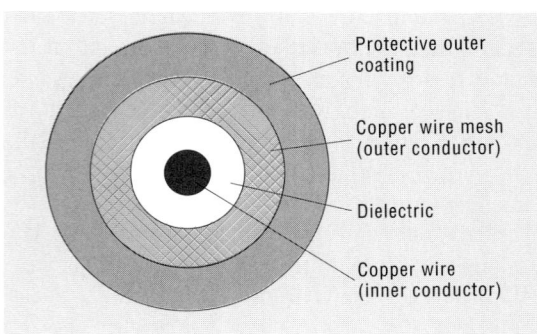

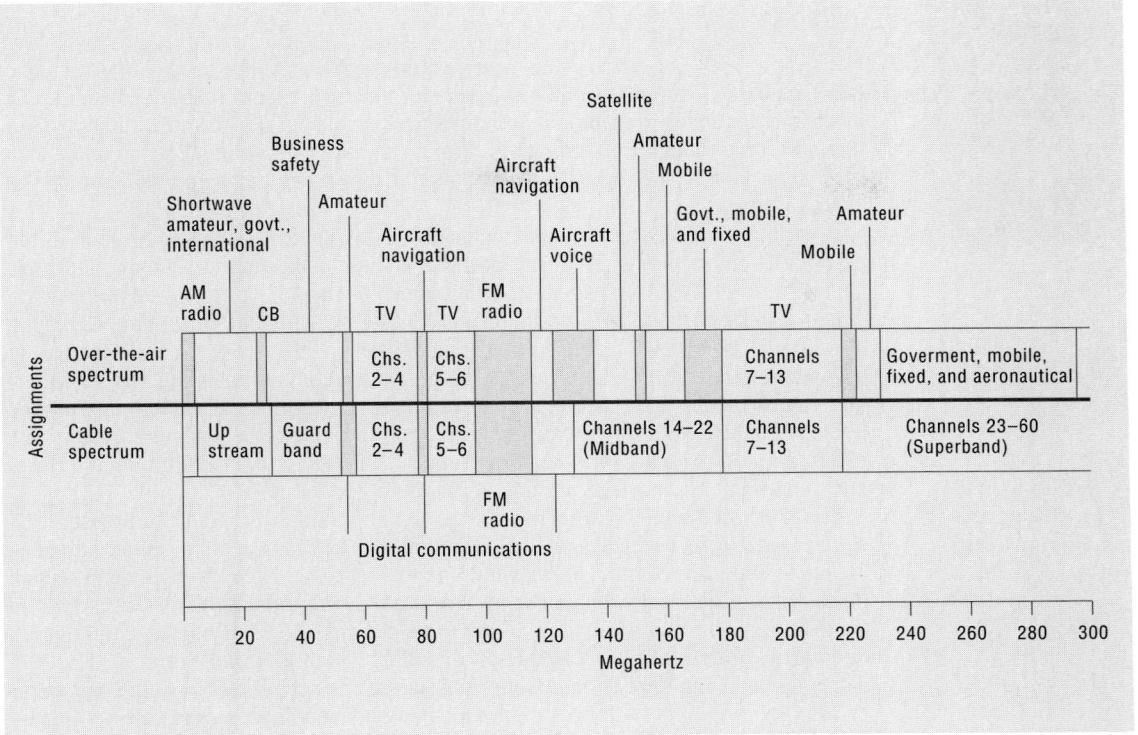

FIGURE 13-9 Cable TV and spectrum use. (Source: Herbert Dordick, *Understanding Modern Telecommunications.* New York: McGraw-Hill, 1986. Used with permission.)

in the reserved frequencies. Thus TV sets and VCRs that are not "cable-ready" require a *converter box* to receive these channels. On some converter boxes channels 14 to 22 are known as channels A to I.

Note that the UHF broadcast spectrum falls outside the range of the 450 megahertz usable by cable. Instead, the cable spectrum uses another region that is off-limits to over-the-air telecasters. Frequencies spanning 216 megahertz and above fall into the cable *superband* (or, with less hyperbole, the highband). On a converter box these channels range from 23 to 36, or from buttons J to W. Note that some of the same frequencies used *within* the cable system are used for important nonbroadcast communication services, such as aeronautical and emergency communication, *outside* the system. Thus it's important that the cable system not interfere with these other services.

As you can see, local cable operators can exert a kind of media power that broadcasters cannot. They—not the FCC—can control where a given TV station is placed on the cable. In addition, cable systems can reuse their own frequencies simply by installing two cables side by side. Thus 120-channel systems are possible. Moreover, a cable company can construct a system that allows communication from the receiver to the headend, enabling two-way, or interactive, communication.

Another advantage of cable is *addressability*, the ability of a cable system to send a program to some houses that request it and not send it to those that don't. Addressability is important in the development of pay-per-view TV where only a portion of subscribers are willing to pay extra to watch a recent movie or sports event.

This flexibility accounts for cable's great expansion in recent years, detailed in Chapter 3. Originally a source of improved reception of only three or four nearby TV stations, the cable has become a broadband source of local and regional broadcast stations, distant superstations, sports, movies, and other special-interest channels to more than one-half the nation.

The structure of a cable system is diagramed in Figure 13-10. The *headend* is the point at which all program sources are received, assembled, and

ECONOMICS

SCRAMBLED SIGNALS

Not that any of you reading this book would do it, but some people have tried to receive cable TV programs without paying for them. To ensure that access to pay TV is limited to paying customers, programmers use several devices. On cable an encoder (or scrambler) electronically alters a signal so that it can be viewed only on a receiver with a specially equipped decoder that takes the scrambled signal and makes it viewable.

Jerrold Communications of Philadelphia has developed a unique approach to the problem of cable piracy. Its technology allows cable operators to send an electronic signal from the head-end to each subscriber household. To legal subscribers the signal is undetectable, and programming is received without interruption. However, Jerrold's "Magic Bullet" blacks out any cable home with an illegal converter box, or a box that has been rigged to receive programming without payment. The TV simply goes black (and the subscriber's address begins to flash at the cable home office).

On March 31, 1991 American Cablevision in Queens, New York, fired the magic bullet. Over 300 cable pirates took a direct hit: they faced criminal prosecution and substantial fines if they did not pay $500 within twenty days.

Those people who own satellite dishes present another problem. To protect their signals from dish owners who are unauthorized to receive them, many services, including HBO, Showtime, The Movie Channel, and Cable News Network, scrambled their signals. The most widely used system is called the VideoCipher II manufactured by M/A-Com, Inc. In a nutshell, here's how it works. A dish owner purchases a decoder for about $300 to $400. Each decoder, which is about the size of a VCR, displays a thirteen-digit serial number that can be punched in M/A-Com's computer in San Diego. When a dish owner pays all the necessary fees to receive a scrambled signal, M/A-Com's computer orders a signal containing the serial number to be bounced off a satellite; when that signal hits the approved user's home dish, it activates the descrambler and the consumer sees a normal picture. If the consumer fails to pay the monthly bill for receiving the service, a phone call from HBO or Showtime or any of the other scrambled services can easily rescramble the signal. One big advantage of VideoCipher II is that it's practically tamper-proof; any fiddling with the device and its microprocessor loses its memory and stops working.

---

processed for transmission. The headend can pick up signals from nearby standard broadcast stations using large antennas and can receive TV signals from more distant stations via microwave transmission. Cable systems are also equipped with one or more satellite dishes that enable them to receive signals sent from superstations, cable-only networks, and premium channels like HBO. Some headends are connected to studios where local-origination programs are produced.

The *distribution system* consists of the coaxial cable, which is strung along telephone poles or buried underground. The *trunk cable* leaves the headend and goes to individual neighborhoods. Since signal strength gets weaker as it passes through a coaxial cable, amplifiers are located at various points along the trunk line. A *feeder cable* is connected to the trunk cable; it is this cable that is actually installed up and down the streets of the various neighborhoods served by the trunk cable. A *subscriber drop* connects a customer's house to the feeder cable.

Satellites

The most important advance in TV transmission technology in the 1970s and 1980s was unquestionably the development of the communication satellite. Satellites changed the relationship between broadcast networks and their local affiliates. They revolutionized the methods of program distribution. Today satellites are changing both the nature and technique of TV news. Before discussing this, let's examine how communication satellites evolved.

Believe it or not, the inventor of the communications satellite was Sir Isaac Newton. In the seventeenth century he theorized that if a can-

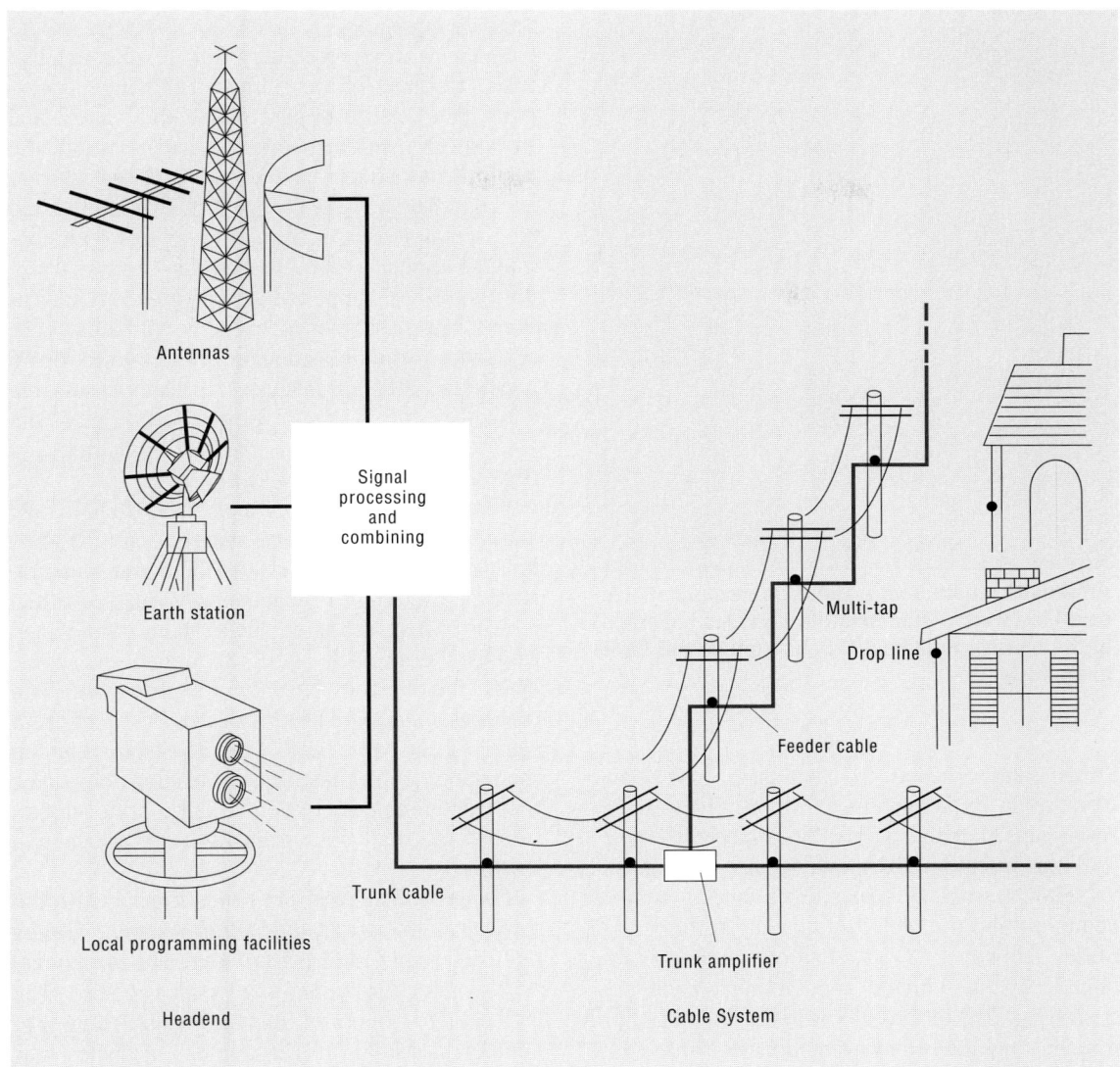

FIGURE 13-10 Structure of a cable TV system.

nonball was shot from a high enough mountain, it would fall back to the ground at about the same rate the Earth revolved away from the original spot of the mountain. That is, it would stay in approximately the same place, never fall back to Earth, and become a "second moon." The modern impetus for satellite development was an article in the October 1945 issue of *Wireless World* in which Arthur C. Clarke envisioned worldwide communications via three satellites positioned in space, in high orbit, equidistant from each other and remaining stationary directly above the equator. In 1957 the former Soviet Union provided the first practical application of Newton's ideas with the launch of *Sputnik*. A flurry of activity followed in both the United States and the Soviet Union. The space race had begun.

In 1962 a satellite expressly designed for communications applications was launched. With financial backing from AT&T, NASA launched *Telstar 1,* the first active communications satellite. (*Echo 1,* an earlier satellite, merely reflected signals from Earth.) *Telstar* was designed with the capability to relay all forms of communication signals, from telephone to TV. Thus by July 1962 it was possible to link simultaneously the United

An artist's rendition of a Galaxy communications satellite in orbit. This satellite provides coverage of the entire continental U.S. as well as the Caribbean basin.

States and Europe via satellite, and David Brinkley in Paris announced directly to watching Americans that there was "no important news."

Within months *Telstar* was regularly being used to relay news and sports events from various points around the globe to viewers in the United States. However, *Telstar* had one serious limitation: it was not in the geostationary orbit suggested by Clarke; thus it was in the line of sight of transmitting and receiving stations for only a few hours each day.

The solution was the launch of geosynchronous satellites, satellites that are "parked" in space. At a height of 22,300 miles above the equator their orbits precisely match the rotation of the Earth. For all intents and purposes they are therefore stationary. By 1966 using communications satellites for simultaneous TV transmissions from distant points to the United States became routine. Today, virtually all satellites used for nonmilitary communications are geosynchronous, and there is considerable traffic in the skies above the equator. As Figure 13-11 illustrates, dozens of communications satellites are parked in space, some as close together as about 1100 miles. Not surprisingly, this crowding prompted the birth of international laws that govern satellite "parking."

Satellite terminology Satellite technology has introduced a new vocabulary into telecommunications. Because of the complexity of their operations, the amount of information they carry, and the power needed to send a signal through the Earth's atmosphere and back, satellites operate in the *super-high-frequency (SHF)* portion of the electromagnetic spectrum. Just as new developments led radio and TV to move up the spectrum, the same has happened in satellite communications. The first geosynchronous satellites were assigned to the area ranging roughly from 4 to 6 gigahertz (gHz, or billions of cycles) in an area known as the *C band*. Newer, more powerful satellites operate from 12 to 14 gigahertz, a region known as the *Ku band*.

Like the situation in TV and cable, there is only a limited amount of space available on a satellite for video and audio services. The rough equivalent of a radio or TV channel is a satellite *transponder*. Most satellites have about two dozen transponders, each about 36 megahertz wide. Thus transponders used for telephone and data transmission can be filled with all sorts of different signals at the same time. Television signals eat up the limited transponder space, leading companies that lease the space to demand high prices and to require much advance planning from prospective lessees.

The ground source that sends signals up to the satellite is known as the *uplink*. Satellite dishes that can receive signals from a satellite are known as *ground* or *Earth stations*. Uplinks require very high power to operate; thus they tend to be few in number in relation to the number of dishes that are capable of receiving satellite signals. Earth stations specially designed just to receive TV signals (those commonly seen in rural areas, atop high-priced condos, and in front of motels and hotels) are known as *Television Receive-Only Earth stations (TVROs)*.

Satellites serving the United States are in an arc 22,300 miles above the equator. They are located at points above a range of 55 and 145 degrees west longitude (southeast of Florida). From there, their signals can be received across the entire country. The area blanketed by a satellite's receivable signal is known as its *footprint*, illustrated in Figure 13-12.

The longitudinal angular position in which the satellite is parked is called the *orbital slot*. Since satellites that are too close together tend to inter-

CHAPTER 13:
VIDEO TECHNOLOGY

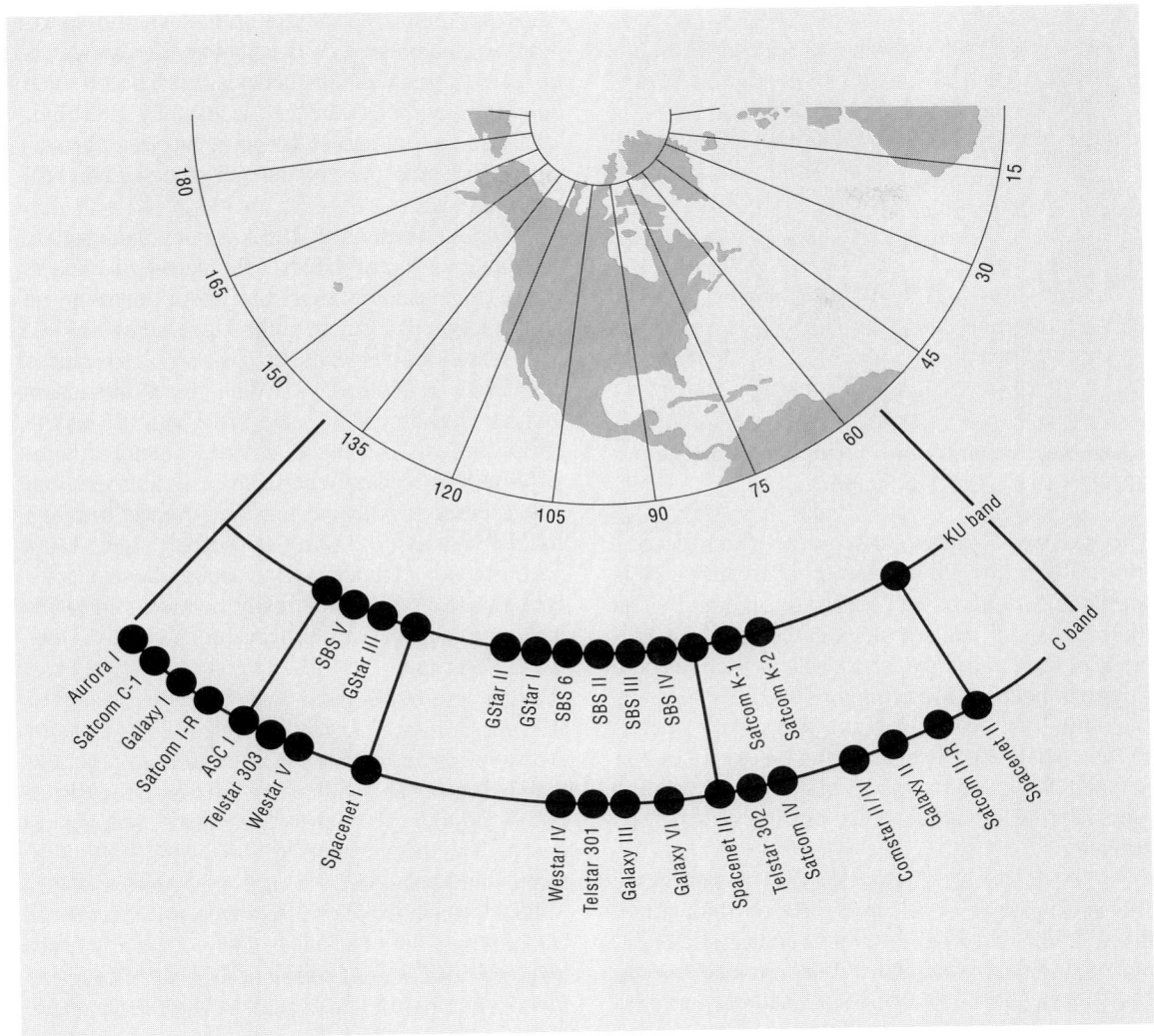

FIGURE 13-11 Orbital positions of geosynchronous satellites. (Source: *Broadcasting & Cable Market Place 1992* and *Broadcasting Magazine.* Used with permission.)

fere with one another, there must be some distance between them. Originally the United States intended to keep its satellites in orbital slots that were 4 degrees (about 1880 miles) apart. Satellites with better directional antennas and more advanced Earth stations, coupled with an increased demand for slots, prompted the United States to reduce the minimum separation between orbital slots to 2 degrees.

To illustrate, in Figure 13-11, *Aurora* I occupies the farthest west orbital slot at 143 degrees. Moving east, its closest neighbor is *Satcom C-I* at orbital slot 139 degrees, a 4-degree separation. A 2-degree separation exists between *G-Star* I at 103 degrees and *G-Star* II at 105 degrees. Clearly, the space available for geosynchronous satellites is limited and there is growing concern, particularly among the developing nations, that all the available slots will be taken before they are ready to launch their own communication satellites.

The range of applications of satellite distribution seems limitless. Businesses use them to exchange data and hold long-distance board meetings, called teleconferences (see Chapter 7). They are also a primary means of sending long-distance telephone calls. Cable systems and TV radio stations use them to receive programming.

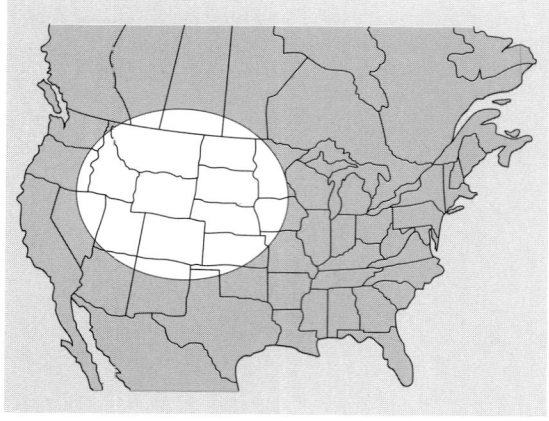

FIGURE 13-12 Satellite footprint.

The major TV networks rely on them to distribute their shows. Groups of local stations use satellites to feed news reports to one another in the process of satellite news gathering. Management of the limited transponder space and orbital slots has become increasingly difficult, especially since the U.S. space program, plagued by failures in the mid-1980s, has been unable to meet the demand for satellite space. As a result, a new type of telecommunications business has been booming.

The development of teleports More than thirty firms in the United States are dedicated to the management of satellite telecommunications. They do this through *teleports,* facilities consisting of uplink ground stations well suited to send a high-grade signal to the satellites parked along the equator. In return for a fee (about $500 per hour in prime times) teleport companies manage the transponder requests of a variety of users, from telephone companies to real estate developers. However, broadcast and cable companies are the primary users of teleport services. In 1989 for example, two-thirds of the revenue of teleport companies came from broadcasting and cable firms. By the mid-1990s, the number of operating teleports may exceed fifty, and investments in this expanding segment of the media business are expected to exceed $1 billion.

Direct-broadcast satellite As better and higher powered satellites are developed and launched, it becomes feasible to beam TV signals to smaller receiving dishes perched atop houses and apartment complexes. This technology is known as *direct-broadcast satellite* service, or *DBS*. If such powerful satellites can be successfully launched, DBS has the potential to replace cable and broadcast TV as the primary means of delivering TV programming.

The development of DBS has been hindered by a number of factors. First, the recent decline in the performance of NASA has led to delays in the construction and launch of the higher-powered Ku band satellites required by DBS. DBS requires significant up-front capital (like the $140 million lost by Comcast in a failed DBS venture in the early 1980s). And there is the problem of competition with cable. Will cable programmers and superstation owners permit their signals to move to DBS satellites, possibly putting their cable partners out of business? Despite these significant hurdles, there are many believers in DBS, including well-known broadcaster Stanley Hubbard (see box).

Microwave relay A microwave relay link, or microwave for short, is usually located atop a very high tower or tall building. At the top of the tower are a number of boxlike structures, known as horns. The horns receive super-high-frequency signals and relay them to the next station.

Like coaxial cable, microwaves are capable of transmitting and receiving the complex signals involved in TV transmission. Microwaves have found a number of applications in the media industries.

Cable systems have used microwaves to import signals from outside their service area. Systems covering a wide geographic region use microwaves to feed their signals from the head-end to substations, called hubs, which become the connecting points for subscriber cable connections. More recently, TV news teams have used small microwave dishes on their remote trucks and vans to beam field reports back to the studio.

Limitations of microwave relay service are both technical and financial. On the technical side, since microwave relies on the line of sight, relay stations must be spaced no more than about 30 miles apart. Further, signals from satellites can interfere with terrestrial microwave systems, a problem that will worsen as DBS gains accep-

PROFILE

STANLEY S. HUBBARD: A BROADCAST BELIEVER IN SATELLITES

As we have seen, most broadcasters have viewed new technologies, like cable and satellites, as a pernicious enemy to be fought at all costs. However, one longtime broadcaster has embraced new technologies and is poised to profit from their increasing proliferation.

Stanley S. Hubbard was born into a broadcasting family. His father was an early broadcasting pioneer; his brother and other family members own and operate nine TV stations and two radio stations from their home city, Minneapolis, Minnesota.

Stan Hubbard was an early advocate of satellite technology for broadcasters. A pioneer in satellite news gathering, Hubbard's KSTP was one of the first local stations to use satellite trucks and related gear. As we have seen in the chapter on TV news, Hubbard owns and operates CONUS, a satellite newsfeed service for TV stations.

Today Hubbard remains a determined advocate of DBS. In 1982 he applied for five orbital slots for DBS. More than a decade later he remains committed to DBS development. In 1991 his United States Satellite Broadcasting (USSB) announced the purchase of transponder space on new satellites to be launched by Hughes. By mid-decade, Hubbard envisions a 150 to 180 channel DBS system, in which consumers would purchase a small satellite receiver (shaped like a square, not a dish), a decoder, and a converter, for about $700. To date Hubbard has committed over $100 million to the project. He expects to sink an additional $180 million to $200 million into the venture to see it come to fruition.

---

tance. In addition, the signal attenuates quickly and must be reamplified at each receiving station.

Multichannel, multipoint distribution service *Multichannel, multipoint distribution service (MMDS)* is short-range microwave transmission from a centrally located transmitter to a number of receiving dishes and tuners, called downconverters. In the early 1980s the FCC adopted rules allowing for two four-channel MMDS systems to be allocated to each major U.S. community. More than 16,000 applications were submitted for this new service. Four-channel MMDS may succeed as a means for transmitting movie and sports programming to urban areas where cable service has lagged. It may be used to deliver data services, like stock information or airline schedules. Some forecasters predict that MMDS may be an appropriate way to deliver the wider bandwidth required by HDTV.

Low-power TV and VHF drop-ins Satellites and microwaves denote big things: big transmitters, big power sources, big bucks. At the other end of the scale of new transmission methods are two modest means of sending TV signals: low-power TV and VHF drop-ins. *Low-power television (LPTV)* stations are local UHF or VHF facilities (about three-quarters of LPTV stations are UHFs), whose signals are concentrated in a small geographic area. An LPTV station operating at 10 watts on a VHF channel can generate a signal covering about 10 miles. The FCC began offering LPTV licenses in 1981, expecting the service to allow neighborhood and community access to the TV medium. The reality of LPTV service thus far has been disappointing. Most stations have tended to emulate independent TV stations, programming syndicated shows or movies to their communities. But they have lacked the economic base for effective operations, with their limited signal coverage hampering their ability to generate advertising revenues.

VHF drop-ins are full-powered stations "dropped in" by the FCC in areas where it ruled that new developments in transmitter design and receiving equipment made new signals feasible without interference to existing ones. As one might expect, there was a great hue and cry against the drop-ins by the broadcast industry, which feared new competition for advertising dollars and audiences. By the 1990s these fears had found little basis in fact. Only a few new stations had been dropped in, and these suffered the same kinds of financial problems that had befallen the fledgling LPTV industry.

Fiber Optics

You may have noted in recent times the fantastic improvement in the quality of telephone service. Long-distance calls sometimes sound as good as or better than calls to your friend next door. The reason is the development of a new type of cable for communication transmissions.

Fiber optic cable is a kind of "wireless wire." Instead of the copper and aluminum elements found in coaxial cable, fiber optic cable contains strands of flexible glass. A piece of fiber optic cable is seen in Figure 13-13.

Fiber optic technology makes use of digital communications. That is, the electrical signals normally transmitted through conventional wire are replaced by light pulses transmitted by a laser light source. These on–off (binary) pulses are decoded at the receiving source by a photodiode, a small light-sensitive device. The process is similar to the way CDs produce high-quality sound (described in Chapter 12).

Fiber optic cable has excellent potential in its applications to telecommunications. Glass is less expensive than aluminum and copper. Electrical interference is nonexistent since there is no real electricity in the wire. Optical fiber is thinner and more flexible than coax. Perhaps most important, the bandwidth is virtually limitless, unlike the relatively narrow band that fits into traditional coaxial cable. On the downside, amplifying fiber optic signals is hard to do, as is connecting and switching fiber optic transmissions.

Fiber optic technology is at the center of a struggle now emerging between the telephone and cable industries. Many regional telephone companies are replacing their copper wires with fiber optics. As regulatory barriers fall and the telephone companies are allowed to offer video services to the home, it's likely that the conversion to fiber will occur much more quickly. Cable companies are also experimenting with fiber. By 1990 twelve of the top-15 cable companies were installing fiber optics in at least part of their systems. The benefits of fiber over the traditional copper cable are significant: clearer video and audio, low maintenance costs, and a virtually unlimited capacity for video services. The huge amounts of signal capacity needed for HDTV, for example, are not a problem with fiber optics.

In the early 1990s the fiber optic market was a bright spot in an otherwise moribund telecommunications economy. Telephone companies and cable systems were racing to replace their trunk lines with fiber optic highways. Like the turtle and the hare, the telcos were way out in front. By 1992 the regional Bell operating companies had laid over 1.5 million miles of fiber optics. By

FIGURE 13-13 Fiber optic cable.

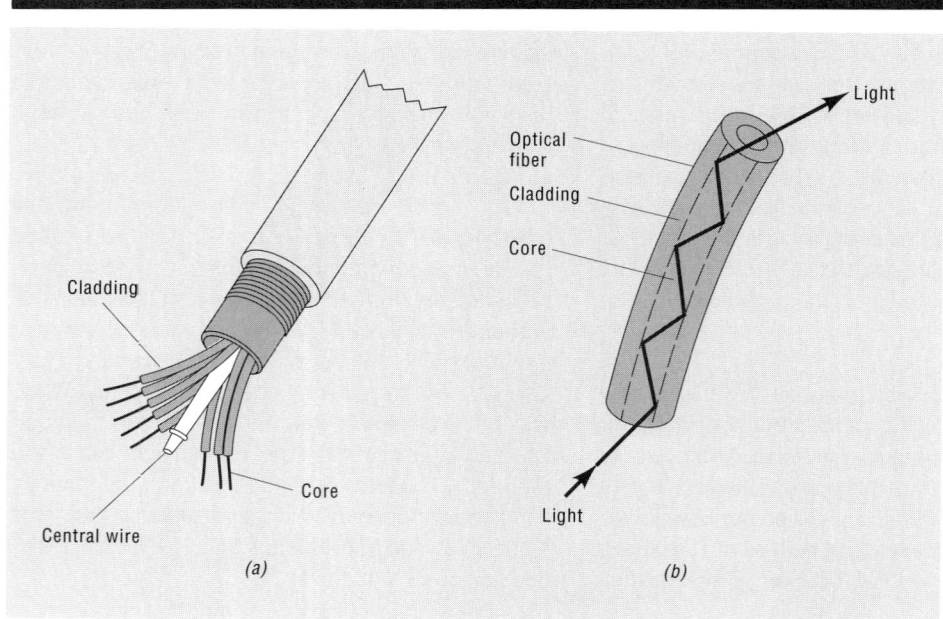

comparison, the leading cable multiple-system operators (MSOs) had rolled out only about 1000 miles of fiber.

Both the telephone and the cable companies still face one big problem: making the inside-the-home switch from copper wire to fiber optics. Such a conversion will take time and cost a lot of money. In any event, by the turn of the century either telephone companies or cable systems will have won the race to bring fiber optics to consumers.

Video Compression

In the 1990s a new digital technique was developed that may have as great an impact on signal distribution as the video toaster has had on video processing. Digital video compression allows for the complex TV signal to be digitally sampled (as a CD player "reads" a CD), stored, and squeezed so that it requires a much smaller bandwidth. Just as a trash compactor enables a lot of garbage to occupy a small space, video compression permits more signals to travel on the relatively narrow pathways now offered by TV channels, coaxial cable systems, and satellite transponders.

Compression has already had a significant impact on cable TV. As we mentioned in an earlier chapter, it is possible for cable programmers to multiplex a number of different versions of their services. Thus by 1992 HBO was offering three different movies each evening to some cable systems. MTV was planning to offer three or four different formats to its subscribing cable systems.

Local cable systems have also experimented with compression. The Time-Warner system in Queens, New York, has succeeded in squeezing 150 channels into its system, giving viewers an opportunity to choose from fifty different movies on a pay-per-view basis.

VIDEO RECEPTION: INSIDE THE TV SET

Let's examine how the scanning process is used to create color images on the TV screen. Basically the scanning process is reversed, in a technique known as retracing. In technical parlance, the TV picture tube is known as a *cathode ray tube,* or *CRT.* Like the pickup tube, the CRT is a glass-encased vacuum tube, with an electron gun at one end emitting cathode rays and an image-collecting screen at the other. The elements of the picture tube are illustrated in Figure 13-14.

The face of the CRT is coated by a phosphorescent material that shines when hit by the stream of electrons from the gun. Black and white sets have a single grayish silver phosphor coating: color sets feature nearly a million tiny dots arrayed on the screen in threes: red, green, and blue. The intensity of the light emitted by the CRT is proportional to the intensity of the electron stream from the gun. That is, a strong stream produces a bright light and a weak signal produces a soft light. The electron stream produced by the gun is controlled by the composite signal that has been sent from the pickup tube. Black and white sets have one electron gun; color sets have three, just like the cameras, each gun calibrated on its corresponding color dots.

The beam scans the face of the tube in exactly the same time sequence as in the pickup tube: thirty times per second, with phases of vertical and horizontal blanking, scanning a total of 525 lines. The synchronization pulse sent with the picture signal makes sure that scanning in the pickup tube is simultaneous with retracing in the picture tube. If all works as planned, the same time the camera sees Dan Rather in the newsroom we see his image as a *raster,* or display of glowing phosphor dots on our home screens.

Improvements in our media devices usually are confined to making them either grossly large or incredibly minute. For example, in radio consider the contrast between nearly piano-size boom boxes and personal stereos custom-fit to the human ear canal. These same developments in the TV medium head the list of new developments in TV reception.

Large-Screen TV

In the early 1970s wall-size TV screens began appearing in bars, nightclubs, and conference facilities. Today they are slowly creeping into the home video environment. In 1991, for example, nearly one-half a million such systems were sold.

Large-screen TVs come in two main types. Projection TV systems beam the red, green, and blue elements that comprise the color signal through special tubes to a reflecting screen suspended at a distance from the video unit. This is the type of system most commonly seen in sports bars.

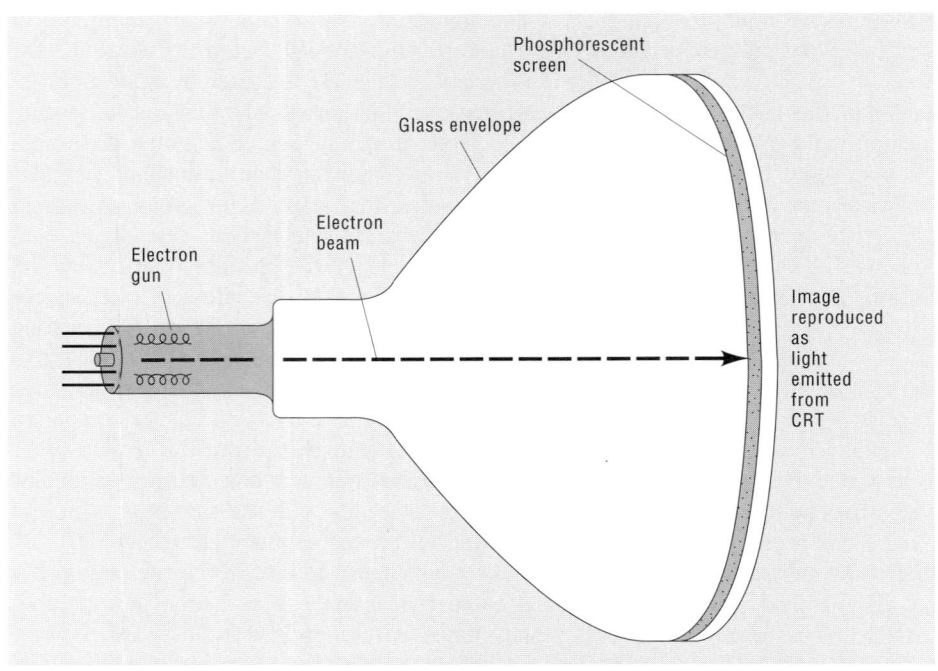

13-14 Inside the TV receiver. (Source: Alan Wurtzel and Stephen Acker, *Television Production*, New York: McGraw-Hill, 1989.)

Solid-state, large-screen systems do not have separate beam and screen components. These receivers make use of new developments in computer imaging, such as the charge-coupled device (CCD) discussed earlier. The picture tube is made up of a matrix of over 200,000 red, blue, and green light-emitting elements. Excellent color, crispness, and detail are achieved. In addition, the screen is flat, free from the type of distortion common in standard curved vacuum tube systems. Although expensive, solid-state, large-screen systems are beginning to move from sports stadiums to the home video market.

Small-Screen TV

For some TV enthusiasts smaller is better. These folks like to take the TV with them to the ball game, the beach, the office, or elsewhere. To satisfy this demand, manufacturers have introduced the visual equivalent of Sony's revolutionary Walkman radio design. Sony and its competitors, Casio and other Japanese manufacturers, have merged microprocessor technology and display expertise with years of experience in the production of transistor radios to produce smaller and smaller TV sets.

To date, a couple of major problems have hindered the diffusion of this technology. For one thing, the screens are so small, they tend to produce very low light levels. Hence they can get drowned out by bright sunlight (a big limitation if you're at a day baseball game or the beach.) Another fault is inherent in the medium. Unlike radio, a TV must be watched (at least some of the time) to be enjoyed. For obvious reaons this has limited the suitability of the Watchman to such proven portable radio activities as driving a car and jogging.

Digital TV

How many times have you tried to watch two or three TV programs simultaneously? Using a remote control or nimble fingers, you flip continuously from one to the other, playing a sort of TV musical chairs—and end up missing the important parts of both programs. Well, if you're an inveterate flipper (like the authors), the digital receiver is for you. The integrated microchips that make tubeless cameras and micro-TVs practical have made it possible for the TV screen to display more than one channel at a time. The technique is known as *picture-in-picture,* or PIP.

Digital TVs also perform other electronic tricks. Suppose, when watching a baseball game, you decide to try to steal the catcher's signs to the pitcher. *Zoom* allows you to fill the entire picture with just one small detail you would like to see, in this case the catcher's hands. To really get the sign, you can stop the picture momentarily, a technique known as *freeze frame.*

Like large-screen TV, digital receivers have thus far been out of reach of most consumers, with a price tag near $1000. Given the popular appetite for gadgetry, expect the cost to come down and these kinds of sets to become increasingly commonplace in the 1990s.

Remote-control devices are now in about three-fourths of all American homes and have significantly changed the way we watch TV. This new way of watching TV is called *grazing.* Viewers peacefully watch a TV program until a commercial or dull spot appears and then they use the remote control to scan quickly the other channels looking for greener pastures. A multipurpose remote control—one that operates both the TV and the VCR, for example—combined with a digital TV set offers viewers the highest degree of control over what they watch, even letting them monitor six channels at once. The TV set of the future will undoubtedly contain other features that accommodate this desire.

VIDEO STORAGE DEVICES

As we have seen, the TV signal is a complex combination of codes, including sound, picture, color, brightness, and synchronization. It should not be surprising, then, that the development of ways to store and retrieve TV signals has perplexed engineers since the beginnings of the medium. Until comparatively recently, there were two primary means of storing TV images for later or repeat broadcast: film and large-format videotape.

The Kinescope Recorder

Many older Americans look back fondly on the days when TV was "live." What happened in the studio appeared simultaneously on the home screen. The experience was immediate, radiating a sense of danger and excitement. More often than not something unexpected would happen: a "dead" character would come back to life when the actor thought the camera was off him or an easy-opening refrigerator door would refuse to budge, despite pulling, yanking, and tugging by an undaunted young model.

There was good reason for TV to be live: no means had yet been developed to preserve a TV program in sufficient fidelity for postproduction or later broadcast. Until the mid-1950s there was only one means to record TV programs and its performance was less than satisfactory. The *kinescope recorder* was a film camera especially equipped to shoot an image off the face of a TV screen.

The quality of a kinescope recording (or "kinny," as it became known) was not very high. For one thing, we know that the scanning rate of the TV tube is $\frac{1}{30}$ of a second. Yet the frame rate of a motion picture camera is $\frac{1}{24}$ of a second. With the kinescope out of sync with the picture tube, a visible bar (the scanning beam itself) often appeared on the recording. In addition, the frame size of the film was different from the shape of the picture tube, leading to recordings with heads, feet, and assorted other body and picture parts cut off. There was the delay involved from recording to replay, owing to the need for processing and drying time for the film. Finally, editing was virtually impossible: who would want to assemble a bunch of fuzzy images together? Because of these limitations, kinescope recording was never an adequate means of preserving TV signals. Its primary use was to make up for the three-hour time difference between the east and west coasts. After one play the kinny was usually discarded, which explains why so few of TV's great early programs can be seen today.

Magnetic Video Recording

In the early 1950s there was much excitement in the radio and recording industries over the development of magnetic tape recording. Magnetic

PRODUCTION

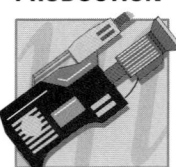

LIFE IN THE KINESCOPE DAYS

Before the VTR, video recording meant kinescope. There were so many problems with this process, however, it was no wonder that the networks eagerly welcomed a substitute. In addition to their poor quality, kinnies used expensive film stock. In 1954 American TV operations used more raw film for kinnies than all the Hollywood film studios combined. NBC alone used more than 1 million feet of film a month to feed programs to stations in the different time zones.

Kinescope recording was also troublesome and a little nerve-wracking. At CBS Television City in Hollywood recording started at 4:30 P.M. to pick up the shows broadcast live at 7:30 in the east. Engineers recorded a 35-millimeter kinescope along with a 16-millimeter backup copy. When the first thirty-four-minute kinny reel was done, a switch was made to a second kinescope machine. Then a courier grabbed the exposed reel and rushed it to a nearby film lab. Meanwhile, another courier took the 16-millimeter copy and, using a different route to minimize the chances that both reels would get caught in a traffic jam, rushed it to the same lab.

As the film came out of the dryers, it was spooled onto reels and packed into cans. The waiting couriers then rushed it back to the CBS projection room. These films were called "hot kinnies" because at air time they were still warm from the dryer. At the same time, another set of couriers was heading back to the lab with the second kinny reels to be developed. If traffic was bad or if the film lab had problems, things could get tense. Veteran engineers recall several times when they were threading up a reel only a minute or so before air time, and there were other occasions when they had to use the 16-millimeter backup copy. Despite these hardships, CBS never lost a show because of a kinescope processing problem. Nonetheless, everyone was relieved when magnetic recording was perfected.

---

sound recordings were high fidelity: except for a slight hiss, the quality of the recording matched the fidelity of the performance. Magnetic tape was erasable and therefore reusable. And the medium was relatively simple to edit, by either recording over and over again (overdubbing) or cutting and splicing tape pieces together.

Naturally there was great interest in applying this technology to the TV signal. But here's the problem. As we have seen, the TV waveform is 6 megahertz wide. Standard sound recording speed of 15 or $7\frac{1}{2}$ inches per second permits storage of up to 30 kilohertz of information. This frequency range is perfectly acceptable for audio but would provide room for only about one-twentieth of the information required to produce an acceptable TV signal.

The two-inch quad transverse video recorder Two things were done to increase the frequency response of magnetic tape. First, wide tape was employed. Instead of tape $\frac{1}{4}$ inch wide, video recording was standardized with tape 2 inches across. Second, the Ampex Corporation developed a revolutionary design for a mechanism that mounted up to four recording heads on a wheel, which spun constantly at right angles to the direction of the magnetic tape. The result was that the amount of information storable in one piece of videotape was increased. The 2-inch tape moved at 15 inches per second: a 12-inch reel could provide just over sixty minutes of TV recording.

Since the picture information was written at right angles across the tape, the method became known as *transverse recording*. And since four record heads were used, it is also known as *quadruplex*, or *quad*, *recording*. Finally, since 2-inch tape was used, the method is sometimes known as *2-inch recording*.

The *videotape recorder (VTR)* made its network debut with "The CBS Evening News with Douglas Edwards" on November 30, 1956. A new age of TV was born. "Live on tape" replaced "live." And reruns became practical for programs that TV itself, and not its rival, the film industry, had produced. Most important to the viewing public, there was no loss in fidelity between a live TV program and its VTR equivalent. For all practical purposes, live and live on tape looked and sounded the same.

Within a few years the color VTR was developed. Transistors replaced tubes, allowing for (slightly) smaller machines. While the process was painstaking and complex, a few intrepid souls (including innovative comedian Ernie Kovacs) were cutting and splicing videotape the way their colleagues in the radio business spliced audio tape. The 2-inch VTR had become the industry standard.

Helical-scan VTRs Even with transistorized components, 2-inch tape was hardly portable. Except for large-scale events like the baseball World Series or the Academy Awards, it was impractical to take the VTRs out of the station. And certainly, having a VTR in the home was out of the question. The next revolution in videotape recording was made by the Japanese. At the forefront of the emerging field of microelectronics, in the late 1960s Sony technicians perfected a means of storing the complex video signal on narrower tape, moving at slower speeds.

The idea was to stretch the magnetic tape around the revolving recording head so that there was virtually continuous contact between the tape and the recording head. For an analogy, imagine that instead of taking a single piece of chalk to one blackboard at the head of the class, your instructor had a piece of chalk in each hand and foot and was surrounded by a huge cylindrical blackboard.

This technique became known as *helical-scan* tape recording, since the heads are positioned obliquely to the tape and the lines of picture information produced on the tape form the shape of a helix. Various helical-scan tape configurations are illustrated in Figure 13-15. Some configurations wrap the tape completely around the heads, in the shape of the Greek letter alpha (α). Thus the pattern is known as *alpha wrap*. More commonly, the tape pattern represents the Greek letter omega (ω); hence omega wrap. To you and me, the shape looks like the letter U: enter U-matic recording.

The birth of a standard: Three-quarter-inch U-matic The helical-scan recorder had many advantages. Tape width shrank from 2 inches to ¾ inch, which had financial as well as technical advantages. The tape could be packaged in a protective case; thus the video-cassette became popular. Most important, the size of the tape recorder shrank. In the early 1970s it became practical to send cameras and recorders into the field for coverage of breaking news and live sporting events. Although a number of helical-scan recorders had been introduced, by 1975 the industry had standardized. The 2-inch quad gave way to the ¾-inch U-matic helical scan.

The ¾-inch VTR was soon a fixture in TV production. After its introduction, recorders became small, portable, and usually dependable. Editing from one machine to another became easy, especially with the development of computer-assisted editing equipment. And the entire field of broadcast news changed: the era of electronic news gathering, or ENG, was launched.

The final set of new developments in TV signal processing is perhaps the most important: the evolution of small-format or home video. In 1976 Sony introduced its first Betamax videocassette recorder (priced at $1200); today the VCR has

FIGURE 13-15 Helical-scan tape configuration.

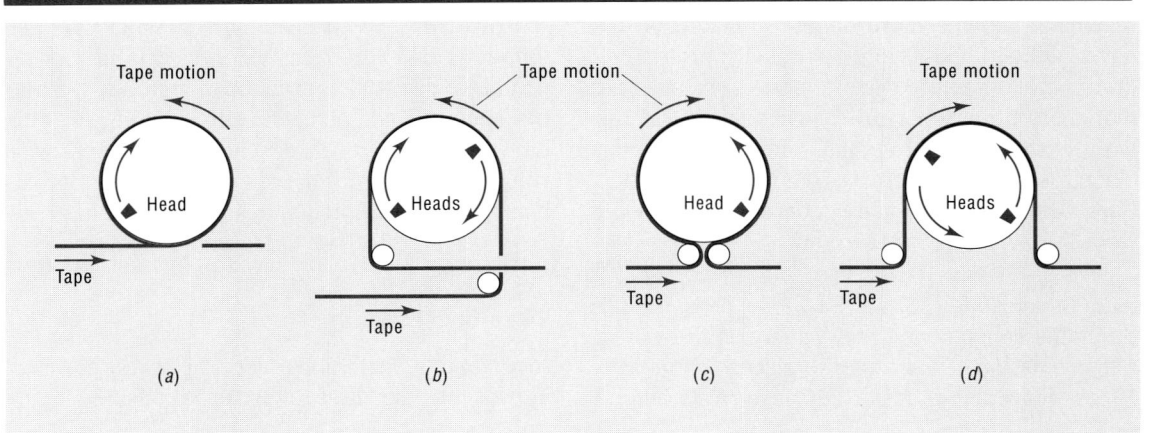

become the centerpiece of a revolution in the way we view TV.

Beta and VHS   The home videocassette recorder was made possible by the development of new means to store TV picture information, so that smaller and smaller (narrower and narrower) tape could be employed. The key was the development of high-density recording. Basically, the difference between broadcast-quality $\frac{3}{4}$-inch U-matic recording and the $\frac{1}{2}$-inch systems introduced in the late 1970s is that the new system eliminated the guard bands on the videotape, providing more room to store the picture signal on a reel of magnetic tape. Essentially, the $\frac{1}{2}$-inch machines sacrificed some picture quality for smaller, lighter, and cheaper recorders with longer recording times. Clearly, the compromise was worth it: in ten years more than one-half the households in the United States acquired VCRs.

Initially there were two competing varieties of $\frac{1}{2}$-inch helical-scan VCRs. Machines produced by Sony and its licensees used the Beta format; machines produced by other manufacturers used VHS tape and electrical components. For ten years, a debate raged about which system produced the better signal, with most technicians and industrial video users concluding that the Beta system had slightly superior performance. However, in the interim the TV consumer had decided on the standard. Since more machines and prerecorded tapes were available in the VHS format, by the mid-1980s the home market had more or less standardized in the VHS format. Even Sony began producing VHS machines.

The Japanese zeal for technological development and the American appetite for gadgets hastened the introduction of a number of features that have become more or less standard on today's VCRs. These range from elaborate programmable timers, special effects from slow motion to stop action, the ability to add words with text or titling programs, and, of course, the ubiquitous wireless remote. This nifty device allows the viewer complete control of the VCR from the sofa across the room.

Beta-SP versus S-VHS   If Sony lost the battle for the home consumer, it may have won the war in the professional video marketplace. In the late 1980s, Sony and its major Japanese rivals (led by Matsushita and JVC) each introduced high-quality equipment in small helical formats. Both Sony's Beta-SP and its rivals' Super-VHS offered broadcast-quality images to stations, networks, and production companies in the smaller packages that are common to consumers. Bulky field cameras (in the old $\frac{3}{4}$-inch format) could be replaced with lightweight camcorders. Editors, special effect generators, and other equipment could be smaller and cheaper, and perform as well or better than the U-matic equipment on hand since the mid-1970s.

After a short battle by the early 1990s it appeared that more professional video shops (including most TV news operations) had chosen Beta-SP over S-VHS. It wasn't long before Sony added another option to its list.

VHS-C, Video-8, and Hi-8   One problem common to Beta and VHS camcorders was the size of the tape cassette. No matter how much the camcorder's electronics were shrunk, the machine still had to hold the comparatively bulky tape cassette. One solution was JVC's compact VHS cassette. Roughly one-half the size of a standard VHS tape, the VHS-C compact cassette can be played back on any standard VHS machine by using a special adaptor. VHS-C can record twenty minutes at the highest speed (SP), up to an hour at its low speed (SLP).

In response, Sony introduced a new tape format. Using tape only 8 millimeters wide, Video 8 can record up to eight hours of broadcast quality images, with camcorders about the same size as snapshot cameras. One limitation of Video 8 is that the tapes cannot be played on a home VHS machine; a special cable connects the camcorder to the VCR or TV set.

However, a high-resolution version of Video-8, known as Hi-8, produces video images as good as or better than those produced by studio U-matic machines. In addition, Sony has found a way to embed both stereo audio and editing time-code tracks into this very narrow format. Thus, today many TV newsrooms and production companies have moved into Hi-8 video in a big way.

Video Disc Players

In 1979 the North American Philips Corporation introduced the first affordable consumer videodisc player (cost: $775). Using a low-powered laser to read microscopic pits embedded in a plastic and aluminum disc, the system featured marvelous visual and audio clarity. Unlike tape, the discs could last indefinitely. Storage capacity was excellent. The potential for special effects seemed limitless. Yet, thus far the videodisc player has flopped as a consumer electronics device, whereas its audio counterpart, the CD player, has been an astonishing success. Why?

For one thing, consumers seem to desire the ability to record as well as play back TV programs. Thus far home videodisc systems have not had recording capability. In addition, it took too many years for a standard to develop. At least five different disc formats have been introduced, each with its own special-sized discs and technical-sounding acronyms, including CLV, CAV, CLD, and CD-V. The ensuing confusion among consumers and the lack of recording capability limited the diffusion of laser videodiscs. They have tended to be most widely employed by libraries and media centers, which use their excellent fidelity and wide storage capacity to archive instructional and classic programs in a virtually permanent form. Some manufacturers, including Pioneer and Magnavox, believe that the success of the audio CD player may help spread the popularity of videodiscs. They have introduced machines that can play all format of videodiscs as well as standard audio CDs. (See Chapter 20.)

The Marriage of Television Set and Computer

The present decade is marked by the merger of home video devices, including the TV set, VCR, and computer. Digital technology now makes it possible for sound, words, and pictures to be stored and displayed by the same device: the common home computer.

We have already discussed one implication of this development: the "video toaster." Consumers with "high end" (more expensive) home computers can produce elaborate special effects that rival the best of MTV. Thus far only those interested in video graphics (like TV commercial producers, newsrooms, and production houses) have experimented to a great extent with this technology.

However, computer manufacturers have introduced a new line of low-cost options for the personal computer marketplace. A "video card" can be installed in any small computer with sufficient memory and a color screen. The video card enables the computer to work like a standard TV set. Plug the card into your antenna or cable, and you can watch TV in a small window while typing away at a term paper. Like any computer file, the images on the TV can be captured and stored on a hard drive or floppy disc.

The Apple company now markets many of its machines with a technology known as Hypercard. Using Hypercard, full multimedia is possible from a MacIntosh. (See Chapter 20.) Stereo sounds, words, and color TV images can be manipulated simultaneously. For example, a Hypercard file on "The Beatles" might begin with a display of the cover of the "Sgt. Pepper" album. By clicking the mouse on any figure, the computer can play a song, reveal a personal biography, play a segment from a news report, roll a music video, and so on.

Thus in the 1990s we are moving from the "tubeless camera: to the "tapeless VCR." All information is being encoded into digital form. In the end, we are all bits and bytes.

SUMMARY

Television is possible because of two characteristics of our perceptual system—persistence of vision and the phi phenomenon—which allow us to perceive motion in a series of rapidly changing still images. Like radio, TV uses the facsimile concept and the principle of transduction of energy.

The first medium to transduce moving images using a mechanical technique was the motion picture. TV uses electronic transduction to achieve the same goal.

Television is based on the notion of scanning. An electron beam scans every element of a picture and the image is then retraced on the TV receiver. Multichannel TV sound provides for stereo or dual-language TV. High-definition TV adds more scanning lines to improve picture clarity and impact.

Before it is transmitted to the home set, the video signal is amplified and mixed with other video inputs using a device known as a switcher. Digital processing of TV signals permits the generation of a large number of special effects.

The traditional systems of transmitting a TV signal are (1) over-the-air broadcasting utilizing electromagnetic radiation and (2) by wire through a cable system. Over-the-air TV channels are located in the VHF and UHF portions of the spectrum. Cable systems use coaxial cable that permits them to carry more than sixty channels of programming. Cable signals are received at the headend of the system and are then sent through trunk cables and feeder cables to a subscriber's home.

Satellites have helped the growth of cable TV. Geosynchronous satellites orbit the Earth in a fixed position that allows them to relay communication signals. Direct broadcast satellites may one day provide original programming directly to the home. Fiber optics may replace coaxial cable as the preferred means of transmitting a video signal through a cable system.

Digital video compression allows for video multiplexing in congested bandwidths, such as satellite transponders and coaxial cable systems.

Inside the TV set, the electron beam rapidly scans lines of phosphors that glow momentarily, thus creating the illusion of motion. Digital TV receivers are capable of many special effects.

Video signals were once recorded on a kinescope, a film of a TV screen, but are now stored on magnetic videotape. The development of $\frac{1}{2}$-inch videotape has helped popularize the videocassette recorder and home video recording.

New developments in video storage include Hi-8 mm tape and video cards for personal computers.

SUGGESTIONS FOR FURTHER READING

Baldwin, T. F., & McVoy, D. S. (1987). *Cable communication.* Englewood Cliffs, N.J.: Prentice-Hall.

Benson, K. B. (1991). *HDTV: Advanced television for the 1990s.* New York: Intertext/McGraw-Hill.

Besen, S., & Johnson, L. (1986). *Compatibility standards, competition and innovation in the broadcasting industry.* Santa Monica, Calif.: Rand.

Harrell, B. (1985). *The cable television technical handbook.* Dedham, Mass.: Artech House.

Ingram, D. (1983). *Video electronics technology.* Blue Ridge Summit, Pa.: Tab Books.

Luther, A. C. (1989). *Digital video in the PC environment.* New York: Intertext (McGraw-Hill).

Michel, S. L. (1989). *Hypercard: The complete reference.* Berkeley, Calif.: Osborne/McGraw-Hill.

Overman, M. (1977). *Understanding sound, video, and film recording.* Blue Ridge Summit, Pa.: Tab Books.

Prentiss, S. (1990). *HDTV: High-definition television.* Blue Ridge Summit, Pa.: Tab Books.

Rainger, P., Gregory, D., Harvey, R. V., & Jennings, A. (1985). *Satellite broadcasting.* New York: Wiley.

Smith, W. A. (1983). *Video fundamentals.* Englewood Cliffs, N.J.: Prentice-Hall.

Wurtzel, A. and Rosenbaum, J. (1993). *Television production* (4th ed.). New York: McGraw-Hill.

PART SIX

WHAT IT DOES:
AUDIENCE IMPACT AND EFFECTS

CHAPTER 14: THE RATINGS: ESTIMATING AUDIENCES

The business of broadcasting/cable is numbers—that is, numbers as debits and credits on ledgers to measure profit and loss, numbers on rate cards to tell advertisers the cost of airing a spot, numbers of people who tune to the station or who listen or view the program. All of these numbers serve as important bits of data to broadcasting stations, cable channels, and advertisers. To understand the broadcasting/cable business—the forces that drive all the other elements of the industry—a person must have a working knowledge of numbers.

You will need to know how to read "the book," industry slang for the latest ratings report, and to be able to interpret the pages of numbers found there so that you can understand how your station or channel and its programming are performing. One of the many complaints heard from professionals is that broadcasting majors who start in the industry can't read the book.

This chapter discusses how to measure and evaluate what people are listening to or viewing; this process is usually called "the ratings" but is more appropriately thought of as one type of audience research. When broadcasters and cable executives discuss research, they usually mean ratings research. This chapter covers how ratings and rating companies developed in the United States. The first part of the chapter should be easy; it is a standard historical review. The second part of the chapter deals with how ratings information is collected, analyzed, and reported. This kind of discussion must include statistics. Assuming that you had at least basic algebra in high school, you should be able to understand the general statistical concepts that are introduced in this chapter and in Chapter 15. We have kept the formulas to a minimum.

Some of you might find this and the next chapter a "tough go," but a radio–TV major needs to know more marketing and audience research techniques than a business major. Most broadcasting/cable majors have aspirations to move into management. This means having the specific knowledge of the industry while also having a grasp of general business practice.

USES FOR RATINGS INFORMATION

The information found in an audience ratings report is an invaluable tool to both the broadcaster and the advertisers. Both are interested in more than just how many people are watching or listening. The various types of information contained in the audience ratings report can be analyzed to make better programming decisions, or the information contained in the report can help a client select the placement for commercials that supplies the best exposure to the selected audience.

Before discussing the history and development of audience estimates, let's examine some of the uses for applying information from ratings services. Ratings are a marketing tool, but they also act as a means for a station to ascertain how well it is doing as compared to the competition. Ratings can provide self-correcting information for a station that does not provide what a particular audience wants.

Audience estimates in a ratings report can be used by stations to determine how they are doing in terms of their total audience. Are more people watching more TV or listening to more radio? The larger the number of people using their sets, the larger the pool of people to sell to advertisers.

These estimates also allow one to determine when people and what type of people are watching. Although advertisers are usually interested in large audiences, many of them are interested in specific kinds of audiences. A show like "St. Elsewhere" survived for five years not because it had a large audience, but because it had the right kind of audience—young urban professionals with a healthy amount of disposable income, which certain advertisers wanted to help them dispose of. Similarly, a local program that has a high proportion of women viewers would make it a good place to advertise products or services that cater to women.

Audience estimates help the station judge its reach—just how many different people view or listen to the show during a week. Some shows with relatively small audiences may be viewed by many different people during a week. A commercial shown at the same time for a week may not be viewed or heard by a large audience at one time, but considered over the week the reach of the program creates a large number of exposures.

News departments can use audience estimates to see how much viewing is occurring in adjacent communities. If there seems to be considerable viewing in neighboring regions, the station's news operations may want to include stories from those areas to encourage continued viewing and better ratings for the news.

Audience estimates of network programming are often used to compare local programming performance with that of the network's programming. Local, not network, programming is the chief contributor to a station's revenues. If local programming is weak, new shows may be needed or a shake-up of the schedule may be in order. Audience estimates during network programming are also used to gauge how well the networks are doing in a particular market. The price of commercial time during network breaks is affected by how well the network is doing in a market.

Average estimated viewing/listening times are advertising selling tools. Station account executives—the sales force of a station—can use these figures to counsel advertisers as to whether a rotation schedule for commercials—spots appearing at different times—or a fixed position—a spot scheduled at the same break every day—is best for the desired audience. The station break average audience provides a good estimate of how many people can be delivered to an advertiser. This knowledge helps the client plan the purchase of time, which is really a purchase of consumers.

Ratings services provide historical audience estimates that usually cover a one-year period. Seasonal trends in audience viewing help the station estimate how well it will do between this survey and the next, thus allowing the station to adjust its rate card to reflect the estimated audience that can be delivered.

These seasonal data can also be used by the programmer to make judgments of how well a particular program is doing. If the total audience remains constant and a program on the station's schedule is losing viewers, changes may be necessary. However, a shrinking audience for a program matched to a shrinking of the total audience may simply reflect seasonal changes, and tinkering with the schedule may be unnecessary.

Program audience estimates show how a particular program is doing against its competition. Here the programmer can make decisions about whether the program is delivering the audience that it should. Remember, it isn't necessary always to have the largest rating to make a profit, but no one would turn down the largest rating. If one station is showing the soap opera "The Bold and the Beautiful" while a competing station is showing "Tiny Toons," the overall rating may be less important than whether or not the target audience is viewing. For example, the percentage of all people with TV sets in their homes who watch "Tiny Toons" might be 10 percent and for "The Bold and the Beautiful," 8 percent. Among women 18 to 49 years old, however, "The Bold and the Beautiful" might be getting 15 percent of those who have TVs in their homes while "Tiny Toons" gets less than 1 percent. This would be a significant fact for advertisers to know. In addition, the flow of the audience or segments of the audience from program to program can be determined from this section.

For the advertiser, audience estimates for specific programs may be used to estimate the average audience for rotated commercials—spots that appear at different breaks in a given program—over a week's viewing period. This is helpful to the client as it gives an estimate of the exposure received.

The various ways listed here suggesting how audience estimates can be used should suggest to you that the ratings process is a handy tool. Ratings provide data that advertisers, broadcasters, and cablecasters may use to help make decisions; however, good programming executives do not make their decisions solely on what a single ratings report says. Remember, ratings are *estimates* of audiences subject to the standard fluctuations one finds in any sampling procedure.

Ratings companies try hard to provide accurate and reliable information for their clients. Ratings companies are constantly trying to improve their product; they have to balance cost versus accuracy. Larger sample sizes could provide estimates with smaller error, but most clients don't want to pay for the larger samples.

EARLY ATTEMPTS AT AUDIENCE MEASUREMENT

Mail was the earliest form of audience measurement. When radio stations were just beginning, listeners were encouraged by station owners to send letters mentioning that they heard the station. Receiving a letter from a listener in Green Bay, Wisconsin, might evoke comments from the people at a Corvallis, Oregon, radio station, but it couldn't tell much about the number of people listening.

Fan mail for a program was another early audience measurement device. The letters that came to a show or a star were tabulated as a rough gauge of the number of people who were actually listening. Fan mail is wildly inaccurate because some shows and personalities evoke a greater response from the public than others. "Twin Peaks" elicited a huge volume of fan mail, but it wasn't the most watched TV program. The people who write fan letters tend not to be your typical broadcasting consumer.

Another means for guessing at the popularity of a radio program was to offer some prize or gift to those who wrote in. Decoder rings were not so much a means to transmit secret information to faithful listeners as they were a way to guess at the number of people who were listening. Giveaways were used as an inducement to have people write and the number of letters received was used as a rough estimate of the listenership. This provided a better approximation of the number of listeners than listener reports or fan mail because the number of people responding was higher. Obviously, this form of audience measurement suffered from many problems.

The information that the station received from these early listeners was of some use—that is, there was someone listening—but it wasn't very useful in estimating the actual audience. Although the station knew how many letters were received, that number couldn't be translated with any reliability into the approximate number of listeners.

Some stations used responses to giveaways as a crude form of determining where the audience was located. Local stations would employ the number of responses from their home city or county to compare to responses from other counties and cities. The volume of fan mail from a city, say, 200 letters, would be compared to letters from other areas, say, 60 letters from rural areas and 180 letters from an adjoining city. The local mail would be assigned a value of 100, with the number of local letters divided into the number of letters received from other areas resulting in a percentage of listenership that could be compared to the home audience; in this case a 30 would be assigned to the rural area and a 90 to the other city. This would generally indicate where the listeners were located.

Radio stations sometimes used other equally crude methods to guess at the listenership. One of the first methods used was to get a map and draw a circle with a 100-mile radius. The station would determine how many people lived within the circle and then announce that figure as the number of listeners. Obviously, any relationship between the number obtained in this fashion and the number of actual listeners was purely coincidental.

Another technique used by the radio networks to compare coverage was to add together the power of all their affiliate stations. A network could then boast that it had so many kilowatts transmitting its signal across America. It was, however, another meaningless number because two 5-kilowatt stations in two different localities would have different coverage areas.

A more exact method was needed to gauge the size of the audience. Advertisers demanded it; they wanted to know if they were getting good coverage for their advertising dollar. Stations wanted to know so that they could make programming decisions and thus could ask for higher rates, that is, more money, for those shows to which more people listened.

RATING SERVICES

The first national radio survey was conducted in 1927. It was paid for by a baking powder manufacturer in an attempt to discover how many people

A census taker at work. It takes a lot of time, money, and energy to do a census. That's why ratings companies typically rely on samples.

heard his network advertisements. Not so many were listening as he had hoped. The survey discovered that a number of stations substituted commercials from local sponsors for network commercials so that the local stations could make more dough.

Cooperative Analysis of Broadcasting

Archibald M. Crossley pioneered the techniques for measuring the size of an audience for particular programs. He undertook his first national survey in 1929. Using randomly selected numbers from telephone directories, he conducted telephone interviews asking people to recall the radio programs that they had listened to during the previous night.

The American Association of Advertising Agencies (AAAA, called "the four A's") and the Association of National Advertisers saw the utility in Crossley's study and formed the Cooperative Analysis of Broadcasting (CAB) in 1930. Crossley was hired by the CAB to continue to produce listener reports. Established as a nonprofit organization by and for advertisers, the CAB was designed to discover how many people listened to programs that were offered by the radio networks. The reports became known as *Crossley Ratings.*

The CAB established offices in thirty-three cities across the United States using the method Crossley had previously pioneered—placing telephone calls to randomly selected numbers taken from the telephone book. The *recall method*—asking the participant to remember previous listening—continued to be used but with refinements. Instead of being asked to recall what was heard last night, calls were made in four-hour cycles and the listener would be asked what shows she or he had heard in the last three or four hours.

One of the early findings by the CAB was that the most popular hours for listening to the radio occurred from 7:00 to 11:00 P.M. In particular, the peak usage hour was from 9:00 to 10:00 P.M. This discovery led to what would be called *prime time.*

Crossley and the CAB attempted to find a more economical way of providing audience measurement. As a cost-saving experiment the CAB attempted to use a system employing postcards. The attempt failed when the number of postcards returned was under 6 percent of all those sent out—a number too small to be reliable. If the experiment had worked, it could have meant a reduction by about 800 percent from the cost of telephone surveys. The cost of each telephone interview was estimated at about 40 cents. It seems cheap now, but remember, in the 1930s one could go to the diner and get dinner for 50 cents and movies were only a dime. The postcard method was abandoned and until its demise the CAB continued to use the telephone interview.

The Crossley ratings were for sponsored network programs only. Those programs that had no sponsor, called *sustaining programs,* were not reported by the CAB, which was primarily paid for by advertisers. Ratings were published every other week based on 3000 daily calls.

There were two major problems with the Crossley ratings. First, they used the telephone and telephone directories. In the early 1930s only about 50 percent of homes had telephones. The telephone directories used by the CAB were for the thirty-three cities in which the CAB had offices, and fourteen of these offices were located on the east coast. If you didn't live in one of these cities, or if you had an unlisted number, you had no chance to be included in the Crossley ratings.

The second problem was the use of the recall method. Although requesting recall of what a person had heard in the last few hours is more accurate than asking what that person had heard the previous night, human memory (as you know when it's test time) is prone to failure.

The CAB continued to operate until 1946. It fell victim to competition from another organization, which had a superior way of measuring the number of listeners. The CAB attempted to adapt by changing its method to match its competitor, Hooperatings, but to no avail.

C. E. Hooper, Inc.

Clark-Hooper, Inc. was an audience measurement firm that began by doing studies of magazine and newspaper advertising effectiveness. In 1934 the company inaugurated its first radio audience survey. The company would split into two separate companies in 1938 with C. E. Hooper, Inc. continuing the audience measurement surveys. The results of these surveys became known as *Hooperatings*.

Hooper's ratings were different from those produced by the CAB. Whereas the CAB was primarily a venture by advertisers, Hooper viewed both advertiser and broadcaster as his clients. Hooperatings were available not only for network programming but also for local programs, an area that had been generally ignored by Crossley because of its sponsorship by national advertising firms.

The methods that Hooper employed in obtaining his measurements also differed from those used by Crossley and the CAB. The Hooper organization worked out of thirty-two cities. These cities were called *areas of equal opportunity*, meaning that stations representing all networks were present in the locality. This provided that each network had an equal opportunity to be heard. The absence of one or more of the networks from a sampling area would bias the results.

Hooper introduced the *coincidental* telephone interview. Respondents were not asked to recall what they had listened to in the past but, rather, to tell the interviewer what they were listening to at the time of the call. The coincidental method is superior to the recall method, as it does not rely on the often fallible memory of the listener.

Since coincidental telephone interviews asked what the listener was listening to at the time of the call, sample sizes were not so large per program as were the CAB's. A new series of telephone calls began every quarter of an hour. Some sample sizes for a given program could be as small as ninety respondents.

The five questions asked of respondents by the Hooper organization were published by C. E. Hooper in his book *Radio Audience Measurement*. The questions were:

1 Were you listening to the radio just now?
2 To what programs were you listening, please?
3 Over what station is that program coming?
4 What advertiser puts on that program?
5 Please tell me how many men, women, and children, including yourself, were listening to the radio when the telephone rang?

The answers to these questions formed the basis for the biweekly Hooperatings. These ratings, like Crossley's, indicated the number of radios that were turned on, or *sets in use;* the percentage of people listening to a specific program as compared to all the radio sets owned, or *rating;* and the percentage of people listening to the show as compared to everyone who was using the radio at the time, or *share*. These concepts are discussed more thoroughly later in the chapter.

The Hooper method suffered from problems that are both common with Crossley and unique: multiple set listening in the home, listening while in the car, small sample size, no rural samples, and a problem with almost all audience research—some people lie. Some interviewers reported that people said they were listening to a particular program but the interviewer could hear another program over the telephone.

The Hooperatings suffered the death blow when TV was introduced into the cities where Hooper was making measurements. The way in which the organization calculated its statistics could not deal with TV viewing and radio listening accurately. Finally, Hooper, Inc. would sell its national sampling to its chief competitor, the A. C. Nielsen, Inc. in 1950.

A. C. Nielsen, Inc.

The best known of all the ratings are those produced by A. C. Nielsen, Inc.—specifically, Nielsen's ratings of network programming. Everyone from your local newspaper to "Entertainment Tonight" reports how network TV programs are doing in the Nielsen ratings. "Nielsen" has become synonymous with "ratings" for many.

Audimeter There are inventions that change the way people live: the light bulb, indoor plumbing, the radio. A. C. Nielsen's invention of the *audimeter,* a mechanical device used to measure listenership, changed the way audience ratings were gathered. The original audimeter was invented by two Massachusetts Institute of Technology professors, Robert F. Elder and Louis Woodruff, in 1936. The audimeter collects data in a different manner from the survey methods employed by its two former competitors. Crossley and Hooper measured people; Nielsen's audimeter measures receiving sets. The audimeter is a device that is attached to the receiving set and keeps track, first using paper tape, later using 16-millimeter film, of the station to which the radio or TV is tuned.

Whereas the previous companies measured listening, Nielsen measured set use. Both Crossley and Hooper asked people about their listening. The audimeter could determine only when the set was on and to what station the set was tuned. The audimeter could not tell if anybody was actually listening when the set was on.

The Nielsen Radio Index was first released to commercial clients in 1942. This was preceded by a number of years of testing and experimentation. The shortage of resources and the effort to win World War II restricted the growth of the Nielsen system until the end of the 1940s. Released from war-time constraints, the Nielsen numbers became paramount, and by 1950 Nielsen's major national competitor, Hooperatings, had sold its national ratings service to Nielsen.

By the early 1950s the audimeter had moved to TV. The device was attached to every set in the selected home. Even portable sets had an audimeter attached, and this provided a record of TV use for an entire home. Once they were installed, the early audimeters were designed to be maintained by the families. About once a week a new recording medium—a strip of film—would arrive in the mail. A member of a "Nielsen family" would take out the old film and insert the new one. The audimeter would then reward the person by dispensing two quarters. The used film would be mailed to the Nielsen company for analysis.

The audimeter has been in a continual process of refinement. Where paper tape and photographic film were once used, magnetic media were substituted. Next, the *storage instantaneous audimeter* (SIA) replaced the old audimeters. This device sent information by telephone lines directly to Nielsen's computers. More recently Nielsen switched from the SIA to the new *Peoplemeter,* a hand-held device that resembles a TV remote control. These devices not only record what shows are being watched when, but also *who* is watching. Peoplemeters are discussed in greater detail later in the chapter.

The old audimeter and the current Peoplemeter note both the station and the time. This measurement can be used for relatively simple analyses, such as what programs are most watched and more complicated viewing behavior such as audience flow. The information can also be used to measure what people don't watch. Channel flipping during commercials can be noted with Peoplemeter information.

When the Nielsen Radio Index was first issued, an attempt was made to sample not just people who lived in cities or who had telephones but a more diverse and representative group of people as well. Nielsen would later claim that homes of all types were represented within their sample. However, college dormitories, hospitals, bars, and the like are not classified as "homes" and are excluded from the sample.

Because of complaints that the Nielsen numbers represented only sets in use and didn't indicate *who* was watching, Nielsen introduced the use of *diaries* in 1955. These diaries were to be filled out by listing the show viewed and who was watching. Diaries were dropped from the Nielsen national sample in 1987 with the introduction of Peoplemeters. More on diaries will follow in the discussion of Arbitron ratings.

Following its predecessors, the Nielsen Radio Index was published every two weeks, as was the original Television Rating Index. Nielsen continues to publish its *Nielsen Television Index* of national audience viewing every two weeks. Nielsen also provides national "overnight" ratings based on viewing behavior in Los Angeles, Chicago, New York, and other cities. These overnights are a rough gauge of how a program did the previous night. However, because the sample is small and urban, it is not so accurate as the weekly reports. The *Nielsen Station Index* is published four times a year and reports on local market viewing.

Peoplemeters The biggest news in TV in 1987 and 1988 was not about the programs but, rather, the way in which the audience was estimated. In September 1987 Nielsen introduced Peoplemeters for use in producing national ratings in response to the introduction of Peoplemeters by AGB Television Research, a company that soon went out of business. Peoplemeters attempt to bypass the problem of the audimeter and its cousins, which record when and to what channel the TV is tuned but can't tell who, if anyone, is watching.

Peoplemeters still are connected to the TV set and automatically note the channel and time, but there is also a remote control, which serves as the heart of the change. Each person in the family is assigned a number. People entering the room to watch TV are supposed to punch their number on the remote control and then punch it again when the person leaves the room. The Peoplemeter box attached to the TV then knows who is watching the program. All this information is sent via telephone lines to central computers, which have stored detailed demographic information about each of the viewers as well as programming information. This demographic information is obtained through personal in-depth interviews with each family member at the time of the placement of the Peoplemeter in the home.

Nielsen began by using 2000 homes as its sample and then increased the number to about 4500. This increase in sample size was an improvement over the older SIA sample size of 1150 homes. In September 1991 A. C. Nielsen changed its estimate of the size of the population that it was sampling. The total number of U.S. households estimated to have TVs was reduced from 93.1 million to 92.1 million. The company said that this reduction was based on the results of the 1990 U.S. census.

The actual identities of Nielsen families are supposed to be a closely guarded secret. If a broadcaster or producer knew who these families were, it would be possible to manipulate the rat-

The Peoplemeter revolutionized television audience measurement. Each family member is assigned a code number that is entered into the handset at right while watching TV. Note also that the Peoplemeter has code numbers for visitors as well.

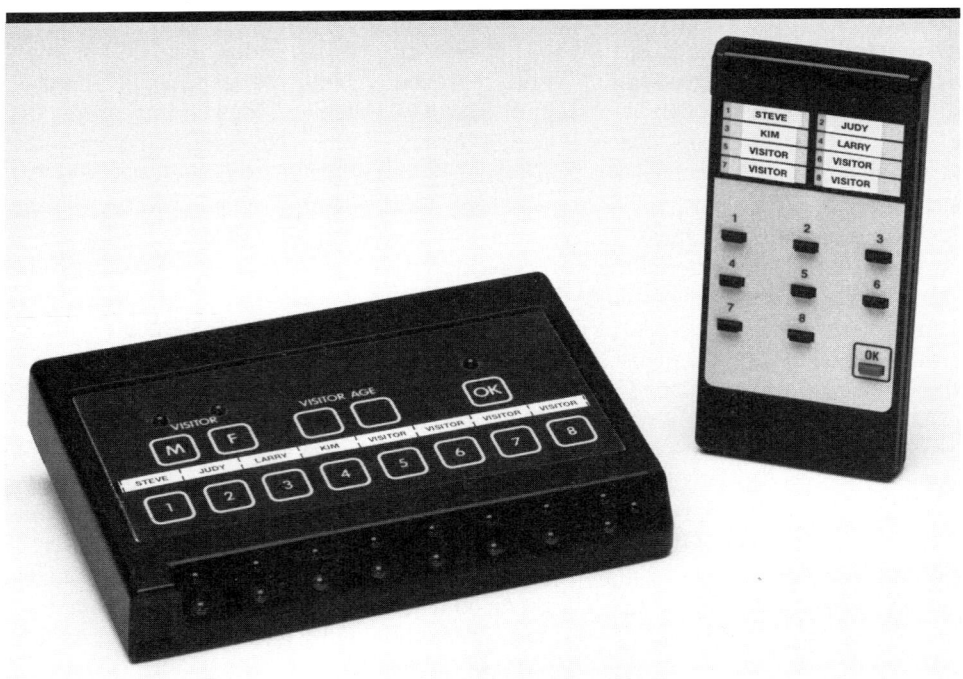

ECONOMICS

HOW TO RIG THE RATINGS

In an advertisement headlined, "This Is a Bribe" KELI, Tulsa, offered to pay $14.30 (their frequency on the AM dial) to anyone who had an Arbitron diary and would fill it out saying that he or she listened to KELI all the time. The person needed only to take the diary to the radio station's office and KELI would give the person a check and even mail the diary for the person.

The reason for such an approach? The ad states, "The problem is that these decisions are made on the listening habits of approximately 900 people. Think about that. The 900 people set the advertising rate and radio programming for a city of over 650,000. . . . Remember here is your chance to get $14.30 free and our chance to prove a point for *justice*."

Although this approach has the advantage of being direct, it was appreciated by neither KELI's competitors, who immediately complained, nor by Arbitron. When it comes to playing with potential ad revenues, the broadcasting industry does not have a very good sense of humor.

---

ings. So Nielsen has attempted to protect, with varying degrees of success, the members of its sample.

The Peoplemeter has a number of advantages. First, it isn't subject to the whims of memory. The machine always remembers, short of a technical breakdown. Second, it monitors set use around the clock. Many time periods were underrepresented in previous ratings because it was difficult to find a respondent who would cooperate when a telephone rang at 1:00 A.M. to ask what show the person was watching. In addition, as long as people remember to enter their codes, Nielsen even knows who is watching.

One of the problems with the Peoplemeter relates to the validity of the responses. A Nielsen family knows that its viewing behavior is being monitored and there is good evidence that people who know they are being observed don't act the way they would normally. These aberrations in behavior probably go away after a period of time, but it is one concern.

Another problem is that the system is expensive to operate. The cost of installing and maintaining a sophisticated technical system is high. This means that changing homes—members of the sample—is difficult for Peoplemeter homes.

Out-of-home viewing Peoplemeters do not reliably measure certain segments of the audience. Coincidental telephone studies indicate that only about one-half of the people visiting and watching TV in a Nielsen home are accounted for.

As mentioned earlier, Nielsen measures home viewing. People who visit someone in a home are supposed to be represented in the viewing numbers. People who watch TV not in a home but, for example, in a dorm or a bar, are usually not accounted for in the Nielsen numbers. Nielsen's own study that was released in 1990 indicates that 2 percent of all TV watching is done outside the home; that is, about 6.3 million people are excluded from the Nielsen estimates. The preponderance of out-of-home viewing takes place at work (33 percent), colleges (21 percent), hotels (17 percent), and bars and restaurants (9 percent). The most popular prime-time network show viewed outside the home is "Monday Night Football" on ABC.

Children ages 2 to 11 remain a measurement headache for Nielsen. A 22 percent loss of viewers age 2 to 11 was reported when the switch was originally made to Peoplemeters. Children's viewing levels have continued to fluctuate from quarter to quarter. The primary problem is that children aren't consistent in logging on to the Peoplemeter. Suggestions that Nielsen reward children for logging on have been met with resistance from the company. It believes that providing rewards (T-shirts, toys, and money have all been suggested) could significantly affect children's viewing behaviors.

Arbitron

The primary competition to Nielsen in reporting ratings was a company called the American Research Bureau (ARB). ARB developed a device similar to Nielsen's called the *Arbitron*. The reports of the ARB were known in the industry as

Arbitrons. The name Arbitron became synonymous with the company, which later changed its name in 1973 to Arbitron. The Arbitron company uses an instantaneous device sample like Nielsen for overnight ratings, but its major emphasis is on a separate diary sample used to measure the viewing of local stations.

Developed in 1958 and similar to Nielsen's SIA, the Arbitron device is connected to all TV sets in a sample of homes to monitor which channels are selected. The information is transmitted by telephone lines to a central site. Arbitron service originated in New York City but was expanded to cover other major cities. In 1961 ARB began to compete against Nielsen as a ratings service. Arbitron, today, offers overnight TV ratings, as does Nielsen, based on the information from the Arbitron devices.

Arbitron, not to be left behind, introduced its own TV Peoplemeter system called Scan America. This Peoplemeter was introduced in Denver in 1987. ScanAmerica has expanded into New York, Chicago, Los Angeles, Philadelphia, and Miami. In late 1992, Arbitron dropped its national ScanAmerica report. The local market reports were kept intact.

ScanAmerica provides more than just viewing information. Along with a device to measure TV viewing comes a data scan wand. This wand is to be run over the universal product code for every product purchased in the home. The bar codes are transmitted to the ScanAmerica device where they are collected for later analysis.

Local television Similar to the national surveys that are designed to make an estimate of the national audience for each network program, local TV ratings estimate the local audience for each program. Arbitron measures TV viewing in 210 TV markets four times a year—February, May, August, and November. In the larger markets Arbitron surveys more frequently. Arbitron uses the diary method in most of its local TV surveys. Medium markets have a sample size of about 1000 diaries that are sent out. The diary itself will be described a bit later in the chapter.

ECONOMICS

THE FALL OF AGB

AGB had great hopes of becoming the premiere national ratings service in the United States just as it had in Europe, but it wasn't to be. The British-based firm seems to have made several mistakes.

AGB's Peoplemeters were first installed in Boston for a two-year test. Although AGB had been warned by Arbitron not to test for so long, AGB ignored the advice and installed its Peoplemeters for a two-year test. In that two years Nielsen, which had relied on its old measuring device and diary supplements, was able to install a nationwide system of its own Peoplemeters. AGB's technological edge with Peoplemeters was history.

One of AGB's greatest problems seems to be that it wasn't ready for the diversity of U.S. programming. Because there is little cable in Europe and syndication is almost unknown, AGB was surprised to find the number of different programs of which it had to keep track. Compounding the large numbers of scheduled programs was the problem of keeping track of changes at local stations that would preempt scheduled programs to cover the last innings of a baseball game or a hockey game that ran long.

Only CBS also signed up with AGB for its 1987 to 1988 ratings. However, CBS made the decision not to renew its contract with AGB for the 1988 to 1989 season. With its only major subscriber gone, AGB shut down its operation in the United States in 1988. The irony of AGB's failure is that if AGB hadn't tried to penetrate the U.S. market, Peoplemeters might still be experimental devices rather than Nielsen's standard method for obtaining its national sample.

Perhaps the final straw for AGB was that it wasn't an American firm. One ad agency executive was quoted in *Forbes* as saying, "We were not about to walk away (from Nielsen) just because somebody came from another country and said they had another way of doing something."

Radio ratings Arbitron radio surveys use the diary method just as Arbitron's local TV surveys do. Arbitron radio surveys became strictly local in 1964. Sample sizes for local studies vary somewhat with the size of the area, or *market,* under study. Sample sizes may range from 500 to 4500 for a local radio market, with medium size markets having a sample size of about 1000 diaries. The number of radio ratings periods depends on the size of the radio market. All 261 radio markets are measured in the spring. The largest half of the radio markets are measured again in the fall. The top-79 markets are measured in the summer and winter. Arbitron is currently working on a radio Peoplemeter, but estimates that it will not be in full service until at least 1995.

Because of the size and frequency of the Arbitron radio and TV surveys, your family has a better change of participating in one of them than in Nielsen's.

THE RATINGS PROCESS

Today's mail includes a letter from Arbitron Ratings, Beltsville, Maryland, addressed to you. Opening the letter, you find that a local representative of the Arbitron company will be calling you soon to "invite you to be in a survey of television viewing." This is it! At last you get a change to be part of the ratings.

The letter is only the first step in a multistep process that Arbitron will go through to recruit, qualify, and inveigle potential respondents to complete a diary and return it. The letter you received is designed to warn you of the forthcoming telephone call and to assure you that the call is not a means of selling you a set of Ginsu carving knives.

Step 2 is a telephone call from a local representative, usually within two weeks of receipt of the letter. The call will be from a woman, as the research companies have found that women are not threatening and have the highest success ratios in telephone sales. This call will be used first to ensure that you don't work for a TV or a radio station and second, to ascertain the number of TV sets in your home and whether you have a VCR.

Step 3 is the arrival in the mail of one diary for every set you claimed to have in your home and a separate diary for your VCR. The diary is where you will keep track of what you view and record for the week. Along with the diary there will be a letter thanking you for your participation and, wrapped in plastic, two quarters.

The money has been found to make you feel indebted to the company and is used to motivate

---

ECONOMICS

BIRCH RADIO: 1978–1991

Birch Radio, a radio ratings company, died December 31, 1991. Born in 1978, Birch grew quickly to cover 273 radio markets. The Birch marketing strategy was to start in the small and midsize radio markets and significantly undersell its major competitor Arbitron.

Unlike Arbitron's seven-day diary, Birch used the telephone interview. Randomly generated telephone numbers in the most populous areas of the towns and cities were called, and respondents were asked to recall when and to which radio stations they had listened the day before and the day before that.

Birch's early growth has been attributed to three factors. First, Birch cost less than rival Arbitron. Second, many people, particularly at advertising agencies, felt that Birch's telephone interviews were superior to Arbitron's diary method. Third, Birch had supplied additional consumer information about its interviewees to clients without additional charge.

The death of Birch was caused by the shrinking number of profitable radio stations in 1991. Radio's financial belt-tightening caused many stations to drop their subscriptions to Birch while keeping their Arbitron service.

Birch is survived by the Scarborough Report, which provides an annual lifestyle profile for the top-55 markets that will now be marketed by Arbitron.

you actually to fill out the diary and to return it. Studies have shown that the return rate, the number of diaries that are returned versus the number mailed, is significantly enhanced by including the money. The research also shows that the return rate is about the same whether the company sends 50 cents or $5.

Your local representative will call to make sure that you received your diary, to answer any questions that you have, and to remind you of when you are supposed to begin writing. Diaries run from Wednesday to Tuesday covering a one-week period. You are supposed to supply information in the diary as to the number of people in the home, their ages, and the TV services/stations that you receive. Time frames in the diary run from 6:00 A.M. to 2:00 A.M. the next day. You are also supposed to list any visitors who drop by and watch with you, giving their sex and approximate age. Additional demographic questions are included at the back of the diary. Information on ethnicity and the number of working women in the home are included here. Questions on whether you have cable TV and the services offered by your cable company are also included in this part of the diary. There is a final back page for your comments and suggestions, but Arbitron does not analyze this information.

Your VCR diary is similar to the TV diary. You fill in the names of people in the home and note during the week what shows you recorded and the date when you watched the playback. Again, as with the viewing diary, there are demographic questions. You are requested to list tapes played back that were recorded before the survey week—old TV shows, video tape rentals, and so on—on the back page of the diary.

The next step is for you to watch TV for a week and to keep up your diary. When you have finished your week of watching, you mail the diary back to Arbitron. And, at last, you can say that you were part of the ratings.

How Is a Sample Selected?

How does one get to participate in these ratings surveys? Asking *some* people to respond means that the final results may not accurately reflect what *everyone* views or listens to. The question of whether a group of people who are questioned actually give the same answers as everyone in the population illustrates the problem of representativeness.

One way to beat the representativeness problem would be to ask everyone. The process of asking everyone—or, more technically, measuring all members of the population—is called a *census*. The U.S. government takes a census every ten years. You may recall a census taker coming to your home and asking questions. But you can see the problem: there are just too many people or homes to be able to ask everyone about his or her viewing preferences. It takes the resources of the federal government to conduct a census, and still the census taker often misses people.

Instead of asking everyone, broadcast and cable researchers ask a select group of people who can represent those who aren't asked. This process is called *sampling*. A *sample* is a careful selection of elements (homes in this case) so that those selected represent the entire group from which they were picked. Sampling is what you do to check the taste of soup. If you are trying to determine whether you have put in enough pepper, it isn't necessary to drink the entire pot. You take a spoonful—a sample—and determine the nature of the whole.

"People aren't soup," we hear you cry. You're probably thinking, "My family's a zoo. No one could represent my 'geeky' sister, Joyce!" Well, in ways, a sample does take geeky Joyce into account. The assumption is that most behaviors, including TV viewing, fit a *normal distribution curve* (see Figure 14-1.) The normal distribution curve is also sometimes called a *bell curve*. Skipping the technical statistical definition, a normal distribution says that behaviors are randomly distributed. For example, some people watch a lot of TV and can be spotted as high scorers on MTV's "Remote Control"; and some people watch very little TV and can be spotted as high scorers on the last test in this course. But the majority of people watch about the same amount of TV. In a randomly distributed behavior, people tend to cluster around the *mean*—another word for the arithmetic average.

A better example than soup would be popcorn. Get an air popcorn popper and pop yourself some popcorn; think of it as a homework assignment. Notice how the popcorn pops. A few kernels will pop after only half a minute or so, and then more

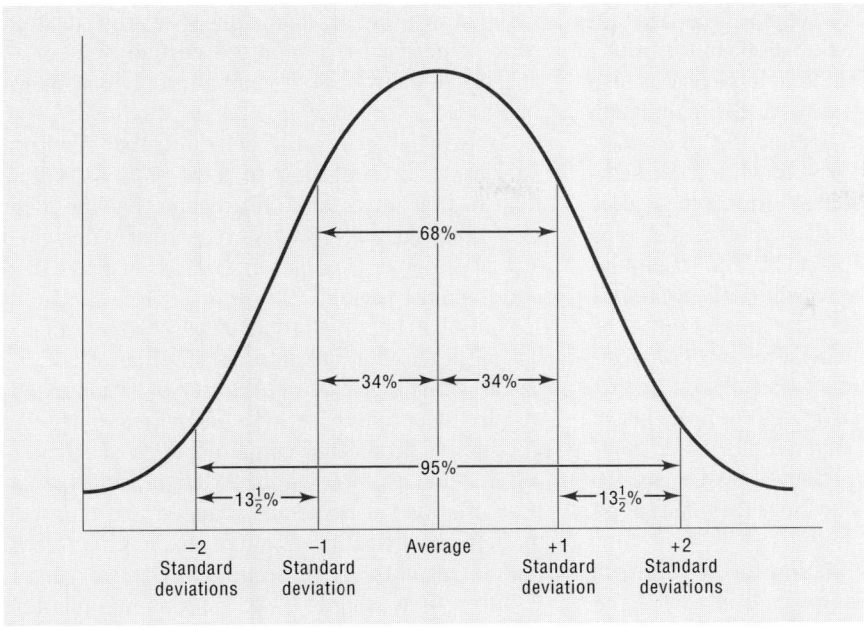

FIGURE 14-1 The normal distribution curve.

and more popcorn will pop. There will be a period when it sounds as if all the corn is exploding and then the activity begins to settle down. Soon you will hear only one or two kernels pop. Think about the normal distribution curve as you watch it pop and you will discover that popcorn popping time is randomly distributed.

Geeky Joyces would be somewhere in this distribution. Average folks cluster around the center and unusual people move out to the ends. Joyce would be at one end.

Drawing our sample from a population that is normally distributed, as the TV audience is assumed to be, we can represent the entire population from the sample. The population of an Arbitron study consists of TV or radio households: any type of residence—apartment, house, mobile home—that has a TV set. This definition does not include dormitories, hospitals, nursing homes, and most other forms of institutional living, although it does count houses on military bases. Arbitron does not determine the number of households in a market. Rather, it contracts with another company to supply the information.

How the sample is selected, called drawing the sample, is important. All homes must have an equal chance to be selected. You wouldn't want to select just those students who sat in the back row as representative of your class as a whole, would you? People who sit in the back row often have different attitudes toward the class than those who sit in the front. Better to find a way to pick and choose people from throughout the class. This could be done by putting everyone's name in a hat and selecting names. When a sample is drawn so that all people have an equal chance, it is called a *random sample*. Arbitron uses a type of random sample called *systematic interval selection*. This process makes a selection from a list every so many times. For example, going through the telephone directory and selecting every hundredth name is a type of systematic interval selection.

Arbitron has its computer go through and select homes from computer tapes that Arbitron purchases from a marketing firm called Metromail. These tapes contain names, telephone numbers, and addresses. A problem with such tapes is that people who have no telephones, mostly the poor, and those with unlisted telephone numbers, such as physicians, police officers, "high profile" individuals, are not included on the tapes.

Arbitron selects a random start point in the list, say, name 53, and then selects a new name at a fixed number, say, 100. A scheme like that described would produce a list of 1000 names from a list of 100,000 people. The actual interval to skip would be affected by the size of the population and the number of people needed in the sample.

The number of households drawn for the sample varies based on a statistical formula that takes into account the number of households that are estimated to be in the market. The minimum number of diaries for the area where the station is dominant is 200. These 200 are spread out over the counties that are contained in the dominant area based on the number of households in that county. In addition, a metro area (defined just a little later in this chapter) minimum is used, ranging from a minimum of 200 diaries for the top-40 markets to 125 for those markets listed as eightieth or smaller.

There is an obvious limitation to the Arbitron methodology, one that has plagued many of the ratings services historically—the reliance on listed telephone numbers. Remember, one of the disadvantages of the Crossley method was its use of telephone directories. Even today, when most homes have telephones, a number of homes aren't in the telehone directory, either because the family lacks a telephone or because it has chosen not to have its number published. Such people would not be represented in the sample. They could never receive a diary because they don't exist in the lists provided by Metromail. This limitation is especially important for people residing in inner cities. Arbitron has developed techniques to compensate in some cases.

The *expanded sample frame (ESF)* was developed in an attempt to include people who would normally be left out of the telephone list. ESF is used in those markets where minority populations may be left out or underrepresented without some additional means to include them. Arbitron generates a theoretical list of all possible home telephone numbers. Excluded from this list are all known home telephone numbers from Metromail. The remaining numbers are then selected using the same systematic interval selection process. When a number is drawn, the household is contacted by telephone and a diary placement is attempted.

In areas with large Hispanic populations, particularly enclaves where English is not the dominant language, Arbitron will send a personal representative to place the diaries. Using a telephone number as a "seed," a starting point, the Arbitron representative then counts a certain number of apartment or front doors and attempts placement there.

Arbitron differs from Nielsen in that Arbitron specializes in reporting to local stations rather than to the networks. This means that the samples which Arbitron draws are primarily designed around groups of competing stations in a specified geographical region.

One of Arbitron's first efforts was to organize TV and radio stations into distinct markets. This is the most general of Arbitron's various geographic descriptive terms. *Markets* are specific geographic areas that contain groups of competing TV stations for whom Arbitron issues a report. Stations in Boise are grouped together as a market because these stations are in direct competition for both viewers' and advertisers' dollars. Arbitron then assigns these stations to a market. The names of markets are taken from the locations served; usually they are city names.

Markets may include more than one city. Sacramento, California, is located 30 miles north of Stockton. Both cities have TV stations assigned to them. Because of their proximity, stations in these cities compete against one another. So Arbitron created a group of markets that it designates as multiname markets. Markets can have as many as four cities in their name; for example, in the midsouthern section of Pennsylvania, Arbitron has designated the Harrisburg–York–Lancaster–Lebanon market.

A subdivision of a market is a *metro rating area*. The metro area is usually the area that contains the highest density of people in the market that Arbitron is surveying. The precise definition of the metro area is a bit more complicated. The metro rating area includes the home counties for stations within a market, in addition to those counties designated as linked together by the U.S. Census Bureau in their census reports. Standard metropolitan statistical areas (SMSAs) are included in the metro area. In some cases Arbitron will include "usage requirements" in the metro rating areas of "historic industry." These metro areas normally comprise the primary shopping areas of the survey. Viewers or listeners in the metro areas are more likely to avail themselves of local stores and services, and so these are the areas where account executives from the station focus most of their attention.

Larger than the metro rating area is a region that Arbitron designates as the *area of dominant influence (ADI)*. The original designation of ADIs occurred in 1965. Arbitron undertook a county-by-county study of TV and radio listening habits to determine which stations were listened to or viewed and assigned the station to that county. If you lived in an area midway between TV stations in two different cities, Arbitron attempted to determine which stations were the ones that were most frequently watched or listened to by people in your area. If people in our Midway City watched stations from city A more than city B, then Midway City would be included within city A's ADI. Arbitron reexamines the placement of cities like Midway City each year to determine whether placement in the assigned ADI is correct.

An ADI is designed to provide an exclusive geographical region. Although ADIs usually are based on counties, they may be broken up if terrain or other circumstance makes reception of TV signals from one part of the county difficult to receive in another county.

The FCC has recognized the designation of the ADI, and it now serves as a standard definition in the TV and radio industries. The ADI is used by national advertisers when they create their marketing plans. Money is often allocated by national sponsors based on ADI and has come to be used by many national companies to define sales and distribution territories for their products.

When Arbitron selects a sample, the sample includes an area that Arbitron calles its *total survey area (TSA)*. These are areas surrounding the ADI that must be included in order to account for 98 percent of the total viewing of commercial TV or radio stations within the market. The TSA is the largest area reported in an Arbitron report. It is a representation of just how far away people may be viewing the station within certain statistical limits. Some areas in the TSA may be shared with other markets.

How Data Are Analyzed

It's Wednesday and you have just spent the week dutifully keeping your diary. You listed every show you watched. You tried to be honest. You didn't write down that you spent the week watching symphonies and operas or William F. Buckley. After all, you are being paid 50 cents for your efforts. Time now to mail the diary back to Arbitron. If you forget, don't worry. The nice lady who called will call again to make sure that you have put your diary in the mail.

Diaries flow into the Arbitron headquarters and are subject to several review processes. The first review is called, predictably, the *first edit*. The first edit is designed to remove those diaries that Arbitron considers to be unusable: diaries mailed before the survey ends without an explanation of why, diaries that arrive too late at Arbitron, diaries that have a blank day without the "set not in use" box checked, diaries that have entries after the postmark date, diaries missing all demographic information, diaries missing station information, diaries listing the head of household as less than 17 years old. These and other anomalies will cause Arbitron to reject the diary.

The second stage review is called the *family edit*. Arbitron employees check that the information entered about family members is correct. They also check to make sure there has been notation of who has been watching what and when.

The third stage is the *can-receive edit*. This is a veracity test. Checkers in this department see if the stations listed on the first page of your diary can actually be received at your home.

Diaries are next sent to the *household edit* for review. Arbitron employees verify that the information marked in the demographics section in the back corresponds to information that you have marked in other areas of the diary.

The *viewing edit* begins a close scrutiny of the programs and times that are marked in the diary. Entries are checked to make sure enough information is available so that viewing credit may be assigned to the proper station.

The final editing step is *station verification*, which ascertains whether the correct channel number or station ID has been entered for the program listed. When in doubt, Arbitron uses call letters or service names for those people hooked to a cable system. For those respondents who don't have cable Arbitron relies on channel numbers. If a name of a show has been entered without channel information, Arbitron attempts to assign the show to the proper originating service. To aid in the assignment of programs, Arbitron sends a program title log to all commercial stations in the survey area. Stations are to complete the log by listing the names of all TV programs that will be on during the survey week.

When all these checks have been made, it's time for the information to be entered into the computer. Once the data from the diaries have been coded and entered into the computer, there are still some additional procedures used before the results are printed out.

One of the more interesting procedures is called *Monte Carlo audience ascription.* In some diaries the information that delineates who watched when is fragmentary. Arbitron has a computer program that assigns viewing based on similar cases that are processed for the market. For missing data the computer enters which persons are watching at a given time based on what similar families are doing. Then the computer randomly decides when actually to count this information, a process known as the Monte Carlo method, for obvious reasons.

Arbitron has also developed a computer program that fills in missing viewing times. This program is used when the information in the diary is ambiguous. The computer can assign times that seem appropriate.

One other special procedure that is common in survey research, *weighting,* also is used. This is a procedure that multiplies responses by some set figure. This multiplication factor is used to offset underrepresentation in the sample.

It is common for many ethnic minorities to be underrepresented in surveys. Hispanics may be 14 percent of the population in the market under study according to the U.S. Census Bureau, but the number of diaries from Hispanics is only 7 percent of the sample. One could multiply, or weight, the results of the Hispanic sample by 2 so that the numbers would be representative of the population as a whole.

Arbitron does weighting for a number of different factors. They weight for geography, weeks reported, subscription to cable, age of household, sex, and for two ethnic groups: blacks and Hispanics. These weighting procedures are used to ensure that the results reported reflect the size of the population and not just the sample. It is insurance that the cable subscribers, for example, who return diaries more frequently than noncable subscribers, are not overrepresented.

After all these checks, controls, and computations the results of the survey are finally printed out. They become part of the Arbitron Television/Radio Market Report.

How Data Are Reported

Arbitron produces market reports that it sells to networks, local stations, advertising agencies, and other interested parties. The report provides information about both the demographics of the survey audience and their viewing habits.

There are four terms that are important in understanding the information that Arbitron supplies in its TV reports. One of the most important is *HUT,* which is an acronym for *households using television.* As are most of the numbers reported by Arbitron and other ratings services, HUTs are expressed as both percentages and numbers of households. HUTs represent the number or percentage of households that have a TV set turned on at a specified time. HUTs are important in interpreting the other collected data and are used in the computation of other percentages that are reported by Arbitron.

The second term is *rating.* This word has been used frequently in this chapter and is the generic word used by many to describe the process of collecting information on what is viewed or listened to and the estimated number of people who are listening to or viewing the broadcast media. Specifically, a rating is the estimated percentage of TV households viewing or listening to a specific program. A rating of 18 means that 18 percent of those in the home that have TV sets were tuned to a specific program. Ratings do not take into account how many homes in the market had their TVs or radios on at the time.

The total number of households watching a station divided by the total number of households that actually have their sets on is called a *share.* This represents an estimated percentage of households with their sets turned on and tuned to a specific program. The difference between share and rating is that rating does not take into account whether the set is turned on while share does take viewing or listening into account.

It is possible for a program like "Geraldo" and "Cheers" to have the same share but have vastly different ratings. Maybe a hypothetical example will help make the distinction clearer. Let's say that in our example there are 1000 households with TV sets. Of these, 100 households have their sets turned on during the day, which gives a HUT of 10; that is, 10 percent of the households have their sets in use at this time. Let us say that 30 households are tuned to the program

"Geraldo." The *rating* for "Geraldo" would be a 3 ($^{30}/_{1000}$), indicating that 3 percent of those people who have a set are watching the show. The *share* for the show would be a 30 ($^{30}/_{100}$), indicating 30 percent of those people watching TV at the time are watching the show. Although the rating isn't good as compared to prime-time ratings, the share is fairly respectable. A few hours later in prime time 900 households have their sets turned on; the HUT is now 90 ($^{900}/_{1000}$). A total of 270 homes are tuned to Sam, Norm, and the people at Cheers. The share for the show is 30 ($^{270}/_{900}$), but the rating is a 27 ($^{270}/_{1000}$). "Cheers" has many more viewers, as indicated by the rating, but it attracts the same percentage of sets that are turned on as "Geraldo," as indicated by the share.

A similar situation occurs with a show like "Late Night with David Letterman." The Letterman show doesn't have a big rating primarily because a lot of people are in bed and not watching TV. But its share is usually in the mid to high 20s because of those people who are up watching TV at the time, a good proportion of them watch the show.

We have tried to avoid formulas in this chapter because we know that these can be troubling to some, but if you followed the last couple of paragraphs, these should be easy. Here are the two main formulas used to compute ratings and shares for TV programs:

Rating = the number of households watching a certain channel/total number of households with TV sets.

Share = the number of households watching a certain channel/total number of households watching TV (HUT).

With a little algebra you can derive the following formulas:

$$\% \text{ HUT} = \text{rating/share}$$
$$\text{Rating} = \text{share} \times \% \text{ HUT}$$
$$\text{Share} = \text{rating}/\% \text{ HUT}$$

For another computational example, see the box "Calculating Ratings and Shares."

The fourth term is *cumulative audience*. This is an estimate of different households/listeners that select a station at least once during the week for the time under scrutiny. Cumulative audience is sometimes known as *cume* or *reach*. Generally more important in radio than TV, reach is a measure of how many different people view/listen at least once during the week.

Average quarter-hour persons is a measure frequently used in radio reports. It is an estimate of the average number of people within a demographic group who are listening to a station within a fifteen-minute period. The average quarter-hour persons is calculated by dividing the estimated number of listeners within a specific time block by the number of quarter-hours (four per hour) within that time block. For example, if there were an estimated 800 women 18- to 24-year-old listeners to a station sometime within the 4:00 to 6:00 P.M. time block, the average quarter-hour persons would be (800 persons/8 quarter hours =) 100 average quarter-hour persons. Even if the estimated 800 women had listened only from 5:30 P.M. to 6:00 P.M., the calculations are the same.

The Book

A TV ratings book covers a four-week period, which is noted on the cover. The reports usually begin with the map page. It contains a breakdown of the area covered by the survey, the stations included in the sample, dates of data collection, and a map. This is the beginning of the *Market Data Section* of the Arbitron ratings book. This information tells the station who its perspective audience is and ultimately who should be considered prime local clients of the station. The counties included in the Arbitron survey are listed, along with the estimated number of homes that have TV in the county and the actual number of homes sampled from that county.

Demographic characteristics are the next major section. This information is broken down by total survey area and ADI. Next is a report on the sample that was used for the current report. A table shows the number of homes that were selected for inclusion, the number of homes from those selected that actually said they would accept the diaries, and then the number of homes that were returned and survived the editing process. This is followed by the number of quarter-hours estimated to have been viewed in a household, by women, men, teens, and children in a week.

APPLICATIONS

CALCULATING RATINGS AND SHARES

Let's use the following hypothetical data in our calculations. A ratings company samples 2400 households (all having TV) in a given market and determines the following for the 7:30 to 7:45 time period.

Station	Number of Sample Households Watching
WAAA	480
WBBB	240
WCCC	120
Other stations	120

Let's figure out WBBB's rating:

WBBB's rating $= \dfrac{240}{2400} = .10$ or 10%

To calculate WBBB's share of the audience, we must first determine the total number of households using TV (HUT) at 7:30 to 7:45. To do this, we add 480 + 240 + 120 + 120 = 960. WBBB's share, then, is

Share $= \dfrac{240}{960} = .25$ or 25%

Also note the following relationships:

(1) % HUT $= \dfrac{\text{rating}}{\text{share}}$

In our example

% HUT $= \dfrac{.10}{.25} = .40$ or 40%

This means that 40 percent of all the TV homes in the sample were watching TV from 7:30 to 7:45:

(2) Rating = share × % HUT

In our example

Rating = .25 × .40 = .10

This is simply another way to calculate the rating:

(3) Share $= \dfrac{\text{rating}}{\text{\% HUT}}$

In our example

Share $= \dfrac{.10}{.40} = .25$ or 25%

Again, this is another way to calculate the share.

Audience Estimates in the Arbitron Market of

Minneapolis-St. Paul

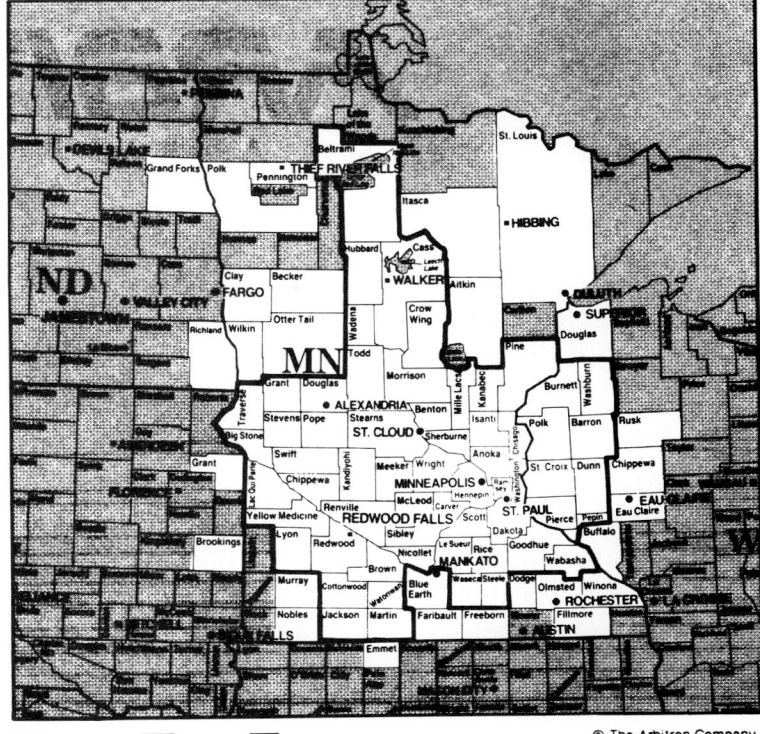

Survey Period: JAN 02, 1991 – JAN 29, 1991

Survey Months:
OCT NOV JAN FEB MAY JUL

This report is furnished for the exclusive use of network, advertiser, advertising agency, and film company clients, plus these subscribing stations -
KITN

Schedule of Survey Dates 1990-1991

October	September 26 - October 23, 1990
November	October 31 - November 27, 1990
January	January 2 - January 29, 1991
February	January 30 - February 26, 1991
March	February 27 - March 26, 1991
May	April 24 - May 21, 1991
July	July 10 - August 6, 1991

☐ Metro ☐ ADI ☐ TSA
• City of license   ■ City of license of satellite station

© The Arbitron Company

The map page in an Arbitron local TV market report. The map shows the total survey area, the area of dominant influence, and the metro area of the market.

The numbers in the demographic section are used by Arbitron and subscribers to Arbitron to make decisions about who is viewing or listening. They are also used in the computation of the statistics—HUTs, ratings, shares, reach—that Arbitron reports in later pages.

Arbitron provides a summary of the ADI marketplace in the third section. Census data, sales information, car registration, magazine and newspaper circulation, grocery stores, and chain stores located in the area are included in this section.

The heart of the Arbitron report is the audience estimates, or who is viewing/listening to what and when. Estimates for TV are broken down by time, called *daypart,* and program. Daytime viewing is broken down into daypart audience estimates that cover nonnetwork viewing—local and syndicated—and network daypart audience estimates, which exclude all local and syndicated programming. These numbers help a station to understand how it is doing in a more general way than looking at specific times, which are reported later.

These pages report HUTs, ratings, shares, and reach, and project the estimated number of people who view a station or a network at a given time. They do not report specific shows but, rather, report viewing times, for example, Monday to Friday, 9:00 A.M. to noon. Trends in viewing over a year's period are also included in these listings so that a station may judge how well it's doing over a period of time. Daypart reports also include a breakdown of the audience by demographics. Thus, it is possible to know, for example, the estimated number of 18- to 34-year-old females who viewed TV during a specific segment of the day.

Time Period Estimates

DAY AND TIME STATION PROGRAM	ADI RTG TN 12-17	CHILD 2-11	CHILD 6-11	TV HH	PERSONS 18+	12-34	WOMEN 18+	18-34	18-49	25-49	25-54	WKG WOM	MEN 18+	18-34	18-49	25-49	25-54	TNS 12-17	CHILD 2-11	CHILD 6-11	STATION BREAK AVERAGES ADI TV HH RTG	MET TV HH RTG	TV HH	TSA IN 000's WOM 18+	MEN 18+
	36	37	38	39	42	41	45	46	47	48	49	50	51	52	53	54	55	56	57	58	5	8	39	45	51
RELATIVE STD-ERR 25%	11	9	11	29	47	48	36	37	35	31	31	33	35	44	36	32	32	40	58	45	1	2	29	36	35
THRESHOLDS (1σ) 50%	3	2	3	7	11	12	9	9	8	8	8	8	9	11	9	8	8	10	15	12	-	-	7	9	9
WEDNESDAY 7:00P- 7:30P																					7:00P				
WCCO 48 HOURS		1	2	144	235	52	124	22	41	37	49	42	111	29	48	40	51	1	14	6					
WAR GLF CBS1	11	6	10	307	504	195	265	64	136	124	134	68	240	98	143	73	91	32	30	30					
--4 WK AVG--	3	2	4	185	302	88	159	32	65	59	70	49	143	46	72	48	61	9	18	12	16	17	223	208	159
WCCO+KCCO (SP)	4	3	5	224	360	105	196	44	82	71	88	58	164	50	82	58	73	11	21	15	18	17	264	244	186
KSTP+ WONDER YEARS	19	14	10	211	245	177	143	62	116	94	98	63	102	52	87	76	80	63	75	32					
ABC SPRP 8PM	11			169	296	103	192	43	133	115	130	62	105	25	91	76	83	34							
--4 WK AVG--	17	11	8	201	258	159	155	58	120	100	106	63	103	45	88	76	81	56	56	24	13	14	187	152	115
KMSP NRTHSTR HCKY		2	2	37	49	15	10		6	6	6	4	40	15	27	27	27		12	6					
7 PM MOVIE				39	34		17		14	14	14		17		15	15	15								
CNN NWS SPCL	2	3	1	57	121	72	55	26	48	35	35	37	67	40	60	35	35	5	15	4					
--4 WK AVG--				43	60	22	25	7	21	17	17	10	35	14	29	23	23	1	7	2	4	4	68	41	57
KARE UNSLVD MYSTR	1	3	3	155	233	67	142	39	67	47	59	35	92	24	37	35	44	4	16	11					
NBC AMER WAR	2	5	8	134	221	84	126	58	103	89	89	24	95	20	60	60	66	6	26	26					
--4 WK AVG--	1	3	5	150	230	71	138	44	76	57	67	32	92	23	42	41	49	4	18	14	8	8	114	102	65
KTMA+ MOVIE AT 700				17	21	7	13	3	5	5	10	4	9	4	4	4	5		1		1	2	18	11	8
KTMA+SATS (SP)				17	21	7	13	3	5	5	10	4	9	4	4	4	5		1		1	2	18	11	8
KITN 29 PRIME MOV	1	2	2	32	56	21	22	8	20	17	17	17	34	9	28	28	28	4	11	7					
TMBRWLV BKBL	3	2	3	45	65	37	17	9	17	17	17	9	48	20	45	35	35	8	13	8					
CNN NWS SPCL				6	11		4				4		7				7								
--4 WK AVG--				29	47	20	17	6	15	13	14	11	31	10	25	23	24	4	9	5	2	2	30	13	23
KTCA PTV	1	1	1	67	107	20	56	8	18	13	15	11	51	9	28	28	29	4	7	4	3	4	46	35	34
KTCI PTV				5	4	2							4	2	4	4	4		1	1			3		2
HUT/PVT/TOT	31	26	27	736	1087	406	600	170	337	276	317	189	489	157	302	257	288	80	120	65	59	55	730	598	490
7:30P- 8:00P																					7:30P				

A sample page from the time period estimate section of an Arbitron local TV market ratings book.

Time period estimates are next. These breakdown the viewing into half-hour blocks. The same type of information reported in dayparts is reported here. However, this section focuses on specific time periods, which make them better for use as sales tools for account executives who must sell time slots. This allows advertisers to buy commercial time based on time of day rather than on a specific program. This section also includes estimates of the audience during station breaks, the periods when the commercials are run.

Arbitron next provides information focusing on how stations have done over a year's period with various viewing audiences. These pages demonstrate how a station does seasonally. Viewing during the summer traditionally declines as people take advantage of the increased daylight hours and as they vacation. These types of trends can be seen and attempts can be made to reduce their impact on viewing. They can also be used as a predictor. Knowing how your station has fared over a year's period allows for speculation about how the station will do in the upcoming months.

The final two sections of the TV ratings book deal specifically with programs. Remember, previous portions of the report dealt with time periods rather than programs. The first of the program sections is the program audience estimate. This section is designed to allow the evaluation of specific programs. Since it is broken down by time periods, station management can see how a program does in head-to-head competition with other programs in its time period.

The next to last section of the report includes an index of programs by title, along with an ADI index of trends. This section helps in locating a show by its time period in the program audience estimate for a detailed view of how the show is faring in the market. This section also includes share and HUT information for the program over a year's period. The final section contains a glossary and a technical section on the statistical methods used in compiling the data.

TECHNOLOGY

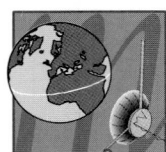

IS BIG BROTHER WATCHING?

In mid-1989 the A. C. Nielsen Company announced plans to introduce a "passive" Peoplemeter system that it is developing with the David Sarnoff Research Center. The passive Peoplemeter is so named because individuals would no longer have to note actively their presence to the machine with any button-pushing. Officials at the major TV networks said that such a system would provide more accurate data than the current Peoplemeter system.

Here's how the new Peoplemeter system works. A remote cameralike sensing system about the size of a VCR watches family members as they watch TV. Each family member's face will be electronically scanned into the Peoplemeter's memory. This digital equivalent of the facial structure of each member of the household will be matched with demographic information about that particular person. In effect, the Peoplemeter scans the area in front of the set, recognizes the faces of those who watch, and matches them up with the profiles stored in its memory. Its developers say that the new system can recognize family members with or without glasses but has trouble distinguishing identical twins. The new passive Peoplemeter will even be able to tell if people are looking at the screen during the commercials.

Will families be reluctant to allow such a device to be installed in their homes? Nielsen doesn't think so. "I don't think we're talking about Big Brother here at all," said one company executive. Time will tell.

SUMMARY

Audience estimates, the ratings, are an important tool for broadcasters, cablecasters, and advertisers. The audience reports published by the different ratings companies make decision making more accurate in the industry.

Early methods of measuring audiences did not accurately reflect who was listening to radio. In 1929 Archibald Crossley began to conduct random surveys by using numbers selected from telephone directories.

In 1930 the Cooperative Analysis of Broadcasting began their own telephone surveys but their technique had many shortcomings.

Hooperatings were the next step in audience analysis. People were asked to say what they were listening to at the moment of the interview, and the results were given to both advertisers and broadcasters.

In recent years the largest ratings companies have been A. C. Nielsen, Inc. and Arbitron, Inc. The Nielsen ratings use Peoplemeters to measure TV viewership. Some problems with Peoplemeters are that people know that they are being measured, the sample may be too small, and the system is too expensive to operate. Arbitron uses its own version of the Peoplemeter as well as the diary method. Arbitron measures both TV and radio.

Companies calculate the ratings, shares, households using TV, and cumulative audience, and then publish their figures in ratings books. The books include maps, demographic characteristics, summaries of ADI marketplace, daypart reports, and reports concerning specific programs. This information is used by stations and advertisers.

SUGGESTIONS FOR FURTHER READING

Arbitron Company. (1980). *Description of methodology*. Beltsville, Md: Arbitron.

Beville, H. M. (1988). *Audience ratings* (2nd ed.). Hillsdale, N.J.: Lawrence Earlbaum Associates.

Chappell, M. N., & Hooper, C. E. (1944). *Radio audience measurement*. New York: Stephen Daye.

Fletcher, J. (1981). *Handbook of radio and TV broadcasting*. New York: Van Nostrand, Rinehold.

Powers, R. (1977). *The newscasters*. New York: St. Martin's Press.

Webster, J. G. (1984). *Peoplemeters*. Washington, D.C.: National Association of Broadcasters.

Wimmer, R., & Dominick, J. (1991). *Mass media research: An introduction* (3rd ed.). Belmont, Calif.: Wadsworth.

CHAPTER 15: BEYOND RATINGS: OTHER AUDIENCE RESEARCH

As the owners and managers of broadcasting stations became more sophisticated, so did their need for information about their audiences and their programming. To compete in a *market*—the area and people served by the broadcasting station—a broadcast manager must have as much information as she or he can get. Information about the station's image and the quality of the programming takes on greater importance in enabling a station to compete more effectively in a market as that market becomes increasingly crowded with broadcasting services, each vying for attention.

Audience measurements, as discussed in the previous chapter, serve as the primary source of information about viewers and listeners. The size of the audience is very important, but information is needed about how to maintain and increase the station's share of the audience. New and better ways are necessary to discover who the audience is, what the audience wants, and how best to provide this information. A subsidiary industry to broadcasting has emerged to find the answers to some of these questions. Consultants and organizations that help define who the station is reaching and radio "doctors" who remedy programming ills at the station have become commonplace.

It has become increasingly important for radio and TV stations to engage in ongoing research activities. Some of these activities may be taken on *in-house*—done by station personnel; other activities are accomplished through the help of research firms and consultants. However the research is performed, successful stations in competitive markets must look to their market, specifically the portion of the broadcasting audience that comprises their market, to find out what the market wants and needs.

This chapter examines four broad categories of research used in broadcasting and cable today: music research, market research, physiological measures, and consultants. Each of these is an important part of the research area in the electronic media.

MUSIC RESEARCH

The bulk of most radio stations' programming is music. The correct musical selections in the *rotation*—the mix or order of the music played—will keep people listening to the station without tuning away for something more desirable.

For a top-40 station to play a "hit" for longer than it appeals to its listeners is called *burnout*, which will drive listeners to another station—or to turn off their radios. Either of these alternatives is a loss to the station and its advertisers.

Music directors at local radio stations make judgments about what should be played on their radio stations to attract the largest audience. Reading *Billboard* or *Spin* supplies information about what is hot at the national level, but this alone isn't usually sufficient for a music director to know what is playing well in her or his market. These judgments are normally based on a combination of experience, reviews, record charts, and gut instinct. The music director will use any source of knowledge about the music to help make a more attractive product that will gain and hold an audience.

Requests generally can't serve as a sufficient indicator of people's desires because requests usually are for either the most popular current songs or songs that have dropped out of rotation. New music that might be placed into rotation will receive very few requests because most people gain their knowledge of new music from the radio. Also, some songs will continue to be requested even after most people are "burned-out" on the song.

Broadcasters at large- and medium-market stations use two methods to test a song's attraction for its listeners: *call-out* and *auditorium testing*. Although these methods use different techniques, the object of both is to discover what people are interested in hearing. These methods give some indication of the public's likes and dislikes, but they alone can't serve as sufficient information. The results of this research detail what songs people like but may not measure how frequently people want to hear a given piece of music. Frequency of play is a decision that must be made by an experienced programmer who has many sources of information.

Call-Out Research

The most frequently used form of music research is a method called the *call-out*. Call-outs may be done by station staff instead of a research firm or a consultant, but many stations do hire outside firms to conduct their call-out research.

Call-outs are accomplished by making a five- to fifteen-second cut of the most identifiable portion of the music, often the beginning of the record. These small portions are called *hooks*. The hooks are then played over the telephone and the potential listener's opinions are solicited. Three methods are used, sometimes together, to select those people who will be called.

First, a member of the radio station's staff may select names of people from the telephone directory.

A second, and better method in terms of the criteria of scientific research, is a random telephone number technique. This technique uses the first three digits of prefixes known in the listening area. These are usually printed at the front of the telephone directory. The last four digits of the telephone number are taken from a list of random numbers found in statistics books, or some other method is used.

The random number technique has advantages and disadvantages. The main advantage of the random number method over the telephone book is that people who have unlisted numbers have the same chance of being called as those of the general population. The main disadvantage is that many of the numbers generated will not be any good, because either they are business numbers or the telephone number isn't in use.

When a number has been called, the next step is to screen for, that is, identify, someone in the household who would generally be an audience member of the radio station. This is also an important step as there is no reason to interview a grandmother if your target audience is 18 to 34 year olds.

The third method of obtaining respondents, the *panel* method, uses known listeners as respondents. These people may be identified through their participation in promotions or contests, or they may be people who call to make requests. The panel method is more efficient than the previously discussed methods because the people on the panel are known listeners and can be contacted easily.

Membership on the panel rotates, with people staying on the panel from four to eight weeks. The longer a person is a member of the panel, the more likely that person is to become dissimilar to the listeners she or he is supposed to represent. Being a member of a music panel often makes the participant much more aware of music than the general public. Since the panel member knows that her or his opinion will be solicited about music, the person may begin to act unnaturally.

Once the right person has been identified, about twenty hooks are played over the telephone. Twenty is usually the maximum number because people tire quickly on the phone. After each hook several responses are elicited. Typical questions concern whether the person likes the cut, whether the person would listen to the music (sometimes we listen to music that we don't like and sometimes music is so objectionable that we change the station), whether the person is tired of hearing it, or whether the song is a favorite.

Scales are frequently used as responses. A person might rate the hook on a scale of, say, 1 to 5, with 1 meaning "I strongly dislike the music" and 5 signifying "I strongly like the music."

Other information about listening behaviors is usually obtained at the end of the telephone session. For example, it might be helpful when a person listens—mornings, lunch hour, driving to work or school—and where—in the car or at work.

The results of these call-outs can be used by the music director to judge how well he or she is doing in the selection and placement of songs. Music directors can identify songs that have peaked in their popularity and highlight the oldies that are still attractive.

Research combined with the music director's knowledge results in programming that is attuned to the tastes and preferences of the target listeners. This leads to more listeners tuning in for a longer time and an increase in profits for the station.

One of the limitations of the call-out method is that it can't test unfamiliar music well. Since call-out uses familiar hooks, music that is relatively unknown doesn't test well. Other methods must be employed in making judgments about new releases. One research method that can be used for new music is auditorium testing.

Auditorium Testing

Auditorium testing is used for more than just music research. It was frequently used in the past by the networks to measure responses to proposed TV programs, but only CBS continues to use it for testing new programs. Advertisers still use auditorium research to measure the effectiveness of their commercials. This section deals specifically with music testing. Advertising's use of the auditorium test will be covered under market research.

In *auditorium testing,* 75 to 100 people are recruited to come together at a central place, usually an auditorium, to listen to hooks (as in call-out studies) for sixty to ninety minutes. Auditorium testing gains two advantages for the music researcher: (1) more hooks may be played in a single session and (2) listeners may better concentrate on the task.

Where the usual call-out research is limited to about 20 hooks, in auditorium testing subjects may be asked to judge up to 600 hooks. Usually after listening and responding to about 300 hooks, most people are tired and the quality of their responses is not good.

This increase in the number of hooks that can be tested is a function of having the respondents out of their homes. A person trying to listen attentively to musical hooks at home is usually plagued with a variety of distractions. By placing the respondents in a controlled environment the quality of the responses improves.

People are generally recruited to participate in the auditorium sessions through a field service company, which specializes in recruiting respondents for research. These companies use various methods to locate potential participants. Some service companies may recruit participants from people standing in line for TV shows or plays. Some firms recruit participants from shoppers at shopping malls. Since recruiting from a line for a TV show or at a shopping mall might produce an unusual sample, some firms recruit using a more random method, such as telephone or mail. The recruitment method depends on the requirements of the company doing the auditorium testing and the client station.

All firms that recruit for auditorium testing offer some sort of reward to those who participate. Rewards of up to $50 are offered to individuals to participate in the sessions, although $20 is the most frequent payment. Auditorium testing is a fairly expensive method as compared to call-outs. Costs usually range from $10,000 to $100,000 dollars for auditorium research.

Although infrequently done, auditorium research can be used to test new songs because the entire recording can be played. The number of recordings that can be tested is limited by time constraints, but auditorium testing of new music is possible.

For radio stations thinking about format changes, auditorium research may be used to test the new sound that the station is contemplating. Rather than hearing just hooks, the participants would be exposed to a sample of the proposed format for their response.

MARKET RESEARCH

No one wants to have a flop. In broadcasting and cable a flop means a loss of listeners or viewers and ultimately a loss of money. It is better to test a product before resources are committed and then to test again to see if the original test results were correct.

A network wants some indication of how well a new TV program will be received. Before committing millions of dollars network executives want to know if they have a potential winner such as "The Cosby Show" or a potential loser such as "Raising Miranda." Market research is one of the ways to obtain information that contributes to informed decision making.

Market research is a method for broadcasters to use in order to gain more knowledge about their products, audience, programming, and image to create the greatest return on investment. Just as manufacturers attempt to learn about their prospective customers and to test new products

PROGRAMMING

RESEARCH DOESN'T ALWAYS GET IT RIGHT

The use of testing can't guarantee a hit program nor can it always identify a loser. The use of cable, auditorium, and other market research testing sometimes misses the mark. "All in the Family" tested so poorly that ABC didn't want it. "thirtysomething" and "Hill Street Blues" also didn't test well but went on to have long, critically acclaimed network runs.

Sometimes testing seems to get it right. Stephen Bochco's "Cop Rock" and Jackie Vernon's sit-com "Chicken Soup" both tested poorly. These shows made it onto the schedule because of contractual agreements but both quickly disappeared. Your best bet for getting a good test is to have a show with animals, children, and/or witches according to an ABC official.

Testing is not well loved by many of the creative people in the TV business. Garry Marshall, the creator and producer of shows "Happy Days," "Mork and Mindy," "Laverne and Shirley," and many others said that he never looked at test results although he would sit in the theatre with the audience. Jay Tarses, who created "Buffalo Bill," "The Days and Nights of Molly Dodd," and "The Slap Maxwell Show," strongly dislikes the testing of new programs. When a new show of his failed in a test and failed to make NBC's schedule, The *New York Times* quote Mr. Tarses as saying, "I don't know who they tested it for. Maybe some kind of plankton."

---

before engaging in the expense of manufacture and marketing, so must broadcasters and cablecasters.

The methods employed in market research are diverse. Some methods require the testing of large groups of people; some test only small groups. Each method is designed to provide broadcast and cable executives with information for better, more reasoned decision making.

Preproduction testing

Before investing huge sums of money in a production, the studio, network, or advertising agency may want to elicit some early response. This testing of initial ideas uses brief descriptions of the program or a brief storyboard of the advertisement. People's general response to these are used as guides in development of the project.

Rough cuts, or models, are used to test the next step before actual production is launched. A rough cut usually is a very simple production meant to give an idea of what the show would be like. Rough cuts have minimal sets and editing and sometimes use amateur actors.

These rough cuts are designed to provide the respondent information about characterizations and approaches. A general sense for the possible quality of the show or ad may be elicited from these tests. But if an audience doesn't seem to like the rough cut, one must remember that the lack of a slick-looking production is probably influencing some of that response.

Auditorium Testing

As mentioned previously, auditorium research is used by advertisers to measure the effectiveness of their commercials. The procedures of gathering an audience together in an auditorium are the same for TV research as for music.

New advertisements may be tested to see how well the audience will accept them. Auditorium testing is a type of high-tech "sneak preview" that film companies pioneered to determine the strengths and weaknesses of a movie before its release.

Special auditoriums are usually used for testing commercials. Each seat in these auditoriums has a hand control. The audience is asked to respond via the hand control to the commercial as it is viewed. The responses from the hand control go directly into a computer, which keeps track of how the audience is responding to the ad. This gives the creators a good idea of what is and what isn't working in an ad.

Advertisers find that auditorium testing of commercials prior to their placement on the networks or local stations is one way to estimate the effectiveness of the commercial. These tests protect advertisers from spending large amounts of money to place commercials that will not sell the goods.

CBS continues to do their own auditorium testing for some of its proposed network programs. While the other networks abandoned auditorium testing for cable testing, CBS uses both. Senior Vice President of Planning and Research at CBS, David Poltrack, says that auditorium testing helps comedies while cable testing favors drama. He feels more comfortable having both types of information.

Cable Testing

The most popular method for testing new programs by the TV networks is through cable testing. Arrangements are made with a few selected cable companies to pay the cable company to show the tested program on a channel that is not assigned to a cable network or broadcasting station. A date and time is set for the showing.

A research firm randomly selects viewers from within the cable company's service area to participate in the test. Calling selected homes, a representative of the research company will screen for the type of viewer that is needed. A household might supply more than one person to participate, a parent and a child, for example. If the person is willing to participate, the person will be told of the time and channel on which the show is to be cablecast. A reminder phone call is usually made the day of the scheduled cablecast to remind the person to watch. At the end of the program the researcher calls back and asks the person a number of questions about the show and the characters.

Cable testing of potential programs has several advantages over auditorium testing. For example, a more heterogeneous audience can be recruited for the study. The greater the diversity in the sample is, the more representative it is of all viewers. Another advantage is that testing takes place in a natural setting. Rather than being in a theatre with a group of strangers, cable testing allows an individual to sit in her or his own home and to watch the program just as she or he would do if the program were scheduled on the network. This natural setting promotes responses closer to what will be experienced when or if the show makes the network's schedule.

Focus Groups

Over the last decade the use of focus groups, or group interviewing, has become very popular. This popularity is due to the apparent ability of these groups to answer the more complicated how and why questions that survey or auditorium research finds difficult to obtain. Focus groups can be a particularly effective method for delving into areas that are difficult or impossible to quantify.

A *focus group* is composed of six to twelve people and a leader. Focus group members are usually paid between $25 and $40 for their participation. Members of specialty groups such as doctors may require a more substantial payment to elicit their participation. As with auditorium testing, members of the focus group are usually obtained by a field service, which also usually provides the setting for the group.

Because of the limited number of people who participate in a focus group, the selection or screening of the participants is of greatest importance. The sample for focus groups must be narrowly defined in order to maximize the value of the group. For example, if a radio station wanted to find out what listeners thought of the station's announcers, the focus group participants might be required to be listeners to the station. To determine if a person is appropriate for the focus group, a series of qualifying questions, called screeners, are used.

The leader, usually called a *moderator,* is the person who attempts to elicit responses from the participants about various topics that are specified by the client. The moderator usually begins the session by summarizing the purpose of the focus group. Next, it is the moderator's responsibility to make the members of the focus group comfortable. Brief self-introductions by focus group members often help.

The discussion, in general, begins with something that all of the members have in common. This may be a response to a question from the screener that was used to select them. Or, as is more likely in broadcast research, the session starts with the listening to or the viewing of a tape.

A single focus group is never enough on which to base good decisions. There is no way to know if the responses from a single focus group represent widespread opinion or are unique to that group. The number of focus groups to be used is a function of how much the sponsoring organization is willing to pay. The minimum number of groups should be two, with the usual number of about six. The greater the number of groups that is used, the more reliable the information will be.

Focus groups are often observed by the clients or researchers either through a one-way mirror or on videotape. People sponsoring the research often like to be present to observe for themselves the reactions of the group. Most focus group sessions are also recorded so that the group's responses may be studied in greater detail later.

In evaluating program material or a commercial in a focus group, members might be asked what type of feelings they had while they watched or what was memorable about the commercial. Questions might focus on the various actors or characters in the programs. These questions and answers provide a more detailed or focused response to questions.

Producers can use these answers as another means to adjust programs so that they will be more effective. If an actor doesn't test well for a role, he or she might be replaced with another actor. A character may be seen by people in the focus group as irritating; thus that character may be rewritten or dropped.

Focus groups can be used on a local level. Television news is often a subject of focus groups. Here the focus group is used to study the perception of the anchors for the station that is paying for the study as well as for those of the competition. What is the image that the news operation projects? How do people respond to the stories covered by the news? A focus group may be used to study a radio station's proposed or present format.

Findings from a focus group may suggest other questions that are best answered by a more representative sample provided through survey research. Decision makers should remember that information from a focus group is information from a very small sample of people. The quality of a focus group is often directly related to how good the leader is. Sometimes members of the focus group may be led to answers that they might not normally give. Some members of focus groups have been known to respond to peer pressure by going along with a perceived majority of the group rather than expressing their own opinions.

It is best to view information from focus groups as a beginning point for more detailed research. Data from focus groups can be thought of as diagnostic, revealing the "whys" behind some behavior.

Survey Research

Surveys of listeners or viewers usually take the form of telephone interviews. Mail surveys have been tried from time to time, but the response rate—the number of returns versus the number sent—is, in general, abysmally low. The small return rate for mail surveys tends to invalidate the results as it hints that there is something special about those people who returned the postcards. Telephone interviewing is the preferred form of media survey work.

Survey research is used when a representative sample is desired. Sometimes the questions that are used in survey research have emerged from focus groups. The station may wish to determine whether the views expressed in the focus group are shared by the population as a whole.

Ratings research is one branch of survey research. The methodological considerations, random sampling, sample size, for instance, are those shared with survey research in general. However, survey research may cover far more than estimating the number of homes listening to or viewing a station.

Some survey work is undertaken in-house but this can lead to problems. Local stations seldom have people trained to conduct survey research. This lack of experience may lead to such problems as biasing answers because of poor question construction, obtaining an unrepresentative sample because the interviewing took place during regular office hours, or receiving subtle cues by staff members who are seeking certain answers. Any of these problems can invalidate the results of a survey.

It is preferable to hire a professional survey research firm. Members of the firm can take the issues of concern to the client and structure questions that address those concerns. "What do you think of TV40?" is probably not a good survey question since it is too broad. Asking a series of more specific questions is a better method in survey research since it allows the respondent to focus on each item and provides manageable answers.

Professional firms either have trained interviewers on staff or hire a field service that has trained interviewers for the survey. The value of trained interviewers is that they have no stake in the answers. Often professional interviewers know nothing of the purpose of the survey. This lack of involvement increases the likelihood that the answers are those of the interviewee rather than an answer the interviewee may have been led to.

The size of the sample is a function primarily of how accurate the client wants the results to reflect the population that is sampled. This accuracy is measured in two ways: tolerated error and confidence level.

The *tolerated error* is the margin of error between the results of the sample and the results elicited if the question was asked of all the population. That is, if everyone was questioned and the response was, say, 65 percent, how close to that number do you want your sample to be? In other words, how much error can you tolerate? If you wish to be within plus or minus 3 percent, which is the tolerated error in national surveys such as Gallup polls, you are saying that you will be satisfied if your sample result in the example is from 62 to 68 percent. If a tolerated error of 7 percent is acceptable, you are saying that the actual population response as a whole could be anywhere from 7 percent lower to 7 percent higher than the one reported, that is, from 58 to 72 percent. Obviously, a 7 percent tolerated error requires fewer people than a 3 percent error does.

The *confidence level* deals with the researcher's degree of certainty that the results reported are an accurate reflection of the population and not just something that happened by chance. The usual accepted level of confidence is 95 percent. This means that 95 times out of 100 times the results reported did not happen by some whim of chance. Put another way, you are betting a 20 to 1 shot that you are correct. Would you feel confident of winning a bet if you knew the odds were 20 to 1 in your favor?

One story of how 20 to 1 came to be the accepted level of confidence says that it originated in the British College of Surgeons. In the nineteenth century the college decided that if a given surgical procedure killed only one person in every twenty, then the procedure was warranted. For a broadcast station to be killed in the ratings is not quite the same thing, but this confidence level seems to be the accepted one.

It is possible to have samples with lower tolerated error, say, 1 percent, and confidence levels set at 99 percent, but the number of people needed for such a sample rises dramatically. The minimum number of completed questionnaires needed in a sample with 3 percent tolerated error and a 95 out of a 100 confidence level is 1067. For a tolerated error of 1 percent and a confidence level of 99 out of 100 the minimum needed is 16,587. Is this increase in confidence and greater accuracy worth it? The answer is "It depends." Each of those responses has to be paid for and the larger the number of respondents is, the larger the bill to the client will be. Frequently the gain in accuracy might not be worth the cost.

Confidence intervals can easily be calculated for ratings data as well. Note that the broadcasting and cable industry has decided that Nielsen's Peoplemeter sample of 4500 represents an acceptable compromise between accuracy and cost. Table 15-1 uses well-known statistical formulas to indicate representative sample sizes needed for various error tolerances and confidence levels. Keep in mind that the accuracy of a given survey sample is a function only of sample size. The size of the population doesn't matter (as long as the population is large, and in broadcasting and cable research it usually is). Thus a sample of 1067 randomly selected from a population of 1 million will produce results just as accurate as a sample of 1067 randomly selected from a population of 90 million.

TABLE 15-1 Random Sample Sizes Needed for Various Degrees of Precision

Confidence Level (%)	Tolerated Error ± 1%	± 3%	± 5%	± 7%
95	9,604	1,067	384	196
99	16,587	1,843	663	339

Station image One subject of survey research for radio stations is the station image study. Station decision makers need to know how the public perceives their station. Is the perception of the public the one that station management has been trying to elicit? If the answer is no, the station must take steps to correct the mistaken image.

A radio station might begin a print or billboard advertising campaign as a means to correct misperceptions. These print-based campaigns are designed to draw listeners back to sample the programming again.

Station image studies are very important after a station has changed formats. At these times stations can determine whether their promotional efforts at alerting prospective listeners of the format change have been effective.

TV Q Who is America's favorite TV personality? Answering this question is a company called Marketing Evaluations, Inc. It provides a list of Q scores, more formally *performer Q*, a measure of a performer's popularity, for over a thousand TV personalities. These scores can be used by producers, networks, advertisers, and others to determine which star appeals to any particular audience segment.

Marketing Evaluations makes their rankings of TV performers by using three panels of 1250 people from across the country who agree to be part of the telephone marketing panel. Members of the panel respond to a list of names, with each member of the panel requested to score each name on a 1 to 6 scale; a score of 1 means "One of my favorites," whereas a score of 6 means "Someone I have never seen or heard before." The ranking for each performer is tallied.

Marketing Evaluations reports a TV Q score for each performer. To calculate this score, Marketing Evaluations first computes a familiarity score for the performer. The familiarity score is based on the percentage of respondents who gave the celebrity a 1 to 5 score. Next, a favorite score is produced based on the percentage of people who gave the performer a 1, "One of my favorites." The performer's TV Q is the favorite score divided by the familiar score. For example, if 50 percent of the sample gave a performer a score of 1 to 5 and only 20 percent rated the performer a 1, then the performer's TV Q would be ($20/50$) or 4. Okay, it is really a .4 but Marketing Evaluations multiplies by 10.

These results are also broken down by the age and the sex of the respondent. This allows the

---

TECHNOLOGY

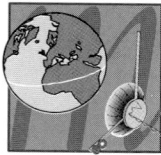

VOX BOX—DOES THE WORLD WANT A BETTER MOUSETRAP?

A system that combined the questions of quality like TV Q with the computer technology of Peoplemeters was the VoX Box. Developed by the R. D. Percy Company, the VoX Box was a way for viewers to respond to how well they liked the program or performers they were seeing on TV. A box with buttons labeled "Excellent, Zap, Dumb, Informative, Credible, Funny, Unbelievable, Boring" allowed the viewer to make judgments on just how good a show or parts of a show were, and a button marked "Person" allowed the people being reviewed to appear on the screen. Like the Peoplemeter, the VoX Box assigned recognition codes to each member of the family so that it could keep track of whose opinion the system was receiving. Programs, as well as performers or characters within a program, could be evaluated with the VoX Box. The VoX Box was connected by telephone lines to a central computer. These computers were able to monitor the responses just as they do for ratings.

The difference between the information supplied by the VoX Box versus the ratings was a matter of perceived quality of the program. Ratings tell approximately how many people are watching a show but don't indicate how many people are enjoying it. It's 3:00 A.M. and you are watching a man demonstrate the Flowbee, a hair-cutting machine that attaches to your vacuum cleaner—Are you enjoying it? VoX Boxes enabled the viewer to provide some response without the use of a Visa, Discover, or Mastercard.

But, alas, the system is history. Primarily financial problems led to the end of the system, but it still seems like a good idea to be able to respond "Dumb" when watching some things on TV.

client to know which performers appeal to which parts of the audience.

Marketing Evaluations also produces a list of performers who have high recognition but low appeal as indicated by a negative TV Q rating. It also lists performers who have low recognition—recognized by a few people—but liked by those who do recognize them.

These lists of performers' TV Qs are used by advertisers in selecting a product spokesperson. Bill Cosby has a very high TV Q, which has helped him secure one of the longest string of endorsements in advertising. Producers trying to cast a hit TV program select actors based on their TV Q score. Also, there are cases where a producer's selection of an actor has been overridden by the network in favor of an actor with a better TV Q.

Performers are not the only commodity that Marketing Evaluations evaluates. It also produces TV Q scores for programs. These program TV Qs are similar to the performer TV Qs. The panel is asked to rank the shows using much the same scale as that used for performers, except that 6 means a "Program I have never seen." Familiarity, favorite, and TV Q scores are calculated in the same fashion as for performers. These program TV Qs are helpful to programmers in deciding whether to take a show off the air or to let it stay on. "Hill Street Blues" was a show that had a high TV Q even though it seldom had high ratings.

Local programmers find program TV Qs helpful. TV Qs can help executives decide which shows in syndication to purchase for their stations. The TV Q scores can indicate not only whether the show is popular but also whether the show fits into the program schedule.

Psychographics and lifestyle surveys Advertisers and program producers sometimes like to classify audiences into readily identifiable groups that help define the audience for market researchers. As noted in the previous chapter, the ratings book breaks down the audience by demographic characteristics such as age, sex, and place of residence. Consumer researchers, however, realize that demographics do not tell the whole story. For example, the viewing, listening, and buying habits of a 20 year old from New York City may not resemble those of a 20 year old from rural Nebraska despite their age. Consequently, some market researchers endorse the notion that psychographic and lifestyle research present a more complete picture of the audience. The terms "psychographics" and "lifestyle" are sometimes used interchangably; however, there is a difference between the two. Psychographics emphasizes generalized personality traits, whereas lifestyles emphasize needs and values that are associated with consumer behavior.

Psychographic variables are measures of such personality traits as leadership, sociability, independence, conformity, and compulsiveness. Psychographic data are collected by asking audience members to rate themselves on a long list of adjectives such as "refined," "tense," "organized," "romantic," using a 5-point scale that ranges from "Agree a lot" to "Disagree a lot."

If the viewers of a particular TV show are psychographically similar, this information could be useful in producing commercials to run in that show. For example, if a psychographic survey noted that the audience for "Designing Women" scored high on independence, ads that stress such themes as "break free from the crowd" or "go your own way" might be efficient selling tools.

Related to psychographic surveys are *lifestyle surveys,* which measure the way a person perceives himself or herself in relation to job aspirations, leisure activities, enduring values, and buying habits. There are several ways to classify audience lifestyles, but the one that is best known is VALS II—values and lifestyles—developed at the Stanford Research Institute. The VALS survey covers such diverse areas as attitudes about life in general, personal goals, media exposure, and consumption patterns. VALS II emphasizes the psychological makeup of consumers and places less emphasis on values than the original VALS (See Figure 15-1).

People are divided into three basic categories: those viewed as principle-oriented, status-oriented, and action-oriented. VALS II also estimates the resources of the individual, which includes such things as a person's health, income, and education. The basic categories and resource information combine to create eight subcategories: affluence ranges from actualizers to strugglers; principle-oriented are broken down into fullfilleds and believers; along the status-oriented dimension the subcategories are achievers and strivers; experiencers and makers are action-

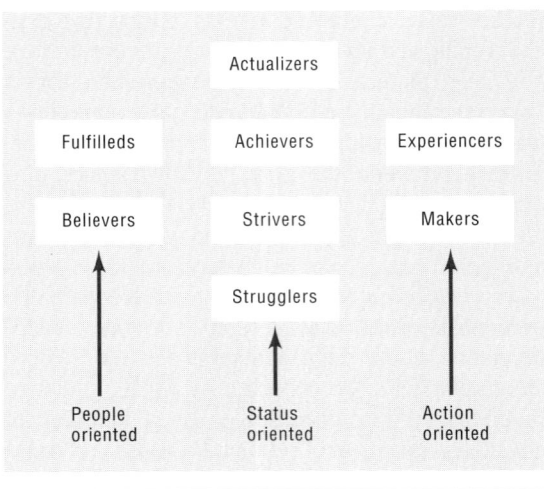

FIGURE 15-1 A simplified diagram of the VALS II categories.

oriented. Other lifestyle surveys used to segment the audience more narrowly are the list of values (LOV) and social style.

Advertisers have used VALS and other lifestyle measurements in developing advertising campaigns, positioning or repositioning a product, and segmenting the audience. Lifestyles measurement helps an advertiser select a compatible medium for the client's advertising or a compatible program within a specific medium. Television and radio stations have only recently started to use lifestyle measurements to differentiate the audiences of their various programs.

PHYSIOLOGICAL MEASURES

Audience researchers have learned how to use machines to measure physiological changes in people. It is assumed that physiological changes are caused by psychological reactions to stimuli. Galvanic skin response and blood pressure are two of the most common types of physiological responses measured by researchers.

Galvanic skin response (GSR) measures the tiny changes in the electrical resistance of your skin. The theory is that as you become excited or interested, your skin loses resistance from changes in perspiration. These tiny changes can be measured through sensors attached to the skin. The same applies to blood pressure. As you become more excited, your heart will pump more frequently and your blood pressure rises. Both measures are used to indicate some form of excitement in the person connected to the machinery. GSR and blood pressure are two of the critical measurements used in lie detectors.

Some sophisticated researchers also take *electroencephalogram (EEG)* readings from subjects. Through the attachment of sensors on the scalp, EEGs measure the four types of brain waves: alpha, beta, delta, and theta. Whereas blood pressure and GSR gauge the results of brain activity, EEGs measure brain activity itself directly. For example, it has been shown that watching TV produces alpha rhythms, which are very close to those the brain has during sleep. This may simply mean that most TV programming makes you want to take a nap, or it may be an indicator that TV works at a much deeper level.

These tests can be used to measure various aspects of the electronic media. Want to know how newly released music will do on the playlist? Let people listen and measure their physiological reactions. In what type of news stories is the public most interested? Show them and see which ones get the greatest physiological response.

The perceived advantage of physiological measurement over other forms of information collec-

---

INTERNATIONAL

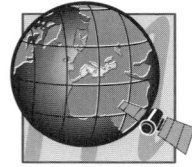

MEASURING THE MOSCOW TV AUDIENCE

As Russia moves toward a free-market economy, its broadcasters are learning a new concept—ratings. Under Soviet rule there was little need to know who was watching. Now, with many Western advertisers considering purchasing commercial time on Russian media, a new service—the Russian Media Monitor—was started in 1992. A panel of 2000 viewers regularly reports what it watches on TV. The first results had to do with how much the panel viewed Western TV shows. What was watched the most? MTV.

One possible arrangement used to collect physiological data in audience research.

tion is that it removes the vagaries of language. Researchers do not have to rely on verbal information, which is often inaccurate or subject to the verbal biases of a person's speech pattern.

If I were conducting research on your likes and dislikes, I might ask if you loved ice cream. I would like you to select from one of these answers: strongly disagree, agree, and strongly agree. You might answer that you strongly agree. Now I ask if you love your mother, using the same scale. You answer that you strongly agree to that, too. Should we assume that you love ice cream as much as you do your mother? Well, probably not. Physiological measurements are meant to help remove the problems created by having to find words to express your reactions.

Entertainment Response Analysts (ERA) was the first major consulting firm to use such techniques. ERA would test programming, especially news programming, for client stations. Using samples of eighty people, ERA would do GSR studies to suggest what worked or didn't work in local newscasts. It studied everything from the anchor to the graphics.

ERA's work was based on early physiological studies that were conducted by Thomas Turicchi on the response of students to music. Turicchi discovered that based on GSR he could predict what music would be successful. He began helping radio stations to select their playlists, with a reported accuracy of about 90 percent. This level of accuracy is unheard of in the music business.

One problem with physiological measures is that one can't always be sure what is causing the change in brain waves or skin resistance. Do you react to the music because it has a beat you can dance to and you like the lyrics, or did it remind you of the time you spent with your date last weekend? Either one of these will give a peak, but how does the researcher know which one it was? It could even be a strong aversive reaction that you are having; the GSR, blood pressure readings, and EEG can't tell the difference.

Another problem is that some people object to the physiological measures, considering them as an invasion of an individual's privacy. Using such measures reduces a participant's right to decide what information to give a researcher. The right of the participant to refuse to provide information to a researcher is considered fundamental.

CONSULTANTS

Stations may find it useful to hire experts from outside their area. These experts, or consultants, can help a station do anything from evaluating the management structure of the station, to providing motivation for sales executives, to suggesting changes in the station's programming, or to help to find new talent.

A consultant usually specializes in one or two areas. Consulting firms may be comprised of one consultant or many consultants. Major consulting firms offer clients a full range of consulting and research services.

Consulting firms have in common the ability to provide experts on a temporary basis to solve immediate broadcasting and cable problems. Consultants provide useful services and expertise to management. They usually charge much more than an average employee at the station is paid. But since the consultant is in town for a limited time only, the cost is less than hiring a new employee to handle the problem.

In general, consultants can be divided into seven types of service categories: financial, management, personnel, programming, public relations, research, and technical.

Financial Consultants

As mentioned before, broadcasting is a business. Understanding the financial ramifications of the business is difficult. Trends in the business marketplace create new opportunities and new pitfalls; but station managers and other station personnel may not be able to keep abreast of these trends. At critical times—when acquiring a new station or cable system, for example—outside help from a consultant is often warranted.

Financial consultants are frequently called in to assist with sales or purchases. Consultants can help with appraisals of a station's assets, suggest methods of financing, and identify brokers; indeed, they may be brokers themselves. Financial consultants may be called in to help with automating the accounts department. The implementation of a new billing or accounting system is often the time to hire an expert to train the staff on the new system. In addition, the consultant is usually placed "on call" for a specified period to answer questions that may arise after his or her departure.

Management Consultants

Department heads sometimes find that they are not able to do their jobs efficiently. Perhaps it's a time management problem—not enough hours in the day to get the work done. Perhaps it's a problem of effectively communicating orders to subordinates or taking orders from superiors. A management consultant might help.

Management consultants usually observe the operation of a business, question management and subordinates, and then make suggestions. They attempt to make the organization run more smoothly and efficiently.

Personnel Consultants

Personnel consultants will do everything from suggesting changes in eye color of the anchors to helping to find an anchor with the right eye color. A different type of personnel consultant can also help with the more mundane matter of how to gain more staff productivity.

There are many consultants who engage in talent evaluation. The hottest area for these consultants is in local news. Frank N. Magid Associates and McHugh and Hoffman are two of the most influential consultants in this area. Talent evaluation does not limit itself to how well a person reads or writes copy, the quality of eye contact, or how well the person dresses. Talent evaluation may also include discussions of whether the person's eyes are too close together or whether the part of the performer's hair should be on the other side.

Some consulting firms maintain files of performers who are looking for a new job. These firms may act as an employment firm for the station and talent.

Programming Consultants

Consultants help local management make decisions about what gets transmitted. Programming consultants are available for every aspect of programming.

At radio stations consultants help select the proper format for the market. If management doesn't wish to change the format, the consultant can help choose playlists or decide the order in which music will be played. Consultants may counsel on the effectiveness of purchasing programming services or program the station themselves.

A program consultant for a TV station will help the station set its daily program schedule. He or she may make recommendations about both the placement of programs that the station presently owns and future acquisitions.

Government/Public Relations Consultants

Consultants help stations deal with external publics. Cable companies and broadcasters deal with Congress, the FCC, and the local governing board of their communities. However, both must also deal with the public as subscriber or viewers. Consultants can help broadcasters and cable companies deal with these groups more effectively.

Mounting an effective promotional campaign, especially in the fall during the beginning of the new TV season, is critical. Planning for the new season may call for a consultant to help map strategy.

If a station is having public relations problems with a segment of its audience, it may be a signal for the services of a consultant who is an expert in this area. The consultant may offer new ideas to improve relations.

Technical Consultants

Technical consultants help make decisions about equipment. They advise the chief engineer in making equipment selection. Some offer specific technical assistance, like finding the best site for a transmitter tower or determining which combination antenna and transmitter will achieve maximum signal coverage. These issues are important to the station, but the knowledge for making these decisions is not always directly in the stations.

Research Consultants

The area that attracts most consulting is in the general area of research. As previously mentioned, most TV and radio stations in the United States don't have their own in-house research people. Research consultants literally come into your community and test for whatever you think is a problem, and if you don't know if you have a problem, consultants will be happy to tell you what your problems are.

Some research consultants specialize in interpreting the results of other research consultants. It is not unusual for a consultant to provide a fairly detailed and thick report. There is usually a summary of the findings, which will be five to ten pages long. At the very beginning there is an executive summary, which is one to two pages. To help understand all of the information within research, you may sometimes need to hire additional help.

Research removes some of the guesswork from decision making. It provides additional information that adds to a programmer's own knowledge and experience. But research is not foolproof. The low survival rate of new network TV programs attests to the fallibility of testing.

A knowledgeable broadcasting or cable executive learns to balance research results with her or his own experiences; experience is often more valuable than research. A balance between what the executive knows and what the research reveals often leads to decisions that are best for the broadcaster or cable system, the audience, and the industry.

SUMMARY

Nonsyndicated audience research includes those forms of research that do not deal with the ratings of particular shows.

Music research is carried out by call-out and auditorium testing. In both methods listeners determine the popularity of established and new recordings.

Market research makes use of preproduction testing to give producers an idea of how an audience will react to a show. Auditorium testing is also used to test commercials and to sneak preview TV shows. Another popular research tool is the focus group. Small groups discuss a series of predetermined topics and give producers insights as to why some things are popular and some aren't.

Some broadcasters and cable operators make use of survey research to gather information about their operations. A popular form of survey research involves delineating the image of a station in the minds of the public. TV Q studies determine the best known and best liked celebrities.

Psychographic and lifestyle research focus on personality traits and values in an effort to understand the audience further. Physiological measures, such as galvanic skin response and blood pressure, are also used in broadcast and cable research.

Finally, some stations hire outside consultants to help them evaluate their operations. These consultants include experts in finance, management, personnel, programming, public relations, technology, and research.

SUGGESTED READING

Backstrom, C. H., & Hursh, G. D. (1963). *Survey Research*. Evanston, Ill.: Northwestern University Press.

Dominick, J., & Fletcher, J. (1985). *Broadcast research methods.* Boston, Mass.: Allyn & Bacon.

Hsia, H. (1988). *Mass communication research methods.* Hillsdale, N.J.: L. Erlbaum Associates.

Lindlof, T. (1987). *Natural audiences: Qualitative research of media uses and effects.* Norwood, N.J.: Ablex.

Marshall, C., & Rossman, G. (1989). *Designing qualitative research.* New York: Sage.

Nichols, J. E. (1990). *By the numbers: Using demographics and psychographics for business growth in the '90s.* Chicago, Ill.: Basic Books.

Stewart, D. W., & Shamdasani, P. N. (1990). *Focus groups: Theory and practice.* Newbury Park, Calif.: Sage.

Wells, W. D. (1974). *Life style and psychographics.* New York: American Marketing Association.

Wimmer, R. D., & Dominick, J. R. (1991). *Mass Media Research: An Introduction.* Belmont, Calif.: Wadsworth.

CHAPTER 16: THE EFFECTS OF BROADCASTING AND CABLE

Do TV and radio have an impact on our lives? Ask yourself the following questions:

1. What brand of gym shoes does Michael Jordan wear?
2. What rights does a person have when arrested?
3. While on a date, have you ever used a line or clever remark that you heard on TV?
4. What kind of a day is it in Mr. Rogers's neighborhood?
5. When you're preparing for an important occasion, do you worry about dandruff, acne, perspiration, bad breath, ring around the collar, yellow teeth?
6. Did you ever consciously dress like a character you saw on TV?
7. How many L.A. lawyers are like the lawyers on "L.A. Law"?
8. Do you vote differently because of televised campaigns for political candidates?

The answers to these questions will tell you how much of an impact TV and radio have had on your personal life.

Now consider the global scale. What impact has the broadcast media had on society? This question is a little more difficult to answer. Nonetheless, it's an important topic. Radio and TV have been, at various times, the alleged culprits behind a host of social ills. Television, it was claimed, made us more violent and antisocial, hurt our reading skills, decreased our SAT scores, fostered sexual stereotypes, and more. From a pragmatic standpoint society needs to know if, in fact, these allegations are valid, and if so, how to correct them. Consequently, this chapter briefly examines how we go about studying the impact of radio and TV, the changing views concerning media effects over the past seventy years or so, and the most current research about the effects of broadcasting in specific areas.

Before beginning we should note that most studies of the topics cited in this chapter examine the effects of traditional broadcasting on the audience. Few studies have concentrated on the social effects of viewing content that is available only on cable TV. Nonetheless, we would expect that most of the research would generalize to cable viewing. At the same time, however, many of the findings mentioned in this chapter might need rethinking in light of the new multichannel environment of TV.

STUDYING THE EFFECTS OF BROADCASTING AND CABLE

There are many ways to examine the social consequences of the electronic media. Some, however, are better than others. A person might, for example, conclude that TV and radio are harmful forces in society because some noted authority says so. All of us have probably seen headlines such as "Noted Psychiatrist Blasts TV" or "Vice President Holds TV Responsible for Morals Decline." If a person takes what these individuals say at face value, he or she is relying on what philosophers call the method of authority for beliefs about media effects.

Alternatively, some people might rely on intuition as the foundation of their beliefs. Intuitive statements are generally self-evident. For example, it is self-evident that all the "garbage" on TV is having a harmful effect on society. Or, since people spend so much time watching TV, it is self-evident that it must have considerable impact on their thoughts, feelings, and actions. Relying on authority or intuition works well in some situations, but each is limited as a foundation for one's opinions. For example, some so-called experts are self-styled and may not know what they are talking about, and what is "self-evident" may not be correct. For many centuries it was self-evident that the sun revolved around the Earth, until Galileo and Copernicus suggested differently.

When it comes to studying the effects of broadcasting, it is safer to rely on evidence derived from using the scientific method. Simply put, the scientific method guards against accepting untested or unfounded assumptions about reality. It is based on testing our assumptions (known as hypotheses) against evidence derived from the real world (a method known as empiricism). Note that Galileo and Copernicus challenged the existing model of the solar system because it did not match what they observed. In other words, they were empiricists; they rejected the conventional wisdom when they saw contrary evidence. Of course, the scientific method may not be best for every situation (it's seldom used to choose a friend,) but it's quite appropriate for analyzing the social effects of radio and TV.

There are, of course, other research approaches. Some scientists employ qualitative methods, in which they make direct observations and in-depth analyses of mass communication behaviors in natural settings. Other scholars use the critical or cultural studies technique that is long popular in the humanities to provide a more interpretive look at the process of mass communication. Both the qualitative and critical/cultural studies approaches suggest new and different ways of explaining and understanding the nature of the impact of electronic media. Those of you who want to know more about these methods should read the books by James Anderson and Richard Adler cited in the list of suggested readings. This chapter, however, focuses more on the traditional and pragmatic social science approach to mass media effects; it emphasizes those research questions that have implications for social policy.

In broadcasting and cable research several techniques are used to gather data about audience effects. These techniques are wide-ranging partly because historically the effects of mass communication have been studied by psychologists, social scientists, political scientists, and others. Not surprisingly, each discipline has relied on the technique most closely associated with it. Thus psychologists use the experimental method, whereas sociologists use surveys. There are, however, many variations on experimental and survey research. In general, it's possible to say that there are three main social scientific methods:

1 Experimental methods, which can take place either in controlled "laboratory" conditions or in more natural "field" conditions
2 Survey methods, which can sample the subjects one time only or continue over time
3 Content analysis, which is a systematic method for analyzing and classifying communication content

Each of these three methods has its own built-in pros and cons. Knowing the advantages and disadvantages of these various techniques is important because they have an impact on the degree of confidence that we have in research results.

Laboratory experiments are done under tightly controlled conditions and allow researchers to focus on the effects of one or more factors that may have an impact on the audience. Usually at least two groups are involved; one group gets treated one way while the other is treated differently. The two groups are then checked for differences. The big advantage of lab experiments over other methods is that they allow researchers to make statements about cause and effect. In an experiment subjects are randomly assigned to experimental conditions and the researcher has control over most external factors that might bias the results, thus making the claim of cause and effect stronger. Their big disadvantage is that they are done under artificial conditions and behavior that occurs in the lab might not occur in real life.

Field experiments occur outside the lab. Sometimes a natural event occurs that creates the conditions necessary for an experiment. For example, suppose one community is about to be equipped with a thirty-five-channel cable system while a comparable community is limited to traditional broadcasting. Such an event would allow researchers to study the effects that the introduction of such a system has on residents. Sometimes field experiments can be set up by the researcher. To illustrate, suppose one program is fed to one-half the homes on a special cable system while the remainder see a different program. The effects of this single program could then be examined. The advantage of field experiments is naturalness; people are studied in their typical environments. The big disadvantage is the lack of control. Unlike the lab, field experiments are subject to the contaminating influences of outside events.

Surveys generally consist of a person's answers to a set of predetermined questions. Surveys are done through the mail, over the phone, or in person and usually involve some kind of questionnaire or other written document. The big advantage of survey research is its realistic approach. People in natural settings are asked questions about their typical behaviors. A big disadvantage with surveys is the fact that they can't establish cause and effect. After a survey a researcher can say only that factor A and factor B are related; the researcher cannot say that A causes B or that B causes A. A survey only establishes a relationship. Another disadvantage is that surveys rely on self-reports. It can only be assumed that respondents give valid and truthful responses.

Surveys can be done one time or they can be *longitudinal* (repeated over time.) A *trend study* is one in which the same question or questions are asked of different people at different times. An example of a trend study would be a poll done six months before an election that asks what presidential candidate people intend to vote for and is then repeated with a different group of people a week before the election.

Panel studies are a special type of longitudinal survey in which the same people are studied at different points in time. The advantage in a panel study is that some evidence of cause and effect can be established, usually through sophisticated statistical analysis. The disadvantages include the fact that panel studies take a long time to do and they suffer from attrition—respondents die, move away, or get bored and are no longer part of the study group.

Content analysis studies segments of TV and radio content in order to describe the messages presented by these media. Such studies have been useful in defining media stereotypes and establishing a gauge of the amount of violence in TV programming. The biggest problem with content analysis is that it cannot be used alone as a basis for making statements about the effects of media content. For example, just because a content analysis establishes that Saturday morning cartoons are saturated with violence, it doesn't necessarily follow that children who watch these shows will behave violently. That might be the case, but it would take an audience study to substantiate that claim—a content analysis by itself would be insufficient.

THEORIES OF MEDIA EFFECTS

These methods have been used to study the impact of mass media since early in this century. Throughout that time our view about the power of media effects has undergone significant changes as social science learns more about the various factors that affect media impact. As we shall see, there were periods in history when the media (including broadcasting) were thought of as quite powerful and other times when they were thought to have little effect. The current thinking seems to represent a compromise between these two extreme positions. The rest of this section briefly reviews the various theories concerning the effects of mass media that have evolved over the years.

Hypodermic Needle Theory

One of the earliest theories of media effects held that mass-communicated messages would have strong and more or less universal effects on the audience they reached. It was thought that the media would "shoot" beliefs into people's minds much the same way a doctor inoculates people with a hypodermic needle. Much of this thinking was due to the apparent success of propaganda before and during World War I (see box). For example, the dominant mass medium of the time was the newspaper, and the papers owned by publisher William Randolph Hearst were credited with pushing the United States into a war with Spain in 1898. In 1914–1916, as World War I broke out in Europe, skillful British propaganda stories were considered responsible for bringing the United States into the war on the side of the Allies. After the war a new medium, radio, further reinforced the hypodermic model. Successful radio rogues, such as Dr. Brinkley (see Chapter 17), and the "War of the Worlds" scare seemed to support the view that mass media can have powerful consequences. In Europe the skillful use of radio and film propaganda helped bring Hitler's regime to power and keep it there.

Reexaminations of the development of this model suggest that it was not so thoroughly accepted as once believed. In addition, some social scientists of this early period also argued that other factors should be considered when discussing the impact of the media. In any case, advances in experimental and survey research began to cast serious doubt on the hypodermic needle model. By the mid-1940s it was obvious that the model's assumptions about the way communication affected the audiences were too simplistic. The pendulum was about to swing in the other direction.

Limited-Effects Theory

Persuasion, especially the political kind that goes on in election campaigns, was the main focus of this new line of research. Several studies indicated that the media did not have a direct effect on the audience, as was previously believed. The newly developed two-step flow theory suggested that media influence first passed through a group of people known as opinion leaders and then on to the rest of the audience. Further research posited that media influence was filtered through a net of intervening factors, such as a person's prior beliefs and knowledge and the influence of family, friends, and peer groups. The mass media were simply one of a great many determinants of how people think or behave.

This view brought some comfort to those who feared that the public might be brainwashed by skillful ideologues and clever propaganda techniques since it suggested that they were unlikely to succeed. It also was appealing to mass media executives, who could use it to counter criticism that the media were the cause of various social ills.

The most complete statement of the limited-effects position (although he doesn't call it that), appears in a book by Joseph Klapper, *The Effects of Mass Communication*, published in 1960. Klapper reviewed the existing research and summed it up in a series of generalizations, the most widely quoted of which held that mass communication alone does not ordinarily cause audience effects but instead functions primarily to reinforce existing conditions.

Klapper's generalizations enjoyed popularity for nearly two decades; in fact, there are some who still subscribe to the limited-effects model even today (for an example, read the essay by William J. McGuire cited at the end of the chapter). Nonetheless, keep in mind that the bulk of the studies reviewed by Klapper were done before TV became the dominant mass medium. In fact, 95 percent of the 275 citations in Klapper's book are pre-1958. About 20 percent of U.S. homes didn't even have TV in 1958, there were only about 400 TV stations, and very few homes could pick up more than three channels. In addition, although the most widely quoted of Klapper's conclusions concerned the reinforcement effect of the media, he also noted that there were occasions when the media could exert direct effects or when the mediating factors that generally produce reinforcement are absent or themselves help foster change. As TV became more prevalent and researchers discovered more areas

BACKGROUND

PROPAGANDA IN WORLD WAR I

World War I was the first modern "total" war. Along with the armies in the field, the entire manufacturing potential of nations was pitted against that of their rivals. This new kind of warfare required the total support of all the citizens in a country so that armies could be raised, production kept at full pace, and money obtained to finance the war. To mobilize the citizenry during World War I, a massive propaganda effort was undertaken by all the belligerents. Messages appearing in newspapers, on posters and billboards, in books, and on records were all designed with common themes: hate the enemy, love your country, and contribute the maximum to the war effort.

After the war was over those involved in these propaganda campaigns revealed that much of what they had told their fellow citizens was false; many of the most effective propaganda stories were fabricated by those involved. For example, in Great Britain, whose effective propaganda campaigns were credited with bringing in the United States and many other countries on the side of the Allies, it was revealed that many atrocity stories, such as German soldiers bayoneting babies, were totally fabricated. On another occasion a British intelligence officer was examining two photos captured from the Germans. One showed dead German horses on their way to a factory where soap and oil were made from their carcasses; the other showed a wagonload of dead German soldiers on their way back behind the lines for burial. The intelligence officer simply switched a caption and under the photo of the dead soldiers ran the inscription "German Cadavers on Their Way to the Soap Factory." The photo and caption were then sent to China, where reverence for the dead is a powerful tradition. The miscaptioned photo was credited as being one of the reasons that China declared war on Germany.

where media effects were direct, the limited-effects model gave way to a new formulation.

Specific-Effects Theory

The most recent theory of media effects represents a middle ground. Researchers realize that media are not all-powerful; they compete with or complement other sources of influence such as friends, family, and teachers. Nonetheless, there are circumstances under which specific types of media content might have a significant effect on certain members of the audience. Although this statement might not be entirely satisfying from a scientific standpoint, in the last few years communication researchers have made great progress in identifying how and when, and sometimes even why, mass media, especially broadcasting, affect individuals and groups. Accordingly, the answer to the question, "What are the effects of the mass media?" has become complex as social scientists continue to define the circumstances, the topics, and the people for whom specifics effects might occur.

The remainder of this chapter examines eight of the most investigated topics in recent broadcasting research: (1) the effect of violent TV programming on antisocial behavior, (2) perceptions of social reality, (3) sex-role stereotyping; (4) TV and politics; (5) TV and educational skills, (6) the impact of broadcast advertising; (7) uses and gratifications and (8) prosocial behavior.

VIDEO VIOLENCE

It is appropriate that we first examine the media violence area. This subject is the most controversial, generates the most research, utilizes all of the four research techniques mentioned earlier, illustrates some of the problems in generalizing from research data, and is the one topic about which communciation researchers know the most.

History

People worried about the effects of media violence long before TV came on the scene. Around the turn of the century parents were concerned that detective stories in "dime novels" would corrupt the morals of the young. A few years later movies took center stage. The crime and gangster films of the 1930s (take a look at *Public Enemy* or *Little Caesar* if you get a chance) were the focal point of an early series of scientific studies that demonstrated that what was on screen had an impact on adolescent attitudes. In the 1950s the spotlight focused on the alleged harmful effects of reading comic books (see box).

During the 1950s and 1960s juvenile delinquency became a topic of national concern. Not surprisingly, the role that TV, then becoming the most popular mass medium, played in this process came under close scrutiny. As early as 1954 a Senate subcommittee examined the small amount of research that existed then concerning the effects of viewing violence on teenagers. The subcommittee reached no general conclusions about the topic but did call for more research. Seven years later this same subcommittee again inquired into TV's impact on antisocial behavior. After holding extensive hearings it ultimately issued a report that was critical of TV industry programming and concluded that watching TV was a factor in shaping the attitudes and characters of young people. The report, however, did not examine the specific effects that TV violence had on its audience.

The urban unrest and the wave of violence that characterized the mid-1960s sparked a new burst of interest in the topic in 1968. A presidential commission, the National Commission on the Causes and Prevention of Violence (NCCPV), again held hearings on the role of the media in times of social friction. After reviewing the available evidence the staff report of the NCCPV concluded with a strong indictment of video violence: "We believe it is reasonable to conclude that a constant diet of violent behavior on television has an adverse effect on human character and attitudes . . . encourages violent forms of behavior and fosters moral and social values about violence . . . which are unacceptable in a civilized society." Several scientists criticized the NCCPV for going beyond their data in drawing these conclusions, but all agreed it was the first time any group studying the issue had come out with a public condemnation of TV violence.

This conclusion was only one of many that appeared in the final NCCPV report. It did not receive much attention outside the social science community. Consequently, it was no surprise that the same controversy surfaced again only a few

BACKGROUND BRIEFING

BEWARE THE COMIC BOOK

Before TV there were comic books. Granted some of today's comic books are bizarre, but they can't hold a candle to the comics of the 1950s. Those were the days of controversial, garish comic books filled with scenes of violence and great comic-book words like ARGHH, THUNK, GLURG, and AANGH. By 1950 some 60 million comic books were being sold every month, the vast majority containing scenes of crime and violence. Some of these illustrations were hideous. One comic book cover featured a close-up of a hanged man with his neck obviously broken and eyeballs protruding. A series of illustrations in another comic pictured a baseball game in which all the equipment consisted of dismembered limbs and other assorted body parts. It was not long before the alleged harmful effects of this medium were a hot topic of debate. In the forefront of the crusade against comic books was a prominent New York psychiatrist, Dr. Frederic Wertham. In *Seduction of the Innocent* he outlined the case against comics.

In the first place, said Wertham, comics were a school for criminals. They pictured the precise details of how to pull off such crimes as arson, car theft, purse snatching, housebreaking, and assault. In the second place, some kids actually imitated the crimes they saw portrayed. Wertham presented at least a half dozen cases, ranging from extortion to murder, where children had copied what they had read. Third, Wertham suggested that comics featuring crime and violence caused a distorted perception of reality. Comics gave unrealistic expectations about violence, sex roles, and minority groups.

Although his research was criticized as selective and impressionistic, it was primarily because of Wertham's writings that congressional hearings were called to examine the relationship between comic book reading and juvenile delinquency. Concerned over possible federal action, comic book publishers cleaned up their act and toned down the most blatant offenders. A comics seal of approval was instituted. Finally, the rise in TV's popularity caused a drastic decline in comic book reading. The same concerns, however, soon surfaced with regard to the newer electronic medium. In fact, many of the same topics highlighted by Dr. Wertham would be the subject of research examining the effects of TV. (A reissue of *Seduction of the Innocent* was planned for 1993. This new edition will contain some of the more gory and provocative examples deemed too controversial for the original book.)

---

years later. This time the impetus for study came from the U.S. Senate, specifically Senator John Pastore of Rhode Island, who urged that the U.S. Office of the Surgeon General undertake a study of the allegedly harmful effects of TV violence on youth. A few years earlier the Surgeon General's Office had released a report on the harmful effects of cigarette smoking on the public's physical health. Pastore reasoned that the same kind of study should be done on the potentially harmful impact of TV violence on the public's mental health. The government earmarked $1 million for the project.

Unlike the NCCPV, which simply reviewed existing data, the emphasis of the Surgeon General's study was the accumulation of new scientific data. Eventually more than fifty new studies were conducted and released when the commission issued its final report in 1972. Unfortunately, there were some questionable tactics used to appoint the members of the advisory committee that wrote the summary report based on the new studies. Several prominent researchers were kept off the committee but two full-time network employees were included. When the summary was written, the committee's main conclusion—that there was a causal link between viewing TV violence and subsequent antisocial acts—was buried in language that was so convoluted and ambiguous that many reporters mistakenly said that the report concluded that TV violence had no adverse impact at all. When the five volumes of technical papers describing the actual studies were made public, however, the link between TV violence and aggressive behavior was more plainly seen. In fact, when the advisory committee appeared at hearings called by Senator Pastore, a strong consensus emerged among the

In 1969, Democratic Senator John Pastore of Rhode Island wrote to the Secretary of Health, Education and Welfare proposing a major research effort regarding the impact of TV violence. This letter eventually led to the formation of the Surgeon General's Scientific Advisory Committee on Television and Social Behavior.

members that TV violence was a significant contributor to antisocial attitudes and behavior, a consensus not apparent in their summary report.

In the years following the release of the Surgeon General's report additional evidence was acquired from field experiments and panel studies, and several groups, including the American Medical Association, the national Parent–Teacher Association, and the American Psychological Association, all campaigned for a reduction in TV violence. In 1982 a ten-year update of the Surgeon General's report was released. The main conclusion of this report, written by social scientists, was to the point: "The totality of evidence . . . supports the conclusion of a causal relationship between televised violence and later aggressive behavior." The update also documented several other areas where TV was having significant impact: perceptions of reality, social relationships, health, and education, to name a few.

The update was criticized by many in the industry and one network even issued a thirty-two-page pamphlet refuting its conclusions. Nonetheless, the majority of social scientists who have studied the topic would probably endorse most, if not all, of the conclusions of the update.

Many studies appearing since the publication of the update have changed their focus to "effect size." This gets away from the simple question of whether TV violence has an effect on the audience and considers instead the more complicated question of how big an effect it has. Again, there is disagreement over the significance of some of the relatively small effects that have been discovered.

So much for the rather lengthy history of this topic. Next, we review specific research evidence, pointing out its strengths and weaknesses, and conclude with an attempt to summarize the research consensus in this area.

Research Evidence

Experiments were one of the first methods used to investigate the impact of media violence. A series of studies done in the early 1960s documented that children could easily learn and imitate violent actions that they witnessed on screen. These studies used the famous "Bobo doll" and demonstrated that young children who viewed filmed violent actions performed by a model were just as likely to imitate those behaviors as were other children who saw a live model.

A second set of experiments from the early 1960s was designed to settle the debate touched off by two competing theories of media effects. On the one hand, the *catharsis theory* posited that watching scenes of media violence would actually reduce the aggressiveness of viewers since their hostile feelings would be purged while watching the media portrayals. Not surprisingly, this viewpoint was popular with many industry executives. On the other hand, the *stimulation theory* predicted that watching scenes of violence actually prompted audience members to behave more aggressively after viewing. The experimental design used to test these competing theories is presented in simplified form in Figure 16-1. The results of these and other experiments showed little support for catharsis. In fact, the bulk of the laboratory research argues for the stimulation effect. This is not to say that the catharsis hypothesis is categorically wrong. There may be some instances where it might occur. Most of the time, however, the likely end product is increased aggression.

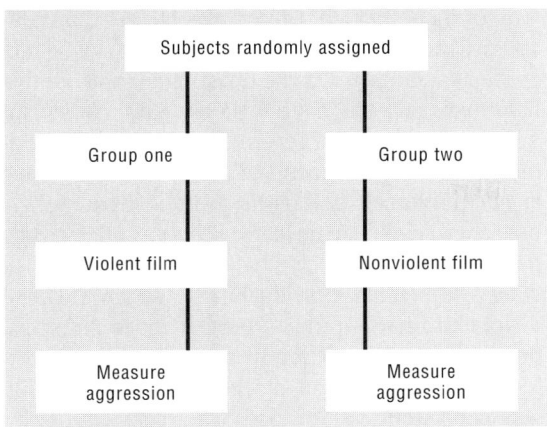

FIGURE 16-1 Simplified diagram for catharsis vs. stimulation experiment.

These early experiments were criticized for their artificiality. They were done in the lab under controlled conditions and used violent segments that were not typical of what everybody saw on TV. Further, the way aggression was measured (hitting a Bobo doll, giving shocks) was not quite the same as aggression in real life. Later experiments used more realistic violent segments and more relevant aggression measures. For example, several experiments used actual programs that contained about the average number of violent acts per hour in prime-time TV (about five or six). Additional experiments used more real-life measures of aggression. Several observed the actual interpersonal aggression of children in play groups or in classrooms. One experimenter showed either violent or nonviolent TV to selected groups of young children and then measured how aggressive they were in a subsequent game of floor hockey. These and other more natural measures confirm that watching violence stimulates subsequent real-life aggression.

Other laboratory studies have demonstrated that this process is complicated and affected by the child's developmental stage. Younger children tend to respond more to dynamic action with little apparent regard as to whether it is violent. Older children's responses are more apt to depend on their understanding of story structure, character motivations, and consequences of the violence.

In sum, the results from laboratory experiments demonstrate that shortly after exposure to media violence, individuals, especially youngsters, are likely to show an increase in their own level of aggression. In fact, since the mid-1970s there has been a marked decrease in the number of lab experiments examining this topic partly because the results have been so consistent. More recent experiments have accepted the fact that exposure to violence facilitates subsequent aggression and have concentrated instead on factors that might increase or decrease the amount of aggression performed in response to media portrayals.

Laboratory studies are important because, as mentioned earlier, they help establish a plausible cause-and-effect pattern and control for the effects of extraneous factors. Still, the laboratory is not real life, and to be more sure of our conclusions we need to examine the results of research that is done outside the lab.

Surveys (also called correlational studies) are done in the real world. Although they offer little evidence of cause and effect, they do not have the artificiality of the lab associated with them. Most surveys on this topic incorporate the design of Figure 16-2. If the viewing of media violence is indeed associated with real-world aggression, then people who watch a lot of violent TV should also score high on scales that measure their own aggressive behavior or attitudes toward aggression. The results from a large number of surveys involving literally thousands of respondents across different regions, socioeconomic status, and ethnic background have been remarkably consistent: there is a modest but consistent association between viewing violent TV programs and aggressive tendencies.

FIGURE 16-2 Survey design examining effects of TV violence.

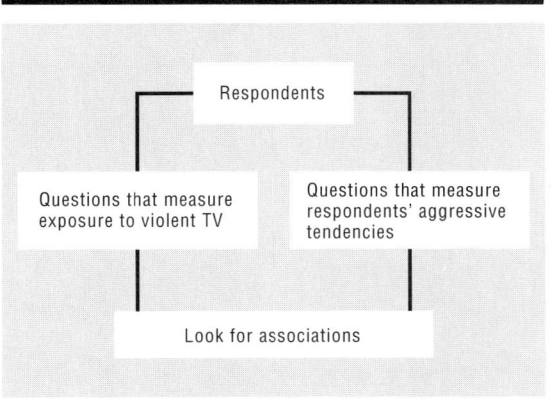

These results, however, are not without problems, like trying to establish cause and effect. Although TV violence viewing and aggression are related, TV viewing does not necessarily *cause* aggression. In fact, it's logically possible that aggressive individuals choose to watch more violent TV, which would mean that aggression could cause the viewing of violent TV. Finally, it's also possible that the relationship might be caused by some third factor. Maybe the real cause of aggression is a history of child abuse, and this in turn is associated with watching violent TV. Survey statistics would show a positive relationship between violent TV viewing and aggression, but the real cause might be something else.

Once again, to sum up, correlational studies provide another piece of the puzzle; they show that viewing TV violence and antisocial behavior are linked in the real world, but they don't tell us anything definitive about cause and effect. Remember, however, that lab studies can determine cause and effect and their results are consistent with the notion that TV viewing causes subsequent aggression. So far we have reason to be somewhat comfortable with that conclusion. But there is still other evidence to consider.

In the past twenty years or so several field experiments were carried out to investigate the potential antisocial effects of TV violence. Recall that field experiments give us some basis for deciding cause and effect but suffer from a lack of control of other, potentially contaminating factors.

The results from field experiments are somewhat inconsistent. At least two done in the early 1970s found no effect from viewing violent TV. One of these two, however, was plagued by procedural problems and its results should be accepted cautiously. On the other hand, at least five field experiments have yielded data consonant with the lab and survey findings. These experiments included children and adults and were done in such diverse settings as a penal institution, a boarding school, a nursery school, and average homes. The main conclusion of these studies seems to be that individuals who watch a diet of violent programs tend to exhibit more antisocial or aggressive behavior. In some studies this effect was stronger than in others, but the direction was consistent.

Figure 16-3 shows the design used in one of these field experiments. In this case the experiment was based on natural circumstances. The researchers were able to identify a Canadian town that was surrounded by mountains and was unable to receive TV signals until 1974. This town was matched with two others, one that could receive only the Canadian Broadcasting Corporation (CBC) and another that could get the CBC plus the three U.S. networks. The towns were studied in 1973 and again two years later. Children in the town that just got TV showed an increase in the rate of aggressive acts that was more than three times higher than those children living in the other two towns.

On balance, the results from field experiments are not so striking as those using the lab and correlational methods. On the whole, though, they tend to support, although weakly in some cases, the notion that viewing violent TV fosters aggressive behavior.

FIGURE 16-3 Canadian field experiment design.

Community	Time one		Time two	
1	No TV	Measure aggression, media exposure, and related variables	One TV channel	Measure aggression, media exposure, and related variables
2	One TV channel		Two TV channels	
3	Four TV channels		Four TV channels	

There is one more piece to be added to the puzzle. Panel surveys, as noted earlier, are longitudinal research projects that examine the same individuals at different points in time. They are not plagued by the artificiality of the laboratory and their design allows us to draw some conclusions about cause and effect. Since the early 1970s the results of several panel studies have become available for analysis.

The panel analysis begins with measurements of both real-life aggression and exposure to TV violence taken at two different times. Next, the researchers determine if TV viewing as measured at time 1 is related to aggression at time 2. At the same time, the relationship between aggression at time 1 and TV viewing at time 2 is also assessed. If early TV viewing is more strongly related to later aggression than early aggression is to later violent TV viewing, then we have evidence that it's the TV viewing causing the violence and not vice versa. Figure 16-4 diagrams this approach.

To illustrate how this technique works, in one study researchers followed sixty-three boys and girls over a five-year period, from ages 4 to 9. From time to time interviews were done with both the children and their parents and data were collected on the children's TV viewing, their aggressiveness, their self-restraint, and how their parents disciplined them. The results showed that later aggressiveness in children was strongly related to heavy viewing of TV violence, particularly the amount of TV violence viewed in their preschool years. In addition, the children who had watched the most violent TV at age 4 were least able to show self-restraint at age 9. The researchers also noted that TV viewing was not the only cause of later aggressiveness since parents who put a lot of emphasis on physical punishment during a child's early years also tended to have more aggressive 9-year-old children. (The original study, by J. Singer, D. Singer and W. Rapaczynski, was reported in the spring 1984 issue of the *Journal of Communication*.)

Several other longitudinal studies done in the United States and Europe, including both panel and trend studies, have found, with some exceptions, similar results. The majority of the studies seem to suggest that the sequence of causation is that viewing TV violence causes viewers to become more aggressive. The degree of the relationship between the two factors is small and, in a few instances, difficult to detect, but it is consistent. In addition, the process seems to be reciprocal. Watching TV violence encourages aggression, which in turn encourages the watching of more violent TV, and so on.

Having reviewed evidence from four basic research methods, what can we conclude about the effects of violent TV on antisocial behavior? Laboratory experiments demonstrate that under certain conditions, TV can have powerful effects on aggressive behavior. Field experiments provide additional, although less consistent, evidence that TV can exert an impact in the real world. Surveys show a consistent but somewhat weak pattern of association between violence viewing and aggression. Longitudinal studies also show a persistent but weak relationship between the two and suggest a pattern whereby watching TV causes subsequent aggression. Unfortunately, few findings in areas such as this are unambiguous. Nonetheless, a judgment is in order. Keep in mind that some might disagree, but the consensus among social scientists seems to be:

1 Television violence is *a* cause of subsequent aggressive tendencies in viewers; it is not *the* cause since many other factors besides TV determine whether people behave aggressively.
2 The precise impact of TV violence will be affected by many other factors, including age, sex, family interaction, and the way violence is presented on the screen.
3 In relative terms, the effect of TV violence on aggression tends to be small.

Does this close the case? Not quite. The third summary statement has been the focus of much current debate. The majority of researchers concede that there is some kind of causal link between viewing video violence and aggression,

FIGURE 16-4 Longitudinal design to study TV violence.

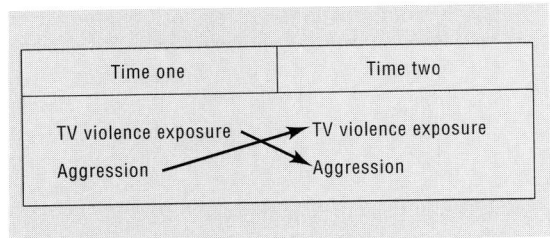

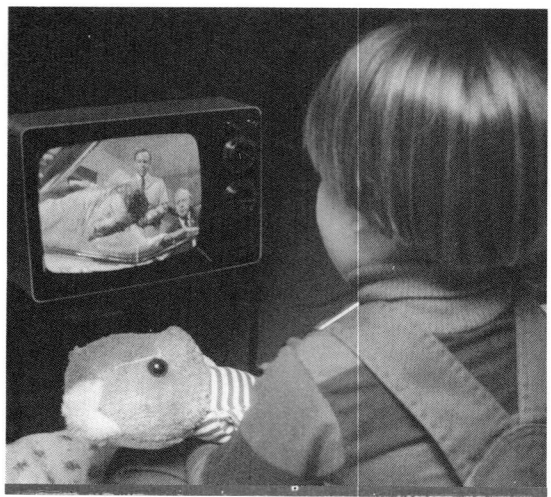

By the time a child graduates from high school he or she will have seen more than 13,000 violent deaths depicted on TV entertainment programs. The effects of exposure to violent TV has been the subject of more than 500 separate studies.

but several argue that the link is too weak to be meaningful. To be more specific, statisticians characterize the strength of any relationship in terms of the amount of variability in one measure that is accounted for by the other. For example, suppose a person was trying to guess your college grade-point average just by looking at you.

Chances are the person wouldn't do too well. Now let's show the person your SAT scores and let the person guess again. Chances are that the person will do a little better. Why? Because there is an association (a moderate one, at least) between GPA and SAT scores. In other words, they share some variation. The more shared variability there is between two measures, the better a person can predict. If two factors are strongly related, changes in one might account for 60 to 70 percent (or higher: 100 percent if perfectly related) of the change in the other. A person could make quite accurate predictions in this instance. If two factors are not associated at all, change in one would explain 0 percent of changes in the other. In this case a person's predictions would be no better than chance. Thus the strength of a relationship is measured by the variability explained. As far as TV violence and aggression are concerned, exposure to televised violence typically explains from about 2 to 9 percent of the variability in aggression. Knowing how much TV violence you watch helps a little, but not much, in predicting your aggression level. Put another way, between 91 and 98 percent of the variability in aggression is due to other causes. Given this situation, is the effect of TV on aggression really that meaningful? Does it have any practical or social importance?

REGULATIONS

SENATOR SIMON, THE TEXAS CHAINSAW MASSACRE, AND REGULATING TV VIOLENCE

As he tells it, Paul Simon, the Democratic senator from Illinois with the bow-tie trademark (and an unsuccessful candidate for the 1988 Democratic presidential nomination), was flipping through the channels on his hotel TV one night and happened to come across a particularly gory scene from what he said was that slash and splatter movie classic *The Texas Chainsaw Massacre*. Not impressed by what he saw, Simon quickly introduced a new bill that would curb the amount of violence shown on TV. The bill would grant an exemption from the antitrust laws to allow the networks, cable, and program producers to form a set of standards regarding the way violence is shown on TV. After several failures, a modified version of this bill was passed in 1990. As of this writing, preliminary talks concerning how to implement this bill were being held.

An interesting footnote to this episode concerns whether or not Senator Simon actually saw *The Texas Chainsaw Massacre* on the TV set in his room. As it turns out, the film had not been licensed for showing to hotels, the networks, cable, or local independent stations on the date when the senator claims he saw it. He also could not recall what hotel he was in at the time so that its records might be checked. Simon maintains it was the famous film but others suggest it was an even lower budget imitation. Whatever it was, it had blood, gore, and chainsaws and was enough to lead to the latest round of the antiviolence fight.

This question is more political or philosophical than scientific, but research can offer some guidelines for comparison. The usual effect size found for TV violence's impact on antisocial behavior is only slightly less than that found for the effects of viewing "Sesame Street" and "The Electric Company" on cognitive skills of the audience. It's also only slightly less than the effect that a program of drug therapy has on psychotic patients. Indeed, several drugs in widespread use have therapeutic effects about as great as the effect size of TV violence and aggression. In sum, although the magnitude of the effect may not be great, it is not that much different from effects in other areas that we take to be socially and practically meaningful. Thus even though the effect may be small, this does not mean that it should be dismissed.

Finally, it should be noted that the bulk of the violence research was done when network TV was still the viewing norm. In today's viewing environment the audience can choose to watch cable offerings that show much more graphic violence than the network ever could before. Further, even youngsters have access to violent movies on rented videocassettes; the *Friday the 13th* series and other slasher/stalker films, for example, contain much more realistically staged scenes of mayhem. One study in Great Britain has already suggested that youngsters who rent and view violent videocassettes also behave more aggressively in real life. The effects of these new distribution channels for violence have yet to be researched adequately. It appears that there will be much more written on the vexing topic of video violence.

A MEAN AND SCARY WORLD? PERCEPTIONS OF REALITY

The preceding section traced the impact of TV on the behavior of the audience. This section is more concerned with the impact of TV on how the audience thinks about and perceives reality.

The media, particularly TV, are the source of much of what we know about the world. The media, however, bring us more than simply information. They also, at least to some degree, shape the way we perceive the outside world. Or, put another way, the media affect the way we construct our social reality. For instance, what is it like to live in southern New Zealand? It seems probable that very few of you reading this book have had much of a chance to experience living in southern New Zealand or to talk to people who have. Consequently, a TV show on the life of the people who live there might have tremendous influence on your perceptions in this area. This isn't much of a problem as long as the media presentations accurately represent reality.

There are, however, many areas where the world presented on TV differs greatly from reality. For example, studies of TV entertainment content have shown that far more TV characters work in crime and law enforcement than do people in the real world. Criminals on TV commit violent crime far more often than do their real-life counterparts. Trials on TV are decided by juries more often than they are in reality. Leading characters in TV shows are almost always American; people from other countries are rarely shown. More people on soap operas have affairs and illegitimate children than do real people. More unscrupulous and treacherous people are shown on TV than exist in real life. Again, this wouldn't be much of a problem if people were able to separate the two worlds, TV and reality, without any confusion. For some, however, separating the two is not easy.

This is the focus of the cultivation theory. The *cultivation theory* was popularized by George Gerbner and his colleagues at the University of Pennsylvania. In simplified form, the theory suggests that the more a person is exposed to TV, the more likely that the person's construction of social reality will be more like that shown on TV and less like reality.

The procedure first used to put the theory to the test requires a content analysis of TV to isolate those portrayals that are at odds with reality. A content analysis is a systematic and objective analysis of the messages portrayed on TV. Gerbner and his colleagues have been conducting such analyses for more than a decade and have come up with many areas where the two worlds diverge (some of these areas have already been mentioned). The second step requires dividing audience members into heavy-viewing and light-viewing groups and then asking for their perceptions of various social events or situations. If the cultivation theory is correct, then a lot of the heavy viewers should give answers more in line with the TV world, whereas a lot of the light viewers should give answers more in keeping with the real world. Figure 16-5 shows the basic model used for analysis.

Gerbner and his colleagues have done several surveys examining the cultivation effect. In one study adolescents were asked how many people were involved in some kind of violence each year, 3 or 10 percent (the TV answer); 83 percent of the heavy viewers gave the TV answer as compared to 62 percent of the light viewers. Another question asked how often a police officer usually draws a gun on an average day. The choices were less than once a day (the real-world answer) and more than five times a day (the answer more in line with the TV world). Three times as many heavy viewers (18 percent) as light viewers (6 percent) said more than five times a day.

Further, Gerbner argues that the world portrayed in many TV entertainment shows is a world filled with violence, duplicity, double-crossing, and insincerity. In short, it's a mean and scary place. If cultivation is occurring, then heavy viewers should also think that their own world is mean and scary. This is basically what surveys disclosed. Heavy viewers were more afraid to walk alone in their neighborhoods at night. They were also more likely to agree with the statements; "You can't be too careful in dealing with people" and "Most people are just looking out for themselves." Other researchers examining soap operas found that heavy soap viewers perceived greater numbers of doctors, divorced people, lawyers, criminals, and illegitimate children than those that existed in the real world, thus reflecting the impact of their viewing.

Cultivation studies have received a fair amount of publicity in the popular press (*TV Guide* even carried an article by Gerbner and his colleagues) and the theory has a certain commonsense appeal. Recent research, however, suggests that the process of cultivation is much more complex than originally thought.

In the first place, remember that most of the cultivation studies relied on the survey approach. Although the survey method can establish a relationship, it cannot be used to rule out other factors that might be causing the relationship. Consequently, more recent studies suggest that when other factors in the process (like age, sex, race, and education) are simultaneously controlled, the cultivation effect either is weakened or disappears.

In response to these findings the cultivation theory was revised to include two additional concepts—mainstreaming and resonance—that account for the fact that heavy TV viewing may have different outcomes for various social groups. *Mainstreaming* means that heavy viewers within social subgroups develop common perceptions that differ from those of light viewers in the same subgroup. For example, among light TV viewers, nonwhites are generally more distrusting than whites and are more likely to report that people will take advantage of you if they get the chance. Among heavy viewers, however, the gap between whites and nonwhites on this measure is significantly smaller. Heavy TV viewing has a homogenizing effect in this instance and brings both groups closer to the mainstream. Similarly, liberals and conservatives who are light viewers show wide variations in their attitudes toward such topics as busing, open housing, and abortion. Among heavy viewers, however, the gap between liberals and conservatives on these issues narrows significantly, once again demonstrating the mainstreaming effect.

Resonance refers to the situation where the viewers get a "double-dose" from both TV and reality. For example, heavy TV viewers who live in a high-crime area have their belief in a scary world reinforced by both TV and their firsthand experience. They should show an exaggerated cultivation effect when compared to light viewers.

Also remember that surveys cannot establish cause and effect. The cultivation theory assumes that TV viewing is cultivating the subsequent perceptions of reality. It is possible, however, that people who are fearful of going out at night stay home and watch more TV, thus making the perception the cause of the viewing rather than the effect. Of course, the best way to sort out cause

FIGURE 16-5 Design for cultivation analysis.

TV viewing	Number giving "TV answer"	Number giving non–TV answer
High	A lot	A few
Low	A few	A lot

and effect would be an experiment, but since cultivation theory talks about the long-term cumulative effect of TV exposure, a definitive experiment would be difficult to design.

Nonetheless, some short-term experiments shed light on the topic. In one of these, college students were divided into low-anxiety and high-anxiety groups on the basis of a psychological test. Students from both the high and low groups were then placed in one of three experimental conditions. One group (the light-viewing group) was asked to watch little TV; the second group (known as heavy viewing/justice) watched a lot of action/adventure shows where the forces of good always won and justice prevailed. The third group (called heavy viewing/injustice) watched a lot of TV, but their shows were chosen so that injustice triumphed (bad people got away with crimes). Six weeks later the students were measured again. As the cultivation theory would predict, the students in the heavy viewing/injustice group showed the greatest increase in anxiety over those in the other groups, and also showed the most increase in their estimate of their likelihood to be victims of violence. The light viewers experienced the least increase in anxiety and fear of violent victimization. In addition, the experiments looked at the voluntary viewing choices of subjects after the experimental exposure was over. They found that students in the heavy viewing/injustice group watched fewer action/adventure shows than those in the other conditions, suggesting that their perceptions don't encourage additional heavy viewing, a finding that supports the reasoning of the cultivation theory that viewing cultivates perceptions rather than the other way around. (This experiment by J. Bryant, R. Carveth, and D. Brown was published in the winter 1981 issue of the *Journal of Communication*.)

Other cultivation studies have tried to specify the conditions that are most likely to foster or inhibit cultivation. Although the results are not entirely consistent, it appears that cultivation depends on the following:

1 *The motivation for viewing.* Ritualistic, low-involvement viewing appears to be more potent than planned and motivated viewing.
2 *Experience with the topic.* Studies have noted that cultivation seems to work best when audience members have only indirect or distant contact with the topic. This seems to contradict the resonance notion.
3 *Perceived realism of the content.* Cultivation appears to be enhanced when the viewer perceives the content of entertainment shows to be realistic.

More recent studies of the cultivation theory have demonstrated that the research itself must be done carefully in order to avoid spurious findings. One study found that the amount of TV viewed that divides the high TV group from the low TV group must be chosen precisely or distortions will appear in the results. Other studies have found that the way that the questions are worded and the precise topics to be evaluated will also have an impact. It should be obvious that a lot of research still needs to be done before we understand the cultivation theory completely. It is also obvious that the area is an intriguing one and represents one of the few topics that have attracted research interest from sociologists, and social psychologists, and mass communication researchers. The influence of TV on the way we construct social reality is likely to be an important research area for the foreseeable future.

SEX-ROLE STEREOTYPING

Somewhat related to cultivation analysis is the research area that examines the impact of stereotyping on viewers' attitudes. The women's movement of the 1960s and 1970s sparked interest in the way males and females were portrayed in TV programs. Early content analyses disclosed that men outnumbered women two to one in starring roles and that men appeared in a far greater variety of occupational roles. When they did appear, women were likely to be housewives, secretaries, or nurses. Female characters were also portrayed as passive, deferential, and generally weak, in contrast to male characters, who, on the whole, were active, dominant, and powerful. More recent content analyses have shown that females are now portrayed in a wider range of occupations but that little else has changed.

A number of studies about the effects of exposure to this material began to surface in the early 1970s. Many of these studies used the correlational approach and although they were not as rigorous as they could be, their findings generally supported what the cultivation hypothesis would predict: youngsters who watched a lot of TV should have attitudes and perceptions about sex roles that are in line with the stereotypical portrayals on TV. In one study heavy TV viewers were far more likely than moderate TV watchers to choose a sex-stereotyped profession (for example, boys choosing to be doctors or police officers) and girls choosing to be housewives or nurses). Another study noted that children who were heavy viewers scored higher on a standardized test that measured sex stereotyping.

Like cultivation analysis, the major problems with this sort of research are establishing causation and sorting out the impact of TV from other sources of sex-role information (schools, peers, parents, books, and so on). In an attempt to clarify the process, panel studies have examined the correlation between viewing and stereotyping. The results suggested that the causal connection evidently works in both directions: TV viewing led to more stereotypical attitudes and people with more stereotypical attitudes watched more TV, thus reinforcing the effects.

In the elaborate Canadian study that has been referred to before, children who grew up in the town that did not have TV held fewer sexually stereotyped attitudes than children who had grown up with TV. After two years of TV viewing, however, the attitudes of the kids in the town where TV was just introduced became more stereotyped. This is an important finding since the influence of other sources, such as peers, teachers, and books, was present in both groups during the entire two-year period of the study. The introduction of TV was itself strong enough to produce an increase in sex-role stereotyping.

A second popular technique for examining the effects of TV on sex-role attitudes usually takes place in the laboratory and consists of showing subjects (usually youngsters) men and women in counterstereotypical roles (for example, a male nurse, a female mechanic) and then seeing if changes in perceptions occur. Results from these studies have found that exposure to this nontraditional content does seem to decrease sex-role stereotyping. In one typical experiment a group of girls saw commercials in which a woman was shown as a butcher, a welder, and a laborer. Another group saw commercials featuring women as telephone operators, models, and manicurists. After viewing, girls who saw the nontraditional roles expressed a greater preference for traditional male jobs than did the other group. Similar results have been found in at least a half dozen other experiments.

Taken as a whole, these studies demonstrate that sex-role beliefs can be affected by the mass media. A qualification is in order here. It also appears that nontraditional portrayals can sometimes boomerang and reinforce the stereotypes they were meant to erase. In one study first and fourth graders saw a tape about a visit to the office of "Dr. Mary" and "Nurse David." When asked immediately after viewing to describe what they had just seen, most of the male students referred to "Dr. David" and "Nurse Mary." Apparently some stereotypes are highly resistant to change.

IMPACT

MUSIC VIDEOS AND STEREOTYPING

Music videos have been one place where sexual stereotyping has been studied closely. Not surprisingly, all studies suggest that music videos constitute a place where females are portrayed in a distinct way. First, few female performers appear. Studies from several different years have found that males make up almost 80 percent of the artists who appear in music videos. Second, when women did appear, either as performers or characters in concept videos, they tended to appear in seductive dress or various forms of undress. Third, several studies rated the portrayal of women in music videos according to a four-step sexism scale, ranging from level 1, "condescending" through level 4, "fully equal." As might be expected, level 1 accounts for most of the video portrayals, ranging from 40 to 50 percent in various studies.

BROADCASTING AND POLITICS

Even the most casual of political observers will concede that the broadcasting media, and TV in particular, have changed American politics. A new term, "telepolitics," has been coined to describe the way politics is now practiced. This section looks at the obvious and not-so-obvious influences of broadcasting on the political system.

Those who have studied the impact of media on politics generally divide the field into two categories. The first has to do with the influences on the ultimate political act—voting. Studies in this category examine how media help shape our election campaigns, our images of candidates and issues, our knowledge of politics, whether we vote, and for whom we cast a ballot. The second category includes studies of how media, TV especially, are changing the basic political structure and how we perceive it. We shall examine each category, but first let's look at the media, electioneering, and voting.

Media Influences on Voting Behavior

The past fifty years have seen rather striking changes in the way political scientists and mass communication researchers have viewed the importance of media in political campaigns. Early fears about the political impact of the media were shown to be unjustified by careful studies done in the 1940s and 1950s. Most people reported that factors other than the media, such as their party affiliation and the opinions of respected others, were the most influential factors determining their vote choice. Since 1960, however, TV has assumed dominance as the most potent political medium and voters have tended to be less influenced by party ties and organizations. Accordingly, the potential for media impact may be on the increase.

One thing is certain. People get a lot of political information from the media during the course of a campaign. Candidates who get extensive coverage also show strong gains in public awareness. Take Jimmy Carter in 1976. At the start of the campaign he was basically unknown (a little less than 20 percent of the voters even knew who he was), but by the end of the 1976 primary season, during which Carter got the bulk of print and broadcast attention, he was known by more than 80 percent. The same pattern occurred in 1992 when Ross Perot went from a regional to a national political figure thanks primarily to TV coverage.

Moreover, the pattern of news coverage during an election campaign can help determine what political issues the public perceives as important, a phenomenon known as *agenda setting*. For example, if the media give extensive coverage to U.S. policy in Central America, audience members may think that this is an important issue and rate it high on their own personal agenda of political issues.

Voters, of course, get more than just issue-related information from the media. Another area of research is concerned with the role of TV and other media in forming voters' images of the candidates. In particular, political ads are designed to project a coherent and attractive image to voters. Studies have suggested that a candidate's image can be the dominant factor in many elections. As party identification weakens, voters tend to rely on a general image to help them make up their minds. This "image effect" seems strongest among uncommitted voters and is most noted during the early stages of a campaign. During the 1988 primary campaign candidate George Bush was plagued by a "wimpy" public image. Especially designed campaign ads and Bush's evidently calculated performance during a heated TV interview with CBS anchor Dan Rather apparently gave Bush's image a needed overhaul at a crucial campaign juncture.

When it comes down to the actual choices of (1) whether to vote at all and (2) for whom to vote, the research is not so definitive as one might think. Voting behavior is a complex activity and many factors—interpersonal communication, personal values, social class, age, ethnicity, party affiliation—along with media exposure come into play. To make it even more complicated, some people are unable to distinguish exactly what factors influence their choice. Existing research, however, does offer some conclusions. Not surprisingly, voters who learn a great deal about a particular candidate are likely to vote for that candidate. Certain kinds of media exposure are also related to voter turnout. Print media readership was found to be related to greater voter participation, whereas people who were heavy TV viewers were less likely to vote. Exposure to TV ads and other political TV programs was also related to turnout and voter choices, but this effect was most pronounced among those who had little interest in the campaign.

TV debates between or among presidential candidates have become a fixture of modern campaigns and their results have been closely studied to determine what, if any, impact they have on voter preference. Numerous studies of the debates of 1960, 1976, 1980, 1984, and 1988 generally agree that the debates reinforced preferences that had formed before the debates. Almost six out of ten voters had made up their minds by the time the debates occurred and most viewers simply have their choices confirmed. Among those who haven't made up their minds, debate viewing generally crystallizes their vague predispositions and many in the audience split according to partisan leanings. Of course, in a close election, such as that of 1960, if even a small percentage of voters are influenced by debate exposure, then the effects can be significant.

A final area that received substantial research interest in the 1980s relates to the influence of *exit polls* on voter behavior. To conduct an exit poll, interviewers stop voters leaving polling places, ask how they voted, and gather other relevant information. Exit polls are used by the news media to predict election results before all the votes are in and all the polls are closed. Do the early calls cause people to stay home or to change their vote choice? Although many critics claim that common sense suggests that people will not turn out to vote if they think their vote is meaningless or will not turn out to support a candidate already declared a loser, the research suggests that the effects are more complicated. In the first place the closeness of a race is an important consideration. If exit polls simply confirm a landslide predicted by preelection surveys, their effect may be quite different from exit polls that confirm the existence of a close race or, conversely, show a sweep when a close race was expected or a close race when a landslide was predicted. Research suggests that exit polls may have their greatest impact in the situation where they change perceptions about the closeness of a race.

Further, only those potential voters who have not voted *and* who hear an election call can be affected. Predictably, many people who haven't already voted by the time a call is made tend to have low interest in the election and are not likely to be exposed to any calls anyway. In addition, in some states the polls are open only an hour or less after a winner is projected, whereas in other states the polls are open much longer. Moreover, other factors such as weather, early concession speeches, and tight local races influence turnout. In sum, the potential pool of people likely to be affected is limited and the influences of many outside factors are hard to sort out.

Nonetheless, taking all the complications into account, perhaps the best summary of the available research evidence is presented by political statistician Seymour Sudman in a 1986 article in *Public Opinion Quarterly*. Sudman suggests that there is the possibility that exit polls may decrease turnout anywhere from 1 to 5 percent in those states where the polls close more than an hour after an election call is made, and in elections where exit polls predict a clear winner in a race that had previously been thought close. In other situations the effect will be hard to find. This may not be a final answer but, given the available data, it is difficult to be more precise. In any case, the whole problem may become academic if Congress passes a uniform poll-closing bill or if news media informally refrain from publicizing poll results until everybody has had a chance to vote.

Media Impact on the Political System

Turning now to the second general area of research, the impact of the media on the political system as a whole, we find that there is less research available but the findings are no less important. One line of research suggests that political institutions and the media have become interdependent in a number of ways. Political reporters need information to do their jobs and the government and politicians need media exposure. Thus public officials hold news conferences early enough to meet media deadlines, presidents have "photo opportunities," and correspondents accompany administration leaders on world trips. Further, many politicians have learned to stage media events that accommodate both sides: the

Televised debates among presidential hopefuls, such as those among the Democratic candidates in the 1992 primaries, receive heavy media coverage, but most research indicates that televised debates seldom change voters' decisions. In most instances, the debates simply reinforce the choice that voters have already made.

politician gets exposure and the TV reporter gets a ten-second "sound bite" for the evening news. The whistlestop railroad campaigns of the 1930s and 1940s have given way to the "airport-stop" campaigns of the 1980s: a candidate flies into an airport, gives a short speech to the cameras, and flies to the next airport where the process is repeated. This interdependence is also demonstrated by the change in political conventions. In prior years conventions actually selected candidates. These days conventions are more like coronation ceremonies, closely orchestrated to maximize prime-time TV coverage.

To sum up this rather broad and complicated area, research concerning the political impact of the media demonstrates the specific-effects viewpoint mentioned earlier. In some areas, such as building political knowledge, shaping political images, and setting agendas, the media effects are fairly direct and evident. In other areas, such as voter turnout and voter choice, the media work along with a host of other factors and their impact is not particularly strong.

TELEVISION AND LEARNING

It seems that every new medium has been indicted at one time or another for lowering the educational level of American youth. During the 1930s critics charged that radio was a negative influence on youngsters' school achievement and made it difficult for them to develop good study habits. In the early 1950s comic books were said to be undermining the reading skills, stunting the imagination, and handicapping the development of an adequate vocabulary of future generations. In the mid-1950s, when TV took center stage, concern about its effects on children mounted quickly, but early studies failed to show convincing evidence that watching TV was related to poor school performance. These surveys did uncover some links between heavy viewing and poor academic performance, but the lack of conclusive evidence pushed the issue out of public focus. In the 1970s, however, the issue was rekindled, primarily as a response to an alarming dip in SAT scores. This section reviews the research on TV's effect on IQ, school achievement, reading, and imagination.

The results of the research on the relationship between TV viewing and IQ are not surprising. Heavy viewers tend to have lower IQs than light viewers, and many critics have argued that watching TV actually lowers a person's intelligence. Further research suggests that the association is not that simple. First, it appears that age is a factor. At least two studies noted that heavy viewing was actually linked to higher IQs in kids up to the age of 10 or 11 but after that the relationship reverses. Data from the field experiment conducted on the Canadian towns (see above) found no relationship between IQ and viewing in the towns where TV had always been available. On the other hand, in the town that just got TV, lower IQ children watched more. Age also appeared to be an important factor. In sum, there appears to be a weak negative relationship between IQ and TV viewing, but age is a complicating factor. And, as is the case with surveys, the direction of causation is not clear.

When it comes to school achievement, the results are not conclusive but they also suggest a slight negative relationship. Youngsters who watch a lot of TV tend not to do so well at school as do their lighter viewing counterparts. In one study of more than 600 sixth to ninth graders high TV viewers tended to score lower on tests of vocabulary and language achievement than did light viewers, even when the effects of IQ were statistically controlled. Interestingly, TV viewing was not associated with math achievement scores. Surveys done among adults confirm that high TV viewers do less well on vocabulary tests, suggesting the negative relationship carries over from youth.

Another study indicated that the type of TV content viewed, along with the amount of viewing, was important in determining the precise relationship between TV and achievement. This survey found that children who watched a good deal of news and educational programs got better grades in school, whereas those who watched a lot of adventure shows got lower grades. A recent panel study among high school students, however, found no evidence of TV's effects on math and reading achievement after controlling for such confounding factors as parents' education level, IQ, race, and school attendance. In any case, virtually all of the studies showed the relationship, if one was found, to be relatively weak. In terms of the amount of variability explained, TV viewing generally accounted for about 1 to 4 percent of the variation in achievement scores, a link even weaker than that discussed between exposure to TV violence and aggression.

The relationship between TV and reading performance is hard to sum up. Although common sense might suggest that TV viewing has a negative influence on reading skills, the research has found no clear relationship. True, a few surveys have found a weak negative relationship between entertainment TV viewing and reading achievement scores, but at least one study has found the opposite. Perhaps the authors of the Canadian study mentioned earlier said it best after their lengthy consideration of the issue. In their study TV had a small negative effect on reading competence, primarily by taking up time that might otherwise have been spent on reading practice. The researchers added that this impact, however, was also affected by IQ, social class, age, and parental attitudes toward reading.

A large-scale study published in 1990 sheds additional light in this area. This research used data gathered from a national sample of about 1750 children over a four-year period. The children were first studied when they were 6 to 11 years old and then studied again four years later. Cross-sectional correlations done at one point in time were as expected: negative relationships between the amount of TV viewing and IQ, reading skill, and arithmetic ability. When the relationships were examined longitudinally, however, the negative relationships disappeared. The best predictors of intelligence and aptitude were parents' educational level, race, size of family, and birth order.

On the more positive side, related research has shown that TV can teach youngsters specific skills that will help them develop reading competence. Research examining "Sesame Street" and "The Electric Company" showed that they helped children develop letter recognition and decoding skills, which promoted reading fluency.

The last area we shall examine in this section deals with TV's influence on the imagination and creativity of youngsters. Has TV brought forth a generation of "couch potatoes" vegetating in front of the screen? Again, the available evidence is not conclusive, primarily because of the difficulty in defining and measuring creativity and imagination.

RESEARCH

TELEVISION INTERFERENCE

Are you watching TV while you are reading this? You might want to turn it off, particularly if you are concerned with doing well in school.

In a study reported in the spring 1990 issue of *Human Communication Research,* researchers Blake Armstrong and Bradley Greenberg were interested in investigating the relationship between TV viewing and academic achievement. They hypothesized that one possible reason for the link between TV watching and poor academic performance is that TV, even if it's just on in the background, interferes with the learning process.

To test this notion, college students were randomly assigned to five different groups. The groups completed several different tests designed to measure intellectual skills: short-term memory, problem solving, reading, and creative thinking. Four of the groups took the tests in the presence of a TV set. The fifth group did the test in silence.

The results disclosed that the groups with the TV set on did worse than the control group on measures of comprehension, problem solving, and creative thinking. The researchers suggest that background TV diverts some of our attention from the task at hand. This might not be a problem for some simple tasks, like short-term memory, but for more complex tasks, such as reading comprehension, TV in the background might have detrimental effects.

Is your TV set still on?

Still, a trend can be detected. Heavy TV viewing has a negative effect on creativity. In the elaborate Canadian study, for example, youngsters in the town without TV scored higher on initial tests of imagination and creativity than did their counterparts in the towns with TV. Two years after TV arrived, however, the scores of these children had declined to the same level as those in the town with TV. One possible explanation, similar to the one about reading skills, holds that TV viewing displaces other activities, like book reading, that are related to creative thought.

Overall, the data offer some qualified support to the notion that TV viewing, at best, has a small adverse effect on IQ, school achievement, and creativity. The results certainly support the conclusion that excluding especially designed educational programs, TV offers no particular educational benefits to its young viewers; all the hours that youngsters invest in it carry no measurable scholastic rewards.

IMPACT OF TELEVISION ADS

It's obvious that the people who run TV ads think they have an impact. Otherwise, why would Procter & Gamble spend $700 million a year touting the virtues of Pampers, Head & Shoulders, and Top Job or the Philip Morris Company spend $475 million advertising products like Miller Beer, Maxim Coffee, and Jell-O? They spend this money because TV and radio commercials help sell products and services. This realization has gone international. The recent decision of the Chinese government to introduce TV commercials was a conscious effort to increase consumer demand and to stimulate mass production.

A great deal of research has gone into determining what makes an effective commercial. Social scientists, however, have been interested in more than the marketing aspect of advertising. Topics that have received a lot of research attention since 1970 include (1) how young children perceive and understand commercials, (2) the effects of common advertising techniques, and (3) how ads influence the development of consumer skills.

Research evidence indicates that young children are able to identify commercials as distinct from the surrounding programs at an early age. About 80 to 85 percent of 5 year olds can distinguish between program content and a commercial. It takes longer, however, for children to understand what commercials are designed to do. Recognition of the selling intent behind an ad varies with age. Kindergarten youngsters, for example, apparently think that commercials are there to entertain viewers or for other unrelated purposes. (In one study a youngster reported that "commercials give the actors a chance to change clothes.") By the second grade, however, most children were able to understand that the real reason for commercials was to sell things, and by the sixth grade most kids were aware not only of the selling intent but also of specific techniques that were used in ads (such as putting a prize in a cereal box to encourage people to buy it) and of the profit-seeking motives of the sponsors.

The amount of credibility that children attribute to commercials is also related to age. Even young children show some skepticism toward commercials; only one-third of a group of 5 year olds believed that commercials always told the truth. By age 12 fewer than 5 percent of the youngsters endorsed commercials as truthful.

Another area of advertising research has investigated the potentially harmful effects of commercials for certain products. For example, analyses of commercials during children's programming revealed that ads for products heavy in sugar comprised about 80 percent of the total. Surveys and field experiments found that children heavily exposed to these ads overestimated the nutritional value of sugared cereals and candy and were more likely to select low-nutrition foods for their own diets. Moreover, many of the ads for sugary cereals carry a statement, called a "disclaimer," that says the product should be "part of a balanced breakfast." Research suggests that most youngsters do not understand the balanced breakfast concept and many even get the impression that the cereal alone is all that's needed for a nutritious meal.

A related line of research examined ads for over-the-counter medicines (such as aspirin, cold remedies, antacids, and sleeping aids) and their potential negative effects: creating misconceptions about health and medicine and a possible connection to illegal drug use. The results of this research indicated that TV drug advertising showed no connection with either legal or illegal drug use. Further, heavy exposure showed only a slight connection with the perceptions that people are sick a great deal and that TV medicines always give quick relief. In short, TV drug advertising did not seem to have a significant impact on youngsters.

More recent studies have looked at the potentially harmful effects of ads for beer and wine. Congressional hearings on this topic disclosed that some researchers found a possible link between exposure to alcohol ads on TV (and for that matter in print) and alcohol abuse among teenagers. One survey found a statistically significant link between ad exposure and beer and liquor consumption among those young people who already drank. Among those yet to start, heavy ad exposure was associated with a stronger desire to drink when older. Like other surveys, this study can't produce conclusive evidence of cause and effect, but the data suggest that such a link is possible and should be studied further. During the hearings on this topic Congress considered removing all alcohol ads from TV (as they did earlier with tobacco ads) but eventually decided against it.

Researchers have also examined how TV ads contribute to parent–child friction and to children's unhappiness. Predictably, children who are more heavily exposed to TV ads tend to make more requests for the purchase of specific products. If parents say no to these requests, disagreements and arguments usually follow. Surveys have shown that kids who see a lot of ads are the ones who protest the most. One study showed that twice as many heavily exposed children argued when their requests were denied as compared to light viewers.

Turning to the final category of advertising research, those who defend children's advertising argue that it makes kids more intelligent and shrewder consumers. This is a difficult area for researchers since it is hard to specify what exactly is meant by effective consumer skills. Surveys suggest that TV ads are one of many things that help contribute to the skills, attitudes, and behaviors that go into consumer behavior. Exposure to commercials has also been related to an awareness of different brands of the same general product and a greater understanding of brand differentiation, two skills commonly associated with being an intelligent consumer. It was unclear, however, if TV viewing or age was the key factor behind these abilities. Perhaps the conclusion that best sums up the research in this area is that no convincing link yet exists between exposure to ads and more skillful consumers. Further study is needed to determine how TV relates to other forces, such as parents, peers, and school, in developing intelligent consumer skills.

USES AND GRATIFICATIONS RESEARCH

Uses and gratifications research is slightly different from the topics discussed so far. Not only does it represent a substantive area of research results, but it also denotes a theoretical model with which to examine the effects of mass communication. Uses and gratifications research takes the point of view of the consumer. It tries to discover the uses to which people put their media consumption and the gratifications they derive from it. To elaborate, uses and gratifications focus on the audience member. We assume that before explaining the effects of mass communication, we must first understand why and how the audience uses the media material. The approach presumes that audience members differentially select various media content to satisfy their needs. In other words, the audience consists of active and motivated individuals. To sum up, the core question in uses and gratifications research is not how the media are affecting attitudes and behavior but how the individual audience member uses them to meet his or her needs.

This approach is not new. Researchers in the 1950s and 1960s were cataloguing why people were listening to the radio, watching TV, or attending to other media. Data showed that some adults watched TV to escape, to have something to talk about with friends, to learn about the world, or just to pass time. Subsequent studies among teens found similar results, but young people also mentioned using TV for companionship and for emotional arousal. Radio was primarily considered a background medium for teens and most used it to help pass the time. More recent studies of the TV audience have produced consistent findings: the audience use of TV tends to be either ritualized or instrumental. *Ritualized viewers* tend to be somewhat passive. They watch TV out of habit or merely for diversion. These individuals would say "I think I'll watch TV; let's see if anything good is on." *Instrumental viewers* are more active and selective. They deliberately plan their TV viewing ("Letterman's on; I think I'll watch his show").

Gratifications differ by age. One study that examined children, young teens, and older adolescents found six main reasons for TV viewing that cut across all ages: because it was habit, to pass time, for companionship, to escape, to relax, and for emotional arousal. Adolescents, however, used TV for an additional reason: to acquire information about the outside world.

Not only do the data provided by the uses and gratifications approach help us understand mass media consumption, but they also illustrate how uses and gratifications can have an impact on other media effects. For instance, one study noted that youngsters who reported that they watched TV to learn were more likely to believe that what they saw on TV was real than were youngsters who viewed for other reasons. Other studies documented that how TV was used was related to the actual viewing of certain programs. People who tended to watch out of habit gravitate more toward entertainment content and stay away from news and talk shows. Studies of media use and politics have noted that different motivations lead to different effects. People who use the media as a learning device about the candidate are more likely to become informed about general campaign issues as well.

The uses and gratifications approach has a practical dimension. For example, a recent study examined the uses and gratifications of "grazing," or rapidly scanning the channels using a remote control device. The most important use reported by a sample of college students was to avoid commercials. Clearly advertisers no longer have a captive audience and must examine new ways to reach the viewer in the age of the remote control. Further, many of the new information technologies, such as teletext, videotex, and two-way TV, have not caught on as quickly as expected partly because consumers have yet to find a compelling use for them. On the other hand, the VCR has become popular because it satisfied the audience's desire to timeshift TV programs and to see recent theatrical movies at home. In the future those who market the new technologies will try harder to educate audiences about the specific gratifications that can be obtained through their use.

The most recent research in uses and gratifications has been directed toward developing a theory that explains the connections among audience motives, media exposure, and gratifications. Some factors important to this process that have been isolated include age, personality, social situations, lifestyles, and the family viewing environment. Perhaps the biggest contribution of

IMPACT

THE GREAT ROCK MUSIC SCARE

Television is not the only medium alleged to have antisocial effects. One of the controversies of the early 1980s and 90s had to do with the influence of rock and rap music lyrics on youngsters. Critics charged that these lyrics encouraged sexual promiscuity, drugs, violence, and even satanism. The Parents Music Resource Center, the creation of Tipper Gore, wife of Senator Albert Gore, Jr. (candidate for the Democratic party's vice-presidential nomination), campaigned for a warning system on rock records much like that for motion pictures.

How much of an influence do rock lyrics have? Well, to begin with, to have any effects at all, they have to be at least understood by their audience. Many rock artists contend that the lyrics are really subordinated to the music, sometimes to the point where they are incomprehensible. In fact, many studies have shown that most adolescents are unable to express the meaning of most rock songs.

To illustrate, in one recent study by Patricia Greenfield and her colleagues that appeared in a 1987 issue of the *Journal of Early Adolescence*, fourth, eighth, and twelfth graders along with college students listened to Bruce Springsteen's "Born in the USA" a song most of us have probably heard at one time or another. (To refresh your memory and maybe surprise you a bit, the song talks about a hard-luck guy who was sent to Vietnam to fight, who lost a brother there, can't find a job when he gets back, and winds up with "nowhere to run" and "nowhere to go." All in all, not a flattering picture about life in America. The music, however, gives it kind of an upbeat feeling.) Even immediately after hearing the song, about 60 percent of the overall sample could not answer all of a simple series of questions about the song's content. Comprehension increased with age. Only 10 percent of the fourth graders as compared to 50 percent of the twelfth graders and 85 percent of the college students understood the meaning of the phrase "yellow man" as used in the song. Even more striking were the results about the overall mood of the song. All of the fourth graders thought the song was about how great it was to be "Born in the USA." Only 30 percent of the eighth graders and 40 percent of the twelfth graders thought the song was about despair and resentment. College students didn't do much better; about half of them didn't get the song's gloomy message.

These results suggest that the concern about the harmful impact of rock lyrics may be somewhat exaggerated. Development appears to influence the understanding of rock lyrics rather than the other way around. Students' views of life in the USA will probably be well formed by the time they comprehend Springsteen's disillusioning message.

Nonetheless, parents continue to have anxiety about the messages in rock music. Witness, for example, two recent court cases. In the first case rock musician Ozzy Osbourne was sued by the parents of a teenager who committed suicide after allegedly listening to Osbourne's recording of a song entitled, "Suicide Solution." In the second case the parents of another teenager who committed suicide claimed a backmasked message (one that is heard when the record is played backward) in a song by rock group Judas Priest was influential in causing the suicide. Both suits were unsuccessful, but they still demonstrate that the concern over antisocial messages in rock music is very much with us.

---

the uses and gratifications approach is that it has demonstrated that many factors contribute to the ultimate impact of mass communication. Further, uses and gratifications remind us that humans are active participants in the mass communication process.

TELEVISION AND PROSOCIAL BEHAVIOR

Let's close this chapter on a positive note. It's tempting, of course, to blame all of society's ills on TV. To be fair, however, we should also point out that TV has had several positive or prosocial effects as well. Prosocial behavior covers a wider range of activity than does antisocial behavior. Usually a prosocial behavior is defined as one that is ultimately good for a person and for society. Some behaviors commonly defined as prosocial include learning cognitive skills associated with school achievement, cooperation, self-control, helping, sharing, resisting temptation, offering sympathy, and making reparation for bad behavior. Many of these prosocial acts are not so obvious or as unmistakable as common antisocial acts; it's a lot easier to see somebody hitting somebody else than it is to see a person resisting temptation. In any case, this section takes a look at the more beneficial results of TV viewing.

First, as noted earlier, especially constructed programs can be effective in preparing young people for school. Without a doubt, the most successful program in this area is "Sesame Street." On the air for more than twenty years, "Sesame Street" is viewed by nearly 6 million preschoolers every week. Further, "Sesame Street" is the most researched TV series ever and the available data highlight the success of the series. To summarize some of the major findings:

1 Children who viewed "Sesame Street" regularly, either in school or at home, scored higher on tests measuring school readiness.
2 The more children watched the program, the better their scores were.
3 Disadvantaged children who were frequent viewers showed gains almost as great as their advantaged counterparts.
4 Frequent viewers also seemed to develop more positive attitudes toward school in general.
5 Children who were encouraged to view the program showed more gains than those who were not encouraged.

The success of "Sesame Street" brought it high visibility and criticism. First, it was noted that "Sesame Street" seemed to enlarge the skills gap between advantaged and disadvantaged children. Although disadvantaged children frequently gained as much as advantaged children, fewer disadvantaged children were frequent "Sesame Street" watchers, leading to the net result that a larger proportion of advantaged kids gained by viewing. To cope with this criticism, educators

Big Bird and his pals on "Sesame Street" are known the world over. "Sesame Street" has been on the air for more than 20 years.

looked for ways to encourage more viewing by the disadvantaged. The second criticism charged that the show's fast-paced style might cause problems when children entered the more slow-paced school environment. It was argued that heavy "Sesame Street" viewers might be bored and/or hyperactive as a consequence. This turned out to be a false alarm, and subsequent research has not linked these problems with frequent viewing.

The success of "Sesame Street" encouraged the production of "The Electric Company," a series whose goal was to teach reading skills to children in the early grades. Much like "Sesame Street," research designed to assess the effects of viewing this program found that viewers did better on tests of reading performance than did nonviewers. The program seemed to do best when viewed in school and supported by related curriculum materials.

Other programs that were constructed to get across prosocial messages about mental health or personal social adjustment have proven helpful.

For example, viewers of "Mister Rogers' Neighborhood" were found to be more cooperative and more persistent than nonviewers. Another program, "Freestyle," was designed to alter sex-role stereotypes among 9 to 12 year olds. An extensive research program disclosed that heavy viewers held less stereotypical attitudes that did light viewers. Two other shows, "Big Blue Marble" and "Villa Allegre," were also shown to have some success in improving attitudes toward minority groups.

All the series mentioned were designed for broadcasting over the public TV system. Although public TV reaches a considerable audience, far more watch the commercial system. Is there any evidence about the prosocial effects of commercial TV? There is and the main conclusions are not surprising. First of all, research sponsored by the TV networks found that children were able to perceive prosocial messages in commercial TV shows. One study showed that 90 percent of the audience for "Fat Albert and the Cosby Kids" were able to verbalize at least one prosocial message after watching specific episodes of the show. Do any of these messages have an impact on behavior? Studies examining the prosocial impact of TV, using both lab and survey techniques, have demonstrated that children who were exposed to specially chosen or constructed episodes of TV shows like "Lassie," "The Waltons," or "Superfriends" were more likely than nonviewers to score higher on a postviewing test of helpfulness. These effects, however, occurred in a tightly controlled environment and involved children performing tasks that were quite similar to what they saw in the program.

Surveys that link the viewing of prosocial TV with the performance of prosocial acts in real life are rare. Those that do exist show a rather weak connection between the two, an association that is even weaker than the correlation between violent TV viewing and aggression. One of the reasons for this might be that prosocial acts on TV are more complicated and easier to miss than violent actions. Also, it is not clear if exposure to prosocial material on TV nullifies antisocial portrayals or if each exerts an independent effect. Further, research has yet to establish if there are any critical periods during the development of children where the effects of prosocial—or antisocial—TV portrayals might have their maximum impact. Finally, no one has yet done an elaborate panel study examining prosocial behavior. Clearly much more work needs to be done before a complete understanding of this area is achieved.

In summary, TV seems to have several effects on behaviors that most would define as prosocial:

1 Television teaches certain cognitive skills that are necessary for school success.
2 Television shows can help reduce gender-related stereotypes.
3 Laboratory experiments suggest viewing commercial TV shows with definite prosocial messages can prompt subsequent prosocial behaviors, but this link has not yet been found to carry over in any significant degree to real life.

Before closing this chapter we should point out that we have merely sampled some of the more prominent issues in research about the social effects of radio and TV. There are many more areas that might have been mentioned. For example, many communications researchers are now devoting increased attention to how individuals process messages in their brain—the study of the cognitive aspects of communication. Moreover, it should be clear by now that the research in many of these areas highlights the thinking embodied in the specific-effects theory discussed earlier in the chapter. Radio and TV usually operate along with a number of other factors to produce an effect, and trying to untangle the unique effects of the media can be quite frustrating. And although the study of media effects is still relatively new, the data are steadily accumulating. It also appears that the electronic media are not necessarily so powerful as many people have charged, although they can exert significant influence in specific instances. Finally, we have a lot more work to do, particularly since new media have transformed the media habits of the audience.

SUMMARY

The social effects of broadcasting and cable have been studied by using experiments, surveys, panel studies, and content analyses.

Over the years various theories have enjoyed popularity as explanations for the effects of the media. The hypodermic needle theory considered the media a powerful persuasive force. This theory held that all people would have more or less the same reaction to a mass-communicated message. The limited-effects theory proposed just the opposite: because of a variety of intervening variables, the media have little effect. Currently the specific-effects theory is in vogue. This theory argues that there are some circumstances under which the media have a direct effect on some people.

The most researched topic in broadcasting is the effect of video violence on the audience. After more than thirty years of research most scientists agree that there is little evidence to support the catharsis theory. Further, most agree that TV violence contributes, although in a small way, to the development of antisocial tendencies.

There is less agreement among scientists about cultivation theory, which states that viewing large amounts of TV will distort a person's perception of reality. Evidence is mixed about this topic and more research remains to be done. Television also seems to play a part in sex-role stereotyping.

Broadcasting has had a major effect on politics. It provides political information, sets voters' agendas, establishes candidates' images, and has some limited influence on voters' attitudes.

Television also seems to have a slightly negative relationship with education. Youngsters who watch a lot of TV do not do so well in school as those who do not watch as much.

As far as TV advertising and children are concerned, attitudes toward ads change as children grow older. More mature children are highly skeptical of ads.

The uses and gratifications approach concentrates on how the viewer uses media material. This approach helps explain some of the varying effects of media content. Finally, TV seems to have a limited impact in prompting prosocial behavior.

SUGGESTIONS FOR FURTHER READING

Adler, R. (1981). *Understanding television*. New York: Praeger.
Anderson, J. (1987). *Communication research*. New York: McGraw-Hill.
Barlow, G., & Hill, A. (1985). *Video violence and children*. New York: St. Martin's Press.
Bradac, J. (1989). *Message effects in communication science*. New York: Sage.
Bryant, J., & Zillmann, D. (Eds.). (1986). *Perspectives on media effects*. Hillsdale, N.J.: L. Erlbaum Associates.
Bryant, J., & Zillmann, D. (Eds.). (1991). *Responding to the Screen*. Hillsdale, N.J.: L. Erlbaum Associates.
DeFleur, M., & Ball-Rokeach, S. (1982). *Theories of mass communication* (4th ed.). New York: Longman.
DeFleur, M., & Lowery, S. (1988). *Milestones in mass communication research* (2nd ed.). New York: Longman.
Graber, D. (1984). *Mass media and American politics* (2nd ed.). Washington, D.C.: Congressional Quarterly Press.
Hearold, S. (1986). A synthesis of 1043 effects of television on social behavior. In G. Comstock (Ed.), *Public communication and behavior* (pp. 66–121). New York: Academic Press.
Huesmann, L., & Eron, L. (Eds.). (1986). *Television and the aggressive child*. Hillsdale, N.J.: L. Erlbaum Associates.
Jeffres, L. (1986). *Mass media: Processes and effects*. Prospect Heights, Ill.: Waveland Press.
Kraus, S. (1988). *Televised presidential debates and public policy*. Hillsdale, N.J.: L. Erlbaum.
Liebert, R., & Sprafkin, J. (1988). *The early window* (3rd ed.). New York: Pergamon Press.
McGuire, W. (1986). The myth of massive media impact. In G. Comstock (Ed.), *Public communication and behavior* (pp. 175–234). New York: Academic Press.
Milavsky, J., Kessler, R., Stipp, H., & Rubens, W. (1982). *Television and aggression: A panel study*. New York: Academic Press.
Palmer, E., & Dorr, A. (Eds.). (1980). *Children and the faces of television*. New York: Academic Press.
Pearl, D., Bouthilet, L., & Lazar, J. (1982). *Television and behavior: Ten years of scientific progress and implications for the eighties* (Vol. 2). Washington, D.C.: U.S. Government Printing Office.
Reeves, B. (1986). *Effects of mass communication*, Chicago, Ill.: Science Research Associates.
Severin, W., & Tankard, J. (1988). *Communication theories* (2nd ed.). New York: Longman.
Signorelli, N. & Morgan, M. (1990). *Cultivation analysis*. Newbury Park, Calif.: Sage.
Singletary, M., & Stone, G. (1988). *Communication theory and research applications*. Ames, Iowa: Iowa University Press.
Williams, T. (Ed.). (1986). *The impact of television*. New York: Academic Press.
Wimmer, R., & Dominick, J. (1991). *Mass media research* (2nd ed.). Belmont, Calif.: Wadsworth.

PART SEVEN

HOW IT'S CONTROLLED: RULES, REGULATION, DEREGULATION

CHAPTER 17: THE REGULATORY FRAMEWORK

The philosophy that guides broadcast regulation has undergone a change in recent years. In the early to mid-1970s, under the Nixon and Ford administrations, the Federal Communications Commission (FCC) reexamined and did away with many regulatory standards that had come to the end of their useful life. In the late 1970s, under the Carter administration, more FCC regulations were eliminated in favor of what became known as the marketplace approach. "Deregulation" became the watchword and it was hoped that the marketplace would achieve the same objectives as regulation. Under the Reagan administration in the 1980s "unregulation" was the operative term. The whole process of regulation was looked upon with some skepticism, many more rules were rescinded, and the validity of the theories that justified broadcasting regulation in the first place was closely questioned. Under the Bush administration the key words were "pragmatism" and "reexamination." Faced with an economic recession and a changing marketplace that have combined to create a dismal economic picture for the traditional broadcasting industry, the FCC was searching for ways to preserve the financial integrity of over-the-air broadcasters while at the same time encouraging a competitive marketplace. Congress meanwhile, confronted with growing public disenchantment with the price and service record of cable TV, introduced measures to reregulate that industry.

In any case, the basic regulatory framework is still intact and Congress has made no serious attempts to rewrite the Communications Act of 1934. Accordingly, this chapter looks at the established forces that shape the overall form of broadcasting and cable regulation. The next chapter will examine in more detail the actual rules and regulations.

RATIONALE

Why regulate anything? Why not let people do what they want? Well, in some industries this is almost the case. Firms that manufacture computers, for example, have to abide by some general business laws, such as those regulating antitrust and fraud; but aside from these, they are pretty much left alone. The government, for example, doesn't decide who gets into the business, how much memory capacity their computers must have, and whether the monitors should be monochrome or color. Broadcasting, however, is a little different. Historically, it has had special requirements and special responsibilities placed on it by government. What makes it different? There have been two main rationales for treating broadcasting as a special case: (1) the scarcity theory and (2) the pervasive presence theory.

The *scarcity theory* notes that the electromagnetic spectrum is limited. Only a finite number of broadcast stations can exist in a certain place in a certain time; too many stations can interfere with one another. (This number may be large—Los Angeles has more than forty stations serving it—but it is still finite. There probably isn't enough spectrum space around to accommodate all those people who might wish to broadcast to the L.A. market.) This means that only a limited number of aspiring broadcasters can be served and the government must choose from among the potential applicants.

An extension of the scarcity theory holds that the spectrum is such a valuable resource that it should not be privately owned. Instead, it is treated like a public resource, owned by all, like a national park. The government simply permits private concerns to use the spectrum for their own purposes. In return for using this scarce national resource, the government treats those fortunate enough to broadcast as trustees of the public and imposes special obligations on them, such as requiring that they provide candidates for public office equal opportunities to use their stations.

By way of analogy, let's look at driving a car. As it stands now, there is no scarcity of cars and everybody who wants to drive can do so as long as they are old enough and pass a basic competency test. But suppose cars were scarce and the number of persons wanting to drive far exceeded the number of cars available. In this situation the government might have to choose those who ultimately must drive. In addition, the government might impose special obligations on them, like taking people to the hospital, transporting the handicapped, and delivering children to school. Thus government regulation would strive to maximize the positive benefits of the auto driving privilege of a few to the general public. A comparable situation exists for those licensed to broadcast. Recent technological advances such as cable TV, which offers would-be broadcasters a large number of channels, have prompted a reevaluation of the scarcity theory. Many critics would argue that it is outmoded. Others, however, point out that as long as the number of people wishing to broadcast exceeds the available facilities, scarcity still exists.

The *pervasive presence theory* is of more recent origin. It holds that broadcasting is available to virtually all of the population and once the TV or radio set is turned on, offensive messages can enter the home without warning, reaching both adults and children. This situation is fundamentally different from encountering an offensive message in a public forum. If you're out in public, you're on your own. You might encounter things that offend your sensibilities. Should you meet someone wearing a T shirt with an offensive word printed on it, your only recourse would be to turn away. The home, however, is not the same as a public place; you should not expect to encounter unwanted offensive messages there. Consequently, the pervasive presence theory holds that you're entitled to some protection. Thus the basic intrusiveness of broadcasting allows the government to regulate it.

Note that the pervasive presence theory emphasizes broadcast content and provides a rationale for the regulation of messages. It is interesting to note, however, that as developed so far by the FCC and the courts, the pervasive presence theory has been used to justify content regulation only with regard to material aimed at children. Its appropriateness as a theory underlying all broadcast regulation has yet to be developed.

HISTORY

How did the government get involved with broadcasting regulation in the first place? It started a few years after Marconi demonstrated the potential of wireless as a maritime communication device (see Chapter 2). In 1910 Congress passed the Wireless Ship Act. Only four paragraphs long, it basically mandated that large oceangoing passenger ships must be equipped with wireless sets. It had little relevance to broadcasting primarily because broadcasting as we know it did not yet exist. Two years later the Radio Act of 1912 was passed. Spurred by the sinking of the *Titanic* and the need to ratify international treaties, the 1912 act required those wishing to engage in broadcasting to obtain a license from the Secretary of Commerce. Further, to prevent interference among maritime stations, it provided for the use of call letters and established the assignment of frequencies and hours of operation. The law also strengthened the regulations concerning the use of wireless by ships. As was the case with the earlier law, the 1912 act still envisioned radio as point-to-point communication, like the telegraph, which uses the spectrum only sporadically, and did not anticipate broadcasting, which uses the spectrum continuously. Consequently, for the next eight years or so, when most wireless communication was of the point-to-point variety, the law worked reasonably well. As the 1920s dawned, however, and radio turned into broadcasting, the 1912 act quickly broke down.

As related in Chapter 2, the number of broadcast stations greatly increased and available spectrum space could not accommodate them. Interference quickly became a serious problem. Attempts by Herbert Hoover, then the Secretary of Commerce, to deny license applications or to limit conditions of operation were struck down by court rulings. Naturally, the interference problem got worse. Hoover called a series of radio conferences from 1922 to 1925. Legislation based on these conferences was duly introduced in Congress but got nowhere, primarily because very few legislators were interested in radio but many were interested in politics and didn't want to see possible presidential candidate Hoover get too much power and publicity. By 1926 the interference problem had gotten so bad that many parts of the country could no longer get a consistently clear broadcast signal. Consequently, after prodding from President Calvin Coolidge, Congress finally passed a new law—the Radio Act of 1927.

The 1927 Radio Act

Although more than sixty years old, the Radio Act of 1927 still demonstrates the basic principles that underlie broadcast regulation. Embracing the scarcity argument, the key provisions of the act were:

1. the recognition that the public owned the electromagnetic spectrum, thereby eliminating private ownership of radio frequencies
2. the notion that radio stations had to operate in the "public interest, convenience or necessity"
3. a prohibition against censorship of broadcast programs by the government
4. the creation of a five-member Federal Radio Commission (FRC), which would grant licenses, make rules to prevent interference, and establish coverage areas with its decisions subject to judicial review

In its initial thinking, Congress apparently never meant for the FRC as originally set up to be permanent. The FRC was expected to complete its tasks in one year, at which time its administrative power would be transferred to the Secretary of Commerce. In light of various political considerations, however, Congress eventually abandoned the idea that the Secretary of Commerce would play a major role in administering broadcasting and gave the job to the FRC.

The FRC immediately set to work to straighten out the interference problem. Its authority under the 1927 act was tested immediately as stations protested denials of their license applications or refusals to renew their licenses. In general, the courts supported the new agency and clarified its role in regulating broadcasting. In one important case, the courts even affirmed that the FRC could examine past station performance in determining if a license could be renewed. The FRC denied the license of KFKB in Kansas because it had broadcast prescriptions for bogus patent medicines recommended by a quack doctor (see box). The courts ruled that the FRC was correct in assuming that such programming was not in the public interest and the failure to renew the license did not constitute censorship. The significance of this decision was that the FRC was not restricted to an examination of solely technical matters and could indeed look at content.

The Communications Act of 1934

The supersession of the 1927 act only seven years after its passage did not take place because of any dissatisfaction with the original legislation. In fact, the law performed quite well and the FRC was able to eliminate the chaos that threatened early broadcasting. Instead, the impetus was a concern in the administration of President Franklin Roosevelt over fragmentation of the various regulatory functions. At the time, regulatory powers over communication were shared by the FRC, the Interstate Commerce Commission, the postmaster general, and the president. In an attempt to streamline matters the new act created a seven-member (later reduced by Congress to five) Federal Communications Commission (FCC) and gave the FCC regulatory power over all forms of wire and wireless communication, except for those frequencies reserved for the military and other federal government services. (Most of the regulation of federal use of the spectrum is done by the National Telecommunications and Information Administration.) The original 1927 act was regarded so highly that much of it was simply rewritten into the new legislation and the underlying philosophy was left unchanged, including the key phrase "public interest, convenience and necessity."

Under this broad and somewhat vague mandate the new FCC was expected to develop its own standards. Like most laws, the Communications Act has been amended and reshaped, usually in response to a current problem or to changing technology. For example, in 1959 Congress amended the act, making it illegal for people to rig quiz shows. New language was added the next year that made payola illegal. In 1962 Congress passed both the Communications Satellite Act, which gave the FCC expanded regulatory powers, and the All-Channel Receiver Bill.

Despite several recent attempts to revise or dismantle it, this 1934 act still remains the organic legislation that controls American broadcasting. Since it is so far-reaching, we shall discuss it at length in a later section.

Cable Regulation

The history of cable regulation is somewhat confusing, marked by indecision, political infighting, and major changes of direction. During the 1950s, when cable was still a novelty (see Chapter 3), the FCC steadfastly refused to regulate it, arguing that cable didn't use over-the-air frequencies. As cable became more widespread and traditional broadcasters began to view it as a potential competitor, it was inevitable that the FCC would get into the picture, if for no other reason than to relieve some of the pressure from over-the-air stations. In 1962 the FCC did an about-face and asserted limited jurisdiction over cable. By 1966 the FCC had instituted comprehensive regulations that protected broadcasters by effectively hampering cable's growth in large markets. For example, cable systems in the top-100 markets could not import a new signal without a hearing on the effects of such a move on existing local stations. (Part of the rationale behind these restrictive rules was to help fledgling UHF stations by keeping cable competition out of the market. Later experience suggested that UHF stations got a big boost by being carried on cable systems. Thus the FCC rules might have hurt rather than helped UHF.) In 1972, now under pressure from cable operators, the FCC issued another set of comprehensive rules. Although still favorable to traditional broadcasters, the 1972 rules were somewhat more palatable to cable operators.

While all of this was going on at the federal level, cable systems were also bound by regulations at the state and local levels. Thus a company that owned cable systems in twenty communities might be faced with twenty different sets of local rules. The most important local rule established exclusivity. Local governments awarded an exclusive franchise to a single cable company to serve a community or neighborhood in exchange for the provision of specialized community channels, payment of a franchise fee, and maintenance of low rates. Many companies, in their zeal to obtain a franchise, overpromised and had to scale down their systems, creating some bad feelings between them and the local governments.

Reexamination

During the late 1970s and early 1980s Congress took a new look at the 1934 Communications Act. The increasing popularity of new distribution technologies, such as cable, satellites, and microwaves, suggested to many policymakers that legislation drafted before TV came on the scene was no longer satisfactory. Accordingly, in 1976 plans were announced to rewrite the act, and several bills to that effect were introduced in Congress. These and other proposals were radical in design. One version eliminated entirely the notion of the "public interest"; another levied a fee on those who used the electromagnetic spectrum. Perhaps because they differed so profoundly from broadcasting's regulatory tradition, none of these bills passed and the 1934 law remains on the books.

Although these attempts failed, they did serve to spark interest in the general area of communications law reform. There was a growing national trend toward deregulation, as reflected in the FCC's initiation of competition for long-distance phone service and congressional toleration of this initiative and the eventual disposition of AT&T's phone companies. Broadcasting deregulation was evident in a number of forums. In 1981 the FCC abolished many regulations pertaining to radio, simplified license renewal applications, and eased operating regulations for TV stations. Eventually even such a longstanding policy as the fairness doctrine was repealed by the FCC. The Supreme Court also supported the FCC's decision that the market rather than the government should regulate radio formats. Congress got into the act by extending the license terms for both radio and TV stations.

On the cable front, the FCC relaxed or repealed many of its 1972 restrictions. By 1977 the FCC was even reexamining its basic rationale for cable regulation: that cable would fragment audiences and harm the public interest by destroying traditional over-the-air broadcasting. The climate of deregulation had firmly engulfed cable as the 1980s began, but the focus shifted from the FCC to Congress. In 1982 and 1983 several bills were introduced dealing with cable. In the final hours of the 98th Congress the Cable Communications Policy Act of 1984, a compromise between cable operators and regulators, was passed. The act put cable firmly under the jurisdiction of the FCC but allowed state and local governments to control franchising. Cable operators were free to set their own rates but franchising authorities could require systems to provide channels for public, educational, and government use (a provision later challenged in court). The deregulatory trend continued the next year when a federal appeals court struck down an FCC provision that dated back to the early days of cable regulation: the "must-carry" rules. These rules required that cable systems carry all local stations and any others significantly viewed in the community. A revised set of must-carry rules drafted by the FCC in 1987 was also declared unconstitutional. Theoretically at least, cable systems are free to carry whatever pro-

PART 7:
HOW IT'S
CONTROLLED:
RULES, REGULATION,
DEREGULATION

PROFILE

DR. BRINKLEY AND HIS GOATS

One of the key cases decided by the new Federal Radio Commission concerned station KFKB ("Kansas Folks Know Best") in tiny Milford, Kansas. Licensed to Dr. John R. Brinkley, KFKB ("The Sunshine Station in the Heart of the Nation") became one of the most popular stations in America during the 1920s, primarily because of the notoriety surrounding its owner. John Romulus Brinkley was the son of a medical doctor. At an early age he decided to follow in his father's footsteps and entered medicine. Young Brinkley graduated, if that's the right word, from the Eclectic Medical University of Kansas City. (He apparently finished his course of study in a few weeks, paid $100, and received a diploma.) Forty states did not recognize degrees from his college, but that left "Dr." Brinkley eight others he could practice in. He eventually wound up in Milford and opened a modest hospital, which he immodestly named after himself. If that wasn't enough, he also set up the Brinkley Research Laboratories and the Brinkley Training School for Nurses.

Brinkley had studied a little bit about the workings of human glands. He also was a shrewd judge of human nature. Consequently, he began to advertise his rejuvenation operations over KFKB. Designed to restore sexual drive in middle-aged men, the actual operation consisted of grafting or injecting material from the sexual organs of male goats into his human patients. The operation was done under local anesthesia and took only fifteen minutes. The cost was a rather significant sum in those days—$750. (The goats, of course, paid a much higher price.) One interesting touch introduced by the doctor: patients could pick their own goat in advance from the doctor's private herd, sort of like picking out a lobster for your dinner from a tank at a seafood restaurant. As time passed, Brinkley was touting his operation as a cure-all for skin diseases, high blood pressure, insanity, and paralysis.

KFKB was the perfect medium for Brinkley's advertising. At the time, thanks to the doctor's investment, it was one of the most powerful stations in North America. Twice a day, people from Saskatchewan to Panama could hear the doctor's own radio talk show in which he promoted his operation, gave precise instructions on how to get to Milford, and even advised prospective patients on how to transport the $750 safely.

People flocked to Milford and Brinkley became a rich man. In an era when legitimate doctors were making an annual salary of about $5000 Brinkley was pulling in about $125,000 a year. He owned his own airplane, his own yacht, and a fleet of cars, and usually wore a couple of 12-carat diamond rings.

In addition to his gland operations, Brinkley also started selling his own prescription drugs. On a show called "The Medical Question Box" Brinkley would read letters people had written him about their ailments. He would then prescribe his medicines by number. Pharmacists all over the region were supplied with the Brinkley formulas, which usually consisted of ingredients such as alcohol and castor oil. For each bottle that was sold, the pharmacist would give a cut to Brinkley.

Not all was rosy, however. Kansas City newspapers attacked the doctor's methods, the American Medical Association exposed Brinkley's quackery, and the Federal Radio Commission eventually revoked his broadcast license and further defined the principle that programming must be in the public interest. Undaunted, Brinkley ran for governor of Kansas as a write-in candidate and nearly won. Since KFKB was no longer open to him, Brinkley bought a station in Mexico, XER, 500,000 watts strong, and kept his messages blanketing North America. Eventually the doctor moved to Del Rio, Texas, across the border from XER, and continued his medical practice there, this time specializing in cures for enlarged prostate glands. When a competing doctor opened up a similar practice in Del Rio, Brinkley moved to Little Rock, Arkansas.

Once in Arkansas, Brinkley's troubles began. Malpractice awards to discontented patients, back taxes, legal fees, the costs of running his Mexican station, and his extravagant lifestyle soon exhausted even Brinkley's sizable bank account. The doctor had to declare bankruptcy. As if that wasn't bad enough, the U.S. government finally persuaded Mexico to ease international radio interference problems by shutting down XER and similar stations. Less than a year later, at 56, J. R. Brinkley succumbed to a heart attack, thus ending the career of one of early radio's unsavory but colorful broadcasters.

gram lineup they choose within copyright law restrictions.

As the 1990s began, on both the traditional broadcasting and cable front, the trend toward deregulation had slowed somewhat and there was a small but noticeable movement toward enacting additional regulations on the industry. In 1990 Congress passed two new laws, the Children's Television Act and the TV Violence Act, that put additional obligations on broadcasters and cablecasters. As 1992 began Congress was debating a new bill to regulate cable TV that would supersede the Cable Communications Policy Act of 1984. Nonetheless, the philosophy of deregulation was still around. The FCC liberalized its regulations limiting the number of radio stations one person or organization could own and issued new regulations that would allow the broadcasting networks to own cable TV systems. At the same time, a decision by the Court of Appeals to allow the regional bell operating companies (RBOCS) to provide video information services threatened to change dramatically the competitive marketplace. All in all, the 1990s will probably contain many regulatory modifications for the broadcasting and cable industry.

REGULATORY FORCES

Broadcast and cable policy results from the interaction of several factors. Legislative, administrative, judicial, political, and economic forces are all important determiners of regulation. Drawing upon and expanding the model used by Krasnow, Longley, and Terry in *The Politics of Broadcast Regulation*, this section examines eight key components in the process: the FCC, Congress, the courts, the White House, industry lobbyists, the public, state and local governments, and the marketplace.

The Federal Communications Commission

The FCC currently consists of five commissioners, one of whom serves as chair, appointed by the president and confirmed by the Senate, for staggered five-year terms. If a commissioner leaves office early, the replacement serves only the remainder of the term, not a new term. Political balance is achieved by requiring that no more than three members can be from one political party.

Serving the commissioners are four bureaus: Field Operations, Private Radio, Common Carrier, and Mass Media. Figure 17-1 shows this arrangement.

The Field Operations Bureau used to be called the Field Engineering Bureau. As its former name suggests, it is concerned primarily with detecting technical violations in the operation of broadcast equipment.

The Private Radio Bureau oversees radio used as point-to-point communication such as citizens' band, amateur radio, and local government radio. It does not regulate the federal government's point-to-point radio services.

The workings of the Common Carrier Bureau need some explanation. *Common carrier* is a legal term that denotes those communications systems that must provide their services to anyone who can pay for them. Unlike broadcasting and cable, common carriers traditionally have no control over the content of the messages that are carried on their systems. Put another way, a common carrier is a neutral conduit for information sent by others. The Postal Service, for example, is a nonelectronic common carrier. If a person can afford the price of a stamp, the Postal Service will deliver the letter. The telephone and telegraph are examples of electronic common carriers. Thus the Common Carrier Bureau regulates both wired (telephone and telegraph) and wireless (for example, microwave and many communication satellites) communication.

The Mass Media Bureau is most relevant to broadcasting and cable. Created in 1983, this bureau is charged with overseeing both over-the-air broadcasting and cable. The Mass Media Bureau has four divisions:

1 Audio Services, responsible for AM and FM radio

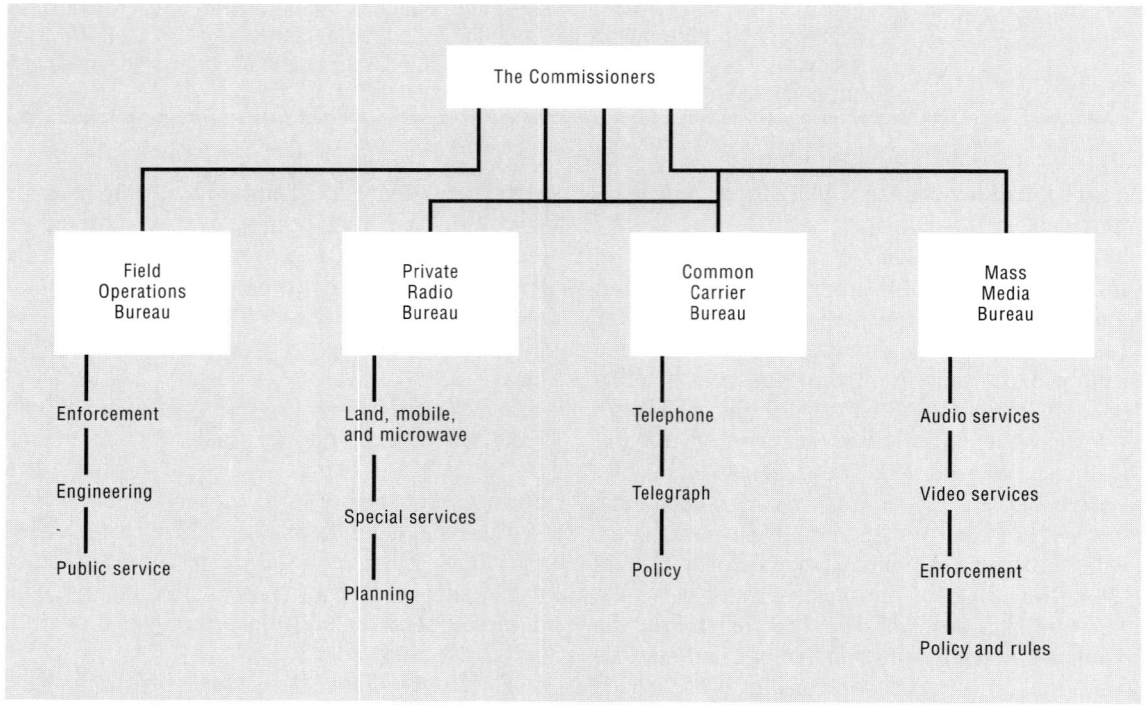

FIGURE 17-1 The Federal Communications Commission.

2 Video Services, which handles cable, UHF, and VHF TV
3 Enforcement, responsible for upholding FCC law
4 Policy and Rules, which handles planning, revising existing laws, and proposing new regulations

The Federal Communications Commission (FCC) regulates more than broadcasting and cable. It also has jurisdiction over such devices as cellular phones.

There are other divisions of the FCC. These include the Office of Public Affairs, which disseminates information about commission actions to the general public; the Office of General Counsel, which advises the FCC on legal matters; and the Office of Administrative Law Judges, whose judges preside over cases designated by the FCC for hearings and issue initial decisions.

The FCC uses a variety of methods to regulate broadcasting and cable. Among the most important are licensing, policy making, and the "raised eyebrow."

Licensing Licensing is the primary function of the FCC and the most important method of regulation. The FCC must approve applications for new station licenses, license renewals, and transfer of ownership to a new licensee. Note that the FCC grants licenses only to stations; it does not license networks. Each of the major networks, however, owns local stations that must have their licenses renewed.

Each radio station must apply for a license renewal every seven years. A TV station license has a shorter term—five years. When a license comes up for renewal, the FCC must determine if the station has operated in the public interest. In past years the FCC required stations to submit sizable amounts of information along with their renewal applications. Stations were required to submit, among other things, program logs, an analysis of programming, and results of community surveys. Not surprisingly, preparing (and processing) an application was a burdensome task. Now, thanks to deregulation, the procedures are simpler. The renewal application for most stations now fits on a postcard (some supporting documentation, however, is also required). Cable TV systems are not licensed by the FCC. Instead, this is the responsibility of the local governments who grant the franchise. The specific regulations and procedures surrounding the granting and renewal of broadcast licenses will be discussed in the next chapter.

Policy making Another regulatory tool is policy making. Under the public interest guidelines, the FCC can enact rules governing some aspect of broadcasting and/or cable. To enact a rule, the commission must announce a proposal and allow for public comments. Moreover, the FCC can't put most new rules and regulations into effect until it publishes a summary, usually in a publication called the *Federal Register*. For example, in the early 1970s the commission enacted the prime-time access rule (PTAR), which in effect took the 7:30 to 8:00 P.M. (EST) slot away from the networks and gave it back to the local stations to program. The rule never quite accomplished what the FCC intended (more local programs), but PTAR did open up the market for syndicated shows. We have also seen how throughout the years the FCC has enacted various rules concerning the operation of cable TV systems.

Not surprisingly, the deregulation climate of the 1980s prompted a trend toward policy unmaking as the FCC dismantled some of its prior regulations or refused to enact new rules. For example, an "antitrafficking" rule that mandated ownership of a broadcasting station for three years before selling was repealed (and is currently being reexamined), as was the FCC's longstanding and controversial fairness doctrine. The FCC also declined to get involved in a decision that would set a single technical standard for AM stereo, preferring to let the marketplace decide.

Nonetheless, the FCC still issues rules and policy statements when necessary. In 1991–1992, for instance, the FCC did the following:

1 It announced a new version of the financial-syndication rules that specify, among other things, how much economic interest a network can have in shows that are in syndication.
2 It changed its rules concerning the cap on owning radio stations from twelve AM stations and twelve FM to eighteen AM and eighteen FM.
3 It enacted new regulations defining the phrase "effective competition" in a cable market as mentioned in the 1984 Cable Communications Policy Act. The revised regulation subjected more cable companies to rate regulation by the local municipality.
4 It proposed to permit regional telephone companies to provide a "video dial tone" that would deliver video programming, information services, and telephone services to homes.
5 It implemented the Children's Television Act of 1990 by imposing limits on commercial time during children's programming and requiring that stations air educational and information programming serving the needs of children.
6 It enacted a new regulation allowing networks to own local cable systems.

Another component of the policy process involves planning for the future. The FCC maintains an Office of Plans and Policy, which analyzes trends and attempts to anticipate future policy problems. In 1991 this office issued a particularly bleak report on the future of conventional TV broadcasting, concluding that the audience and the revenues of that industry were in an "irreversible, long-term decline." In turn the FCC asked broadcasters to submit possible changes in FCC rules that would help the broadcasting business to be more competitive in the future.

Raised eyebrow Regulation by raised eyebrow is an informal technique of suggesting policy without resorting to the long and involved process of making rules. Instead, an FCC commissioner will make an important speech or the FCC will issue a press release or an interview might appear in the trade press. No matter what channel, the industry gets the word about how the commission feels and acts accordingly. This technique is somewhat controversial because it bypasses the traditional rule-making procedure and is not subject to judicial review.

In the early 1970s the FCC issued a statement that radio stations which played songs that apparently glorified drug use would not be operating in the public interest. Most stations got the message and pulled several questionable songs from their playlists. The 1980s and 1990s have not seen the FCC raise its eyebrows very often. Nonetheless, the implied threat of future action with regard to license renewals makes this an effective tool of broadcast regulation.

Enforcement. The FCC has several different methods it can use to enforce its decisions, ranging from a slap on the wrist to a virtual death sentence. At the mildest level, the FCC can issue a letter of reprimand that scolds a station for some practice that is not in the public interest. Usually, unless the station persists in provoking the FCC, that ends the matter. Other legal remedies include a cease-and-desist order (which makes a station halt a certain practice) or a fine (called a forfeiture). A station can be fined up to $25,000 a day to a maximum of $250,000 for each separate violation. Under the chairmanship of Albert Sikes the FCC has taken a more get-tough attitude about enforcement. FCC field inspectors have been busier in recent years than they have for a decade as they scrutinize stations' compliance with technical and content regulations.

The FCC, in effect, can put a station on probation by failing to renew a license for its full term. This is called short-term renewal and usually ranges from six months to two years. The idea behind short-term renewal is to give the commission a chance to take an early look at a station to determine if past deficiencies have been corrected. Infractions likely to bring short-term licenses involve deceptive promotions, lack of supervision of station facilities, and violation of equal employment practices.

The FCC can sound a death knell for stations by refusing to renew a license or revoking one currently in force. Although this action is used sparingly, it does occur. Since 1989 the FCC has denied the renewals of a TV station in San Francisco and a radio station in Ohio for misrepresentation. In addition, the commission has started revocation proceedings against a radio station in Georgia and one in South Carolina.

Congress

The FCC regulates broadcasting with Congress looking over its shoulder. The FCC is a creature of Congress. Congress created the FCC and, if it wished, Congress could abolish the FCC. (In fact, in 1982 Congress changed the status of the FCC from a permanent agency to one that had to be reauthorized every two years.) The U.S. Congress is an important part of broadcasting and cable regulation because of its power to enact and amend laws. It was Congress that passed the Communications Act of 1934, which created the FCC. Throughout the years Congress has maintained its involvement by amending or adding to the act, as it did in 1984 with the Cable Communications Policy Act.

In addition, Congress has enacted new legislation that has had an impact on broadcasting. In 1969 it prohibited cigarette advertising on TV and radio. A 1972 law required broadcasters to provide reasonable access to their facilities for candidates for federal office. In 1990 Congress passed two new laws, one dealing with children's TV and the other with TV violence. Another piece of legislation regulating cable TV was still under consideration as of early 1992.

By and large, Congress has preferred to enact narrow amendments to the 1934 act rather than reformulate its underlying principles. Many of these amendments cover political broadcasting, an area of great concern to politicians; others represent compromises between industry and lawmakers. In the latter category a series of 1982 amendments extended the license terms of radio and TV stations but also gave the FCC the option of using a lottery weighted in favor of women and minorities to decide among several qualified applicants for new broadcast stations. The FCC has used this technique to assign low-power TV stations but until recently had hesitated to use it with full-power stations. In 1989, however, the FCC proposed using a weighted lottery system rather than comparative hearings to choose among competing applicants for new AM, FM, and full-power TV channels.

Congress can also send a message to the FCC by enacting laws that have nothing to do with the Communications Act. In 1980 Congress passed a law that directly affected the Federal Trade Commission (FTC), a sister regulatory agency to the FCC. In the minds of many legislators the FTC had gotten a little too activist in its efforts and had

ECONOMICS

WHAT IT MIGHT COST TO MISBEHAVE

In 1991 the Federal Communications Commission (FCC) announced a new schedule of fines for infractions of various FCC rules. The amounts listed below are "base" amounts for some selected examples and are subject to revision upward and downward due to the circumstances. Nonetheless, they represent good ball-park figures.

Violation	Base Fine ($)
Misrepresentation/lack of candor	20,000
False distress broadcast	20,000
Inadequate tower lighting and marking	20,000
Failure to respond to FCC communications	17,500
Exceeding power limits	12,500
Broadcast of indecent programming	12,500
Operation at unauthorized location	10,000
Failure to identify sponsors	6,250
Broadcasting telephone calls without permission	5,000
Failure to identify station when on air	2,500

---

angered many in the business community. Consequently, Congress passed a law that required the FTC to submit all final rules to Congress for approval, thus instituting a legislative veto. This legislative veto was ultimately declared unconstitutional by the Supreme Court. Nonetheless, the whole episode probably made the FCC aware of what might happen if it aroused similar congressional ire.

Moreover, although it sounds odd, Congress can affect regulatory policy by doing nothing. If it so chooses, Congress can terminate pending legislation or postpone it indefinitely by tabling it. This is exactly what happened during the 1980s as various rewrites of the Communications Act were ignored and eventually left to die. Rather than go through the full-scale political battle that sometimes surrounds the drafting of new or amended legislation, Congress has been content to exert regulatory pressure in less obvious ways.

Some of the nonlegislative methods used by Congress include committee hearings, budgetary review, and approval of FCC commissioners. The House and Senate Committees on Interstate Commerce (whose jurisdiction includes broadcasting) and their subcommittees periodically hold hearings on matters of policy. These sessions highlight, at least in general terms, congressional feelings on a certain topic. The FCC in turn usually gets the message and may shape its policies accordingly. In recent years committee hearings have focused on TV violence, children's programming, and public broadcasting. In 1987 the House Telecommunications Subcommittee held unprecedented hearings on the impact of network news budget cuts on the quality of TV news.

Congress also controls the budget of the FCC. All commission requests for money, such as salaries and operating funds, must be approved by Congress. Sometimes the money comes with strings attached. In 1980 the FCC got nearly a half million dollars extra, but the money was to go toward a study of AM spectrum separation. At the other extreme, in 1988 Congress reduced the FCC's budget by $7.4 million, forcing the commission to cut back on some of its activities. The Congress also can simply refuse to give an agency a regular budget, as it did during its squabble with the FTC in the late 1970s and early 1980s. At one point the FTC actually ran out of money and had to shut down for a day.

In addition, all presidential appointments to the FCC must be approved by Congress. This gives the Congress an opportunity to delay or block a particularly sensitive appointment. It also allows important members of Congress to sponsor or champion the cause of their own favorites for appointment to the FCC. A nominee with the backing of one or more influential members of Congress will generally win easy confirmation.

Finally, Congress, of course, is a political animal and responds to various political pressures. Broadcasters, because they control electronic media exposure—something very dear to the hearts of politicians—are an important constituency and are likely to be listened to by the representatives. By the same token, special-interest and citizens' groups are made up of voters and they also command attention. Broadcasters and citizens' groups may try to influence Congress to amend or pass new legislation or they might try to get Congress to exert pressure on the FCC.

The Courts

If Congress looks over one shoulder of the FCC, the federal court system looks over the other. If a broadcaster, cable operator, or a citizen disagrees with an FCC decision, he or she appeals it to the federal courts. Most cases concerning broadcasting are decided by the U.S. Court of Appeals for the District of Columbia, whose decisions are subject to review by the Supreme Court only. Courts, however, do not act on their own. They wait for others to initiate actions. If a complaint is raised, courts do not normally reverse an agency's decision if the agency acted in a fair, nonarbitrary way and if the FCC actually has jurisdiction over the matter in question.

Challenges regarding FCC decisions are typically examined on two grounds. The first is constitutional. The courts decide if the commission violated the First Amendment protection of freedom of speech. The courts, particularly the Supreme Court, tend to be extremely careful in deciding constitutional issues, preferring to make their decisions on other grounds if possible. But on occasion, constitutional issues may be addressed. For example, the U.S. Court of Appeals struck down the FCC's must-carry rules on grounds that they violated the First Amendment rights of cable operators.

Cases can also be decided on statutory grounds. This means that FCC decisions are examined to see if they are in keeping with the standards of existing laws. In broadcasting this usually boils down to determining if the FCC has followed the Communications Act of 1934 or the Administrative Procedures Act. In cable it usually consists of determining if the Cable Communications Act of 1984 has been followed.

Although it does not get involved in many of the decisions of the FCC, when it does get a chance to speak, the judicial system has shown itself to be an important factor in determining broadcast policy. It was a court decision that articulated the pervasive presence rationale behind broadcasting regulation and affirmed the FCC's right to regulate indecent programming. Another court decision totally revamped the renewal hearing process. Before 1966 only those who would be affected by technical interference or economic hardship caused by the renewal or granting of a station license could appear and offer evidence and testimony at a hearing. Then, in the WLBT case, the Court of Appeals ruled that private citizens had the right to participate in the process and citizen involvement grew rapidly in succeeding years. Deregulation has somewhat diminished citizen involvement, but the WLBT case established their right to intervene.

The courts can also be influential in other ways as well. In 1991 a district court judge removed a restriction on the regional telephone companies that barred them from providing video services, such as video newspapers and video ads, outside their service areas. (The 1984 Cable Communications Act prohibits the regional phone companies from providing video services in their own service area.) If the decision survives appeals and review by higher courts, it would mean more competition for broadcasters and cablecasters.

Of course, courts do not exist in a vacuum; they interact with many of the other forces active in regulating broadasting. Federal judges are appointed by the president with the approval of the Senate. Congress writes and rewrites the laws that the courts interpret. Court decisions, as with WLBT, alter the ease of access for some groups into the regulatory process. Broadcasters and citizens can bring lawsuits to attract judicial attention. All in all, courts play a pivotal role in the regulatory process.

The White House

If it were possible for the FCC to have a third shoulder, the White House would be watching over it. The influence of the executive branch may not be as visible as that exerted by the Congress and the courts, but it is nonetheless potent.

REGULATIONS

BROADCASTING AND AIRPLANES

The Federal Communications Commission (FCC) is not the only federal agency that regulates broadcasting. The Federal Aviation Administration (FAA) also gets into the act. Why should the FAA be concerned with broadcasting? In the first place, broadcast towers are tall structures that might present a hazard to airplanes. Consequently, the FAA has rules that restrict where broadcast towers can be located and how tall they can be. Second, airplanes and control towers use the electromagnetic spectrum to send radio messages back and forth and any interference from broadcast stations might pose a danger to airline safety. In that connection the FAA drafted tough new rules governing interference reduction in 1991. The National Association of Broadcasters, the Association for Maximum Service Television, the Land Mobile Communications Council, and the Cellular Telecommunications Industry Association all criticized the new rules as overly restrictive and announced their opposition. For its part, the FCC called for more interagency discussion and cooperation. Veteran Washington observers, however, described the new rules as part of a power struggle between the FCC and the FAA. In any case, the new interference rules demonstrate one of the difficulties involved in regulating an industry that crosses traditional regulatory boundaries: there is the potential to get caught up in bureaucratic conflicts.

In the first place, the president has the power of appointment over both FCC commissioners and judges. With regard to the FCC, since many members do not serve out their full terms, the president can fairly quickly establish a slate of congenial commissioners. Shortly after George Bush took office in 1989 there were three vacancies on the commission, affording the new president the means to establish an immediate influence. Although the Communications Act limits the number of commissioners from one political party to no more than three, most presidents are able to find people who express the administration's prevailing political philosophy. Furthermore, the president can designate any of the sitting commissioners as chair at any time, without congressional intervention or approval. As a result, the president can set the regulatory tone of the FCC.

Second, the White House has its own agency that specializes in the broad field of telecommunication—the National Telecommunications and Information Administration (NTIA). Housed in the Commerce Department, the NTIA allocates radio frequencies that are used by the federal government, makes grants to public telecommunications facilities, advises the administration on telecommunications matters, and represents the administration's interests before the FCC and the Congress. The NTIA is one of those organizations that will play as big a role in broadcasting and cable policy making as the president desires. It has the potential to exert a powerful political influence, as did its predecessor, the Office of Telecommunications Policy, during the Nixon era. On the other hand, should a president decide to take a more detached position with regard to regulation, the NTIA could be left to languish. Under the Bush administration the NTIA has taken an active role. In 1991 the agency argued for the auctioning of spectrum space that would serve new technologies, such as cellular telephones, to the highest bidder.

Third, the legal arm of the executive, the Department of Justice, has had a strong influence on broadcast policy, particularly through its antitrust division. The Justice Department was influential in the breakup of AT&T. It has gone so far as to challenge FCC decisions in the courts. It was the Justice Department's opposition that was partly responsible for blocking an FCC-approved merger between ABC and ITT in the late 1960s. It was also the Justice Department that sued the three networks for having too much control over prime-time programming. In 1979 the Justice Department went after the National Association of Broadcasters (NAB), charging that the commercial time limits in the NAB Code of Good Practice artificially reduced the amount of commercial time available and drove up prices. The NAB ultimately abolished its codes. As is probably obvious, the Justice Department is not afraid to take anyone to court, and its actions exert a strong influence on regulatory policy.

Another cabinet division that has an influence is the State Department. When the United States enters into international negotiations over such things as spectrum use, satellite orbital slots, and any mass communication-related trade topic, the United States is represented by the State Department. The State Department's role will become more important in the future as more domestic policy making is influenced by the international context. For example, companies in Japan, Europe, and the United States have developed different systems of high-definition TV (HDTV). A decision to adopt one of the foreign systems would cost the United States billions of dollars and worsen the trade imbalance.

Further, the president can initiate communication legislation. Although the president cannot personally introduce a bill in Congress, the White House or the NTIA can draft a possible law and the president can usually find a friendly member of Congress to introduce it. For instance, in 1991 the Bush administration, looking for new ways to raise revenue, proposed levying spectrum use fees on broadcasters and cable operators. Legislation to that effect was quickly introduced in Congress where it faced heavy opposition from the industry.

The White House can also make its views known on matters that are pending before the FCC. In 1991, when the FCC was embroiled in drafting new rules to control the networks' financial interest in the syndication market, White House Chief of Staff John Sununu wrote a letter in which he underscored the president's desire to avoid unnecessary government interference in business matters.

Finally, the president has the power to veto or threaten to veto legislation enacted by Congress. The potency of this instrument was clearly demonstrated in the recent controversy over retention of the fairness doctrine. Congress favored writing this doctrine into law; President Reagan opposed it. In mid-1987 proponents of the fairness doctrine succeeded in passing a bill that would give the doctrine the status of law. The president promptly vetoed the bill and Congress was not able to override the veto. Several months later the fairness doctrine was again written into law, this time as part of a catchall appropriations bill. When the president threatened to veto this bill, even if it meant shutting down the government for lack of money, its backers removed this provision from the spending bill, promising to revive it at some later date.

Industry Lobbyists

A lobbyist is a person who represents a special interest and tries to influence legislators' voting behavior. Some people regard lobbyists as a negative force in policy making, but it should be pointed out that they serve a necessary information function. Many lawmakers regard lobbyists as an asset; they help the lawmaker learn about the impact of pending bills on various segments of society. Lobbying has gone on in Washington for more than 200 years and is a deeply ingrained part of the political process. It comes as no surprise, then, that the broadcasting and cable industries maintain extensive lobbying organizations whose task is to make the industry's wishes known to the FCC, Congress, the courts, and the president. In fact, some observers of the Washington scene suggest that the recent technological advances in communication and the trend toward deregulation have made lobbyists among the most important forces shaping broadcasting and cable policy.

The lobbyists themselves may be long-time Washington lawyers, political staffers, former politicians, or former regulators. There are numerous examples of ex-FCC chairpersons going on to become highly influential lobbyists. Richard Wiley (1974–1977) has done work for CBS, the National Association of Broadcasters, and the National Religious Broadcasters Association; Charles Ferris (1977–1981) represented Turner Broadcasting and clients in the cable TV industry. Lobbyists are generally backed by teams of lawyers, economists, public relations experts, and specialized consultants. The lobbyists represent trade associations and single companies. Each of the major networks maintains its own lobbyists, as do a host of trade associations: the National Association of Broadcasters (traditionally the most influential for the broadcasting industry), the National Cable Television Association (the NCTA—the NAB's rival in the cable industry), the Association of Independent TV Stations, the National Association of Public TV Stations, the

National Religious Broadcasters Association, and the National Association of Farm Broadcasters—to name just a few. No matter who they are or whom they represent, lobbyists have one thing in common: they are expensive. To maintain an aggressive and effective lobby in Washington might cost several million dollars a year. Although this may seem expensive, if a lobbyist is effective, his or her sponsor might save ten times the cost. (Not surprisingly, organizations with the most money generally have the most-effective lobbies.)

A recent tool used by many lobbyists is the political action committee (PAC). PACs attempt to elect candidates who are sympathetic to their industry's needs by donating money to such candidates' campaigns. Both the NAB and the NCTA maintain PACs and both have donated to the reelection campaigns of members of the congressional committees that regulate the two industries. Compared to other industries, however, broadcasters and cablecasters raise only a modest amount of PAC money.

Many regulatory battles can be seen as contests between one or more of the industry lobby groups, with many of the recent ones involving the NCTA. In 1989 the NCTA and the NAB squared off over the issue of syndication exclusivity rules, which would force cable systems to black out syndicated programming on distant stations if the same programming appeared on local stations. Broadcasters favored the rules but cable operators didn't; both sides mobilized a substantial lobbying effort to influence the FCC. The broadcasters finally prevailed and the rules were enacted.

In 1991 Congress was debating a new bill to regulate cable. A section of this proposed law gave broadcasters the right to charge cable companies a fee to carry their stations. This provision was the target of a massive lobbying campaign by the NCTA. Television ads, newspaper ads, and inserts mailed to cable subscribers along with their monthly bills all warned that the proposed regulation would significantly raise cable rates. Subscribers were urged to contact their senators and to make known their opposition to this provision. The NAB responded with a countercampaign that urged viewers to support the new legislation and to end "cable's free ride." The next year or two may decide which lobbying effort was more effective.

The National Association of Broadcasters (NAB), the chief lobbying organization for radio and TV stations, has its headquarters in Washington, D.C. About 5000 radio and 950 TV stations are members of NAB.

At about the same time, the FCC was the focus of an extensive lobbying effort between the networks and the Motion Picture Association of America over the financial interest and syndication rules. The motion picture industry favored some restrictions over network involvement in syndicating TV shows; the nets opposed it. Interestingly, when the FCC finally announced its new rules, both groups were disappointed.

There are times when industry lobbying groups work with one another. Such a case occurred when the cable industry and the broadcast industry together came up with a compromise cable bill they could support that eventually became the 1984 Cable Act. The NCTA has also tried to work out its differences over cable copyright fees with the chief lobbying organization for the movie industry, the Motion Picture Association of America.

Lobbying tends to work best in a negative sense. It seems to be easier for lobbyists to stop something they consider bad from happening than to get something good to happen. Some of the biggest lobbying successes stem from the blockage of proposed legislation. For example, one of the NAB's major successes of 1991 was the defeat of a bill that would have placed a special tax on the use of the electromagnetic spectrum. This tax would have financed public TV.

Further, the lobbying arena has been complicated by the recent fractionalization of the broadcasting and cable industries. There are now far more associations involved in lobbying than there were twenty years ago. This means that the aims of many of these groups often conflict. Networks and affiliates can disagree on rules that would limit the amount of network programming; syndication companies are at odds with the networks over who should have a financial stake in syndication profits; the interests of TV stations frequently differ from those of radio stations. Consequently, it's difficult for one organization, like the NAB, to represent the needs of broadcasting in general. Cable had just begun to see this fragmentation as the 1980s drew to a close.

Finally, the expense and energy that must be expended to support lobbying groups mean that one key player in the policy arena is underrepresented—the general public. Most citizens' groups lack the expertise, time, and money to maintain full-time lobbyists. As a result, policy making tends to be influenced far more by the industry than by the average citizen. There are, however, a few ways in which citizens get involved in the process and this is the focus of the next section.

The Public

After the WLBT decision opened the door, citizen involvement in broadcasting policy increased dramatically during the 1970s. An outgrowth of the civil rights movement, citizen interest groups or media reform groups were in operation in thirty states by 1975. FCC commissioner Nicholas Johnson, always somewhat of a maverick, even published a book, *How to Talk Back to Your Television Set*, which was a guide for citizens to challenge the renewal of local stations' licenses. The primary strategy was to file a *petition to deny*. Most petitions alleged that the station had been deficient in its programming toward minority groups or was delinquent in complying with the equal employment laws. At one point in 1974 petitions to deny had been filed against 280 stations. Although these petitions were rarely granted, they did force stations into lengthy and expensive legal proceedings. Eventually stations sought to avoid petitions to deny by entering into direct negotiation with consumer groups, which resulted in citizen–station agreements.

Some citizens' groups were remarkably successful in achieving their goals. Station WXYZ in Detroit signed an agreement with the National Organization for Women to program ninety minutes about women's issues in prime time every year. A Texas TV station agreed to program more minority shows and to hire more black reporters in return for the withdrawal of a petition to deny. Action for Children's Television (ACT), taking a different tactic, petitioned the FCC directly for rules regarding children's programming. The commission ultimately issued a set of policies and guidelines that covered the area.

By 1980, however, consumer groups had lost most of their clout in the policy-making arena. Why? First of all, private sector support for citizens' groups was drying up. Major foundations were unwilling to support media reform groups (the Ford Foundation was providing 57 percent of all public-interest funding before its decision to withdraw from the arena) and no government funding source could be found. Second, the FCC refused to honor any citizen–broadcaster agreement that abdicated the station's responsibility over its own programming. In effect, this was an escape clause for most stations; it meant they could not enter into any agreement that specified a fixed amount of time devoted to a particular issue. Third, many recordkeeping rules were abolished. In the past, stations had been required to keep detailed programming logs. Public-interest groups used these logs as the basis for many of their petitions to deny. Without them, the citizens' groups had to engage in their own long and costly monitoring, something many groups were not equipped to do. Fourth, the terms of broadcast licenses were lengthened, making the petition to deny a less potent weapon. Finally, the general atmosphere of consumer activism had faded, and citizens' groups were somewhat out of step with the prevailing deregulatory philosophy.

This decline does not mean that members of the general public are shut out of the process. In the first place, a few public-interest groups are still functioning. The Telecommunications Research and Action Center (TRAC) serves as a source of consumer information about policy matters. The National Coalition on Television Vio-

Peggy Charren of Action for Children's Television (ACT). Until its dissolution in 1992, ACT was one of the most effective citizen's groups to bring about changes in broadcasting policy.

lence rates TV programs, music videos, and movies according to their level of violence. The National Black Media Coalition campaigns for increased minority participation in broadcasting. Until it disbanded in 1992 ACT actively pursued better programming for children.

Although petitions to deny have lost their power, consumer groups are finding ways to attempt to influence policy in other arenas. For example, a conservative group announced plans to buy CBS so that it could fire Dan Rather and change the allegedly liberal bias of the network (the group was unsuccessful). Some groups publicize the names of advertisers associated with violent or sexist programs in the hope that consumers will boycott their products. The head of Accuracy in Media purchased stock in the companies that own the networks and used the annual stockholders' meeting as a forum to criticize network news coverage. In the future, with the emphasis on regulation by the marketplace, consumer groups might shift from legal to economic pressures.

Further, the Administrative Procedure Act requires all federal agencies to allow public comment on proposed changes in rules and regulations. The FCC allows interested persons to file "informal" comments with the commission by means of a letter or postcard. One study found that the FCC received more than 1900 informal comments about its decision to deregulate radio. Although the FCC claims to read and categorize each comment received, the influence of these comments in shaping policy is unclear. Nonetheless, this is an easy way for members of the general public to at least make their views known to the FCC.

Finally, the public influences policy in a more general manner. Both the president and members of Congress are elected by the public and they in turn help shape communications policy.

State and Local Government

Federal laws supersede state and local regulations but many states have legislation that covers areas that are not specifically touched on in federal communications law. For example, almost every state has a law dealing with defamation (injuring a person's reputation) and/or reporters' rights to keep sources confidential. Many states have their own laws covering lotteries, the conduct of contests and promotions, and statutes covering fraudulent or misleading ads. Public broadcasting, moreover, is regulated by many states through acts that spell out the ownership, operation, and funding of educational radio and TV stations. And, of course, broadcasters must comply with local law governing taxation and working conditions.

The Cable Communications Policy Act relegated some regulatory power to states and local communities. In general, this act relies on state and local franchise rules as the principal means of regulation but also contains rules that shape and limit this process. For example, the law allows cities to determine the number of cable systems they will license. Implicit in this is the fact that a city might choose to grant an exclusive franchise to a company and exclude others. This raises some First Amendment issues that have yet to be definitely decided by the courts. It seems likely, however, that a Supreme Court decision on the constitutionality of cable franchising will be forthcoming.

The cable act also gave cities the power to require that some cable systems set aside channels for public, educational, or government uses (called *PEG channels*) and that the operator must make additional channels available to others on a leased basis. This provision also raised some constitutional questions. A 1987 decision by a California federal court judge held that these rules violated the First Amendment. At about the same time, however, a court decision in Pennsylvania held just the opposite. Again, a Supreme Court test may be in the future.

Further, local communities can prohibit obscene programming as part of their franchise agreement. (Subsequent to the 1984 Cable Act Congress also made obscene programming unlawful.) States and cities can also enact laws that protect the privacy of subscribers to local stations. Information about personal viewing habits cannot be released to unauthorized persons or organizations (this prevents, for example, a political candidate from obtaining from a cable company the information that his rival subscribes to an adult movie channel). Communities can also collect a franchise fee, subject to limitations, from a cable operator and determine the duration of the franchise agreement.

The Marketplace

In its move toward deregulation the FCC has often used the phrase "let the marketplace decide." It follows, then, that the marketplace has become an important factor in determining broadcasting policy. But what exactly is "the marketplace?" Broadly speaking, it's a place where buyers and sellers freely come together to exchange goods or services. Obviously, with regard to broadcasting and cable, there is no single place where buyer and seller are physically present. In this circumstance the marketplace notion refers to general economic forces, like supply, demand, competition, and price, that shape the broadcasting and cable industry. More and more, the FCC appears to be turning toward the marketplace as the ultimate determiner of the public interest. For example, the FCC refused to issue a single technical standard for AM stereo, propose rules for the TVRO (television reception only) industry, or get involved in the selection of radio program formats. In each case the FCC deferred to the marketplace.

This shift reinforces the basic change in recent broadcast regulatory philosophy. In the past broadcasters had been viewed as public trustees of the airwaves, obligated to offer public service to the community in which they are licensed. Under this rationale the best broadcasters are not the ones that compete best in an economic sense but the ones that provide the best service as determined by the FCC. There were times when the FCC attempted to coerce the industry into making greater public service commitments than economic incentives would dictate. The marketplace notion views the spectrum as an economic asset that would be utilized best if the government steps back and lets the laws of economics take over. In short, the marketplace will encourage those services that the public wants.

In principle, this is an appealing choice since the marketplace idea incorporates the traditional American values of free enterprise, capitalism,

---

REGULATIONS

STATE TAXES

Is it legal for a state to impose a special tax on one mass communication medium and not on others? In 1987 Arkansas extended its 4 percent sales tax to cable services. (About twenty-five other states also had some form of tax on cable.) Representatives of the Arkansas cable industry sued the state arguing that cable was protected speech under the First Amendment and not subject to the tax. In addition, since the tax was levied only against cable and not on other purchased media (for example, newspapers, magazines, motion pictures), it also violated the equal protection clause of the Fourteenth Amendment.

The Supreme Court issued a ruling on the case in 1991 and maintained that the tax was not discriminatory. The Court explained that the tax was a general one, extending to natural gas, water, telephone, and other utilities, and did not single out the cable industry. In addition, the Court found that the tax was not content-based nor designed to censor any expressive activities of cable programs. The decision was seen as a green light for states to use taxes on cable services as an additional source of revenue.

freedom of choice, and efficiency. Like most things, however, this approach has pluses and minuses. Let's quickly examine some of the advantages and disadvantages of relying on the marketplace to decide communication policy. First, the good points.

The marketplace promotes efficiency. According to economists, a competitive market is generally believed to keep costs and prices to a minimum. It rewards those who efficiently run their operations and satisfy the needs of customers. For example, in the case of AM stereo, the marketplace should ultimately reward the company that produces the best system at the most reasonable price. Second, the marketplace should create opportunities for new services. If, for example, radio stations in a market are not meeting the needs of the elderly because most stations are programming for a younger audience, another station owner might be encouraged to develop a format to serve this group. Third, the marketplace works automatically. The laws of supply and demand usually need no monitoring. The FCC doesn't have to check on station performance and broadcasters are freed from onerous recordkeeping and reporting rules. Finally, since there are no artificially created barriers to entry, an open marketplace encourages diversity by allowing new competitors to enter the process. Those who favor the marketplace approach argue that the government should broadly define the rules of the game so that the largest possible number of people can play, create as many outlets as possible, fashion a few rules that assure open entry, and then get out of the way.

On the downside, those who do not favor the marketplace approach point out that the marketplace responds only to economic forces; it is insensitive to external social values (called externalities by economists) and cannot achieve nonmarket objectives. Profit becomes the highest standard. For example, let's suppose society places a high value on the education of its members. An educational radio station can be used to help achieve this goal. In the marketplace, however, the owner of an educational station would probably be tempted to switch to a more commercially successful format to maximize profits. This is because the added value that the educational station gives to society is not measurable in economic terms. Further, the FCC recently repealed its antitrafficking rules, which stated that a broadcast property had to be held three years before selling. The external values that apparently were behind the original enactment of the antitrafficking rules were stability and localism, as expressed by the station's roots in the community. After repeal, broadcast stations were traded much like baseball cards, as companies turned them over quickly to make a profit. Stability and localism did not fit into the economic equation. Consequently, critics of the approach suggest that some things are too important to be left to the marketplace. We don't, for example, let the marketplace decide which drugs are unsafe. Critics argue that localism, diversity, and freedom of expression are important externalities that should not be decided by economic forces.

Next, these opponents point out that a free marketplace approach is used selectively by broadcasters and cable operators. For example, the NAB has lobbied for must-carry rules for cable systems. These rules take the choice of what channels are to be carried out of the marketplace and put it in the hands of regulators. By the same token, cable operators have argued for legislation to dictate the amount of money they pay to program suppliers rather than let the marketplace set the price. In short, the marketplace notion often is endorsed only when it preserves profits.

Finally, challengers of the marketplace notion argue that the real marketplace seldom approaches the perfect model posited in economics texts. For example, entry into the market is difficult since TV stations and cable systems require high start-up costs. Further, once in the marketplace, not all participants are on equal footing. Big media companies exert far more clout than small companies. In sum, foes of the marketplace theory admit that FCC regulations were far from perfect, but they also point out that flawed markets are no better than flawed regulations.

As the 1990s began, both Congress and the FCC were taking a closer look at the marketplace concept. In some areas, such as children's programming, the notion was abandoned in favor of regulation. In others, such as the FCC's proposal to relax its rules and to let the broadcast networks own cable systems, the impetus seems to be heading in the other direction. In any event, it seems likely that the marketplace will continue to be an important factor in broadcasting and cable for the foreseeable future.

THE REGULATORY NEXUS

A nexus is a series of linked connections and it's a word that seems to fit the system of broadcast regulation. The eight components that we talked about are interrelated and interconnected as shown in Figure 17-2. Laws passed by Congress concerning broadcasting and FCC decisions are ruled on by federal courts. Members of these courts are appointed by the president and approved by Congress. Lobbyists try to influence state and federal legislators, who in turn are elected by the public. Citizens' groups representing members of the public try to influence the FCC and Congress. In addition, the whole system responds to the pressures that are exerted in the marketplace. Obviously, the system that produces policy is complex; no one component exists in a vacuum.

In the next chapter we shall take a more detailed look at the laws and regulations that this system has produced.

FIGURE 17-2 The regulatory nexus.

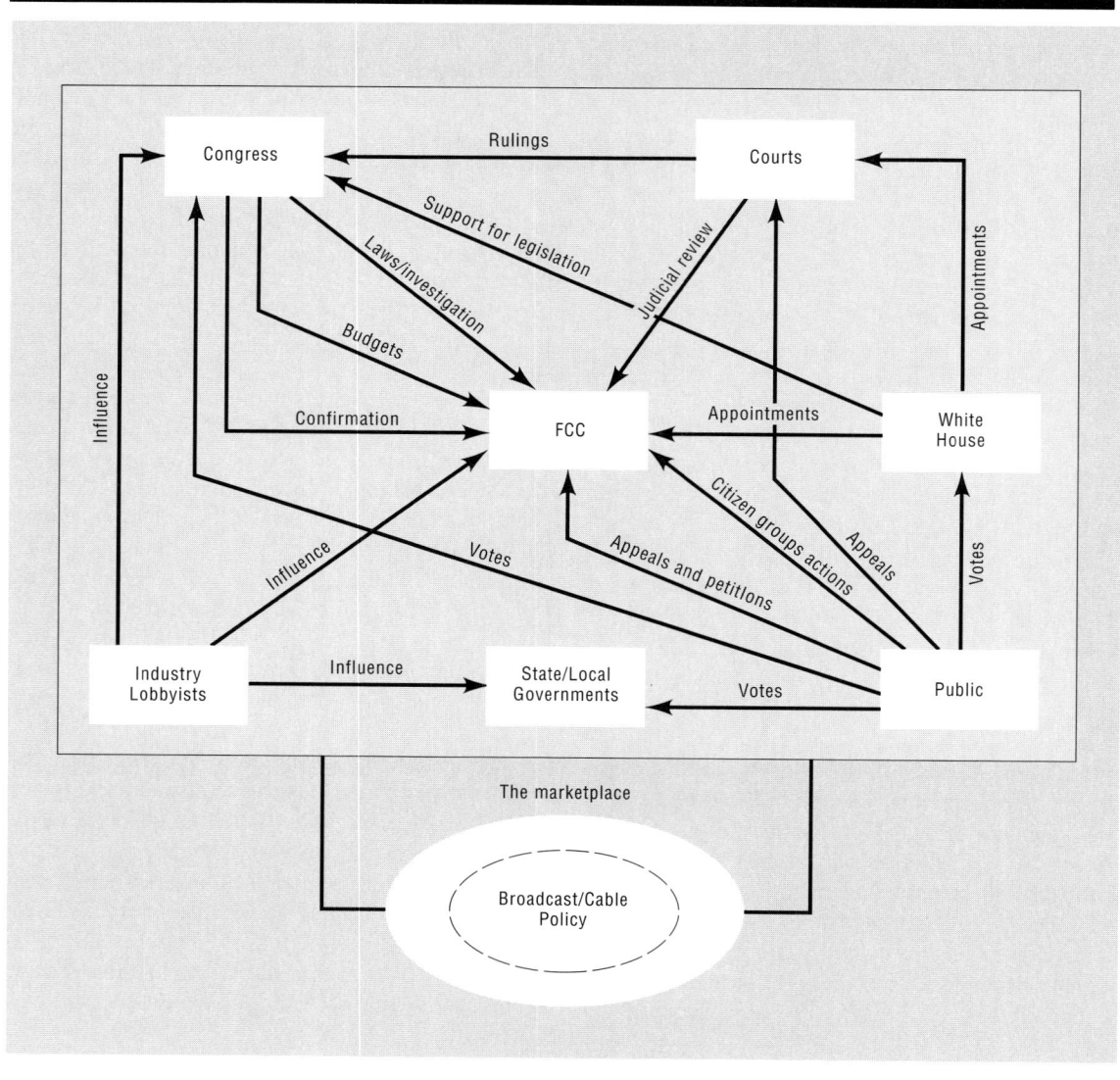

SUMMARY

The regulation of broadcasting is based on two rationales: the scarcity theory and the pervasive presence theory.

The Radio Act of 1927 was established to solve the problem of interference among radio stations. The Communications Act of 1934 replaced the 1927 Radio Act.

In 1984, the Cable Communications Policy Act was passed. This act allowed state and local governments to control franchising and permitted companies to set their own rates.

Legislative, administrative, judicial, political, and economic forces are important in determining regulation.

Regulation of broadcasting and cable is influenced by the FCC, Congress, the courts, the White House, industry lobbyists, the public, state and local governments, and the marketplace. Each is interconnected with the others.

SUGGESTIONS FOR FURTHER READING

Brenner, D., & Price, M. (1986). *Cable television and other nonbroadcast video: Law and policy*. New York: Clark Boardman Co.

Carter, T., Franklin, M., & Wright, J. (1988). *The First Amendment and the fourth estate* (4th ed.). Mineola, N.Y.: Foundation Press.

Cole, B., & Oettinger, M. (1978). *The reluctant regulators*, Reading, Mass.: Addison-Wesley.

Kahn, F. (1984). *Documents of American broadcasting*. Englewood Cliffs, N.J.: Prentice-Hall.

Krasnow, E., Longley, L., & Terry, H. (1982). *The politics of broadcast regulation*. New York: St. Martin's Press.

LeDuc, D. (1987). *Beyond broadcasting*. New York: Longman.

Tunstall, J. (1986). *Communications deregulation*. New York: Basil Blackwell.

CHAPTER 18: BROADCASTING AND CABLE LAW

One result of deregulation: the rules that are left take on added significance. This chapter examines (1) the Communications Act of 1934, as amended; (2) the FCC's regulations covering broadcasting, cable, and related technologies; (3) other federal law that applies to broadcasting and cable; (4) laws pertaining to electronic journalism; (5) advertising regulations; and (6) international rules. A note of caution: this is a turbulent area; by the time you read this book, things may have changed. Like industry professionals, students, publishers, and textbook authors face the task of keeping up with this dynamic area.

THE COMMUNICATIONS ACT OF 1934

Still around after all these years, the Communications Act has survived numerous rewrite attempts. Anything that has endured this long in the volatile world of broadcasting deserves a closer look.

The Communications Act of 1934 and related pieces of legislation are found in Title 47 of the U.S. Code. Title 47 is divided into chapters. The chapter most pertinent for our purposes is Chapter 5, "Wire and Radio Communication." Chapter 5 is further divided into subchapters which incorporate the 1934 Act. For example, subchapter 1 presents the general purposes of the act and creates the Federal Communications Commission (FCC). Other sections in subchapter 1 define the terms and the organizations and functions of the FCC. Subchapter 2 deals entirely with telephone and telegraph regulations and is not a concern here. Subchapter 3 contains provisions relating to radio and TV, and it is this part of the act that has most relevance for broadcasters. Some of the key sections most often mentioned in the trade press are:

- *Section 301.* Users of the electromagnetic spectrum must be licensed by the FCC.
- *Section 312.* Candidates for federal office must be given reasonable access to broadcast facilities.
- *Section 315.* Use of broadcast facilities by candidates for public office is outlined. (This provision is so important an entire section is devoted to it later in this chapter.)
- *Section 326.* The FCC is forbidden to censor the content of radio and TV programming.

Subchapter 4 of the act, among other things, specifies how decisions of the FCC may be appealed. Subchapter 5 spells out the sanctions and penalties that the commission can use against offending stations. Subchapter 6 contains miscellaneous provisions.

When Congress passed the Cable Communications Policy Act of 1984, rather than doing an extensive renumbering job, the act was incorporated into Chapter 5 as subchapter 5A. Some key provisions in this subchapter are:

- *Section 542.* Sets a cap on cable franchise fees.
- *Section 543.* Covers the setting of rates by a cable system.
- *Section 559.* States that the transmission of obscenity over cable is illegal (more about this later).

In early 1992 Congress was debating a new law that would significantly change the regulations concerning cable TV. If it passes in a form similar to its original version, the bill would give local municipalities more power over the rates charged by cable companies, permit broadcasters to charge cable systems for retransmitting their signals, and require the cable industry to make its programming available to competitors such as direct broadcast satellites or multichannel, multipoint distribution services (see below). Not surprisingly, the cable industry was lobbying fiercely against this bill.

THE ROLE OF THE FCC

Licensing

The previous chapter contained a general introduction to the FCC's licensing function. This chapter concentrates on the specifics. Applying for a license to operate a new station is a fairly complicated process. (In fact, there are only a limited number of radio and TV frequencies available that would not cause major interference problems with existing stations. Most persons who want to get into broadcasting generally do so by purchasing an existing facility. Many rules discussed below would also apply in this situation.) After finding an available radio frequency or TV channel all applicants must meet some minimum qualifications. First, the applicant must be a U.S. citizen or an organization that is free from significant foreign control. Second, the applicant must meet certain character qualifications, but the law is vague on what exactly constitutes a "good" or "bad" character. In practice, the FCC usually looks at things that are directly relevant to the potential future conduct of a potential broadcaster, although recent developments suggest that the commission has begun to look more closely at general character qualifications as well. In 1990 the FCC required broadcasters to report all convictions for felonies, serious misdemeanors, any adverse civil judgments involving antitrust or anticompetitive activity, and any cases of misrepresentation to government agencies. Such activity, said the commission, could jeopardize the broadcaster's ability to hold or acquire a license. In one case the FCC revoked the license of a radio station in South Carolina after its majority owner was convicted of drug trafficking.

Next, the new applicant must assert the financial capability to build and operate a new station. In addition, as might be expected, all applicants must demonstrate that they can meet all of the technical requirements set forth in commission rules concerning equipment operation. The applicant must also propose an affirmative action employment plan to assure the hiring of minority group members and women.

The FCC is committed to the diversity of media ownership. In this connection it has established rules limiting the multiple ownership of media properties. Recent developments, however, suggest that these rules might be relaxed.

First, until recently no person or organization could own more than one station of the same type (AM-FM-TV) in the same community. In December of 1988 the FCC voted to liberalize its restrictions on the ownership of radio and TV stations in the same market. The commission announced that it would allow common ownership involving stations in the top-25 markets with thirty or more separately owned broadcast licenses. Ownership of an AM and FM station in the same market has always been permitted. (Many readers can probably think of several communities where one company owns an AM, an FM, and a TV station in apparent violation of this rule. The explanation is that these combinations were in existence before the rules were drawn up and thus were allowed to continue. In legal jargon, the combinations were "grandfathered.") In addition, some combinations occasionally come about if the FCC can be persuaded to waive its rules. The FCC may do this if the overlap doesn't affect many people or if justified by the financial condition of one of the stations.

Second, there is an absolute limit on the total number of stations that one person or organization can own. In 1992, the FCC raised this number for radio from twelve AM and twelve FM stations to eighteen AM and eighteen FM stations. In addition, the new rules would allow one person or organization to own up to four stations in the largest radio markets. Some critics charged these new rules would increase concentration and decrease diversity in the industry but others felt the new caps would bring economic stability to the industry.

In television, the maximum ownership cap is twelve stations but no single owner can reach more than 25 percent of the national TV audience no matter how many stations are owned. There are also special provisions in this rule that encourage minority and UHF ownership. A recent report by the FCC's Office of Plans and Policy suggested that the twelve-station cap on the ownership of TV stations and the rule limiting a broadcaster to just one TV station in a market be repealed. These proposals had the endorsement of FCC Chair Alfred Sikes, but it was unclear if the full commission would vote for them.

Third, newspaper-broadcast ownership in the same community is limited. A daily newspaper may not own AM, FM, or TV stations that serve essentially the same markets as those served by the newspaper. (Again, this rule was grandfathered. Most newspaper-broadcasting combinations that existed before this rule was passed, in 1975, were allowed to continue operation unless such operation jeopardized the public interest.)

Fourth, a TV station cannot own a local cable TV system serving the same area as the TV station. Television owners can own cable systems (many do), but they must own systems in places other than where they own TV stations. In mid-1992, the FCC erased a 22-year ban and allowed TV networks to own cable systems as long as the cable systems served no more than half of any local market and no more than 10 percent of all homes nationwide.

If any applicant attempts to apply for a license in violation of these rules, the application will not be considered.

What happens when there is more than one applicant for a license? This can occur in two separate situations: when several applicants desire a new frequency that has never been used before or when a license for an existing station is challenged by one or more competitors.

In the first situation the FCC insists that all applicants meet the minimum qualifications listed previously and then examines additional criteria such as diversity of control (all other things being equal, applicants who possess no other media interest in the area will be preferred over those who do), participation in the station's operation by the owner (applicants who agree to get involved with the station would get preference over those who simply want to hire others to manage the station), and past broadcast record (considered only if the past record is unusually good or unusually bad).

In practice, deciding between competing applicants takes a long time and costs the FCC, the applicants, and ultimately the taxpayer a lot of money. Consequently, the FCC has examined other options for deciding between or among competing applicants where qualifications were basically the same. One alternative was modified random selection, also called the lottery approach. This process was used to award licenses for low-power TV stations (see Chapter 5), but in 1990 the FCC scrapped the lottery system as a means of choosing among applicants for a broadcast license. The lottery approach was still used, however, in assigning licenses to multichannel, multipoint distribution services (MMDS) or "wireless cable" operators (see below).

The second situation, where the request for renewal of an existing station with a record of past performance is being challenged by a party with no record who promises to do more for the public interest, poses a vexing problem for the FCC. On the one hand, giving preference to the current licenseholder could close out newcomers who might do a better job. On the other hand, ignoring past performance could be unfair to a station that served the public well and ought not to be replaced by an untested challenger. In addition, granting licenses to challengers could be unfair to a station that has spent a lot of money investing in its operation, which would be lost if the license renewal was denied. In fact, it would be difficult to encourage investors to back broadcasting operations if there was a significant chance that the station might not be around after five or seven years.

The FCC has wrestled with this problem for years. Several FCC proposals were struck down by the courts, which eventually forced the FCC to come up with some precise guidelines on how to handle this comparative renewal problem. For now the FCC is forced to compare the incumbent with the challenger when competing applications are filed. If the incumbent has provided "substantial, sound, favorable past services, not characterized by serious deficiencies" (violations of FCC rules), it gets what is known as "renewal expectancy." If this occurs, challengers have a hard time winning no matter how good they look on paper.

This may not be the way the FCC prefers to do business. The commission has asked Congress several times to change the Communications Act to create a so-called two-step renewal system. Under this system, when renewal time came, only the incumbent would apply at first. The FCC would then decide if the incumbent should be renewed. If the renewal was granted, there would be no chance for a challenger to ask for a license. If renewal was denied, however, new applicants could file. Congress has not acted on this proposal partly because broadcasters have been unable to agree with Congress on what a station would have to do to guarantee its license renewal.

Some changes in the system were made in the early 1990s. One problem was that the current standard for "renewal expectancy" involved how well a station has served its community in the past. That standard worked until the early 1980s, when the FCC started requiring less programming information from stations. Today, without such items as program logs, it's difficult for a station to prove it has done a good job. Perhaps, as a result, the last part of the 1980s saw an increase in the number of "sham" competing applications filed against stations in the hope that the station would agree to a cash settlement or make other concessions to the competition in order to avoid the costly process of a hearing. In 1990 the FCC attempted to stop these abuses by enacting strict new rules governing the amount and nature of such payments. In addition, the commission announced plans to streamline and speed up the comparative hearing process. Although these changes may help, it appears that the comparative renewal process may still be a source of difficulties in the years to come.

What does the FCC look for in a license renewal? The postcard form gives a pretty good idea. The form makes sure that the station's most recent Annual Employment Report, which documents numbers of minorities and females in various job categories, and its Ownership Report, which lists owners and shareholders, are on file with the commission. The FCC is thus interested in promoting fair employment and the diversity of ownership. The postcard form also makes sure the licensee is an American citizen, checks for the extent of foreign control, and asks if the station has maintained a file for public inspection that contains those documents required by commission rules.

What must a station do to have its request for license renewal denied? One big reason for denial is lying to the FCC. If a licensee knowingly gives false information to the commission (such as concealing the actual owners of the station), the license is in jeopardy. Unauthorized transfer of control is another serious offense. Programming violations alone, such as violating the indecency rules, have rarely led to nonrenewal. Moreover, the FCC is unlikely to deny a license renewal on its own initiative. Cases that are not routinely passed (perhaps 2 to 3 percent of the total) are brought to the commission's attention by private citizens who file a petition to deny. Of these, only a tiny fraction of licenses (less than 1 percent of the total) are not approved at renewal time. Nonetheless, the threat of nonrenewal is perceived as real by many broadcasters and is a potent weapon in the FCC's enforcement arsenal.

The FCC and Cable

Cable TV systems are not licensed by the FCC. Instead, state and local governments grant franchises. The length of the franchise period is also set at the local level, usually at some period between ten and fifteen years. (Many franchises are not exclusive and some cities can authorize *overbuilds*, allowing two cable companies to serve the same area.) When a franchise comes up for renewal, or earlier, if the cable operator asks for early renewal, the cable operator's service record is examined. Unless the operator has failed to live up to the franchise terms or has provided inferior service, there is a strong presumption that the franchise will be renewed.

Although the FCC does not license cable systems, the Cable Communications Policy Act of 1984 does give the FCC jurisdiction over cable, but it also sets limits on what the FCC (or local governments) can control. Practically speaking, this means that programs originated by a cable system must comply with the equal opportunities rule for political candidates (see below) and, thanks to the syndicated exclusivity rule, must black out programming carried by a distant station or an imported superstation (like WGN or WTBS) if a local station on the cable system has exclusive rights to that programming.

Two other content regulations in force at the time that the act was passed were in limbo as of 1992. As we have seen, the FCC's must-carry rules were struck down as unconstitutional and most of the fairness doctrine (see below) was repealed. Both of these, however, might eventually resurface in some other form. The FCC has jurisdiction over several nonprogramming areas as well. First, as noted earlier, it regulates cross media ownership. Local newspapers were permitted to own cable systems in their market areas but local TV stations were not. In late 1991 the FCC ruled that long-distance telephone companies such as AT&T, Sprint, and MCI may enter the cable business. Local phone companies are prohibited by federal law from providing cable service in areas where they also provide telephone service. The FCC, however, recently ruled that local telephone companies should be allowed to provide a "video dial tone" service in which the phone company would act as a common carrier for the content supplied by other programmers. This decision, along with a federal court decision that cleared the way for regional phone companies to move into cable (see Chapter 17), seemed to suggest that the phone companies will play a greater role in providing video to the home.

In addition, the FCC defines certain terms contained in the 1984 Cable Communications Act. One such term is "effective competition." If a cable system in a community faces effective competition, the Cable Act prohibits local communities from regulating the rates charged by a cable system. The FCC originally defined effective competition as three broadcast signals—a standard met by most cable systems, and, consequently, most cable systems escaped local rate regulation. In 1991 the FCC redefined effective competition, making it tougher for systems to be exempted from municipal rate stipulations. Finally, the FCC also has jurisdiction over certain technical standards that must be met by cable operators.

The FCC and Emerging Technologies

Several recent forms of communication technology don't fit nicely into traditional molds and the FCC has made special rules for them. In general terms, the FCC has taken a light touch approach toward regulation in order to encourage experimentation and development.

Multichannel, multipoint distribution service (MMDS) is free from most FCC content regulation. MMDS sends programming to receivers equipped with special antennas via microwaves. Thus far MMDS has been used mostly in urban areas to transmit pay-TV services (HBO and Showtime) to apartment buildings, hotels, and private homes. After the FCC changed its rules in 1983 and permitted operators to offer up to four channels per community thousands of applications for MMDS licenses were received by the FCC. As of 1992 the number of MMDS channels per community was raised to thirteen and more than 135 markets in thirty-two states had an MMDS service. These systems were recently classified as point-to-multipoint nonbroadcast services, which removed them from the application of various broadcast rules, including section 315.

Teletext is a form of communication in which text and graphics are sent over broadcast frequencies or through cable. Most teletext systems have several hundred pages of information that are transmitted in a continuous cycle. Subscribers with decoders can "grab" individual pages as they appear. The FCC has generally taken a hands-off approach to teletext, arguing that it is more akin to the print media than to broadcasting. The only statutes that apply to teletext are those that prohibit cigarette advertising and obscenity.

Videotex is a two-directional information service that uses computer terminals in combination with cable or telephone lines to receive and send information between the user and the data bank. Since most videotex systems use telephone lines, FCC regulation is limited to that which covers the telephone. In general, it is the least regulated of all the technologies discussed so far.

Equal Opportunities: Section 315

From radio's earliest beginnings Congress recognized that broadcasting had tremendous potential as a political tool. A candidate for political office who was a skilled demagogue might use the medium to sway public opinion. By the time the 1927 Radio Act was written, Congress had already seen how some skilled communicators could use the medium for their own advantage. Consequently, what would eventually be known as section 315 was incorporated into the 1927 act.

The crux of section 315 in the current act is the following:

If any licensee shall permit any person who is a legally qualified candidate for any public office to use a broadcasting station, he shall afford equal opportunities to all other such candidates for that office in the use of such broadcasting station.

Sounds simple enough but in operation section 315 can prove complicated. Note that section 315 talks about equal *opportunities* as opposed to equal *time*. If a station provides a candidate with thirty minutes of prime time, it cannot offer an opponent thirty minutes at 3:00 A.M., nor can it offer the second candidate thirty one-minute spots throughout the day. The station would have to provide thirty minutes in prime time. Note further that section 315 does not obligate a station to provide free time to a candidate unless free time was first offered to an opponent.

Also note that the section is not self-triggering. A station is under no obligation to tell opponents that a candidate has used its facilities. The station is required to keep political files, however, and a candidate can easily examine these to see if any of the other candidates had made use of the station. The opponent must request equal time within a specified time interval. Also keep in mind that section 312 of the Communications Act requires broadcasters to provide reasonable access for candidates for federal office. They can, however, require candidates to pay for this time.

In addition, there are rules governing how much stations can charge candidates for purchased time. Broadcasters are prohibited from charging more for political time than they do for other kinds of commercials. The rules also make sure that a station doesn't charge candidate A $1000 for a spot while charging candidate B $5000 for an equivalent spot. Section 315 does not force stations to stop selling time to one candidate simply because his or her opponent cannot afford to buy the same amount. The FCC determined that the equal opportunities rule was not meant to equalize differences in campaign funds.

Other complications also arise. Who is a legally qualified candidate? According to the commission, there are three criteria:

1 The candidate must have announced publicly an intention to run for office.
2 The candidate must be legally qualified for the office (to run for president, you must be a U.S. citizen and at least 35 years old).
3 The candidate must have taken the steps spelled out by law to qualify for a place on the ballot or have publicly announced a write-in candidacy.

This means that if a highly popular president nearing the end of a first term in office, generally expected to seek reelection, appears on TV and harshly criticizes a likely opponent, Section 315 will not be triggered because, technically speaking, the president has not announced publicly for the office. Similarly, suppose the 30-year-old leader of the Vegetarian party announced plans to run for the presidency. The law would not apply since a 30 year old is not legally qualified for the office.

Section 315 applies only among those candidates actually opposing each other at that moment. This means that it works differently during a primary than it works during a general election. During a primary an appearance by a candidate for the Democratic nomination to an office would not create equal opportunity rights among those running for the Republican nomination. Once the primaries are over, however, an appearance by the Democratic nominee would mean the Republican nominee would probably be entitled to an equal opportunity to appear.

Section 315 raises a question regarding the definition of "use" of a broadcasting facility. Broadly speaking, a use occurs when the candidate's voice or picture is included in a program or commercial spot. A program or commercial about a candidate in which the candidate does not appear would not qualify. Most of the time, a use is fairly easy to recognize. Sometimes, however, it's trickier. What happens if the candidate appears in a role that is totally different from that of candidate? How about the candidate who makes a guest appearance in a skit on "Saturday Night Live"? Or, what about a candidate for the U.S. Senate who appears on a wild-life program and discusses his love for fishing? Could his opponents claim equal time under section 315? What about a station that showed an old Ronald Reagan movie while he was campaigning for the presidency? The commission has ruled that a candidate's appearance is a use even if the candidate is appearing for a completely unrelated purpose and never mentions his or her candidacy. Thus if the host of a children's show on a local station becomes a legally qualified candidate, each time he or she appears on the kiddie show the opposition is entitled to a comparable amount of time for free.

But what about the situation where an incumbent president who has already announced for reelection holds a press conference? Does this mean that all the other candidates are entitled to equal time? A similar situation occurred in 1959 when a minority party candidate for mayor of Chicago requested equal time because the current mayor who was running for reelection was shown in a series of film clips used on the evening news. Congress reacted by amending section 315 and providing these exceptions. Section 315 does *not* apply to:

1 bona fide newscasts
2 bona fide news interviews
3 bona fide news documentaries where the appearance of the candidate is incidental to the subject of the documentary
4 on-the-spot coverage of bona fide news

Bona fide newscasts and bona fide news documentaries usually are easy to identify. A bona fide news interview is one that deals with a newsworthy topic and is under the control of the station. This means that the moderator guides the direction of the interview; the candidate is not simply allowed to give a speech. The meaning of on-the-spot coverage is a little less clear. Most of the time the FCC defers to the station's judgment that the event was newsworthy if the coverage was not obviously designed to favor one candidate. This last exception also provides a loophole for the telecasting of debates among the major candidates for public office. A debate is a newsworthy event and stations are covering it on the spot. Therefore, it does not trigger section 315 and stations do not have to give equal time to a host of minor-party candidates.

Questions can also be raised about exactly how much time an opponent is entitled to. Usually if the candidate appears or the audience hears the candidate's voice in a thirty- or sixty-second spot, opponents are entitled to thirty or sixty seconds, even if the candidate was on screen for a few seconds only. But, if the candidate was on screen or on mike for only a portion of an interview show, opponents are entitled to an amount of time equal only to that featuring the candidate. For example, if a candidate appears for fifteen minutes on an hour-long "Late Night with David Letterman," opponents are entitled to fifteen minutes, not a whole hour.

Finally, there are limits on what content the broadcaster can control in messages in which political candidates appear. Broadcasters cannot censor the material even if it's in bad taste, defamatory, vulgar, or racist. Obscene material, however, may be censored. Since the station has little control over the content of these messages, the courts have held that the stations cannot be held liable for what is said.

The Fairness Doctrine: To Be or Not to Be

As of mid-1992 the fairness doctrine was dead—or, to be more precise, mostly dead, but there was talk in Congress of resurrecting it. At any rate, its future is murky, to say the least. Such a situation poses special problems for a textbook. Consequently, this section briefly reviews the history of the fairness doctrine with special regard for its constitutional implications and presents a brief summary of its provisions. We suggest that readers check the trade press and other timely publications for an up-to-date status report on this controversial issue.

REGULATIONS

YOU MAKE THE CALL: SECTION 315

(A)

Your newscaster has become a legally qualified candidate for the U.S. Senate. You inform her that she will have to be reassigned to a position that doesn't require her being on the air since every time she appears it would trigger the equal time provisions of section 315. She argues that since bona fide newscasts are exempt from section 315's provisions and since her job consists of being a part of a bona fide newscast, she should be exempt from section 315's provisions? Is she right?

(No. The courts have held that a newscast consists of reports of newsworthy events and people. Those who report the news aren't part of the news. The news events that she reports while performing as anchor would have taken place without her. Merely reporting them on a newscast does not qualify her for an exemption.)

(B)

Your station donates time to a candidate so that the candidate can show a five-minute videotaped "personal history." The candidate's opponent demands equal time under section 315 and you reply that you would be happy to donate five minutes of comparable time so that this candidate might also show a personal history tape. The candidate replies, "I don't want to show a stupid tape. I want to give a five-minute speech." Must you grant the five minutes for speechmaking?

(Yes. Remember that the station can't censor the remarks of the candidate appearing under section 315. Trying to dictate the format or the content of the candidate's time constitutes censorship.)

Back in 1941 the FCC declared that stations should not editorialize. Eight years later a different commission with a different philosophy reversed itself and encouraged broadcasters to comment on controversial issues of public importance provided that they fairly covered all sides of these controversial issues. Thus the fairness doctrine was born.

Many broadcasters thought the doctrine was unconstitutional. In the late 1960s the whole constitutional issue came to a head. Court cases stemming from complaints over the fairness doctrine reached the Supreme Court.

Two of the arguments made to the Court by those opposing the fairness doctrine merit special mention. A lower court had noted at the time that there were more than three times as many commercial radio and TV stations in the country as newspapers and concluded that the scarcity of the spectrum argument could not be used to justify more regulation for broadcasting than for newspapers. Broadcasters also argued that the fairness doctrine actually curtailed the First Amendment rights of the broadcasters by encouraging self-censorship. Many stations, they argued, would be fearful of the consequences of the doctrine and wouldn't cover controversial issues in the first place.

In its decision (formally known as *Red Lion Broadcasting vs. The FCC*) the Supreme Court rejected both of these contentions. It first noted that the scarcity argument was still valid as long as there were more people who wanted to broadcast than there were available frequencies. Thus the total number of broadcast stations was irrelevant as long as demand exceeded supply. Consequently, broadcasting could be regulated differently from print media. In addition, the Court argued that the First Amendment protected the public's right to receive information as well as the right of the broadcasters to send it. In truth, said the Court, it was the right of the viewers and listeners to receive information that was paramount and took precedence over the right of the broadcasters. In sum, the fairness doctrine, at least as applied in these cases, was constitutional.

Broadcasters still had problems with the doctrine, however; and the FCC issued reports in 1974 and 1976 in order to clarify it. In the deregulatory climate of the 1980s, though, the fairness doctrine became a favorite target for those who argued for fewer rules. In 1981 the FCC requested Congress to repeal the doctrine; Congress did not respond. (Many members of Congress personally favor the doctrine. Moreover, members of Congress are especially sensitive to interest groups among their constituencies. These interest groups also favor the fairness doctrine.)

The FCC released a detailed study of the fairness doctrine in 1985, which concluded it was not serving the public interest. Congress was again asked to abolish it; again, Congress did nothing. Meanwhile the FCC was still enforcing a doctrine it was fervently denouncing. Finally, a federal court cleared up a legal ambiguity by ruling that the fairness doctrine was not part of the statutory law but was simply a regulation of the FCC. This meant that the FCC no longer had to ask Congress to repeal it; the commission could do the job itself. The FCC did just that in its 1987 *Meredith* decision.

The FCC did not contest the *Red Lion* contention that the rights of viewers and listeners to be informed was paramount. The commission concluded that the fairness doctrine should be abandoned because it didn't enhance that right; in other words, it didn't work. In actual practice, said the commission, the fairness doctrine actually hampered the expression of diverse points of view. The best way to achieve a multiplicity of viewpoints in the marketplace of ideas was to minimize government intervention and to stop enforcing the doctrine. In 1990 the Supreme Court refused to hear an appeal of a lower court's decision that upheld the FCC's right to do away with the doctrine.

Since its abolishment in 1987 Congress has tried several times to resurrect the fairness doctrine by making it a federal law. These efforts have been unsuccessful because of a presidential veto or the threat of a veto. In 1991 Congress introduced yet another bill that would reimpose the fairness doctrine. Should such a bill ever get through Congress, most likely it would be vetoed by President George Bush. If someone else occupies the White House, then the outcome might be different. Time will tell.

Since any resurrected version of the doctrine might resemble the original version, here's what the fairness doctrine did when it was in force. It forbade broadcasters from using their facilities to promote only one particular point of view on a controversial issue. Further, it imposed a duty on broadcasters to identify controversial public issues and to present programming that dealt with those issues. This programming could include the station's newscasts, special panel shows, documentaries, and other formats decided upon by the station. On any issue the broadcaster needed to make a good faith effort to present significant opposing viewpoints. This didn't have to be achieved in one program, but the broadcaster was expected to achieve balance over time. Note that the fairness doctrine never said that opposing views were entitled to equal time; the doctrine simply suggested some reasonable amount of time should be granted. In other words, the fairness doctrine required two things: (1) reasonable attention to opposing views and (2) reasonable attention to some (not all) controversial issues of public importance in the community.

Most of the time, decisions concerning the fairness doctrine were subjective at both the local station level and the commission level. Keep in mind that the doctrine deals with controversial issues, those that already have caused emotions to run high. What appears to be balanced and fair coverage to station management might be perceived as biased and unfair by a proponent of the issue. It's not surprising that both broadcasters and the commission were not too fond of the fairness provisions.

OTHER FEDERAL LAWS COVERING BROADCASTING AND CABLE

In addition to the Communications Act of 1934, broadcasters and cable system operators must abide by other relevant federal statutes that relate to their operations. This section discusses five topics where such laws apply: children's TV, TV violence, copyright, obscenity, and lotteries.

Children's Television

The Children's Television Act of 1990 imposed an obligation on TV stations to serve the informational and educational needs of children through programming designed especially to serve those needs. The legislation also put a cap on the number of commercial minutes allowed in children's programs (ten and one-half minutes per hour on weekends and twelve minutes per hour on weekdays) and established a multimillion dollar endowment for the funding of children's educational programming. In interpreting the specifics of the act, the FCC stated that the programming provided for children would be evaluated every five years during a station's license renewal process and that commercial stations must keep a record of all their programming efforts directed at children.

The act raised some constitutional questions. President George Bush refused to sign the bill into law (it became a law anyway since any bill that the president declines to sign becomes law ten days after going to the White House), and his action was taken as encouragement for those who wished to challenge the law on constitutional grounds. As of this writing, a test case has yet to reach the courts.

Television Violence

Despite continuing pressure from citizens' groups (see Chapter 19), the FCC has been unwilling to pass regulations restricting the portrayal of violence in TV programming. In 1990, however, Congress passed a law that was designed to encourage the industry to reduce violence. Normally members of the broadcasting, cable, and film industries could not meet together and adopt industrywide guidelines without fear of triggering an antitrust investigation by the Justice Department. The 1990 law allows a three-year exemption from antitrust action so that members of the broadcasting, cable, and film industries can collaborate on a code to limit violent content.

As of mid-1991 representatives from the networks, National Association of Broadcasters, and program production companies had begun discussions about forming new guidelines. Representatives from the cable industry declined an offer to meet with the broadcasters, preferring instead to come up with their own standards.

Copyright: Trying to Keep Up

Copyright law protects intellectual property; it allows creative people, such as writers, photographers, and painters, to control the commercial copying and use of intellectual property, such as books, films, phonograph records, audio and video tapes, and sculptures, that they create. The assumption of copyright law, written into the

U.S. Constitution, is that people will not be encouraged to create original works unless they can be guaranteed access to any profit from such works. The way to do that is to declare such creations "intellectual property" and to give the creators control over how they are used. New technological advances have made it difficult for the law to keep pace with new means of copying and disseminating various creative works. Consequently, the law has been revised twice this century, the last time in 1976, to cover computers, videocassettes, and cable.

In its simplest terms, the current copyright law protects works that are "fixed in any tangible means of expression." This includes phonograph records, dramatic works, motion pictures, TV programs, computer programs, and sculptures. Ideas and news events are not copyrightable (you could, however, copyright your particular written or recorded version of a news story). Copyright protection lasts for the life of the author plus fifty years. Frequently the creator of a work transfers the copyright so that in some cases the owner is not necessarily the creator. To be *fully* protected a work must contain notice of the copyright (usually consisting of the letter C in a circle, the copyright owner's name, and the date of origination) and must be registered and deposited with the Copyright Office in Washington, D.C. Nonetheless, any fixed form of an idea is now granted some protection from the moment that it is fixed in a tangible form.

REGULATIONS

YOU MAKE THE CALL: COPYRIGHT PROBLEMS?

(A)

You're the manager of Vern's Video Rental Store. Increased competition has hurt your business and you're looking for ways to increase revenue. Your assistant manager comes up with an idea. How about taking all of the currently unused storage space on the second floor of your shop and turning it into a number of small, private viewing rooms? Families who don't own VCRs can then come to your store and rent a tape, a recorder, and a viewing room for less money than it would cost to go to the theater. The assistant manager even suggests you sell soft drinks and popcorn. What do you think? Is there a copyright problem involved?

(Yes. The courts have ruled that even when families and/or friends are watching a tape in a private room, since the video store in effect charged admission, this constitutes a public performance.)

(B)

Vern's Video Rentals is still facing bad financial times. Once again your assistant manager comes up with a brilliant idea. You have two or three VCRs sitting around idle at any one time. How about taping all the local newscasts and starting a TV Newsclip Service, just like companies that do the same thing using newspapers? Individuals or firms that want to see how they're portrayed on the news would pay you a fee and you would send them copies of the news stories that you taped off the air. Is there a problem with this new scheme?

(Probably. Selling TV newsclips taped off the air from a commercial station without permission and without payment is not a fair use of the material. Courts have ruled in similar circumstances that TV news clipping services are not like newspaper clipping services since the newspaper services actually buy the papers they clip from. In addition, the service would harm the value of the clips if the local stations tried to sell them. Note, however, that the above reasoning applies to local stations. In a 1991 case CNN filed suit to stop a clipping service from selling tapes of its reports. The court found that CNN could not copyright its entire broadcast day since it was not the original copyright owner of some of the segments it aired.)

Once a work is copyrighted, the owner has the right to authorize works derived from the original (for example, a novelist could authorize a movie script based on the book), to distribute copies of the work, and to display and/or perform the work publicly. Even though a work has been copyrighted, others can borrow limited amounts from the material under the doctrine of *fair use*. Critics, for example, can quote from a work without needing permission.

Let's take a closer look at how copyright laws apply to broadcasting and cable, starting first with performance rights. An author is entitled to a royalty payment when his or her work is performed publicly. When a copyrighted work is broadcast over the air, that is, when a radio station plays a record, that constitutes public performance. The broadcasters pay royalties for such performances. The audience for the material pays nothing.

Music licensing There are more than 10,000 radio stations and 1500 TV stations in the United States and it would obviously be difficult for artists and performers to negotiate a royalty agreement each time their songs were played on the air. Accordingly, private music licensing organizations were established to grant the appropriate rights and to collect and distribute royalty payments. In this country performing rights are handled by the American Society of Composers, Authors and Publishers (ASCAP), Broadcast Music Incorporated (BMI), and the Society of European Stage Authors and Composers (SESAC, which, despite its name, handles U.S. clients).

The major licensing firms grant what is known as *blanket rights*, whereby media firms pay a single fee to the licensing agency based on a percentage of their gross revenues. In return the stations get performance rights to the agency's entire music catalogue. These rights do not come cheaply. Radio stations in large markets might pay around $75,000 annually for music rights. Stations in medium-sized markets might pay $15,000 to $30,000. In 1990 royalty payments to ASCAP and BMI amounted to more than $100 million. The licensing agencies distribute this money to composers and publishers. The amount of the licensing fee has been a source of friction between broadcasters and music licensing organizations since the 1920s. New licensing rates are negotiated every several years, so the fee will probably continue to cause problems in the future as well.

Although TV networks supply programs to their affiliates with the rights included, syndicated shows pose special problems. If a station buys a syndicated show, it doesn't get the performing rights to that show and still must have a license from ASCAP or BMI. Television stations have argued for a system of "source licensing," where the syndicated program producers acquire the music rights when they initially produce the programs. As of this writing, such a system had yet to be worked out, primarily because BMI and ASCAP have been reluctant to grant such an arrangement.

Copyrights and cable Cable systems retransmit the signals of local broadcast stations and distant stations outside the market. Must cable systems pay licensing fees to all the program production companies and to the originating broadcast stations in order to rebroadcast this material? The Copyright Act as revised in 1976 gives special consideration to the situation of cable systems. The law states that cable systems need not pay royalties to independent or network-affiliated local stations that are carried on the system. Cable systems may carry the programming of distant stations (superstations like WTBS or WGN) without the copyright owner's consent in return for a compulsory license fee. This fee is fixed by law and its amount depends on the size of the cable system and whether the distant station is commercial or educational. Cable networks, such as HBO, do not come under the compulsory license system. The Copyright Act also created a Copyright Royalty Tribunal, an impressive-sounding name for a body whose job is to collect and distribute royalty payments derived from the compulsory licenses. About $175 million in fees were collected in 1991. Most of the license fee goes to program suppliers (production and syndication companies) and a minor share goes to sports organizations (such as major league baseball). Broadcasters receive only a small amount of the total.

The compulsory license is also a point of disagreement between broadcasters and cable operators. In general, the cable industry favors retention of the license while broadcasters would prefer another arrangement. The license has helped the cable industry prosper. In addition to getting local stations' programming for free, the license allowed the cable industry to save money on some forms of programming. For example, the fees paid by cable under the compulsory license system to major league baseball teams carried on superstations such as WGN and WTBS are far less than that which would be paid if cable systems had to purchase the rights on the open market. Broadcasters don't like the fee because it bars them from receiving revenue from local cable systems that carry local stations. Abolishing the license might provide local stations with another revenue stream if stations were able to charge cable systems for local programming.

In late 1991 both houses of Congress began a review of the compulsory license arrangement. Several alternatives were considered, including one that would link a local licensing system with a new version of the "must-carry" rules. Action on this issue might proceed slowly since it involves many parties and several expensive and complex issues.

VCRs and copyright As is probably obvious, the tremendous popularity of videocassette recorders (VCRs) has created new copyright problems. For example, is it a violation of copyright law for a person to tape a program off the air and watch it at a more convenient time? This question was decided in 1984 by the Supreme Court in what is popularly referred to as the *Betamax* case. The court held that VCR owners could record copyrighted broadcast TV programs for their personal, noncommercial use. Such activity, said the Court, represents fair use of the protected material.

What about renting prerecorded movies on videocassettes? Must a person who rents a tape pay a copyright fee to the copyright holder? No. Viewing a prerecorded movie on tape does not constitute a performance of the work. When a videocassette is purchased, by either a rental store or an individual, a royalty is paid to the author. If, however, that videocassette is sold again or rented, the copyright holder does not receive further payment. This arrangement is known as the "first-sale" doctrine. (A similar situation exists in the textbook market. Copyright holders receive royalties when they first sell a book to a bookstore, but should the bookstore buy back the book and sell it again, no further royalties are paid.) Not surprisingly, the first-sale doctrine has caused friction between motion picture companies and video rental stores as the production companies search for a way to get a cut of the lucrative rental business.

What about renting or buying a movie on cassette and then copying it at home? This is illegal since such copying takes away potential sales and rentals and hurts the market for the copyrighted material. In fact, the motion picture companies estimate that they are losing millions of dollars a year because of illegally copied tapes.

Obscenity, Indecency, and Profanity

Section 1464 of the U.S. Criminal Code states that anybody who utters profane, indecent, or obscene language over radio or TV is liable to fine or imprisonment. Both the FCC and the Department of Justice can prosecute under this section. If found guilty, violators face a fine of up to $10,000, possible loss of license, or even jail. This seems clear enough. A couple of problems, however, quickly surface. First, remember that the FCC is prohibited from censoring broadcast content. Second, how exactly do you define obscenity, indecency, and profanity?

Let's take the easy one first. Profanity is defined as the irreverent or blasphemous use of the name of God. Practically speaking, however, the FCC has been unwilling to punish stations who air an occasional curse. Particularly in the era of deregulation, it is unlikely that profanity will become a major issue.

Obscenity and indecency are another matter. The struggle to come up with a workable definition of obscenity has been long and tortuous and will not be repeated here. The definition that currently applies to broadcasting is the definition spelled out by the Supreme Court in the *Miller* vs. *California* (1973) case. To be obscene, a program, considered as a whole, must: (1) contain material that depicts or describes in a patently offensive way certain sexual acts defined in state law (if you really want to know what some of these acts are, read F. Schauer's book, *Law of Obscenity*); (2) appeal to the prurient interest of the average person applying contemporary community standards ("prurient" is one of those legal words the courts are fond of using; it means tending to excite lust); and (3) lack serious artis-

CHAPTER 18: BROADCASTING AND CABLE LAW

A Florida court ruled that controversial rap group 2 Live Crew's album "As Nasty as They Wanna Be" was obscene. The decision was later overturned, but in the meantime a record dealer was convicted for selling a copy of it to an undercover police officer.

tic, literary, political, or scientific value. Obviously, most radio and TV stations would be wary of presenting programs that come anywhere near these criteria for fear of alienating much of their audience. Consequently, the FCC tends not to issue too many decisions based on the obscenity criterion alone.

But what about cable? Some of the movies commonly shown on certain cable channels go much further than what is shown on traditional TV. The 1984 cable act made it a crime to transmit obscenity over cable. Further, state and local franchise agreements can prohibit obscene programming. Even though cable content is far more daring than over-the-air TV, subscribers are unlikely to see the kind of movies that would be defined as legally obscene. In fact, most hardcore or XXX-rated films are released in two versions, one for theatrical or videocassette release and a less explicit version for cable. In practice, then, obscenity, as legally defined, is seldom at issue in broadcasting and cable.

This leaves the area of indecent programming—the area in which the FCC has chosen to exercise vigilance. Indecent content refers to content that is not obscene under the *Miller* standards but still contains potentially offensive elements. To be more specific, here's the common legal definition of broadcast indecency:

Something broadcast is indecent if it depicts or describes sexual or excretory activities or organs in a fashion that's patently offensive according to contemporary community standards for the broadcast media at a time of day when there is a reasonable risk that children may be in the audience.

For example, a program that simply contains nudity is not obscene, although some people might be offended. Further, four-letter words that describe sexual or excretory acts are not, by themselves, obscene. They may, however, be classified as indecent.

In the past the FCC has acted against stations broadcasting what the commission thought were indecent programs. In 1970 the FCC fined a station for broadcasting an interview with rock musician Jerry Garcia that contained four-letter words.

Some stations objected to the commission's actions in this area, citing the First Amendment and the prohibition against censorship in the Communications act. It wasn't long before a pivotal case arrived before the Supreme Court.

The seven dirty words case George Carlin is probably the only comedian ever to have his act reviewed by the Supreme Court. Here's how it happened. On the afternoon of October 30, 1973 WBAI-FM, New York City, a listener-sponsored station licensed to the Pacifica Foundation, announced that it was about to broadcast a program that would contain sensitive language that some might find offensive. The WBAI DJ then played all twelve minutes of a George Carlin routine entitled "Filthy Words." Recorded live before a theater audience, the routine analyzed the words that "you couldn't say on the public airwaves." Carlin then listed seven such words—five of them were four letters long, one nine letters long, and one twelve—and used them repeatedly throughout the rest of his monologue.

A man and his teenaged son heard the broadcast while driving in their car and complained to the FCC: the only complaint the commission received about the monologue. The FCC decreed that the station had violated the rules against airing indecent content and put the station on notice that subsequent complaints about its programming might lead to severe penalties. Pacifica appealed the ruling and an appeals court sided with WBAI and chastised the FCC for violating Section 326 of the Communications Act, which prohibits censorship. Several years later, however, the case made its way to the Supreme Court and the original FCC decision was reaffirmed. The Court said, among other things, that the commission's actions did not constitute censorship since they had not edited the monologue in advance. Further, the program could be regulated because the monologue was broadcast at a time when children were probably in the audience. Special treatment of broadcasting was justified because of its uniquely pervasive presence (see Chapter 17) and because it is easily accessible to children. Thus the FCC could regulate indecent programming.

In the years following the *Pacifica* decision the FCC generally followed a liberal policy toward possible indecent programming. Isolated swear words were not enough to cause action; indecent language might be broadcast in news shows; such content might be permitted if it had literary, artistic, political, or scientific value.

This reticence may have encouraged broadcasters to push the limits a little too aggressively. Many stations adopted a new format called "raunch radio" or "shock radio," which featured sexual innuendo and vulgarity. In 1987 the FCC warned several stations to clean up their acts or face possible fines for indecent broadcasts. Most stations heeded the warning. Later that same year the commission suggested that programming aired after midnight, when fewer children would be in the audience, might have greater latitude in presenting potentially offensive material. In late 1988, however, Congress passed legislation that would require the FCC to enforce its anti-indecency restrictions around the clock, and the FCC announced a twenty-four-hour ban on indecent programs. Broadcasters immediately questioned the constitutionality of this ruling, and in 1989 the U.S. Court of Appeals issued a stay, preventing the FCC from enforcing a twenty-four-hour prohibition of indecency, pending further justification for the rule.

In 1991 the U.S. Court of Appeals rejected the FCC's basis for its ruling and declared that the twenty-four-hour ban was unconstitutional. The Court ordered the FCC to establish a "safe harbor," a time period during which the number of children in the audience is small and during which broadcasters might air indecent material without fear of recrimination. In turn the FCC announced that, pending further study, it would limit its ban on indecent material to the period from 6:00 A.M. to 8:00 P.M.

Indecent content and cable Cable channels generally have a much wider latitude with programming that might be considered indecent. In fact, the 1984 cable act did not authorize the regulation of indecent material. The act differentiated between obscene and indecent material and permitted regulation of the former only. In addition, federal and state courts have found that the FCC's *Pacifica* ruling does not apply to cable. (George Carlin was able to perform his "Filthy Words" routine on HBO without incident.) The courts argue that cable is not so pervasive as broadcasting (you have to order it and pay additional fees for it) or so easily accessible to children since subscribers could buy lock boxes, which they could use to limit viewing.

The courts have also overturned state attempts to bar indecent content from cable. A Utah law authorized fines for cable companies whose programs displayed bare buttocks or breasts. The Supreme Court affirmed a lower court decision that struck down the rule since it did not meet the requirements of the *Miller* test of obscenity. The courts have also found similar laws in other states to be unconstitutional.

Lotteries

Federal laws concerning the advertising of lotteries have recently undergone a drastic overhaul. Federal restrictions on lotteries have been around since 1895. In 1934 Congress enacted restrictions on the broadcasting or mailing of advertisements or information about lotteries. After several states established state-run lotteries in the 1960s and 1970s, however, Congress exempted these games from the advertising ban.

Further reflecting the changing attitude toward lotteries, a 1988 law passed by Congress removed federal restrictions on the advertising of all legal lotteries. The law, however, did not take effect until mid-1990 in order to give individual states the opportunity to put on the books their own laws regulating lottery advertising. The practical effect of all of this will be that broadcasters and cable system owners will have to look to state law more than before to determine what they can and can't do. In practice, most states have some sort of restrictions on lottery advertising.

Assuming a state chooses not to restrict lottery ads and that lotteries are legal within that state, it would be permissible for broadcasters and cablecasters to carry lottery advertising by certain privately run, nonprofit gambling enterprises, such as church-sponsored bingo games. Advertising of lotteries by commercial organizations, such as a travel agency advertising a raffle for a free trip to a resort, would also be authorized. To be allowed, this advertising by a profit-making organization must be conducted as a promotional activity and must be ancillary (subordinate) to the primary business of the commercial organization running the ad. Further, the 1988 bill expanded advertising rights for state-run lotteries, permitting them, for instance, to advertise lists of prizes and also allowing advertising for games of chance held on Indian lands.

But what if you are in a state that chooses to restrict lottery advertising? What exactly constitutes a lottery? Three elements must be present: (1) There must be a prize. Prizes may be money, merchandise, vacations, or anything of value. (2) There must be chance. The winner of the prize is selected at random; no skill or talent is involved in being selected. (3) There must be consideration, which generally causes the most problems. Consideration is present if in order to win the prize, contestants must purchase something or furnish something of value. This consideration must go to the person or organization giving the prize. What exactly constitutes consideration in a given case is sometimes difficult to define. A contest in which postcards are sent to a station to be drawn for a cash jackpot generally would not constitute a lottery since the act of sending the card does not constitute sufficient consideration. Having to travel a short distance to a station or to a store to sign up for a contest has also been ruled as not qualifying as consideration. Having to pay an entry fee or having to purchase a product, however, would probably qualify as consideration.

Stations run into problems with the lottery law when they run their own promotional contests and when they run commercials for promotions from their advertisers. Some cases can be extremely difficult to decide (see box). One safeguard that a station can adopt is to make the award of the prize contingent on skill as opposed to chance. For example, contestants might be required to write a seventy-five-word essay on a given topic; the winner would be the person whose essay is judged best. Or contestants might have to answer a question or two to win the prize. In questionable situations many stations will consult with their attorneys before airing any material.

Equal Employment Opportunities

Broadcasters and cablecasters are bound by federal law prohibiting job discrimination based on race, color, sex, religion, or national origin. Fur-

REGULATIONS

YOU MAKE THE CALL: LOTTERY OR NOT?

Let's assume your state has prohibited the advertising of all lotteries.

(A)

Arnie's Auto Emporium wants your radio station to carry an ad touting Arnie's big spring promotion. Anybody who buys a car during April and May is invited to put a copy of his or her bill of sale in a big urn; come June 1 Arnie himself will draw a winner, who will receive $500 worth of free gas. Should you carry the ad?

(You better not. This has all the elements of a lottery: consideration, buying the car, chance, a drawing, and a prize—$500 worth of gas.)

(B)

Buck's Bait Shop is sponsoring a bass-fishing tournament at a private lake. The lake has been newly stocked with 200 bass, each sporting a little metal tag with a dollar amount printed on it. Amounts range from $1 to $500. Aspiring fishing experts can go to Buck's shop, pay a $15 entry fee, and get a one-day admission ticket to the lake. Each contestant who catches one or more fish is entitled to the cash amount printed on the tags. Buck wants you to advertise this little venture. Should you?

(A tough call. This promotion certainly has consideration—the $15 entry fee—and a prize. But does it have chance? Probably not. In similar situations courts have ruled that there is some skill involved in fishing and catching fish is not necessarily a matter of chance. Therefore, it's probably not a lottery.)

---

ther, they must adhere to regulations set down by the Equal Employment Opportunity Commission. Although not directly responsible for enforcing all equal employment opportunity (EEO) provisions, the FCC has made it clear that it will assert jurisdiction over this area in broadcasting and cable.

Speaking in general terms, the commission's EEO program spans two separate concepts: (1) ensuring there is no discrimination and (2) enforcing affirmative action rulings. The first concept requires cable and broadcast operators to maintain nondiscriminatory policies in recruiting, hiring, training, promotion, pay, and termination. The affirmative action provision places an obligation on broadcasters and cablecasters to seek out, hire, and retain qualified women and minority applicants. The commission has also developed a system of guidelines that stations should follow. Each year stations with five or more employees must file detailed reports listing employees by job category, race, and sex, Even with the current deregulatory atmosphere, it is unlikely that these regulations will be modified or repealed.

Stations with continuing patterns of halfhearted or unsuccessful fulfillment of EEO plans are often given short-term license renewals. Since the FCC doesn't license cable systems, about the only way for the commission to spur a cable operator to greater efforts in this area would be a fine.

Antitrust

Broadcasting and cable are businesses and, as such, are subject to antitrust laws. These laws date back to the 1890s and protect business against unlawful restraints and monopolies. Although much of the antitrust litigation in recent years has concerned newspapers, the antitrust laws also have relevance for the electronic media. Over the years, as we have seen, the FCC has enacted rules that tried to maximize the diversity of ownership of broadcast facilities. At the same time, there has been a trend in the industry toward greater concentration in media ownership and growing conglomerate involvement. For example, TCI now has about 17 percent of cable subscribers. In addition, TCI owns part of Turner

Broadcasting, half of the American Movie Classics channel, and parts of the Discovery Channel, Black Entertainment Television, and the Fashion Channel. The top 5 radio station groups control about 25 percent of the radio listening market. In 1986 Capital Cities Communications merged with ABC and General Electric purchased RCA, owner of NBC. The federal government has generally taken a hands-off policy toward these developments.

Some antitrust challenges, however, may be on the horizon, particularly in the cable industry. One question that still persists is whether a cable system should be a regulated local monopoly. Further, some cable companies are now getting involved in the production, distribution, and exhibition of TV programs (a process called "vertical integration"), an arrangement that traditionally has raised antitrust concerns. Since policy makers are relying more and more on the marketplace as a substitute for regulation and antitrust laws govern the marketplace, look for more reliance on these laws in the decade to come.

THE LAW AND BROADCAST JOURNALISM

There are several areas of law that are most relevant to the news-gathering and reporting duties of stations and cable channels. Some of these laws cover topics that are common to both print and electronic journalists (such as libel and invasion of privacy). Other laws are more relevant to video reporters (such as the restrictions on cameras in the courtroom). Further, many of these laws vary from state to state but some have raised constitutional issues that have been addressed by the Supreme Court.

This section covers four areas that are highly relevant to the broadcast press: (1) defamation, (2) privacy, (3) protection of confidential sources, and (4) use of cameras and microphones in the courtroom.

Defamation

The law regarding defamation is concerned with the protection of a person's reputation. A *defamatory statement* injures the good name of an individual (or organization) and lowers his or her standing in the community. The law of defamation makes sure that the press does not act in an irresponsible and malicious manner. As we shall see, erroneous stories that damage the reputation of others can bring serious consequences.

There are two kinds of defamation: libel and slander. *Slander* comes from spoken words; in other words, slander is oral defamation. *Libel* is defamation stated in a tangible medium, such as a printed story or a photograph, or in some other form that has a capacity to endure. Since libel exists in a medium that can be widely circulated, the courts treat it more seriously than slander, which evaporates after it is uttered. Broadcast defamation is usually regarded as libel rather than slander.

Broadcast journalists have become particularly wary of suits alleging libel because they can carry sizable cash awards to the victims if the media are found at fault.

Wayne Newton, for example, was recently awarded $19 million by a jury that decided he was defamed by an NBC news report falsely linking him to organized crime figures. An appeals court eventually overturned that judgment, but another jury found that a Chicago TV journalist had defamed the Brown & Williamson Tobacco Corporation and awarded the company more than $5 million in damages. The Brown & Williamson award was reduced on appeal to $3.05 million, with $50,000 having to be paid personally by the TV journalist. Such awards are enough to give broadcast journalists something to think about.

Elements of libel If someone brings a libel suit against a station or cable channel, he or she must prove five different things to win. First, the statement(s) in question must have actually defamed the person and caused some harm. The material must have diminished the person's good name or reputation or held the person up to ridicule, hatred, or contempt. Harm can be such things as lost wages or physical discomfort or impairment in doing one's job. A prime example of a defamatory statement would be falsely reporting that a person has been convicted of a crime. The second element is publication. This usually presents no ambiguities; if a story was broadcast, then it fulfills the publication criterion. The third element is identification. A person must prove that he or she was identified in a story. Identification does not necessarily have to be by name. A nickname, a cartoon, a description, or anything else that pinpoints someone's identity would suffice.

The fourth element is fault or error. To win a libel suit, some degree of fault or carelessness on the part of the media organization must be shown. The degree of fault that must be established depends on who's suing, what they are suing about, and which state's laws are being applied. As we shall see, certain individuals have to prove greater degrees of fault than others. Finally, in most instances the individual has to prove the falsity of what was published or broadcast.

Defenses against libel When a libel suit is filed against a broadcast reporter and/or station, several defenses are available. The first of these is truth. If the reporter can prove that what was broadcast was true, then there is no libel. This sounds a lot easier than it is since the burden of proving truth, which can be an expensive and time-consuming process, falls on the reporter. Moreover, when the alleged libel is vague and doesn't deal with specific events, proving truth may be difficult. Finally, recent court decisions suggest that in many circumstances people who bring libel suits against the media must bear the burden of proving the defamatory statement false. All of these factors diminish the appeal of truth as a defense.

Note that libel suits concern defamatory statements of fact only. There is no such thing as a libelous *opinion* since no one can prove that an opinion is true or false. In 1990, however, the Supreme Court ruled that expressions of opinion can be held libelous if they imply an assertion of fact that can be proven false. Consequently, reporters and commentators must pay close attention to the assertions that they broadcast. Labeling a statement as an opinion does not necessarily make it immune from a libel suit.

A second defense is privilege. Used in its legal sense, privilege means immunity. There are certain situations in which the courts have held that the public's right to know is more important than a person's reputation. Courtroom proceedings, legislative debates, and public city council sessions are examples of areas that are generally conceded to be privileged. If a broadcast reporter gives a fair and accurate report of these events, the reporter will be immune to a libel suit even if what was reported contained a defamatory remark. Keep in mind, however, that accurately quoting someone else's libelous remarks outside an official public meeting or official public record is not a surefire defense against libel.

The third defense is fair comment and criticism. People who thrust themselves into the public arena are fair game for criticism. The performance of public officials, pro sports figures, artists, singers, columnists, and others who invite public scrutiny, is open to the fair comment defense. This defense, however, applies to opinion and criticism; it does not entitle the press to report factual matters erroneously. A movie critic, for example, could probably say that a certain actor's performance was amateurish and wooden without fearing a libel suit. If, however, the critic went on and falsely reported that the bad acting was the result of drug addiction, that would be a different story.

In 1964, in the pivotal *New York Times v. Sullivan* case, the Supreme Court greatly expanded the opportunity for comment on the actions of public officials. The Court ruled that public officials must prove that false and defamatory statements were made with "actual malice" before a libel suit could be won. The Court went on to state that actual malice meant publishing or broadcasting something with the knowledge that it was false or with "reckless disregard" of whether or not it was false. In later years the Court ruled that public figures also must prove actual malice to win a libel suit.

In effect, two different degrees of fault are used as standards in libel cases. In many states a private citizen must simply prove that the media acted with negligence. For some states negligence means that accepted professional standards were not followed. Other states take negligence to mean that a reporter did not exercise reasonable care in determining whether a story was true or false. A few states define negligence in relation to whether the reporter had reason to believe that material that was published or broadcast was true. Public figures and public officials are held to a higher standard; they must also prove actual malice. (Note that the really big financial awards in libel cases come from what are called "punitive damages"—awards levied by juries to punish the offending media outlet. To win punitive damages, even a private citizen must prove actual malice.)

Invasion of Privacy

Most people have seen reporters thrust microphones and cameras into the faces of relatives of victims killed in some tragedy or have seen videotape taken by hidden cameras that were smuggled by reporters into someone's house or place of business. Many of you have heard about reporters staking out the home of a political candidate to catch him or her in some extramarital activity or publishing a list of the videocassette rentals of a nominee to the Supreme Court. Such practices have led to a growing concern over the media's intrusion into the privacy of the people they cover.

The right to privacy is a relatively new area of media law. Stated simply, it means that the individual has a right to be left alone—not to be subjected to intrusions or unwarranted publicity. The right to privacy is not specifically found in the Constitution, although many have argued that the right is implied in several amendments. Like defamation, laws concerning privacy differ from state to state.

Nor surprisingly, invasion of privacy is a complicated topic. In fact, legal experts suggest that it actually covers four different areas. The right of privacy protects against (1) unwarranted publication of private facts, (2) intrusion upon a person's solitude or seclusion, (3) publicity that creates a false impression about a person (called "false light"), and (4) unauthorized commercial exploitation of a person's name or likeness.

Private facts Disclosure of private facts occurs when a broadcast station or cable channel reports personal information that the individual did not want to make public. Disclosure resembles libel in that the individual must be identifiable and the facts must cause the individual humiliation or shame. Unlike libel, however, the facts revealed are true and may not necessarily damage a person's reputation. For example, suppose that a TV station shot footage of you in a hospital bed after cosmetic surgery when your face was swollen, black and blue, and generally ugly looking. The station then ran the tape without your consent on the six-o'clock news with your name prominently mentioned as part of a series it was doing on the hospital. It's possible you might have a case for invasion of privacy. To qualify as the basis for a suit, private facts must be highly offensive to a person of "reasonable sensibilities."

When faced with an invasion of privacy suit, the media have several different defenses. A reporter might argue that he or she received the consent of the individual to release the private information. Consent can be explicit, such as a signed form approving release of the information, or it can be implied. People who know that they are talking to reporters and who know that what they say might be published have given implied consent.

Reporters usually do not have to fear an invasion of privacy suit if what they publish was obtained from a public record. In the mid-1970s the Supreme Court struck down a Georgia law that prohibited publicizing the name of a rape victim. An Atlanta TV station brought the suit to the Court's attention after broadcasting the name of a 17-year-old rape victim. The station argued the name was obtained from a public record and the Court agreed with the station's position. (Of course, reporters are under no obligation to publicize everything that is on the public record. Many broadcast stations, as a matter of ethics, voluntarily withhold the name of rape victims.)

The problems involved with publicizing the name of a victim of an alleged rape were underscored by the 1991 trial of William Kennedy Smith. Despite a Florida law banning the publication of a rape victim's name, the tabloid paper, the *Globe*, along with NBC News and the *New York Times*, identified the woman involved in the *Smith* case. Only the *Globe* was charged under Florida law. A court eventually ruled that Florida statute unconstitutional, but both CNN and a cable network, the Court Channel, which televised the trial, edited out mentions of the woman's name and used an electronic matrix to screen out her face. After Smith was found innocent the woman involved in the trial chose to identify herself, but the whole episode prompted lengthy debates over the proper way to report such trials.

Finally, the station could argue that the private information was released in connection with a newsworthy event. If individuals are involved in

REGULATIONS

YOU MAKE THE CALL: TELEVISION NEWS?

(A)

During your six-o'clock newscast one of your anchors reads a story about a woman who has just opened a new auto repair shop. Unfortunately, your anchor misreads the script and instead of saying, "the 32-year-old mechanic," actually says "the 52-year-old mechanic." The woman, of course, was watching the story and quickly calls your station. "You idiots," she yells. "You've just added twenty years to my age. I'm totally embarrassed. My friends will make fun of me. People will think I've been lying about my age. How humiliating! You should have been more careful. I'm going to sue your station for libel." Does she have a case?

(No. There's nothing defamatory about being 52. The woman may have been hurt, annoyed, and embarrassed, but she was not defamed. And she would probably have a difficult time proving fault. Nonetheless, the professional thing to do would be to have your newscaster correct the mistake and apologize.)

(B)

Your TV station has broadcast an investigative report about a local nightclub. In the report a police detective described the club as a "sleazy bar" where you could get just about anything, including "drugs, sex, and contract murders." Your report also shows an interior shot of this club with a nude female dancer in the background. Although you electronically mask the more private parts of her body, her face is clearly seen in the report. After the segment has aired, the female dancer sues the station for invasion of privacy citing the public disclosure of her private life as a reason. She states that her friends and classmates at college discovered from the report that she was a nude dancer and that this fact has irreparably damaged her reputation. Does she have a case?

(Probably not. The station was merely giving publicity to an activity in which the young woman was dealing with the public. Any person, of legal age, with the price of admission to the club could see her perform. Since her dancing was a public activity and not part of her private life, no privacy invasion could occur.)

---

crimes, fires, or other disasters, it is unlikely that they can sue for invasion of privacy. For example, a person who is photographed in the nude while fleeing a fire might have difficulty recovering damages for privacy invasion if the fire was deemed a newsworthy event. The determination of whether an event is newsworthy is generally left to the courts, and they have been rather liberal in their definition.

Intrusion and trespass *Intrusion* is an invasion of a person's solitude or seclusion without consent. The intrusion may or may not include *trespass*, which is defined as physical presence on private property without the consent or approval of the property's owner. Thus using a telephoto lens to secretly take pictures through a person's bedroom window could constitute intrusion. Secretly sneaking through the yard to take pictures through the bedroom window with a regular lens would probably constitute trespass and intrusion. Problems with intrusion and trespass generally crop up in the news-gathering stage. In fact, a reporter can be sued for trespass or intrusion even if he or she never broadcasts the information that was gathered.

Whether some news-gathering technique violates a person's right of privacy depends on how much privacy a person should normally expect in the particular situation involved. For example, people in their living rooms have a right to expect that they will not be secretly photographed or have their words recorded. People walking down a public street have less expectation of privacy.

As a general rule, the more public the surroundings are, the less is the chance that an invasion of privacy suit will be successful. If a scene is available to be viewed by the general public, then it's probably safe for a camera crew to televise it. For example, the courts have ruled it is not intrusion to take a picture of private property from a public sidewalk or to photograph from outside a pharmacy people inside who could be seen from the street.

There are several places, however, where a good deal of privacy is expected. People do not expect to have microphones secretly hidden in their bedrooms, their phones tapped, or their mail opened. In one famous case two *Life* magazine reporters intruded on the right of privacy of the subject of their story when they secretly took pictures and made recordings inside their subject's house. Bugging and wiretapping are considered intrusions.

Broadcast journalists can generally avoid charges of trespass by securing the consent of the owner of private property before entering it. In some limited instances during fires and natural disasters police or fire officials might control property and can grant consent to reporters to enter. Reporters, however, do not have the right to follow an unruly crowd onto private property without the consent of the owner or the police. In addition TV reporters might be guilty of trespass if they enter, with cameras rolling, private property that is open to the public—like a place of business.

False light A reporter can cast a person in a false light in several ways: omitting pertinent facts, distorting certain information, falsely implying that a person is other than what he or she is, using a photograph out of context. A suit that alleges invasion of privacy through false light is similar to one brought for defamation (in fact, the two are often brought at the same time). A person can't be put in false light by the truth. Thus, like libel, there must be a false assertion of fact. In addition, the misinformation must be publicized. Finally, to win a false light suit, some people might have to prove actual malice on the part of the media. Unlike libel, in false light suits the false assertion is not defamatory. In fact, suits alleging false light have been brought over the publicizing of incorrect information that was flattering to the person. People who bring false light suits seek compensation not for harm to their reputation—how others feel about them—but for the shame, humiliation, and suffering *they* feel because of the false portrayal.

Television stations often run into problems with false light suits when using film or tape of people walking down a sidewalk if offensive or antisocial characteristics are associated with an identifiable person in the shot. For example, if a newsclip singling out a passer-by was accompanied by audio that said, "One out of every seven city residents has a venereal disease," the person shown in the clip might have cause for a false light suit.

False light can also crop up when stations use file footage. Although not a problem in its original context, when used in a different story, the context gets lost and false light becomes more likely. For example, in reporting a story about a doctor who allegedly used an AIDS-infected cotton swab on a patient, a Chicago TV station illustrated the story by running file footage of another doctor, who could easily be identified, performing a similar procedure. The other doctor had no connection with the infected swab case and the file footage came from a story that had been done on the doctor more than two years earlier. The doctor who was seen in the file footage sued the station claiming that the report put her in a false light. Other problems can crop up with attributing false quotes or actions to real individuals in news or dramatic presentations. Docudramas, which blend fact and fiction, are another trouble spot.

Commercial exploitation This area tends to be more of a problem area for advertisers, promoters, agents, and public relations practitioners. Basically, it prohibits the unauthorized use of a person's name or likeness in some commercial venture. Thus it would be an invasion of privacy if you walked into a supermarket and found that a pickle manufacturer had, without your permission, put your face on a whole shelf of pickle jars. The best way for broadcasters to avoid privacy invasion suits based on unauthorized exploitation is to obtain written consent from subjects.

Well-known individuals, of course, can have more of a stake in these matters. In fact, over the last several years an offshoot of the commercial appropriation area known as the "right of publicity" has been articulated by the courts. In contrast with the traditional privacy area, which is based on the right to be left alone, the right of publicity is concerned with who gets the financial rewards when the notoriety surrounding a famous person is used for commercial purposes. In this regard the right of publicity resembles a property right rather than a personal right and bears a little resemblance to copyright. In essence, the courts have ruled that famous people are protected against the unlawful appropriation of their fame. For example, Johnny Carson was successful in his suit against a manufacturer of portable toilets that were sold under the name "Here's Johnny!" Similarly, a California superior court ruled that the producers of *Beatlemania*, a show that used Beatle lookalikes and soundalikes to perform twenty-nine Beatles' songs, had appropriated the Beatles' persona. More recently a California court ruled that the Ford Motor Company and its ad agency had violated singer Bette Midler's right of publicity when it used a soundalike singer and instructed her to imitate Midler's voice for the soundtrack of a TV ad.

Protecting Sources

Since 1950 a conflict has developed between courts and reporters over the protection of news sources, notes, and news footage. Some background will be useful before we discuss the specifics.

There are two basic types of litigation. In a criminal trial the government seeks to punish someone for illegal behavior. Murder, rape, and robbery are examples of proceedings that would be covered by criminal law. Civil law involves a dispute between two individuals for harm that one has allegedly caused the other. Breach of contract, claims of overbilling, disputes between neighbors over property rights (the kinds of things decided by Judge Wapner on "The Peoples' Court"), along with defamation and invasion of privacy are examples of civil suits. In a criminal or civil trial it is necessary for the court to have at its disposal any and all evidence that bears directly on the outcome of the trial in order to assure a fair decision.

Courts have the power to issue subpoenas, official orders summoning witnesses to appear and testify, in order to make sure all who have information come forth and present what they know. There are only a few exceptions to this principle, called privileges. No person is forced to testify against himself or herself. Husband–wife communication is also considered privileged. Other common areas of privilege are lawyer–client communication and doctor–patient conversations. For many years journalists have argued that their relationship with a news source qualifies as the same type of privilege. They argue that without a promise of confidentiality, many sources would be reluctant to come forward and the news-gathering process would be severely hindered. In fact, many journalists have been fined and sent to jail for failure to disclose the names of sources or for failing to turn over to the court notes, photos, and videotapes.

Before the 1970s the courts were reluctant to grant any privileges to journalists. Since then, however, journalists are in a better position thanks to several court decisions and the passage of state "shield laws," which offer limited protection for journalists. The landmark decision was handed down by the Supreme Court in 1972 in *Branzburg v. Hayes*. Originally regarded as a defeat for the principle of journalistic privilege, the decision actually spelled out guidelines that covered when a reporter might legally refuse to testify. To make a reporter reveal confidential sources and information, the government must pass a three-part test. The three-part test listed below is often used in civil proceedings but seldom used in criminal trials. The government must prove:

1. that the journalist has information that bears directly on the case
2. that the evidence cannot be obtained from any other sources
3. that the evidence is crucial in the determination of the case

The government's success is passing this test varies with the legal context. Reporters are most likely to be required to testify before grand juries, particularly if the reporter witnessed criminal activity; they are least likely to be compelled to testify when the defense in a civil trial is trying to obtain information. Civil libel actions where the medium is the defendant create special problems. Some states don't permit the protection of a shield law if it would frustrate a plaintiff's effort to show actual malice.

Twenty-five states granted limited protection to journalists by enacting shield laws. These laws, of course, vary from state to state. In some

Brian Karem was reporter for KMOL-TV in San Antonio, Texas, when he was jailed for not revealing his sources.

states protection is given to a journalist who refuses to reveal a source but protection does not extend to notes, tape, and other media used in the news-gathering process. Moreover, some states have strict definitions as to who exactly is a journalist and is covered by the shield law. Several states require the journalist to prove that a source was actually promised confidentiality. Some states require that the information be published before the shield law is triggered. Other state laws list exceptions where the protection is not granted.

For the broadcast journalist, the practical effect of all of this should be increased caution in promising confidentiality to any source. At the time a pledge of secrecy is given the reporter can't be sure in what legal context he or she may be asked to violate the pledge. Many news organizations advise journalists to obtain an editor's approval before promising to keep a source secret. This approval is granted only in those circumstances where the reporter proves the information couldn't be obtained otherwise. Despite these suggestions, reporters continue to run into trouble with promises of confidentiality. A San Antonio TV reporter was sentenced to a six-month jail term in 1990 for his refusal to reveal the name of a source (he was ultimately freed when his source released him from his pledge). In 1992 legal proceedings were begun against a Peoria, Illinois, TV reporter who also refused to reveal his sources for a story on a local murder case. All in all, clearly journalistic privilege is not absolute and reporters should act accordingly.

Cameras in the Courtroom

The controversy over cameras in the courtroom highlights an area where two basic rights come into conflict: the right of a defendant to a fair trial and the right of a free press to report the news.

In 1932 the infant son of national hero Charles Lindbergh was kidnapped and later found murdered. A German immigrant named Bruno Hauptmann was arrested for the crime and put on trial in a small courtroom in Flemington, New Jersey. Not surprisingly, media attention was intense. Reporters and photographers swarmed over witnesses, jurors, and anybody even remotely associated with the case. Before long the proceedings degenerated into near chaos and, in hindsight, may not have constituted a fair trial for Hauptmann, who was found guilty and later executed.

In the trial's aftermath a special committee of lawyers was set up to study how to cover judicial proceedings. As a result of this committee's recommendations, canon 35 was added to the code of conduct of the American Bar Association. Canon 35 prohibited the taking of photographs in court or the broadcasting of court proceedings. Many states and the federal courts took the language of canon 35 and enacted it into state law.

The issue surfaced again in 1962 when the Texas trial of Billie Sol Estes, an accused swindler, was televised over his objections. (Texas and Colorado were the only states which at that time allowed cameras in the courtroom.) Estes appealed his subsequent conviction to the Supreme Court, which ruled in his favor and ordered a new trial. (Estes didn't gain much from this reversal; he was convicted again minus TV.) Although the Court ruled that TV had prejudiced Estes's trial, several of the justices wrote opinions in which they conceded that this might not always be the case. They noted that subsequent advances in technology that might make broadcasting less obtrusive might also make it permissible to have cameras in the courtroom. Thus the courtroom door was opened a tiny crack to broadcasters.

As the Court predicted, cameras and microphones did become smaller and less obtrusive. Nonetheless, canon 35 remained in effect until 1972, when it was retitled canon 3A(7). The new canon still prohibited broadcasting, telecasting, or taking pictures of court sessions but did allow for

Live TV shots form the courtroom are becoming familiar sights to viewers as almost every state has a permanent or experimental plan that allows coverage. Here William Kennedy Smith is questioned by prosecutor Moira Lasch.

the use of tape and film for the presentation of evidence and permitted limited electronic recording for purposes of education and instruction.

Despite the American Bar Association's hostility, many states started experimental programs with cameras in court. By 1980 at least twenty-two states had some experience in televising trials. It was one of these pilot programs that led to another significant Supreme Court ruling. A portion of a Florida trial was televised over the objections of the defendants. The defendants were convicted and appealed all the way to the Supreme Court, arguing that the cameras had prevented a fair trial. The Court rejected their arguments. The Court's opinion, ironically enough, was written by Chief Justice Warren Burger, an opponent of cameras in the courtroom. The gist of the Court's ruling held that the mere presence of cameras in the courtroom does not automatically mean that the participants were denied a fair trial. Instead, those who appeal a decision on this basis must prove that the outcome of their trial was somehow adversely affected by the presence of cameras and microphones. States were free to pursue their experiments with cameras in the courtrooms. The door to the courtroom was opened even wider to electronic journalists.

Following this ruling the American Bar Association softened, at least a little, its canon 3A(7). The revised canon still prohibited recording, televising, or photographing in a courtroom, unless an appropriate authority rules otherwise. Then a judge can allow coverage, providing it is unobtrusive and does not interfere with a free trial.

As of 1991 thirty-five states had permanent rules concerning TV coverage; nine had experimental rules in place, whereas six states allowed no coverage at all. A cable network, Court TV, which carried televised court proceedings, made its debut in 1991, suggesting that cameras in the courtoom, at least at the state level, will become increasingly common.

Cameras are still banned from most federal proceedings, but a limited experiment that started in 1991 permitted TV in some federal courtrooms. Radio and TV, however, are still barred from the Supreme Court.

REGULATING ADVERTISING

Advertising is big business: about $28 billion was spent on broadcast and cable advertising in 1990. Since it is such an influential industry, the government has enacted laws and regulations that deal with advertising messages. For our purposes there are two relevant areas. The first deals with the question of whether advertising (or "commercial speech," as it is called in legal circles) qualifies for protection under the First Amendment. The second has to do with how false and deceptive advertising is regulated. Let's first examine the constitutional issue.

Advertising and the First Amendment

The relationship of commercial speech to the First Amendment is rather complicated and has recently undergone significant reinterpretation. For many years the issue never came up. Everybody simply assumed that advertising was different; it was simply hawking one's wares. As such it merited no protection under the Constitution. The Supreme Court had not considered the issue until 1942, when it ruled that commercial speech was not protected under the First Amendment. States were free to pass laws regulating the availability and content of ads.

This was the prevailing philosophy for about twenty years until the famous *New York Times v. Sullivan* decision in 1964. Among other important points in this landmark ruling was the notion that advertising that promoted an idea, as opposed to a product or service, did have constitutional protection. The rationale behind this decision foreshadowed the *Red Lion* decision discussed earlier. The Court recognized the First Amendment rights of the public to receive truthful and accurate advertising about lawful goods and services.

This rather neat division was in vogue for about a dozen years until the courts expanded the protection. A Virginia law prohibited advertising for abortions. A newspaper that carried such an ad was found guilty of breaking the law, and the managing editor was fined. The case was ultimately appealed to the Supreme Court, which ruled against the state. The Court argued that even though it advertised a service, the ad also contained ideas and information that were entitled to protection. Similar to its reasoning in the *Red Lion* case, the Court claimed there was a public right to receive accurate information about lawful products or services. Within less than a year First Amendment protection was extended even further. Again, it was a Virginia law that was at the center of attention. Like many other states, Virginia had a law that prohibited advertising the price of prescription drugs. The Supreme Court declared that law unconstitutional and seemed to extend constitutional protection to all commercial speech. Subsequent decisions extended protection to price advertising by lawyers, "For Sale" and "Sold" signs on private homes, and contraceptive advertising.

The Court, however, was unwilling to grant commercial speech the same degree of protection it granted political speech. A 1980 decision spelled out a four-part test that determines if states can regulate commercial speech:

1 Is the commercial speech eligible for protection? Ads that are false and misleading or promote illegal products or services do not qualify for protection and the rest of the test is inappropriate.
2 Does regulation of the commercial serve substantial government interest? For example, a state may prove that regulation of advertising is necessary to preserve the health or safety of its citizens.
3 Does the regulation itself advance the state's interest?
4 Are the regulations narrowly drawn? A regulation that is more extensive than necessary to serve the government's interest would probably not be legal.

There is some question about the precise boundaries of the fourth provision. In a recent case the Supreme Court ruled that the government of Puerto Rico could discourage gambling by native Puerto Ricans by banning *all* casino ads in Puerto Rico, even though gambling there was legal. (Ads for Puerto Rican casinos running on mainland media were not affected.) The Court found this blanket prohibition was not overly broad. Advertisers on the mainland noted with some trepidation that this ruling seemed to suggest that Congress could ban all ads for liquor and cigarettes, even though it chooses not to ban the underlying product.

To sum up, the most recent rulings suggest that at least some commercial speech is entitled to First Amendment protection. The level of protection given to commercial speech, however, is less than that given to political and other forms of noncommercial expression.

FTC Rules and Regulations

We have considered ads that are accurate and truthful, but what about false and deceptive advertising? How is that controlled?

First, consumers are protected from false advertising claims at the state and local level by both state laws and various state departments that specialize in consumer protection. The attorney general of a state, for example, might start an action against an ad that was deemed to be deceptive. Second, several federal agencies, such as the Postal Service, the Food and Drug Administration and the Securities and Exchange Commission also have the power to regulate advertising in print and on radio and TV. However, the agency that has been the most visible is the Federal Trade Commission (FTC), an organization that we have encountered before. (See Chapter 17.)

Congress set up the FTC in 1914 to regulate unfair business practices. In 1938 its power was broadened to include jurisdiction over deceptive advertising. Nonetheless, as far as advertising was concerned, the FTC remained an obscure institution until the consumer movement of the 1960s and 1970s when it became highly active in regulating questionable ads. (It even ordered some companies to run "corrective" ads to counteract any misunderstanding caused by their original ads.) After the move toward deregulation in the 1980s the FTC assumed a much lower profile.

Like the FCC, the FTC is an independent regulatory agency with five commissioners appointed by the president for renewable seven-year terms. The FTC's Bureau of Consumer Protection handles advertising complaints. The FTC is charged with regulating false, misleading, and/or deceptive advertising and may investigate questionable cases on its own or respond to the complaints of competitors or the general public. The FTC may hold formal hearings concerning a complaint and issue a decision. Again, like the FCC, the FTC's decisions can be appealed through the federal courts.

The FTC guards mainly against deceptive advertising, and over the years the commission has developed a set of guidelines that define "deception." Taken as a whole, the ad must have deceived a reasonable person. Any kind of falsehold constitutes deception. For example, a broadcast ad promising buyers a genuine diamond is deceptive if, in fact, what the consumer receives is a zirconium. Even a statement is literally true might be judged deceptive. Wonder Bread ads claimed that the bread was fortified with vitamins and minerals that are necessary for healthy growth. Although literally true, the ad was ruled deceptive because it did not point out that every fortified bread contained the same vitamins and minerals.

Televised product demonstrations can also raise the ire of the FTC. The classic case in this area involved the claim by the makers of Rapid Shave that it could soften even sandpaper. To demonstrate, a TV commercial showed a razor shaving clean what looked like a sheet of sandpaper. It turned out that what was used in the commercial wasn't sandpaper at all; it was a sheet of plexiglass that had been covered with sticky sand. The FTC claimed the ad was deceptive. The company argued that the claim was literally true: Rapid Shave could shave sandpaper—as long as you first soaked the sandpaper in the shaving cream for about an hour and a half. The Supreme Court sided with the FTC and ruled the ad deceptive.

In late 1990 Volvo Cars of North America faced an FTC inquiry because of a deceptive TV ad. In the ad a Volvo apparently survived intact after a "monster" truck drove over it while other vehicles were crushed. The Volvo's frame, however, had been reinforced by special steel braces to help it withstand the ordeal. Volvo pulled the commercial and ran several corrective ads, claiming that the commercial production company had modified the car without the company's knowledge. The company's forthrighteous approach apparently headed off any serious action by the FTC.

Testimonials are another problem area. If a celebrity or expert endorses a product, then he or she should actually use the product. Singer Pat Boone learned this lesson the hard way when it was revealed that not all of his daughters used Acne-Satin skin medication as the ad claimed. Boone had to contribute his share of Acne-Satin

"Informercials," programs that look like legitimate informational shows but are really commercials, have caused special problems for both the FTC and FCC. This is Tom Skerrit in an infomercial for Time-Life's "Air Power."

sales as restitution to persons who purchased the product.

In recent years comparative ads—those that claim the superiority of one product over its identified competitors—have become popular. Making false comparisons, however, can be deceptive. Likewise, the presentation of surveys or test results can be false and misleading. The statement that 75 percent of the doctors surveyed recommended a particular pain reliever might be literally true, but if the survey was done among only four doctors who also happened to be related to executives at the ad agency, it might be regarded as misleading.

The FTC, however, does recognize that there is room for reasonable exaggeration in advertising. Consequently, the commission permits "puffery" in subjective statements of opinion that the average consumer will probably not take seriously. Thus it would probably be OK for a restaurant to claim that it serves "the world's best coffee" or for a service station to advertise the "friendliest service in town." Puffery crosses over into deception when exaggerated claims turn into factual assertions of superiority.

The FTC has several enforcement means at its disposal. As the mildest level the FTC can simply express its opinion about the questionable content of an ad, as it did in 1988 when it notified aspirin manufacturers that they should not promote aspirin as a preventive for heart attacks. Further, it can require that certain statements be carried in the ads to make the ads more accurate. For example, when some Listerine ads claimed that the product killed germs that caused sore throats that came with a cold or fever, the FTC ordered the makers of Listerine to include a statement in their ads to the effect that Listerine was not a cure for sore throats associated with colds or fever. An appeals court upheld the right of the FTC to order advertisers to engage in such corrective campaigns even though the advertiser objected.

The FTC also can notify an advertiser that its ads are deceptive and ask that the advertiser sign a consent decree. By agreeing to a consent decree, the advertiser removes the advertising in question but does not admit that the ad was in fact deceptive. More than 90 percent of all FTC cases are settled by consent decrees. A stronger weapon is a cease-and-desist order. In this situation the FTC issues a formal complaint against an ad and a formal hearing is scheduled. If after the hearing it is ruled that the ad was deceptive, a judge issues a cease-and-desist order and the company must stop airing the offending ad.

REGULATIONS

YOU MAKE THE CALL: DECEPTIVE ADVERTISING?

(A)

The manager of a local ad agency wants to introduce a new TV campaign for Very Berry Soda Pop on your station. You watch a few TV ads and all of the ads emphasize the fact that Very Berry contains natural fruit juices. Curious, you ask the ad executive just how much natural fruit juice is in every serving. The executive replies that every eight-ounce serving has exactly six drops of natural juice. The ad executive goes on to say that a version with 10 percent real juice tasted awful so they dropped it but kept the natural fruit juice campaign because market research suggested this was what consumers wanted. In any case, argues the exec, the ads are true; the product does contain real fruit juice. What's your reaction?

(Be careful. Even an ad that is literally true can still be deceptive, and these ads certainly sound deceptive. The FTC would probably consider the effect of the ad as a whole on reasonable people before making a final determination. The odds are pretty good that this campaign is false and misleading.)

(B)

The TV ad campaign for new Follicle Shampoo states that the product will "work miracles for your hair" and make it "look like it's never looked before." Do you think Follicle needs some evidence to substantiate these claims?

(Probably not. These phrases would most likely be looked upon by the FTC as permissible puffery. If the ad had said "Dermatologists say that new Follicle is healthier for your hair than any other shampoo," then some documentation would probably be needed.)

Failure to do so results in a fine. In rare cases, when an ad might have injurious effects on the public, the FTC can seek an injunction to stop quickly the offending ad.

In practical terms, broadcasters and cable system operators should be encouraged to know that very few of the ads aired in any given year are likely to raise legal questions. In addition, since the advertising industry conducts most of its business in public, it is concerned with the harmful effects that might follow the publication or broadcast of any false or misleading ad. Consequently, the advertising profession has developed an elaborate system of self-regulation (see next chapter) that prevents most deceptive ads from being released. Networks also have departments that screen ads before they are accepted. Nonetheless, an occasional problem ad might crop up, which makes a knowledge of FTC rules and regulations helpful.

INTERNATIONAL OBLIGATIONS

American broadcasters also follow the rules of the International Telecommunications Union (ITU). The ITU is a United Nations organization that is responsible for coordinating the broadcasting efforts of its member countries. The ITU tries to minimize interference between stations in different countries, regulates radio spectrum allocations, assigns initial call letters to various countries (U.S. stations begin with W or K, Canadian stations with C, and Mexican stations with X), and works to improve telecommunications services in new and developing countries. The ITU is not a regulatory body like the FCC. It does not license stations; it has only the power that its member nations allow it to have. The ITU's main purposes are to encourage cooperation and efficiency and to serve as a negotiating arena where equitable policy regulation can be worked out.

The supreme authority of the ITU is given the somewhat ungraceful name of the Plenipotentiary Conference. Composed of representatives of all member nations, it meets every five to eight years. The World Administrative Radio Conferences (WARC) are another important part of the ITU. General WARCs, held at twenty-year intervals, discuss and develop global communications policies. The last general WARC enlarged the shortwave broadcasting band and established policies covering satellite communications. The International Frequency Registration Board is a division of the ITU that maintains a master list of frequency use throughout the world. It allows individual countries to use certain frequencies without creating interference.

Over the years the ITU has managed to concern itself largely with technical matters. Recently, however, politics have crept into ITU operations. As more third world nations joined the organization, the role of the ITU was broadened from merely dealing with technical matters to providing technical assistance to those countries that request it. In addition, the most recent plenipotentiary conference of the ITU in 1989 was characterized by disagreements between Arab states and Israel. In 1991 friction arose between the United States and the ITU over the operation of TV Martí, a TV service beamed from the United States to Cuba. The ITU claimed that TV Martí was causing interference on Cuban TV stations. The State Department denied that the U.S. service was causing problems. In any case, involvement of the United States with the ITU is periodically subject to review.

SUMMARY

The Communications Act of 1934, which established the FCC, provides the groundwork for the regulation of broadcasting. Some of the major tasks of the FCC are to assign licenses to new radio and TV stations, to renew licenses, and to decide between competing applicants for the same license. The FCC does not license cable systems but does have some regulatory power over the industry. The Cable Communications Policy Act of 1984 is the most important legislation for cable.

Section 315 of the Communications Act of 1934 provides for equal opportunities for political candidates on broadcasting stations. Bona fide news coverage is an exception to section 315. The controversial fairness doctrine is currently in a state of limbo. Although it was repealed by the FCC, Congress has made new efforts to reinstate it as federal law. When it was in force, the doctrine made stations present balanced programming regarding controversial issues in their communities.

Copyright laws are important in broadcasting and cable. Music licensing is the method by which performers and composers are paid for the use of their work by broadcasters and cable operators. Cable systems also pay a fee for transmitting the signals of distant stations into their local markets. VCR owners can tape a program off the air for their own personal use without violating copyright laws.

Federal laws pertaining to obscenity apply to broadcasting and cable. In addition, the FCC has drafted special provisions that deal with indecent content on radio and TV. The specifics of these provisions are still undergoing legal review.

Federal law dealing with lotteries has been liberalized. Federal restrictions on the advertising of legal lotteries have been removed, but individual states can restrict lottery advertising.

Cablecasters and broadcasters are bound by the regulations set down by the Equal Employment Opportunity Commission. Both the broadcasting and cable industries are also subject to antitrust laws.

Several legal areas touch upon the practice of broadcast journalism: defamation, invasion of privacy, protecting sources, and using cameras and microphones in the courtroom.

Advertising qualifies for protection as free speech under the First Amendment. The FTC is the main agency that regulates false and deceptive advertising.

Finally, American broadcasters follow international rules and regulations as set down by the ITU.

SUGGESTIONS FOR FURTHER READING

Brotman, S. (1987). *The telecommunications deregulation handbook.* Boston: Mass. Artech House.

Carter, T. B., Franklin, M. A., & Wright, J. B. (1986). *The First Amendment and the fifth estate.* Mineola, N.Y.: Foundation Press.

Carter, T. B., Franklin,, M. A., & Wright, J. B. (1988). *First Amendment and the fourth estate.* Westbury, N. Y.: Foundation Press.

Dill, B. (1986). *The journalist's handbook on libel and privacy.* New York: Free Press.

Ferris, C. D., Lloyd, F. W., & Casey, T. J. (1987). *Cable television law.* New York: Matthew Bender.

Francois, W. E. (1986). *Mass media law and regulation.* New York: Wiley.

Holsinger, R. H. (1990). *Media law.* New York: McGraw-Hill.

Middleton, K. R., & Chamberlin, B. F. (1993). *The law of public communication.* New York: Longman.

Powell, J., & Gair, W. (1988). *Public interest and the business of broadcasting.* Westport, Conn.: Quorum Books.

Rowan, F. (1984). *Broadcast fairness.* New York: Longman.

Spitzer, M. (1986). *Seven dirty words and six other stories.* New Haven, Conn.: Yale University Press.

Zuckman, H. L., & Gaynes, M. J. (1983). *Mass communications law.* St. Paul, Minn.: West Publishing.

CHAPTER 19: SELF-REGULATION AND ETHICS

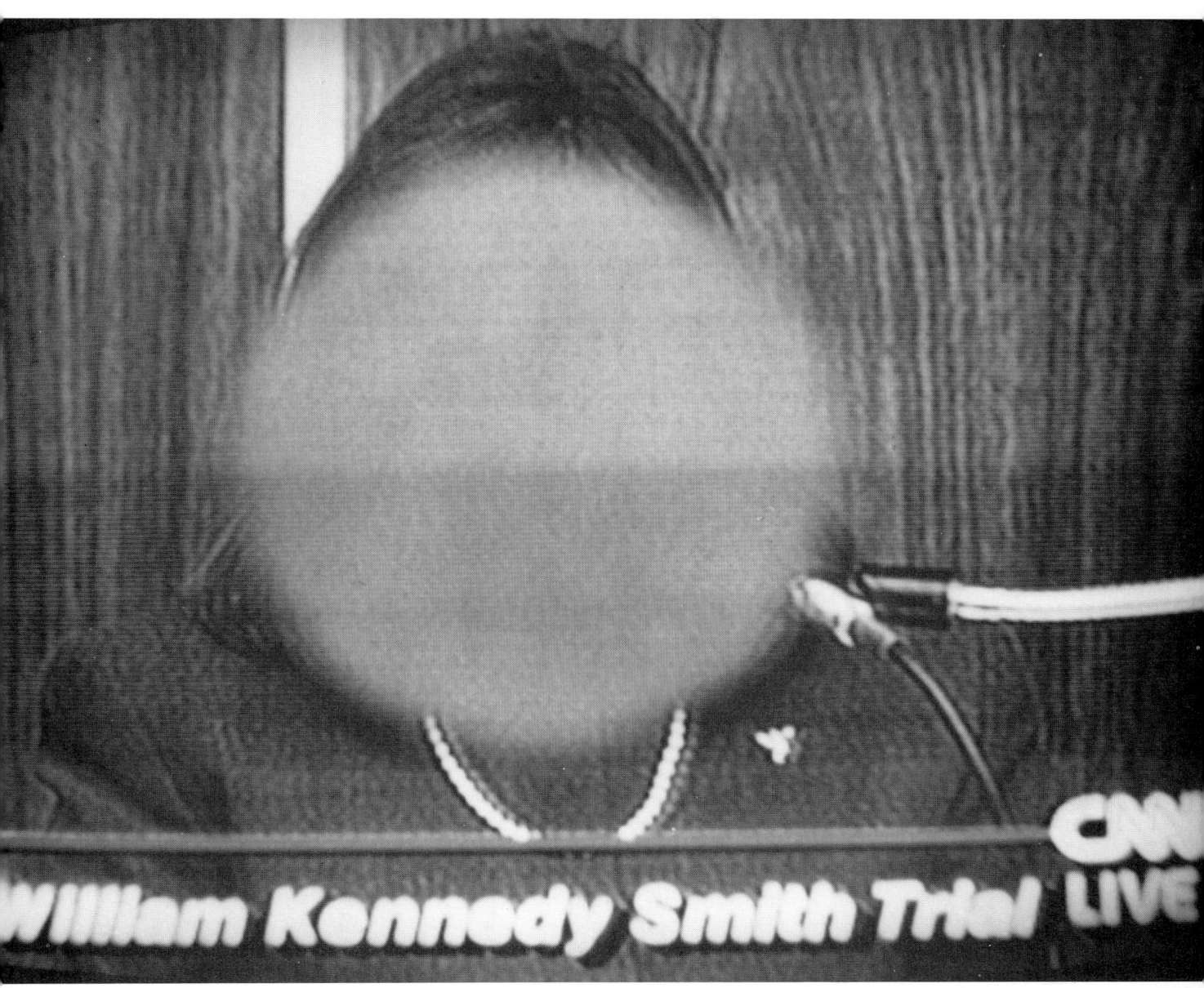

The preceding two chapters dealt with laws regarding broadcasting and cable. There are many situations, however, that are not strictly covered by law or FCC regulations. In these circumstances broadcasters and cable professionals must rely on their own internal standards and/or organizational operating policies for guidance.

Consider the following example. You're the manager of an independent TV station in a medium market in the southwest. Your station has acquired the rights to the all-time cult classic *Texas Chainsaw Massacre*. With the right promotion campaign, you figure that a prime-time showing of the film will get your station its best ratings ever and make a bundle in advertising revenue. But wait, says your program director, this film contains scenes of graphic violence and is likely to be too intense for children. Wouldn't it be smarter to show it after midnight when few children will be in the audience? But wait, says your sales manager, if shown after midnight, the film will not make nearly as much money from the sale of commercial spots. Well, says your program director, if you're going to leave it in prime time, how about cutting out the goriest scenes and flashing "edited for television" on the screen before it starts? But wait, says your sales manager, when people see "edited for television" they get angry and will probably tune out, thus hurting the rating. You remember that when a Florida station showed an edited version of *The Breakfast Club* the switchboard was swamped with protest calls. Well, says your program director, if you're going to show it uncut, at least display a warning that the film might be offensive. What do you do?

In this situation the main consideration is not what is legal but what is the right thing to do. There is no law that says *The Texas Chainsaw Massacre* cannot be shown on TV. This and similar decisions boil down to a consideration of professional standards and personal ethics. In short, it's a matter of self-regulation. A consideration of self-regulation is important because the trend toward fewer formal regulations means that more and more decisions are being left to the discretion of broadcasters and program producers. Accordingly, this chapter focuses on (1) the major forces that influence self-regulation in broadcasting and cable and (2) ethics and its relationship to self-regulation.

SELF-REGULATION IN BROADCASTING AND CABLE

Broadcasters and cable operators use several methods to achieve self-regulation. Some are familiar because they are also factors that influence laws and policies (see Chapter 17); others will be new to you. Specifically, this section examines (1) codes of responsibility and/or good practice, (2) departments devoted to maintaining proper standards, (3) professional groups and organizations, and (4) citizens' groups.

Codes

Codes are written statements of principle that guide the general behavior of those working in a profession. Code statements can be prescriptive—"Thou shalt do this"—or proscriptive—"Thou shalt not do this." At one end of the spectrum they imply the minimum expectations of the profession (journalists should not plagiarize); at the other they embody the ideal way of acting (journalists should tell the truth.) Professional codes are common in medicine, law, pharmacy, and journalism.

Although many group-owned broadcasting stations had strict codes, probably the most famous of all in broadcasting was that developed by the National Association of Broadcasters (NAB). The NAB established the first radio code in 1929, just two years after the FRC was founded. In 1952, as TV was growing, the NAB adopted a code for it, too. By 1980 the two codes had been amended many times to keep up with the changing social climate. Adherence to the code was voluntary and figures from 1980 show that about one-half of all radio stations and two-thirds of all TV stations were code subscribers. The NAB had a Code authority with a staff of about thirty-three people who made sure that stations followed the codes. The only punishment, however, the NAB could dish out to the violator was revocation of the right to display the seal of good practice, a penalty that hardly inspired fear among station owners.

A glance at the codes themselves showed that they covered both programming and advertising. Included in the programming area were such diverse topics as news presentation, political broadcasting, religion, community responsibility, and programming aimed at children. The code provisions were basically innocuous. They suggested special care in the presentation of mental and physical deformities and afflictions, responsible presentation of marriage and relationships, and humane treatment of animals. The advertising section of the codes contained, among other items, guidelines about what products were acceptable, rules for the presentation of broadcast ads, and time standards suggesting limits on the time per hour that should be devoted to commercials.

The time standard provisions ultimately got the codes into trouble with the Department of Justice (see Chapter 17), which claimed in an antitrust suit that broadcasters were keeping the prices of ads high by artificially restricting the amount of commercial time available. The NAB, after a negative court decision, suspended the advertising portion of its code. Later, upon advice of lawyers, the NAB revoked the programming sections as well. By 1983 the codes ceased to exist.

Nonetheless, the codes still have some lingering impact. Although the courts said it was anticompetitive for groups of broadcasters to get together to produce common codes, it was still permissible for individual stations and group-owned stations to maintain their own codes. As a result, many stations still informally follow their provisions; other stations have not, as evidenced by the 20 percent increase in the number of broadcast commercials from 1985 to 1990, the growth of raunch radio, and the continuing controversy over children's TV.

In 1990 the NAB's Executive Committee developed new voluntary programming principles. In order to avoid any of the legal complications that surrounded the original code statements, the NAB's legal department generated the principles and emphasized that there would be no enforcement of these provisions by the NAB or other groups. The new principles were entirely consistent with the philosophy of the original code but were restricted to four key areas: children's TV, indecency and obscenity, violence, and drug abuse. They urged special care in children's programming and that violence should not be portrayed excessively or in a gratuitous manner. Further, the guidelines noted that obscenity was never acceptable for broadcast and that all sexually oriented material should be presented with particular care. Finally, the principles recommended that glamorization of drug use should be avoided.

In addition to these NAB principles, there are other industrywide guidelines that deal with broadcast journalism and advertising.

ECONOMICS

NEWS VERSUS COMMERCIALS

Pay close attention the next few times you watch a local station's local TV newscast. How much consumer reporting is presented? Are any of the stories critical of local stores and businesses that advertise on the local station?

At several stations across the United States consumer reporters have charged that legitimate news stories have been killed because station management feared that the stories would cause local firms to cancel advertising on the station. At one newsroom a consumer reporter planned a story about cars being stolen from a local auto dealer's service lot and what recourse the consumer had if faced with such a situation. The story was scrapped apparently because station executives feared that the negative publicity would discourage auto dealers (big purchasers of TV ad time) from buying spots. Other reporters complained that the names of local businesses that have been the target of many consumer complaints have been deleted from their reports. Some reporters have begun to censor themselves, deliberately avoiding stories likely to cause problems with the sales department. One reporter admitted that she no longer does stories on how to negotiate for the best car price or how to get compensation for a lemon simply because they would never air.

This conflict is always present at TV stations, but during a recession, when stations are after every dollar they can get, the problem is aggravated. The dilemma also illustrates the ethical problems involved in local news: What is the proper balance between news and commerce? So far, at least, it appears that most stations have been deciding in favor of commerce. There are half as many TV consumer reporters on the air now as there were in 1980.

The Radio and Television News Directors Association (RTNDA) has an eleven-article Code of Broadcast News Ethics that covers items ranging from courtroom coverage to privacy invasions. The Society of Professional Journalists (SPJ) also has a code of ethics that covers all media including broadcasting (see box). This code covers such topics as fair play, accuracy, objectivity, and press responsibility. The language of both codes is general and far reaching. From the SPJ code: "The news media must guard against invading a person's right to privacy" and "Truth is our ultimate goal." Before 1988 the SPJ code ended with an interesting paragraph: "Journalists should actively censure and try to prevent violation of these standards." This was an obvious attempt to suggest a means of enforcement, but cases where journalists actually censured other journalists are virtually impossible to find. In any event, the censure provision was dropped from the code.

In the advertising area, the American Advertising Federation and the Association of Better Business Bureaus International have developed a nine-item code that deals with such topics as truth in advertising ("Advertising shall tell the truth . . ."), taste and decency ("Advertising shall be free of statements . . . which are offensive to good taste"), and responsibility ("Advertisers shall be willing to provide substantiation of claims made"). This code has been endorsed by the NAB and many other industry groups.

The absence of a general code of behavior for broadcasting and cable has given station managers and program directors a great deal of discretion in such matters as children's programming, artistic freedom, religious shows, programs devoted to important local issues, and acceptable topics. Management must be sensitive to the political, social, and economic sensibilities of the community; otherwise, a bad decision can cost a radio, TV, or cable organization credibility, trust, and, in the long run, dollars.

To help guard against bad decisions, many stations have developed their own policy guidelines covering sensitive issues. These station policies generally complement and expand the industrywide codes of conduct. Not surprisingly, the broadcast newsroom is the place where a written policy is most often found. A mid-1980s survey noted that about two-thirds of stations surveyed reported a written policy that dealt with problems of broadcast journalism. Another survey of TV stations and production companies revealed that only one in five trained their employees in the ethics of business. There was, however, a sentiment toward more specific guidelines. Fifty percent of the companies surveyed favored the creation of a comprehensive code of ethics.

Written codes, however, are not without controversy. On the one hand, proponents of codes argue that they indicate to the general public that the media organization is sensitive to its ethical duties. Moreover, written codes of conduct ensure that every employee understands what the company views as proper or correct behavior; ethical decisions are not left to the whim of each individual. On the other hand, opponents argue that any company policy statement that covers the entire workings of the organization will have to be worded so vaguely that it will be of little use in specific day-to-day decision making. Further, opponents fear that a written code or policy statement might actually be used against a radio or TV station in court. For example, a private citizen who's suing a TV station for libel might argue that the station was negligent because it failed to follow its own written guidelines.

Many of the company codes that exist today tend to be proscriptive in nature and narrow in focus. Basically, they tend to spell out the minimum expectations for an individual employee. For example, many TV station policies forbid journalists from taking gifts from their news sources or program directors from accepting gifts from program suppliers. Stations also have policies that bar employees from accepting free sports tickets and junkets or from holding outside employment that might conflict with their job at the station.

It's apparent that written guidelines can't do the whole job. In fact, most policy statements and codes fail to address the number one ethical problem revealed by a survey conducted by *Electronic Media*: the conflict between making money and serving the public. Three out of four media executives agreed that the goals of making money and public service were sometimes in conflict. This is not surprising since most policy statements and guidelines are directed at the individual, whereas ethical problems are increasingly originating at the corporate level. For example, many stations and cable systems are owned by groups. How much of the profit made by a local operation should be reinvested back in the station and how much should go to shore up unprofitable operations in other communities or to finance new acquisitions? Should a TV station run sensational stories on its TV newscasts during sweep weeks to inflate its ratings? How much local news should be carried on a radio station if it means losing some money in the process? These are questions that codes seldom deal with.

Departments of Standards and Practices

The major TV networks maintain staffs to make sure that the networks' commercials and programs do not offend advertisers, affiliated stations and cable systems, the audience, and the FCC. In prior years these departments, usually called standards and practices (S&P) or something similar, were large and influential. Their primary role seemed to be that of a moral guardian, shielding the audience from words and portrayals that mainstream society might find embarrassing or offensive. In the 1990s, as network audiences and profits declined, these departments were cut back. Their influence remained strong, however, since a recession made advertising dollars scarce and networks became increasingly reluctant to offend their advertisers. Their mission, though, seemed to change. The networks appeared less interested in guarding morals and more absorbed with the bottom line.

At the same time, the limits of acceptability among viewers have changed dramatically over the years. Back in the 1960s the networks S&P departments decreed that Barbara Eden of "I Dream of Jeannie" couldn't reveal her navel on network TV and that Rob and Laura of "The Dick Van Dyke Show," couldn't be shown in bed together even though they were married. Liberalization of standards began in the early 1970s with such shows as "All in the Family" and "Maude." In addition, cable, with its uncut and uncensored films, conditioned viewers to more permissive standards. By the 1980s network S&P departments were grappling with programs about rape, homosexuality, racial prejudice, and abortion. By the 1990s, however, networks became more cautious about subject matter. Although the restrictions on language, provocative dress, and violence have been relaxed, the broadcast networks have become more vigilant about certain controversial topics. In 1990, for exam-

Many decisions about taste are left to the producers of prime time series rather than to network executives. Some producers, such as those on Fox's "In Living Color," have used this freedom to expand the limits of acceptability.

ple, NBC lost $500,000 when advertisers pulled their ads from an episode of "Law & Order" that dealt with the bombing of an abortion clinic. ABC lost $1 million when sponsors canceled advertising in an episode of "thirtysomething" that focused on a gay relationship. Similarly, NBC asked producers to revamp an episode of "Quantium Leap" that dealt with a gay character in order to head off advertiser defections. Also fearing an advertiser backlash, the producers of "L.A. Law" changed a lesbian character into a bisexual. NBC also toned down the sexually explicit dialogue in the opening scene of "Sisters." As Bruce Paltrow, award-winning creator of such shows as "St. Elsewhere" and "Civil Wars," put it: "What we're seeing is the whitebreading of television." Some producers, aware of what will or will not arouse the ire of the networks, have started practicing what amounts to self-censorship, avoiding those scripts that might conceivably cause acceptability problems.

A network's competitive position also influences how many risks it will take. Fox, for example, in an attempt to catch viewers' attention,

REGULATIONS

SOCIETY OF PROFESSIONAL JOURNALISTS' CODE OF ETHICS

The Society of Professional Journalists, Sigma Delta Chi, believes the duty of journalists is to serve the truth.

We believe the agencies of mass communication are carriers of public discussion and information, acting on their Constitutional mandate and freedom to learn and report the facts.

We believe in public enlightenment as the forerunner of justice, and in our Constitutional role to seek the truth as part of the public's right to know the truth.

We believe those responsibilities carry obligations that require journalists to perform with intelligence, objectivity, accuracy, and fairness.

To these ends, we declare acceptance of the standards of practice here set forth:

RESPONSIBILITY

The public's right to know of events of public importance and interest is the overriding mission of the mass media. The purpose of distributing news and enlightened opinion is to serve the general welfare. Journalists who use their professional status as representatives of the public for selfish or other unworthy motives violate a high trust.

FREEDOM OF THE PRESS

Freedom of the press is to be guarded as an inalienable right of people in a free society. It carries with it the freedom and the responsibility to discuss, question, and challenge actions and utterances of our government and of our public and private institutions. Journalists uphold the right to speak unpopular opinions and the privilege to agree with the majority.

ETHICS

Journalists must be free of obligation to any interest other than the public's right to know the truth.

1 Gifts, favors, free travel, special treatment or privileges can compromise the integrity of journalists and their employers. Nothing of value should be accepted.
2 Secondary employment, political involvement, holding public office, and service in community organizations should be avoided if it compromises the integrity of journalists and their employers. Journalists and their employers should conduct their personal lives in a manner which protects them from conflict of interest, real or apparent. Their responsibilities to the public are paramount. This is the nature of their profession.
3 So-called news communications from private sources should not be published or broadcast without substantiation of their claims to news value.

---

permitted shows like "Married . . . with Children," "In Living Color," "Get a Life," and "The Simpsons" to venture into areas that the three older networks might have avoided.

Even cable networks, which do not have the same amount of advertiser pressure as do their broadcast counterparts, keep a close eye on content. In one of the most publicized actions MTV declined to play Madonna's "Justify My Love" video because of its controversial sexual content. MTV was also hit with complaints after it premiered the Michael Jackson "Black or White" video. An edited version was eventually accepted for airplay.

These conflicts emphasize the uneasy marriage between the creative and the commercial side of TV. Producers and writers want artistic freedom; advertisers want programs that are conducive to

4 Journalists will seek news that serves the public interest, despite the obstacles. They will make constant efforts to assure that the public's business is conducted in public and that public records are open to public inspection.
5 Journalists acknowledge the newsman's ethic of protecting confidential sources of information.
6 Plagiarism is dishonest and is unacceptable.

ACCURACY AND OBJECTIVITY

Good faith with the public is the foundation of all worthy journalism.

1 Truth is our ultimate goal.
2 Objectivity in reporting the news is another goal which serves as the mark of an experienced professional. It is a standard of performance toward which we strive. We honor those who achieve it.
3 There is no excuse for inaccuracies or lack of thoroughness.
4 Newspaper headlines should be fully warranted by the contents of the articles they accompany. Photographs and telecasts should give an accurate picture of an event and not highlight a minor incident out of context.
5 Sound practice makes clear distinction between news reports and expressions of opinion. News reports should be free of opinion or bias and represent all sides of an issue.
6 Partisanship in editorial comment which knowingly departs from the truth violates the spirit of American journalism.
7 Journalists recognize their responsibility for offering informed analysis, comment, and editorial opinion on public events and issues. They accept the obligation to present such material by individuals whose competence, experience, and judgment qualify them for it.
8 Special articles or presentations devoted to advocacy or the writer's own conclusions and interpretations should be labeled as such.

FAIR PLAY

Journalists at all times will show respect for the dignity, privacy, rights, and well-being of people encountered in the course of gathering and presenting the news.

1 The news media should not communicate unofficial charges affecting reputation or moral character without giving the accused a chance to reply.
2 The news media must guard against invading a person's right to privacy.
3 The media should not pander to morbid curiosity about details of vice and crime.
4 It is the duty of news media to make prompt and complete correction of their errors.
5 Journalists should be accountable to the public for their reports and the public should be encouraged to voice its grievances against the media. Open dialogue with our readers, viewers, and listeners should be fostered.

---

sales and networks want to maintain peace between the two. The S&P departments are at the leading edge of this conflict.

The networks also prescreen commercials and reject ads that they find to be too suggestive, too controversial, or in bad taste. There is a little more leeway here since most networks will allow their owned-and-operated stations to accept ads that might be banned at the network level. For example, the three major networks have a long-standing ban against condom advertising but have let their owned-and-operated stations make their own decisions about their acceptability. In 1992 the three networks were reexamining their policies in light of Fox Broadcasting Company's decision to accept condom ads. Standards in commercials have become more liberal over the years, but every once in a while an ad will be

There are fewer content restrictions on material produced for cable, but MTV initially decided that Madonna's "Justify My Love" video was inappropriate for its viewers and declined to carry it unless it was first edited.

rejected by the nets. Such an event happened to a spot for No Excuses jeans starring Marla Maples. In the spot Maples dumped copies of tabloid newspapers into a trash can. The nets objected because the titles of the actual newspapers were seen in the spot. In another example the 1991 movie title *The Pope Must Die* was deemed too controversial. The producers of the film overcame that problem by renaming it *The Pope Must Diet*.

Many cable services run programs that were originally carried by the networks. They can usually be confident that little offensive content is present since the episodes have already been screened. On the other hand, producers of original series for cable generally have more creative freedom. Several series that have appeared on cable networks would be unacceptable to the broadcast networks.

At the local level both broadcasting and cable operations pay close attention to questions of taste and appropriateness but they do not have formalized departments that handle the task. Some group owners of stations have codes that they try to apply to all the stations they own, but most standards and practices decisions are made by managers or program executives. Further, not all decisions will be the same. The acceptability of certain TV or radio messages depends on several factors: (1) the size of the market, (2) the time period, (3) the station's audience, and (4) the type of content involved.

To elaborate, what's acceptable in New York City might not be appropriate for Minot, North Dakota. Standards vary widely from city to city and from region to region. The local radio, TV, or cable executive is usually the best judge of what his or her community will tolerate. For example, several TV stations have a problem with running uncut R-rated films. A Rochester, New York, TV station conducted a viewer poll in its market before making a decision. The result: 23,000 viewers said it was okay; 5200 objected. The station ran the movie without incident. On the other hand, a Kansas City station ran an uncut R film and prompted viewer complaints and possible sanctions from the FCC. Although most markets see nothing offensive in the "Donahue" show, a Denver TV station dropped it because of the program's apparent fascination with topics such as bisexuality, transvestism, and pedophilia.

In cable Playboy at Night hardly raised an eyebrow in some markets but was the target of spirited protests in others. Multiple-system cable operators face a special problem in this regard since their services reach many different types of communities. For consistency and ease of management decisions on what to carry on their systems may be made at a corporate office rather than at the local level. Such an example occurred in the wake of the Jim Bakker/PTL scandal in 1987 when two big cable companies removed "The PTL Club" from all their systems, even though some local managers might have pre-

ECONOMICS

CALL NOW . . . SUPPLIES ARE LIMITED

You have probably seen one—a late-night TV show with a name like "Incredible, Amazing Breakthroughs, Discoveries and Inventions," a format that resembles a news or talk show and a pitch for products such as hand mixers, spot removers, car wax, steam ovens, juicers, exercise equipment, anticellulite treatments, self-improvement tapes, diet plans, skin care products, and so on. Usually thirty minutes long, these programs, which look like legitimate TV shows, are really "infomercials"—program length commercials produced entirely by advertisers whose goal is to get you to call the 800 number on the screen and to order their merchandise.

In years past infomercials might be found only after midnight on marginal stations that are desperately in need of advertising dollars. More established stations shunned them, thinking that infomercials cheapened their image. Now times have changed and infomercials are cropping up everywhere, on quality stations and in all time periods. Why? First, the weak economy has made advertising money a scarce commodity and stations are tapping into sources that they previously might have disregarded. Second, the infomercials themselves have gotten better. They sometimes use celebrities, like John Ritter, Fran Tarkenton, and Ali MacGraw, and have tremendously improved their production values. Whatever the reasons, infomercials are big business, generating almost $1 billion in consumer sales in 1991. Stations like them because they get cash up front and can sell air time (such as after midnight) that would normally not produce much revenue. Merchandisers are impressed with their effectiveness. A thirty-minute infomercial for a hand-held kitchen mixer cost about $125,000 to make and has so far produced $55 million in sales.

Nonetheless, infomercials raise some ethical questions. Are the programs exploiting and deceiving the audience? The formats are designed to make the programs look less like a commercial and more like a straight information show. As a result, do they prey upon unsuspecting viewers? What if the products don't live up to the sometimes hard-to-believe claims made by the participants? Most TV stations apparently don't see many ethical problems. The number of infomercials on TV has increased from about 2500 a month in 1985 to about 21,000 in 1991.

---

ferred otherwise. The biggest problems faced by radio stations these days have to do with the acceptability of song lyrics and the amount of latitude given to so-called "shock jocks."

The time period also has a lot to do with acceptability. Stations tend to be more careful if there's a good chance that children will be in the audience. Programs and films with adult themes are typically scheduled late in the evening when few children are presumed to be listening. Cable movie channels generally schedule their racier films after 10:00 P.M.

The audience attracted by a station's programming is another important factor. Radio stations that feature a talk format with controversial topics usually spark few protests because their listeners know what to expect. Listener-sponsored Pacifica radio stations routinely program material that other stations would heartily avoid. Although they do get into occasional trouble (for example, the *Pacifica* case), the audience for Pacifica stations rarely complains.

Finally, the kind of program also makes a difference. Rough language and shocking pictures are sometimes okay for a news or public affairs program. For example, sexually explicit language was heard during two highly rated news stories in 1991—the confirmation hearings of Supreme Court Judge Clarence Thomas and the William Kennedy Smith rape trial. Similar content in an entertainment program would probably not be allowed.

Professional Groups

Professional groups are trade and industry organizations that offer their members advice about research, technical, legal, and management issues. The best known of these for broadcasters is the National Association of Broadcasters. The NAB has about 5000 radio station and 940 TV station members along with 1500 individual members in the broadcasting industry. We have already examined the lobbying component of the

NAB; its other services include designing public service campaigns, offering legal advice, conducting technology research, and maintaining an information library. In cable, as we have noted, the National Cable Television Association (NCTA) is the most influential group, with a range of services that parallels the NAB's. Other professional groups include the Radio Television News Directors Association, the National Association of Television Program Executives, the National Association of Farm Broadcasters, the Association of Maximum Service Telecasters, the National Association of Public Television Stations, and the National Association of Black Owned Broadcasters, to name a few. A recent issue of *Broadcasting-Cablecasting Yearbook* listed about 175 national associations concerned with cable and broadcasting.

Professional groups contribute to self-regulation in both formal and informal ways. On the informal level, each group sponsors conventions and meetings where members exchange relevant business and professional information. These meetings offer ways for managers from one part of the country to learn how managers from other regions are handling similar problems. A station in Maine might find that a policy or set of guidelines used by a station in Ohio solves its problem. Organizational meetings let professionals get feedback from their peers. Professional organizations also set examples and demonstrate standards of meritorious behavior that members can emulate. For example, recently the NAB participated in public service campaigns for AIDS awareness, safe driving, and against drug abuse.

Now that the NAB codes have been repealed, neither the NAB nor the NCTA is involved with formal procedures of self-regulation. The code enforcement section of the NAB no longer exists and practices of member stations are not monitored. In fact, most professional associations do not monitor subscriber performance. A few professional groups, however, do enforce codes of ethics. The National Religious Broadcasters, for example, recently adopted a code of ethics that requires members to divulge financial details of their operations.

The most elaborate system of formal self-regulation by a professional group deals with national advertising. Basically, the process works like this. If a consumer or a competitor thinks an ad is deceptive, a complaint is filed with the National Advertising Division (NAD) of the Council of Better Business Bureaus. Most complaints are filed by competitors, but the NAD monitors national radio and TV and can initiate action itself. The NAD can either dismiss the complaint or contact the advertiser for additional information that might refute the complaint. If the NAD is not satisfied with the advertiser's response, it can ask the company to change or stop using the ad. If the advertiser agrees (and most do), then that's the end of it. On the other hand, if the advertiser disagrees with the NAD, the case can be taken to the National Advertising Review Board (NARB).

The NARB functions much like a court of appeals. A review panel is appointed, the case is examined again, and the complaint is either upheld or dismissed. If the complaint is upheld, the advertiser is asked once again to change or discontinue the ad. If the advertiser still refuses, the case can be referred to the FTC or another government agency for possible legal action.

One of the problems with this system is that it takes a long time. In many cases, by the time the NAD has weighed the facts of the case, decided that the ad in question was truly deceptive, and notified the company to stop, the advertising campaign has run its course and the ad is already off the air. Nonetheless, although it doesn't have any legal or formal authority, the NAD/NARB system and the implied threat of legal action is taken seriously by most people in the industry.

Lastly, scholarly and academic organizations, such as the Broadcast Education Association and the Association for Education in Journalism and Mass Communication, contribute to self-regulation through their work with students. These and similar organizations urge colleges and universities to emphasize ethical and professional responsibilities in their curricula. Academic organizations also study and help clarify the norms and standards of the profession so that practitioners can have some guidance in making ethical decisions.

Citizens' Groups

In addition to shaping the legal and policy environment of broadcasting (Chapter 17), citizens' groups (or pressure groups) exert a force for self-regulation by communicating directly with broadcasters. In recent years citizens' groups have been most vocal in three areas: (1) portrayal of minorities, (2) presentation of violence, and (3) children's programming.

The concern over the portrayal of minorities began during the civil rights movement of the 1960s. African-Americans correctly noted that few blacks appeared in network prime-time programming and those few that did were usually shown in menial occupations. Their pressure on the networks for more balanced portrayals ultimately led to more important roles for blacks, such as Bill Cosby's in "I Spy" and Diahann Carroll's in "Julia."

The success of these efforts prompted Latinos to campaign successfully for the removal of a commercial character known as the Frito Bandito, which many found offensive. Ethnic stereotyping in a series called "Chico and the Man" was also a target of Mexican-American citizens' groups. Other groups, such as American Indians, Italian-Americans, the Gray Panthers, and Asian-Americans, have also pressured the networks to eliminate stereotyped portrayals. Most recently the National Gay Task Force has lobbied the networks for a balanced portrayal of homosexuals. In 1991 the National Federation of the Blind picketed ABC over the portrayal of a blind character on the short-lived sitcom "Good and Evil." Sensitive to this pressure, many networks and production companies regularly consult with members of appropriate citizens' groups to make sure their portrayals are not offensive.

The portrayal of violence is also a recurring topic of attention for citizens' groups. The National Coalition on Television Violence brings public attention to bear by periodically publishing a list of the most violent programs on broadcast TV and cable. Other groups have gone to more radical means, threatening a boycott of those sponsors who advertise in shows with unacceptable levels of violence. For example, in late 1990 Christian Leaders for Responsible Television (CLeaR-TV) urged a boycott of Burger King because its ads were judged to be supporting TV shows with excessive amounts of sex and violence.

Children's programming is another recurring concern of many citizens' groups, including the national Parent-Teacher Association and the National Education Association. Until it disbanded in 1992 (see box), Action for Children's Television (ACT) was probably the most effective citizen's group in this area.

Mindful of their obligations toward impressionable children, many networks and production companies hire child psychologists as consultants to help them judge the appropriateness of content.

Some citizens' group may not get national exposure but are nonetheless highly influential in a particular community. For example, Japanese-Americans in the Los Angeles area have formed citizens' groups that are active in shaping local broadcasting practices.

Citizens' groups pose special problems for the self-regulatory attempts of the industry. They increase the sensitivity of programmers toward potentially offensive material, but at the same time, they severely restrict the creative freedom of writers and producers. Their cumulative effect thus may lead to a kind of self-censorship. In

PROFILE

ACTION FOR CHILDREN'S TELEVISION

Action for Children's Television (ACT) was founded in 1968 by a group of concerned parents. Under the leadership of Peggy Charren the group became the most influential and effective citizens' group concerned with TV programming directed at children. ACT skillfully brought pressure to bear on both politicians and broadcasters; some of its more notable achievements include the removal of drug and vitamin ads on kids shows, the elimination of host-selling, reduced advertising time in children's TV, and increased prominence for kids shows at the networks. Never content simply to criticize, ACT also presented awards to those programs of most value for the young audience.

ACT disbanded in 1992, in part because of the passage of the 1990 Children's Television Act, legislation strongly supported by ACT. Under the new law broadcasters can lose their licenses for failing to serve the educational needs of children. As Peggy Charren pointed out, "People who want better TV for kids now have the Congress on their side."

The explicit sexual references that occurred during the Supreme Court confirmation hearings of Clarence Thomas raised ethical considerations for broadcast news executives.

addition, program producers are forced to walk a thin line between alternatives. For example, gay citizens' groups are calling for increased portrayals of homosexuals in TV drama. If the producers accede to their demands, they might risk offending a substantial part of their viewing audience. Similarly, networks and production companies are careful not to allow potentially harmful acts of violence to remain in a show, but even seemingly innocuous acts of aggression may still raise the ire of some citizens' groups. Satisfying everybody is not an easy task.

ETHICS

All of us deal regularly with ethical problems. For example, suppose one of your teachers returns your test with a grade of 95 on it. As you look through the test, you realize your teacher made an addition mistake and your real score is 85. Do you tell your teacher? Or suppose the cashier at the bookstore gives you $25 in change when you were supposed to get only $15. Do you return the extra $10? Personal ethical principles are important in the self-regulation of broadcasting and cable since many decision makers rely on them in situations that are not covered by codes or standards and practices. And since the many crucial ethical decisions of broadcasters and cablecasters are open to public inspection and criticism, ethical behavior is critical to those working in radio and TV. This section takes a brief look at the formal study of ethics and then examines how ethics might apply in the day-to-day world of TV and radio.

The Greek word *ethos* originally meant an abode or an accustomed dwelling place, a place where we would feel comfortable. In like manner, the study of ethics can be thought of as helping us make those decisions with which we are comfortable. In formal terms, ethics addresses the question, "Which human actions are morally permissible and which are not?" or, in more familiar terms, "What is the right thing to do?"

Ethics and law are related; both limit human activities. Laws, however, are enforced by sanctions and penalties, whereas ethics are enforced by our moral sense of the proper thing. In many cases the law and ethics overlap—the legally correct action is also the morally correct action. In other cases a decision that's perfectly legal may not be the best decision from an ethical standpoint.

Perhaps a brief and basic tour of the philosophical groundwork that underlies ethics might help.

Ethical Theories

Without getting too bogged down in the technical language of the philosophy of ethics, we first need to define two key concepts: teleological theories and deontological theories. A *teleological* (from the Greek, *teleos*, an end or a result) *theory* of ethics is one that measures the rightness or wrongness of an action in terms of its consequences. To say that it is unethical for a TV station to show a violent film in prime time when a lot of children are watching because some of them might imitate the film and hurt other people is a teleological judgment. A policy that forbids reporters from accepting gifts and consideration from the people they cover because it would hurt their journalistic credibility is another teleogical judgment.

A *deontological* (from the Greek, *deon,* duty) *theory* does not concern itself with consequences; instead, it spells out those duties that are morally required of all of us. The source of these moral obligations can come from reason, society, the supernatural, or human conscience. For example, a policy that forbids investigative TV journalists from assuming a false identity while researching a story because such actions constitute lying, an action forbidden by one of the Ten Commandments, is based on supernatural deontological grounds. Note that it doesn't matter if the consequences of the lying are beneficial to society. A lie is prohibited since it is counter to God's will as expressed in the commandments.

With that as a background, let's examine some specific ethical principles and see how they might relate to broadcasting and cable.

Utilitarianism Probably the most popular and clearest ethical principle is a teleological theory, utilitarianism. *Utilitarianism* holds that a person should act in a way that produces the greatest possible ratio of good over evil. In other words, one should make the decision whose consequences will yield the greatest good for the greatest number (or the least harm for the fewest number). Under this principle the ethical person is one who adds to the goodness and reduces the evil of human life.

A broadcaster who chose to follow utilitarianism as an ethical standard would first have to calculate the probable consequences that would result from each possible course of action (this is not easy; sometimes it's impossible to predict all the consequences of a given decision). Second, the broadcaster would have to assign a positive or negative value to each probable consequence and then choose the alternative that maximizes benefit and minimizes harm.

To illustrate, suppose the sales manager asks the news director (let's assume that's you) at a given station to kill a story about health code violations at a local restaurant. The restaurant is one of the biggest advertisers on your station and has threatened to cancel if you run the story. From the utilitarian perspective you first define your possible actions: (1) run the story, (2) kill the story, and (3) run an edited version of the story without naming the restaurant involved. The consequences of killing the story are easy to predict. Your station continues to get money and stays in business, thus keeping everyone on your staff employed. It might also help the restaurant stay in business, since consumers who are health conscious might have stayed away had the story been aired. On the other hand, failing to run the story or running it minus the name of the offending restaurant might have significant negative effects. First, it might hurt the credibility of your news operation, particularly if other local media identify the restaurant. In the long run this might cause people to stop watching your newscast and hurt your revenues. Second, the morale in your news department might be negatively affected. Third, unwary consumers who eat at the restaurant might have their health jeopardized by the unsanitary conditions. Carrying the full story would help credibility, improve station morale, and warn unsuspecting members of the general public. In short, our cursory utilitarian analysis suggests that it would be better to carry the story.

Egoism Another teleological theory is known as *egoism*. Its basic premise is simple: act in the way that is best for you. Any action is judged to be right or wrong in terms of its consequences for one's self. If decision *A* is more in your self-interest than decision *B*, then decision *A* is right.

Egoism was popularized by writer-philosopher Ayn Rand, who argued that a person should not sacrifice self to others. Rather, a person should have the highest regard and owe the greatest obligation to his own self. (Rand's book *The Fountainhead* is a forceful interpretation of this view. Incidentally, a movie of the same name, starring Gary Cooper, was based on this book, making it probably the only motion picture ever to be based on a formal ethical theory.)

Egoism sounds like an intellectual rationalization for doing whatever you please. But it goes deeper than that. Egoism doesn't preclude kindness or thoughtfulness toward others. In fact, it requires the individual to make a thoughtful and rational analysis of each choice to determine what exactly is best for the individual. Doing what is best for you frequently also entails doing what is best for others. In any case, egoism is an interesting example of an individualistic ethic that many have criticized as being paradoxical and inconsistent (see box).

In our restaurant example the news director goes through the same analysis as discussed earlier but now he or she is not concerned about the consequences for the station, the restaurant, the public, or the news staff. The news director is interested only in what will be best for himself or herself. In this case the news director might well conclude that killing the story would harm his or her reputation as an aggressive, probing journalist and run the story since in the long run it would be beneficial to a news director's career.

Categorical imperative Probably the most famous deontological theory was developed by Immanuel Kant. Our proper duty, not the consequences of our actions, must govern our decisions. Kant argued that our proper course of action must be arrived at through reason and an examination of conscience. How do we recognize our proper duty? By subjecting it to the *categorical imperative*; act only on those principles that you would want to become universal law. This philosophy is close to the famous golden rule in Western cultures. An ethical person should perform only those behaviors that he or she would like all to perform. What is right for one is right for all. In this context, "categorical" means without exception. What's right is right and that's the end of it.

Someone following the categorical imperative might handle the restaurant news story in this manner. Since holding back important information is not something that I would want everybody to do, the story must run. In this instance there is a categorical imperative against squelching the publication of important information. Therefore, it should not be done in this or other situations. Period.

The golden mean Another deontological theory stems from the writings of Aristotle. A natural scientist, Aristotle grounded his theory of ethics in natural law. He noted that too much or too little of something was harmful. A plant would wither if given too little water but would drown if given too much. The proper course of action is somewhere between these two extremes. Moderation, temperance, equilibrium, and harmony are concepts that are important in his philosophy.

When faced with an ethical dilemma, a person using this theory begins by identifying the extremes and then searches to find some balance point or mean that lies between them—the *golden mean*. This may be easy or hard to do, depending on the situation. In the restaurant example one extreme would be to publish everything; the other, to publish nothing. Perhaps a compromise position between the two would be to air the story completely but also to provide time for the restaurant owner to reply to the charges. Perhaps the story might note that these are the first violations in the restaurant's history or include other tempering remarks.

Cultural ethics *Cultural ethics theory*, which is grounded in society as opposed to nature, holds that an individual is shaped by his or her culture. Our moral judgments are shaped by those we experienced when growing up and when entering other socializing environments such as school or work. Cultural ethics suggests that the individual accepts the discipline of society, adjusts to its needs and customs, and finds ethical security within it. There are no universals by which each culture is judged; each is self-legislative, determining its own rules of right and wrong. Moreover, many different subcultures within society may be relevant. An ethical problem encountered on the job might be assessed in the context of the workplace with all of its relevant norms.

In the broadcasting area the practical application of this theory suggests that a station or cable system owner consult the norms of other media or of other businesses or of the general community before making an ethical decision. In the restaurant example the station involved might see how other stations are treating the story or have treated such stories in the past. Similarly, the way that other media, such as newspapers, handle the situation should be noted. Or the station might consult the managers of other stations throughout the area to get advice about the proper course to follow.

Situational ethics The theory of situational ethics also examines ethics in relative terms. Unlike

DEVELOPMENTS IN TECHNOLOGY

VIDEO TOASTER

So-named because it will make TV editing as easy as making toast, the video toaster is an inexpensive and user-friendly computerized editing and special effects system for television. ▶

PANASONIC RADIO RECEIVER

Car radio meets computer: The Panasonic "ID Logic" car stereo stores the frequency, call letters, location, and programming format for more than 10,000 radio stations in the U.S. The listener can push one of the six format buttons on the bottom of the radio—classical, country-western, rock, jazz, easy, and talk—and the radio will automatically find the six strongest stations in that format anywhere in the U.S. ▼

HDTV SPLIT SCREEN ▲

A comparison of an HDTV picture (left) with current TV. The FCC has set 1993 as a target date for selecting technical standards for HDTV. One proposal allocates new channels in the UHF spectrum for HDTV and proposes a 15-year phasing in of the new system.

DIRECT TV BROADCASTING ▲

Those large and ungainly backyard satellite dish antennas may be replaced by this 18-inch dish mounted outside the window. This new system picks up signals broadcast directly from a satellite and sends them to the family TV. Many big companies are betting that direct broadcasting by satellite (DBS) will be a strong competitor to cable TV.

8MM CAMCORDER

A TV camera in the palm of your hand. Fisher's new 8mm camcorder sports a horizontal design making it easier to hold. About 25 percent of U.S. homes had camcorders in 1992.

INTERACTIVE TV ▲

Interactive television has many applications in the workplace and in the classroom. These children use a keypad to communicate their choices to Teacher Bear—actually an interactive videodisc.

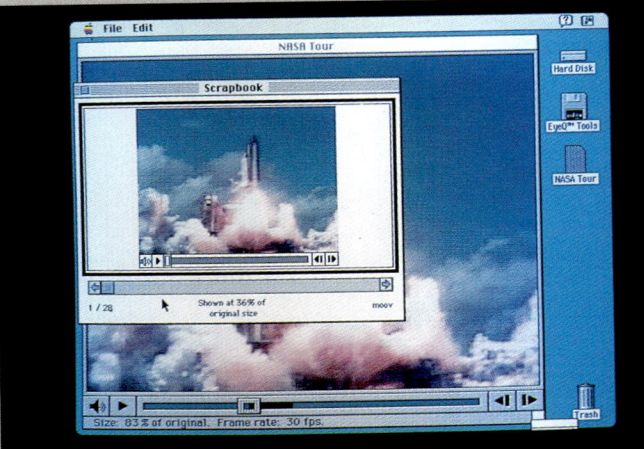

MULTIMEDIA ▲

Multimedia systems blend the computer, TV set, compact video discs, and telephone into one system. Apple's system above shows how a video frame can be frozen, reduced, and stored in the computer's memory.

VIRTUAL REALITY ▲
Reality as we know it? Virtual reality systems put the viewer inside a computer-generated world. The person shown above is playing a two-person virtual reality game in which the participants try to ambush each other. The joystick and data glove allow them to move about the computer environment and "fire" at one another.

cultural ethics, however, *situational ethics*, or *situationism*, argues that the traditions and norms of society provide inadequate guidance because each individual problem or situation is unique and calls for a creative solution. Decisions regarding how we ought to act must be grounded in the concrete details that make up each circumstance. Universal ethical principles like the categorical imperative might provide some guidance, but they are merely hints as to the correct course; they do not apply in every situation. A general principle that we endorse, such as "TV newscasters should always tell the truth," can be violated if the situation calls for it (such as broadcasting a false story to gain the release of hostages).

Of course, the way we see and analyze situations varies from person to person. No two individuals will define the same situation in exactly the same terms. How we perceive things depends on our moral upbringing and our ethical values. Consequently, different answers are possible depending on how one assesses the situation.

In our restaurant example the entire situation must be examined before making a decision. Was the Board of Health on a vendetta against this particular restaurant? Has the restaurant recently changed management? How much money will actually be lost if the restaurant cancels its advertising? Could this revenue be replaced from another source? Have all the violations been corrected? These are some aspects of the situation that might be considered before a decision is made. From a situationist point of view there is no one "correct" answer to this problem.

"Doing" Ethics

A knowledge of ethical principles is useful for all those entering the radio and TV profession. First, it is important to have some predefined standard in place when a quick and difficult ethical call is required. Second, a personal code of ethics assures an individual of some measure of consistency from one decision to the next. Third, media professionals are often asked to explain or defend their decisions to the public. A knowledge of ethical principles and the techniques of moral reasoning can give those explanations more credibility and validity. Fourth, and most pragmatically, good ethics and good business generally go hand in hand, at least in the long run. Although short-term economic gains may result from making ethical compromises, one study done in the late 1970s showed that companies that ranked high in ethics also tended to rank high in growth and in earnings per share.

BACKGROUND

THE PARADOX OF EGOISM

Suppose that you are an advocate of the ethical position known as egoism. You believe that a person should always act in a way that promotes that person's own self-interest. Let's further suppose that you're a salesperson for a local radio station. You have been working for months to sell a large package of ads to a big company in your town. Your colleague comes to you one day to chat.

COLLEAGUE: I have an ethical problem.
YOU: Act in a manner that promotes your own self-interest.
COLLEAGUE: I'm glad you said that. My brother-in-law knows the guy at that big company you've been trying to sell. My brother-in-law says that the guy trusts him and will buy advertising from anyone my brother-in-law recommends. I know how long you've been working on this one and I was going to get my brother-in-law to recommend you. But since you told me to act in my own self-interest, I'll get myself recommended, make the sale, and pocket a big commission. I knew you'd understand.
YOU: Something is wrong here.

In this case recommending egoism to someone else contradicts your own self-interest. It's not to your advantage to have everybody else pursue their own advantage. You're in the interesting position of having an ethical stance that you can practice yourself but cannot recommend to others.

Ethical problems at the corporate level are exemplified by ads such as this one for a "dating line." Ads such as these earn money for TV stations but raise questions about exploitation.

How can a person develop a set of ethical standards that will be relevant and useful in day-to-day situations? If you have read this far, you already have gotten a head start. The first step is to become familiar with enduring ethical principles, such as those just enumerated, or others that may be germane to your profession. For example, if the classical ethical theories do not seem helpful, you might want to consider the set of ethical principles proposed by Edmund Lambeth in *Committed Journalism*. Although designed for working journalists, his principles have relevance for everyone employed in broadcasting and cable. Lambeth suggests that an individual's decisions should adhere to the following principles:

1 Tell the truth.
2 Behave justly.
3 Respect independence and freedom.
4 Act humanely.
5 Behave responsibly.

Another model for journalists endorsed by many in the profession has seven principles or moral duties:

1 Don't cause harm.
2 Keep all promises.
3 Make up for previous wrongful acts.
4 Act in a just manner.
5 Improve your own virtue.
6 Be good to people who have been good to you.
7 Try to make the world better.

It matters little if you subscribe to Aristotle's golden mean or principles suggested by Lambeth; the point is that you need to develop some underlying set of ethical principles.

Next you need to develop a model or plan of action that serves as a guide in making ethical choices. When a person is confronted with an ethical dilemma, the model serves as a blueprint for thinking. One possible model might be the following:

Stage one. Determine the situation. Compile all the facts and information that are relevant to the situation. Learn all you can about the circumstances that prompted the problem.

Stage two. Examine and clarify all the possible alternatives. Make sure you are aware of all your possible courses of action. (If you subscribe to a teleological theory of ethics, you may also have to assess the possible consequences of each action.)

Stage three. Determine what ethical theories or principles you will follow in the situation. Try to develop some consistency in the choice of theories or principles over time.

Stage four. Decide and act accordingly.

Ethics in the Real World

We have now seen some theories and models promulgated by philosophers who have the luxury of time to reflect on their various ramifications. In the real world of TV and radio, however, rarely do professionals have the time to reflect on their decisions. For example, in New York City a convict with a gun threatened to kill hostages if his demands weren't broadcast immediately by a local TV station. The police on the scene urged the station manager to grant this request. The station involved in this dilemma had about two minutes to decide what to do. That's probably not enough time to do a thorough utilitarian or Kantian analysis of the situation. (In real life the TV station agreed to the request and bloodshed was averted. Subsequently, however, the station was soundly criticized by local newspapers for giving a forum to a gunman.)

Moreover, many broadcasters and cablecasters enter their profession without a personal code of ethics. Academic courses in media ethics became popular only in the early 1980s. Many current media executives were trained in selling, programming, or news but had to pick up whatever ethical standards they have informally along the way. In fact, many executives would probably be hard-pressed to articulate exactly where and how they developed their ethical standards.

Third, although some broadcast and cable organizations have codes of conduct or good practice, they are generally written in broad, general terms and may have little relevance to the individual. Companies rarely conduct education and training sessions in ethics. In a 1987 survey done by *Electronic Media*, only 19 percent of a sample of 144 reported their company conducted training sessions concerning ethics. A new employee who expects the company to provide him or her with a ready-made set of ethics will probably be disappointed.

Finally, recent experience suggests that many ethical problems exist at the corporate level as well as at the individual level. The view that the only responsibility of business is to make a profit has come under attack by those who argue that since corporations have such a significant impact on the environment, their surrounding communities, and the individual consumer, they have an ethical obligation to serve the needs of the larger society. It is not surprising, then, that the number one ethical problem that was cited in the *Electronic Media* survey mentioned earlier was balancing corporate profits against public service. This suggests that employees of broadcasting and cable companies will be faced with ethical decisions throughout their careers. New employees will probably have to make choices about their own personal behaviors while veterans who have risen to the managerial level will be making decisions regarding proper corporate policy. Personal codes of ethics must also accommodate the corporate situation.

What were some common ethical problems uncovered by *Electronic Media*'s survey? These were: sales clients trying to manipulate the news department, devoting news time to wealthy friends of the station manager, inflating the ratings during sweep weeks, deciding whether or not to report a supervisor who was apparently taking payola, misrepresenting the station's ratings and performance to potential advertisers, and running questionable religious programming that emphasized raising money. In addition, TV and radio news operations typically confront the same type of ethical problems faced by their print counterparts: privacy, deception, conflicts of interest, and confidential sources. A couple of other interesting facts were turned up by the survey. More than two-thirds of the executives felt that ethical standards in TV were lower now than they were ten years ago and about the same percentage indicated that some generally accepted business practices in their industry were unethical. The survey also highlighted the importance of cultural norms in guiding the ethics of the industry. When faced with a tough ethical question, almost one-half of the media professionals indicated that they would turn to a working colleague for advice.

In sum, a theoretical knowledge of ethical principles and analysis is helpful, but it must be tempered by an awareness of the nature of the day-to-day pressures in TV and radio.

Ethics: A Final Word

Many texts on media ethics present case studies that raise ethical problems for discussion. Students spend much time debating alternatives and find that there may be little agreement on the proper choice. Different ethical principles do indeed suggest different courses of action. Many students become frustrated by this and look upon ethics as simply a set of mental games that can be used to rationalize almost any decision. This frustration is understandable. It helps, however, to point out that ethical problems are not like algebra problems. There is no one answer that all will come up with. Ethical theories are not like a computer; they do not print out correct answers with mechanical precision. Abstract ethical rules cannot anticipate all possible situations. They necessarily leave room for personal judgment.

Ethical decision making is more an art than a science. The ethical principles and models presented here represent different perspectives from which ethical problems can be viewed. They encourage careful analysis and thoughtful, systematic reflection before making a choice. A system of ethics must be flexible, and it ought to do more than merely rationalize the personal preferences of the person making the choice. As Edmund Lambeth says in *Committed Journalism*, an ethical system "must have bite and give direction." In sum, this brief discussion of ethics is not intended to give everybody the answers. It was designed to present some enduring principles, to encourage a reasoned approach to ethical problems, and to foster a basic concern for ethical issues, recognizing all the while that many ethical decisions have to be made in an uncertain world amid unclear circumstances with imperfect knowledge.

SUMMARY

Personal ethics have become an increasingly important issue for broadcasters as the competitive atmosphere, fostered by deregulation, heightens. Station managers now must be accountable for regulating themselves. Codes, departments, professional groups and organizations, and citizens' groups all help promote responsibility.

The acceptability of a message depends on the size of a market, the time period, the station's audience, and the type of content involved.

Ethics and law share common threads. Both are restrictive measures. The difference lies in the fact that one is enforced by the state, whereas the other is enforced by personal judgment.

There are numerous ethical theories that attempt to explain how a person determines right from wrong. Some major theories are utilitarianism, egoism, the categorical imperative, the golden mean, cultural ethics, and situational ethics.

SUGGESTIONS FOR FURTHER READING

Bittner, J. (1982). *Broadcast law and regulation.* Englewood Cliffs, N.J.: Prentice-Hall.

Blum, R., & Lindheim, R. (1987). *Primetime.* Boston, Mass.: Focal Press.

Christians, C., Rotzoll, K., & Fackler, M. (1991). *Media ethics* (3rd ed.). New York: Longman.

Conover, C. (1967). *Personal ethics in an impersonal world.* Philadelphia: Westminster Press.

Davis, L. (1987, July/August). Looser, yes, but still the deans of discipline. *Channels,* pp. 32–39.

Ethical dilemmas. (1988, February 29). *Electronic Media,* p. 1.

Fink, C. (1987). *Media ethics.* New York: McGraw-Hill.

Frankena, W. (1973). *Ethics.* Englewood Cliffs, N.J: Prentice-Hall.

Goodwin, H. (1987). *Groping for ethics in journalism* (2nd ed.). Ames, Iowa: Iowa State University Press.

Klaidman, S., & Beauchamp, T. (1987). *The virtuous journalist.* New York: Oxford University Press.

Lambeth, E. (1992). *Committed journalism.* Bloomington, Ind.: Indiana University Press.

Merrill, J., & Lowenstein, R. (1971). *Media, messages and men.* New York: David McKay.

Merrill, J., & Odell, S. (1983). *Philosophy and journalism.* New York: Longman.

Meyer, P. (1987). *Ethical journalism.* New York: Longman.

National Association of Broadcasters. (n.d.). *This is the NAB.* Washington, D.C.: National Association of Broadcasters.

National Cable Television Association. (n.d.). *Fact sheet.* Washington, D.C.: National Cable Television Association.

Phelan, J. (1980). *Disenchantment: Meaning and morality in media.* New York: Hastings House.

PART EIGHT

WHERE IT'S GOING

CHAPTER 20: THE FUTURE

One difficulty in predicting what the electronic media will be like in the future is that it is easy to succumb to the notion of "technological determinism." Simply put, this concept means that technology is the driving force behind political, cultural, social, and economic change. In reality, however, the simple fact that we have the technology that is sufficient to cause change doesn't necessarily mean that change will happen. In the future, as in the past, it will be people, not the machines, who will be the driving force of change. Unfortunately, people are quite unpredictable, so what follows in this chapter is a "best guess" as to what the next two or three decades might bring. The chapter starts with a look at the probable industry outlook in the future, moves on to an examination of three emerging communication technologies, and closes with an examination of the social concerns most likely to be faced.

THE INDUSTRY OUTLOOK

The last decade has seen fundamental change in the ownership and operations of electronic media businesses. Like much of American enterprise, the late 1980s and early 1990s were characterized by a flurry of economic activity, including deregulation, "mergermania," and the shock effects of a downturn in the economy.

Things have slowed down enough now to begin to predict the shape of the media industry in the foreseeable future. The next two decades are likely to be marked by three related trends: *vertical integration, globalization,* and *privatization.*

Vertical Integration

All businesses, including the media, have four major components: product development, production, distribution, and consumption. Somebody must invent a good, produce it, and deliver it to retailers, and there must be someone willing to purchase and use it. For the first seventy-five years of its existence, most media companies were prohibited from participating in each phase. Fearing a monopoly of information (first, by the major movie studios; later, by the three TV networks), the FCC and other regulatory bodies forbade such vertical integration.

However, the stampede of deregulation and the economies of scale needed to compete in a growing worldwide market (see below) have led to the growth of diversified media companies involved in all phases of media businesses. For example, media giant Time-Warner is involved in product development (including the new Court TV and Comedy Central cable networks), production (movie studios, TV production centers like Lorimar), distribution (it owns ATC cable, among other systems), and consumption (Time-Life books and videos, to name a few). The trend toward vertical integration is likely to continue and to trickle down from the major media conglomerates to the local level.

Globalization

One major element fueling the fire of vertical integration is the globalization of electronic media. Thirty years after media theorist Marshall McLuhan predicted that electronic media would unite the world as a "global village," his prophesy is coming true.

To find evidence of this trend, one doesn't have to look further than Hollywood, once the "all-American city." Today such studio stalwarts as Columbia, Universal, MGM, and Fox are owned by international media firms (SONY, Matsushita, Paretti, and Rupert Murdoch's News Corporation, respectively). The West German firm Bertelsmann A.G. is the parent of RCA records, among other labels. The British firm Thorn-EMI controls labels like Capitol, and Matsushita (MCA) is the parent of Motown and Geffen. The Dutch company Philips is the parent of Polygram.

The media environment of the next century will undoubtedly be marked by an increasingly international character. Remaining American-owned firms (like Time-Warner, Paramount, Capital Cities/ABC, CBS, and GE/RCA) will have to globalize to remain competitive. Students anticipating a career in the mass media would do well to take coursework in international business, plus a foreign language (Japanese, German, or Spanish would be a good choice).

Privatization

Probably the major reason for the new globalism is the changing character of media consumers. Perhaps you have heard President George Bush and others proclaim a "new world order." The end of the Cold War, marked by the reunification of Germany, the demise of the Soviet Union, and the potential democratization of China have very specific implications for the electronic media.

As we have documented in Chapter 8, until recently in much of the world the media were controlled by the parties in power. This was particularly true in socialist countries, where the press was dominated by the ideology of a strong central government.

Today many of these governments have been abolished; others are in disarray. The new democracies of central and eastern Europe, even the nondemocratic leadership of China, are looking for new ways to fund and operate their domestic media. Many are looking to the advertiser-supported, commercial model developed in the West, especially in the United States.

The reasons for this trend are clear. As the research literature shows, most people do not want ideology and propaganda from their media. Rather, they seek entertainment and reliable information. Think of the number of East Germans who crossed the border to buy VCRs; or the youth of Moscow who turned out for heavy-metal concerts by the Scorpions and Guns N' Roses; or the phenomenon known as *karioke,* where Japanese youth jump onstage in bars and nightclubs to lip-synch to American pop tunes; or the appeal of CNN to citizens and governments worldwide.

The mass media businesses of the coming decades will be increasingly owned by private companies, funded through some form of advertising. This presents great opportunity to students of American media (like yourselves). If we have any expertise at all to export worldwide, it's how to produce commercial radio and TV!

EMERGING TECHNOLOGIES

Multimedia

Imagine a single machine that's a combination TV set, computer, telephone, and CD player and you have the essence of the multimedia concept. At its simplest, multimedia combines audio, still and moving images, graphics, and text into a system that allows the user to call up specific information and to customize it in a way that fits the user's needs.

Some examples show the potential for such a system. One multimedia setup manufactured by Philips hooks up to a conventional TV and stereo system and uses an interactive video disc. Viewers can watch an interactive "Sesame Street" that lets them tour Bert and Ernie's apartment and, with the help of joystick, examine objects in the room. Point to a book and Ernie will read it aloud. Point to a picture of a farm and Bert will explain how crops are grown, using a video of tractors and harvesters to illustrate his point.

Another system provides a spreadsheet program that allows the user to make voice-over comments and explanations and even to include a picture or a video sequence to emphasize a point. Microsoft has published a special software reference library for multimedia systems that contains an encyclopedia, complete with many animated illustrations, a world atlas, and a dictionary that pronounces the words you look up. Other examples of available software for multimedia machines include *Treasures of the Smithsonian* in which you can visit 150 exhibits complete with sound effects and musical accompaniment; *Multimedia Beethoven,* which contains pictures, text, and graphics along with the music of the entire *Ninth Symphony;* and a program from Kodak that allows you to edit video from your VCR, add titles, music, graphics, and commentary to your home videos.

Although the potential for multimedia is large, many of its uses seem to capitalize on the system's educational or business capacities. Will multimedia have an impact on the informational and entertainment functions performed by conventional TV and radio? This area is still being developed by most experts who see a blend between conventional electronic media and multimedia. In the not so distant future, for example, a multimedia computer could be programmed with your personal tastes in entertainment and news. While you were out working or playing, the computer would scan the offerings of the 150 or so cable channels available to you for content that fits your personal menu, record the material, and prepare a master list to choose from when you were ready to view.

Some developmental problems face multimedia systems. First, they tend to be somewhat expensive for home users. The hardware costs about $2000 to $3000 and the software can range anywhere from $200 to $1000. Second, there is the compatibility difficulty. Software designed for one system will not necessarily run on another system. Industrywide standards have yet to be developed. Other skeptics note that the consumer demand for multimedia has not been explored. Some see it as a solution for which there is no problem.

Nonetheless, the big computer manufacturers are betting heavily on multimedia. In 1991 fierce rivals, IBM and Apple, announced a deal in which they would cooperate to produce computer products. A key part of that agreement was their intention to develop a compatible multimedia technology.

VIRTUAL REALITY

Back in the 1960s, when computers were huge contraptions that filled up entire rooms, the thought that someday people would actually wear one would have been preposterous. With virtual reality (VR), however, that is exactly what a person does. In technical terms, VR is the name given to a computer interface that enables users to move through, and interact with, computer-generated images in three dimensions. To experience VR a person must wear a pair of goggles that are actually two liquid crystal color computer screens and a pair of pressure-sensitive gloves that detect hand motion and position. Since the two viewing screens send slightly different images to each eye, the result is a three-dimensional picture that puts the person in the middle of the scene. The sensation that most first-timers

TECHNOLOGY

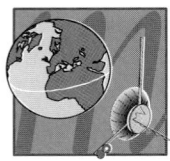

TECHNOPHOBIA

Vexed by your VCR? Confused by your computer? Mixed up by your microwave? Confounded by your camcorder? Perplexed by your programmable phone? Dumbfounded by your digital watch? Don't feel bad. Along with thousands of others, you're suffering from the malady of the 1990s: technophobia, the fear of any device that comes with a bunch of buttons and a long instruction manual.

Psychologists say technophobia comes from a deep-seated fear of being considered inadequate. True technophobes are afraid they will blow up their new computer or melt the microwave if they make a mistake. They fear disapproval or criticism if they do something that makes them look bad in the eyes of their technologically more advanced friends.

Technophobes are everywhere. Some people are afraid to change the film in their advanced cameras. They have them loaded and unloaded at the camera shop. Some people make sure they have the salesperson set the time on their new digital watch before they leave the store. Some people never use the memory buttons on their new phones. Others use only the defrost setting in their microwaves.

The one device, however, that prompts the most technophobic anxiety is the VCR. Many people are afraid to hook them up. Others aren't quite sure what they can do. The parents of one of the authors, after two years of VCR ownership, finally realized that they could tape programs off the TV set and play them back later. They had thought VCRs could only play movies rented at the local video store. Other people tape and play back regularly but aren't able to set the VCR timer for automatic recording. The biggest problem by far, however, is trying to get the VCR clock to quit blinking. The blinking 12:00 drives a lot of people crazy—and try as they may, they can't get it to stop. One clever Californian placed a classified ad that began "Flashing VCR?" and offered to come by and set the clock—for a small fee. One couple figured out a less costly, if somewhat primitive, solution. They simply took a piece of masking tape and covered the clock.

experience when wearing the VR gear is one of "My gosh! I'm inside this picture." The view shifts as the individual turns his or her head just as it would if the person were looking at a real scene, hence the term "virtual reality."

The potential of VR is mind-boggling. Consider what some organizations have already done with this technology:

- NASA has used VR systems to mimic space travel. Astronauts are put in the middle of lunar or other planetary terrain to help with mission planning.
- Architects are experimenting with VR to allow them to "tour" virtual buildings under design to look for flaws in the layout.
- Physicians use VR as a learning tool. Apple Computer is developing a VR system that would allow new physicians to practice surgery on virtual patients.
- Researchers at Boeing are designing virtual airplanes. By taking the plane on a virtual flight, engineers are able to tell if there are any bugs in the design.
- Matsushita in Japan has opened a virtual kitchen showroom. Customers don VR headgear and "tour" a custom-designed virtual kitchen in which they can actually pick up virtual dishes, turn on virtual faucets, and open virtual cabinets. If they find something that they don't like, the plans are redone before the actual kitchen is installed.

As far as the media are concerned, however, the educational and entertainment potential of VR is of most interest.

On the educational side, programmers are at work developing a library of Shakespeare in VR. Students would be able to walk back and forth among the characters (sort of like an invisible man or woman) and watch them perform. (Imagine standing next to Hamlet while he mused about "To be or not to be . . ."). Other researchers are working on ways to make students part of a VR algebra world. A student would actually become the "$2x$" in the equation $2x + y = 961$ and see how various solutions bring the equation into balance. Microbiology students could actually walk

It may not make your hair stand on end as it does for the model in this picture but virtual reality will probably have many applications in the future.

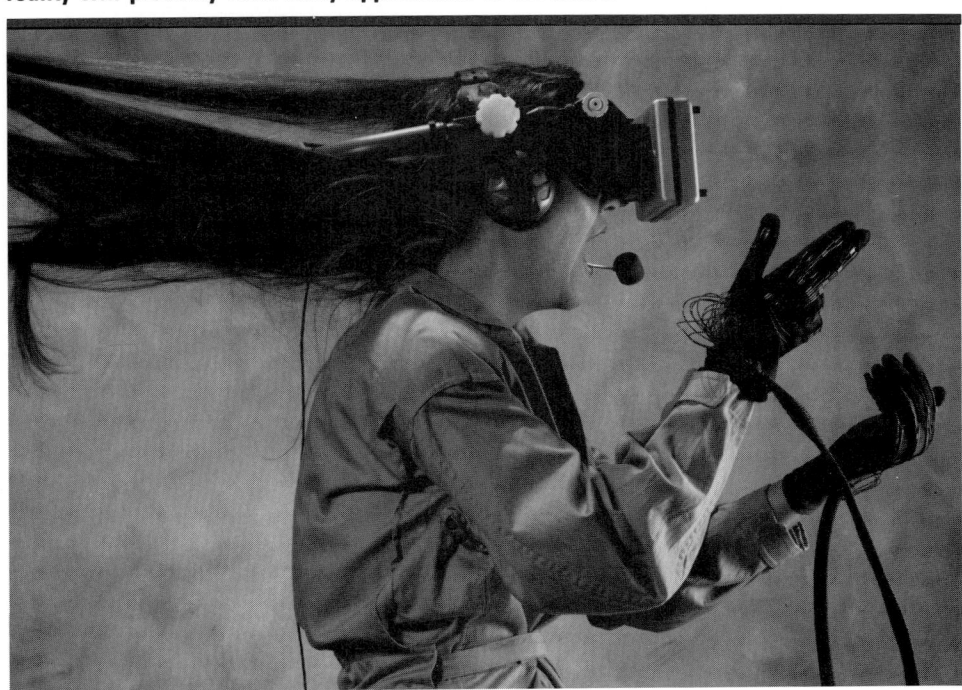

through a VR cell. All of this sounds far more exciting than reading books and going to class.

On the entertainment side, many experts predict that VR will compete with TV. One obvious application of VR would be video games. In fact, a British company has already developed prototypes that are suitable for arcade play. One game, "Battlesphere" is an interactive game in which players actually see each other (more precisely, they see computer-generated images of each other) in the virtual world. The cost of playing one episode of Battlesphere (and most episodes last under one minute) was about $2.00. A division of the big entertainment conglomerate, MCA, is working on plans for a series of VR test theaters. Rather than simply watching a movie, VR viewers would actually take part in it.

Keep in mind that VR in the 1990s is about at the same point in development that TV was during the 1930s. A lot of kinks have to be worked out before VR becomes a mass medium. One problem is the cost. The computer screen goggles cost about $10,000, the sense-gloves cost about $9000, and a person would need about $250,000 worth of hardware and software to run the system. Nonetheless, research is underway to bring down the costs into a range that more people could afford. A second problem is picture quality. Many of the images in VR have a rather cartoonlike look to them, a problem that can be overcome with bigger and faster computers. Most experts, however, suggest that it will take another ten to twenty years before the level of quality of the images in VR approach those of photographs.

In any case, given the tremendous potential and wide range of applications of VR, more sophisticated and powerful versions will be available in the future. It's a virtual certainty!

HOLOGRAPHY

Although still several years off, the next quantum jump in electronic media technology might well be *holography*. The technique of holography creates *holograms*—laser-generated three-dimensional images that appear to hang in the air. (Most people have seen primitive versions of holograms. They are routinely used on credit cards as protection against counterfeiting.) If the projection system is sophisticated enough, a viewer can inspect a hologram from any angle just as if the image was the object itself.

To construct a hologram, one bathes an object in laser light. The laser light reflected from the object strikes a photographic film at about the same time as another beam of laser light that shines directly at the film. The light waves coming from these two sources reinforce and cancel each other out to form what is known as an interference pattern. A close-up look at the pattern on the film would reveal an incomprehensible pattern of loops, whorls, and bulls-eyes. When this film is properly illuminated, however, the process that created the interference pattern is reversed and the pattern recreates the actual light waves reflected from the object, producing an image, which, like the original object, is three-dimensional.

Some holograms are reproduced behind the film, making the hologram look like a window through which the objects are viewed. Others appear in front of the film, jutting out toward the viewer. No matter where they appear, the viewer is almost compelled to touch the image—a frustrating experience since a hologram exists only in the optical nerve and brain, and fingers simply pass through empty space.

The principal research into holography is currently being conducted at the Media Lab at MIT. There researchers are working on holographic TV, a system that would project a three-dimensional image into the middle of a viewing room. Although theoretically possible, the problems are immense. For example, in order to transmit the enormous amount of data necessary for holographic TV (HTV?), a person would need more than the entire block of electromagnetic spectrum space now devoted to all of TV broadcasting. Nonetheless, researchers are making breakthroughs. In 1991, scientists at the Massachusetts Institute of Technology (MIT) announced that they could now make "synthetic" holograms, images produced purely from computer data. Although it may take some time, specialists in holography predict that the day will come when Americans will turn on their HTV and watch twenty-two 6-inch-tall figures play the Super Bowl on the living room rug.

SOCIAL CONCERNS

So far all of this sounds like fun but there's a downside as well. All this new hardware may continue to increase the distance between the media-haves and the media-have-nots. New technologies slowly filter down to lower socio-economic groups. The people who could most benefit from easy access to information are usually the last to get it because the rosy picture painted above is an expensive one. For example, multimedia computers that teach preschoolers how to read and how to count will be a tremendous education aid. They will be expensive, however, and perhaps only available to advantaged children. When these children enter school the gap between their skills and those of disadvantaged children may be even greater than it is today.

Most of the services mentioned will cost the consumer. The more money a family has to devote to entertainment and informational services, the more that will be available to them. At some point in the development of new technologies there will have to be a debate over the social issue of how to distribute equitably the informational wealth that these new technologies will bring.

With information comes political enfranchisement. The problem will not be whether a family has only thirty channels to watch or 300 but whether an individual has the ability to gather and control information. People who have little access to the electronic data bases will have less power than those who do.

Then there is the problem of information overload. Information overload hardly needs to be defined to college students who are studying nine or ten different subjects a year. The term generally refers to the feeling of confusion and helplessness that occurs when a person is confronted with too much stimulation in a limited time period. With 200 or more TV channels, video games, multimedia computers, and virtual reality systems, the consumer of the future will have more information available than he or she could ever use. Let's just take one isolated example. Suppose the house of the future has 200 TV channels available. Let's further suppose that the viewer decides to flip rapidly from one channel to the

TECHNOLOGY

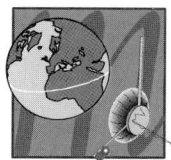

PERSONAL COMMUNICATION SERVICES OR "HELLO SPOCK? THIS IS CAPTAIN KIRK."

You are probably already familiar with one of the new technologies touted for the near future. Remember the small hand-held speakerphone used in the original "Star Trek" series? The "communicator" used by the crew of the *Enterprise* will be available soon, only now it's known as personal communications services, or PCS.

PCS merges the technologies of the cellular phone, satellites, and existing local communication services to the home, like telephone lines and cable TV. Under plans being promulgated by leading media companies, including the "Baby Bell" phone companies, TCI, and Cox Cable, among others, it will soon be possible to send messages from a small phone that will fit in a shirt-pocket. Beamed to nearby microwave towers, up to new, powerful satellites in low Earth orbit, and back down to local phone companies or cable systems, these calls could be transmitted and received anywhere in the world.

Already in development are PCS receivers with the ability to translate languages, receive fax messages, or interface with a personal computer. Even the Star Trek "Trichorder" will soon be a reality. This was the machine Spock and Bones used to measure temperature, atmosphere, and the presence of Klingons. One wonders if PCS will be able to detect any visible signs of intelligent life on this strange planet. . . !

other to see if something good is on. Assuming an average of ten seconds viewing per channel, it would take more than thirty minutes to complete the circuit. By that time the programming on the channels scanned first might have changed completely, necessitating another spin around the dial. Will viewers of the future be able to cope with all these choices? Will we have enough time to plan our viewing choices intelligently? Or, will we simply confine most of our viewing to our four or five favorite channels and, if that's the case, why would we want 200?

Does the availability of all this information mean that we will be better informed in the future? Not necessarily. Note that as the media become more specialized and people use multimedia computers to create personalized menus of their own media choices, it becomes easier to avoid certain types of information, such as political news. Currently about one-half of Americans vote in presidential elections. In the future, when it would be much easier to see and read only what we want to, would half of the American people even know who was running for president? Further, even if you were interested in political news, it matters little if your multimedia computer tapes twenty programs for you to watch if you don't have time to view more than one program.

Lastly, there is the problem of escapism and isolation. The term "couch potato" came into vogue during the 1980s to describe a type of person who turned away from personal interaction and merely vegetated in front of the TV. How many people will become couch potatoes in the future when 200+ channels of TV are routinely available? And, what about the VR technology? Some people currently spend four or five hours in front of the TV set escaping into the virtual reality of sitcoms, game shows, and action/adventure programs. Others spend equally significant amounts of time trying to master the latest video games from Nintendo. How much more time will they spend in the far more beguiling and seductive world of VR? Indeed, some critics of VR have labeled it "electronic LSD" because many people find it addictive.

The new communication technologies will bring promise and peril. Just like the telegraph, telephone, radio, and TV that preceded them, they will bring fundamental changes to our lives. They will require difficult decisions from policy makers and from consumers alike. They will require us to learn more about them and what they can do. We hope that this chapter is a first step toward a greater knowledge of their capabilities. As Loy Singleton pointed out in his book, *Telecommunications in the Information Age*, the new technologies will best serve those who know how to use them.

SUMMARY

The mass media industry of the future will be characterized by vertical integration, globalization, and privatization. New media developments to watch for include (1) virtual reality—a computer-generated environment; (2) multimedia—the blending of the computer, TV, telephone, and CD player; and (3) holography—three-dimensional photography using a laser beam. New media developments will be accompanied by concerns over costs, information overload, and isolation.

SUGGESTIONS FOR FURTHER READING

Dizard, W. (1989). *The coming information age.* New York: Longman.

Gross, L. S. (1990). *The new television technologies.* Dubuque, Iowa: William C. Brown.

Klapp, O. E. (1986). *Overload and boredom.* New York: Greenwood Press.

McCavitt, W. E. (1983). *Television technology: Alternative communication systems.* Latham, Md.: University Press of America.

McLuhan, H. M. (1964). *Understanding media: The extensions of man.* New York: McGraw-Hill.

Rheingold, H. (1991). *Virtual reality.* New York: Summit Books.

Roth, J. P. (1988). *CD-ROM: Applications and markets.* Westport, Conn.: Meckler.

Sigel, E. (1980). *Video discs.* White Plains, N.Y.: Knowledge Industry Publications.

Singleton, L. (1986). *Telecommunications in the information age.* Cambridge, Mass.: Ballinger.

Syms, R. R. (1990). *Practical volume holography.* New York: Clarendon Press.

GLOSSARY

Account executive Salesperson who visits local merchants to sell them broadcast advertising.

Addressable converter Device that allows pay-per-view cable subscribers to receive their programs.

Advanced television (ATV) Improved resolution TV that is compatible with existing TV receivers.

Affiliate Local radio or TV station that has a contractual relationship with a network.

Aftermarket Alternative markets for TV series after they run on the major networks; syndication and overseas markets are examples.

Alternator Device that generates continuous radio waves; necessary for the broadcasting of voice and music.

American Society of Composers, Authors and Publishers (ASCAP) Group that collects and distributes performance royalty payments to various artists.

Amplifier Device that boosts an electrical signal.

Amplitude Height of a wave above a neutral point.

Amplitude modulation (AM) Method of sending a signal by changing the amplitude of the carrier wave.

Analog signal Transduced signal that resembles an original sound or image; for example, a phonograph record contains analog signals.

Area of dominant influence (ADI) In ratings terminology, that region of a market where most of the viewing or listening of that market's TV and/or radio stations occurs.

Average quarter-hour persons In radio, average number of listeners per 15-minute period in a given daypart.

Audience flow Movement of audiences from one program to another.

Audimeter Nielsen rating device that indicates if a radio or TV set is in use and to what station the set is tuned. *See also* Storage instantaneous audimeter.

Audio board *See* Mixing console.

Audion Device invented by Lee De Forest that amplified weak radio signals.

Auditorium testing Research technique that tests popularity of records by playing them in front of a large group of people who fill out questionnaires about what they heard.

Barter Type of payment for syndicated programming in which the syndicator witholds one or more minutes of time in the program and sells these time slots to national advertisers.

Beam splitter Optical device that dissects white light into its three primary colors: red, green, and blue.

Bicycle network A network that distributes programs physically by shipping tapes to various stations.

Blacklist List of alleged Communists and Communist sympathizers circulated in the 1950s; contained the names of some prominent broadcasters.

Blanket rights Music licensing arrangement in which an organization pays BMI or ASCAP a single fee that grants the organization the right to play all of BMI's or ASCAP's music.

Blanking pulse Signal carried inside the TV camera that shuts off the scanning beam to allow for persistence of vision.

Block programming In radio, programming to one target audience for a few hours and then changing the format to appeal to another group. Used by many community radio stations.

Bumper Segment that introduces a newscast after a commercial break.

Burnout Tendency of a song to become less popular after repeated playings.

Call-out research Radio research conducted by telephone to evaluate the popularity of recordings.

Carrier wave Basic continuous wave produced by a radio or TV station; modulated to carry information.

Catharsis theory Theory that suggests that watching media violence relieves the aggressive urges of those in the audience. There has been little scientific evidence for this position.

Cathode ray tube (CRT) Picture tube in a TV set.

C band Satellite that operates in the 4–6-gigahertz frequency range.

Channel Frequency on which a station broadcasts.

Channel capacity In cable, the number of channels that can be carried on a given system.

Charge-coupled device (CCD) Solid-state camera that uses computer chips instead of tubes.

Chromakey Process by which one picture is blended with another in TV production.

Clandestine radio services Unauthorized broadcasts, usually political in nature, conducted by groups in opposition to the current government.

Claw In a motion picture projector, the device that grabs each frame of film by the sprocket holes and holds it in place in front of a light.

Clearance Process in which an affiliate makes time available for a network program. In syndication, refers to the number or percentage of TV markets in which a program is carried.

Clone Digital copy made from a digital master.

Clutter Commercials and other nonprogram material broadcast during program breaks.

Coincidental telephone interview Method of audience research in which a respondent is asked what radio or TV station he or she is listening to at the time of the call.

Common carrier Communication system available for public use such as the telephone or postal service.

Community-service grants Money given by the government to public TV stations to support programming of special interest in the station's broadcast area.

Compensation Fee paid by a network to a local station in return for the local station's clearing time for network shows.

Compulsory license Fees paid by cable systems that use distant nonnetwork signals from other markets.

Condenser microphone Microphone that uses an electrical device to produce the equivalents of sound waves.

Cooperative advertising Arrangement in which national advertisers assist local retailers in paying for ads.

Corporate video Usually done in a business setting, video production intended for a specific audience and usually not for public use.

Cross-licensing agreement Agreement made between or among companies that allows all parties to use patents controlled by only one of the parties.

Cultivation theory Theory suggesting that watching a great deal of stereotyped TV content will cause distorted perceptions of the real world.

Cumulative audience In ratings, the number of different households that watch or listen to a program in a specified time period. Also called the unduplicated audience.

Deflection magnet Device that directs the scanning beam inside a TV camera.

Demographics Science of categorizing people based on easily observed traits. Age and sex, for example, are two common demographic categories.

Dichroic mirror Mirror that separates white light into red, green, and blue light. Used inside a color TV camera.

Dielectric Insulated middle portion of a coaxial cable.

Digital audio broadcasting (DAB) Broadcasting a radio signal using a binary code (0s and 1s).

Digital audio tape (DAT) High-quality audio tape that uses the same digital technology as a CD.

Digital compact cassette (DDC) Digital audio tape packaged in a cassette format.

Digital signal Transduced signal that consists of binary codes (0s and 1s) that represent the original signal.

Direct-broadcast satellite (DBS) Satellite transmission designed to be received directly by the home.

Drop In cable, that part of the system that carries the signal from the feeder cable into the house.

Duopoly (1) System of broadcasting in which two systems, one public and one private, exist at the same time, as in Canada. (2) Owning more than one AM or FM station in the same market.

Dynamic microphone Microphone that uses a diaphragm and electromagnets to change sound energy into electrical energy.

Effective radiated power (ERP) Amount of power a radio station is permitted to use.

Electroencephalogram (EEG) A physiological measure of brain waves used in broadcasting and cable research.

Electroluminescence (EL) Method of providing a flatscreen TV receiver.

Electron gun Device in a TV camera that produces a stream of electrons that scans the image to be televised.

Electronic news gathering (ENG) Providing information for TV news with the assistance of portable video and audio equipment. Also called electronic journalism (EJ).

Encoder Device that combines the red, green, and blue information in a color TV signal with the brightness component.

Equalizer Electronic device that adjusts the amplification of certain frequencies; allows for fine tuning an audio signal.

Erase head The part of a tape recorder that returns the metal filings to a neutral position, thus erasing any signal on the tape.

Evening drive time In radio, a peak listening period that extends from 3:00 to 7:00 P.M. when many people are commuting home from work or school.

Exclusivity deal In cable, an arrangement whereby one premium service has the exclusive rights to show the films of a particular motion picture company.

Exit poll Survey in which voters are asked about their voting decisions immediately after they leave the voting booth; used to predict the outcomes of elections before the polls close.

Expanded sample frame In ratings, a technique by which a sample is increased to include more minority groups.

Fair use In copyright law, a small portion of a copyrighted work that can be reproduced for legitimate purposes without the permission of the copyright holder.

Fairness Doctrine Currently defunct policy of the Federal Communications Commission that required broadcast stations to present balanced coverage of topics of public concern.

False light A type of invasion of privacy in which media coverage creates the wrong impression about a person.

Feeder line In cable, that part of the system that transfers the signals from trunk lines to house drops.

Fiber optic Cable used for transmitting a digital signal via thin strands of flexible glass.

Fidelity Degree of correspondence between a reproduced signal and the original.

Field Half of a complete TV picture; one field is scanned every sixtieth of a second.

Financial interest and syndication rules (fin-syn) FCC regulations limiting network participation in ownership and subsequent syndication of programs produced for the network.

Focus group Small group of people who discuss predetermined topics, such as a TV newscast.

Footprint Coverage area of a communications satellite.

Format The type of music or talk that a radio station chooses to program. Formats are usually targeted at a specific segment of the population. *See also* Demographics.

Formative evaluation In corporate video, testing the storyboard, script, and rough cut of the program to determine if these elements are designed as originally planned.

Frame Two fields or one complete TV picture; one frame is scanned every thirtieth of a second.

Franchise agreement A contract between a local government and a cable company that specifies the terms under which the cable company may operate.

Frequency Number of waves that pass a given point in a given time period, usually a second; measured in hertz (Hz).

Frequency modulation (FM) Method of sending a signal by changing the frequency of the carrier wave.

Frequency response Range of frequencies that a radio set is capable of receiving.

Future file Collection of stories to be used in upcoming newscasts.

Galvanic skin response (GSR) Measure of the electrical conductivity of the skin. Used in broadcast and cable research.

Globalization Tendency for mass media firms to have interests in countries all over the world.

Grazing Method of TV viewing in which the audience member uses a remote-control device to scan all available channels during commercials or dull spots in a program.

Headend In cable, the facility that receives, processes, and converts video signals for transmission on the cable.

Height above average terrain (HAAT) Measurement of the height of a transmitter tower; used to classify FM stations.

Helical-scan tape recording Method of videotape recording in which the signal is recorded in diagonal strips.

High-definition television (HDTV) Improved resolution TV system that uses approximately 1100 scanning lines.

Holography Three-dimensional lensless photography.

Homes passed Number of homes that have the ability to receive cable TV; that is, homes passed by the cable.

Hook Short, easily identifiable segment of a recording.

Households using television (HUT) Number of households that are watching TV at a certain time period.

Hue Each individual color as seen on color TV.

Hypodermic needle theory Early media effects theory stating that mass communicated messages would have a strong and predictable effect on the audience.

Independent Station not affiliated with a network.

Interactive television System in which TV viewers respond to programs using a special keypad.

International broadcasting Broadcast services that cross national boundaries and are heard in many countries.

International Telecommunications Union (ITU) Organization that coordinates the international broadcasting activities of its members.

Keying Process by which one video signal is electronically cut out of or into another.

Kinescope Early form of recording TV shows in which a film was made of a TV receiver.

Ku band Communications satellite that operates in the 12–14-gigahertz frequency range.

Laser optical media (LOM) General term of CDs, videodiscs, and related devices that use a laser to read information.

License agreement Arrangement between a syndicator and a TV station specifying the number of times a movie or TV show may be shown in a given time period.

Limited-effects theory Media effects theory suggesting that media have few direct and meaningful effects on the audience.

Liquid crystal display (LCD) Flat-screen display system being developed for use in TV receivers.

Live assist Form of radio production where local announcers and DJs are used in conjunction with syndicated programming.

Local origination Program produced by a local TV station or cable system.

Lowband That part of the cable occupied by channels 2 to 6 and FM radio.

Low-power television (LPTV) Television stations that operate with reduced coverage and have a coverage area only 12–15 miles in diameter.

Luminance Degree of brightness of a TV picture.

Market Specific geographic area served by radio and TV stations. The United States is divided into approximately 210 different markets.

Midband That part of the cable signal occupied by TV channels 7 to 13 and cable channels 14 to 22.

Minidisc (MD) A more compact version of the CD, about one-fourth the size of a standard CD.

Minidoc Multipart reports that generally air Monday to Friday on local TV stations. Each minidoc segment may only be three or four minutes long.

Mixing console Master control device in an audio studio that selects, controls, and mixes together various sound inputs.

Modulation Encoding a signal by changing the characteristics of the carrier wave.

Morning drive time In radio, 6:00 to 10:00 A.M. Monday through Friday when large numbers of people are listening in their cars.

Multichannel, multipoint distribution system (MMDS) System using microwave transmission to provide cable service into urban areas; also called wireless cable.

Multichannel television sound (MTS) Stereo sound and a second audio channel are multiplexed in the audio portion of the TV signal.

Multimedia System that combines TV set, computer, CD player, and telephone.

Multiple-system operator (MSO) Company that owns and operates more than one cable system.

Multiplexing Sending different signals within the same channel.

Must-carry rule Regulation which stated that cable systems had to carry the signals of all broadcast TV stations seen in the market served by cable. As of mid-1992, the rule was not in effect.

National Association of Broadcasters (NAB) Leading professional organization of the broadcasting industry.

National Cable Television Association (NCTA) Leading professional organization of the cable industry.

National representative (rep) Organization that sells time on a local station to national advertisers.

National spot sales Advertising placed on selected stations across the country by national advertisers.

National Television System Committee (NTSC) Group that recommended the current technical standards for color TV. Also refers to the North American standard for television broadcasting.

Network compensation Money paid by a network to one of its affiliates in return for the affiliate's carrying network shows and network commercials. *See also* Clearance.

Network programming Programs that are financed by and shown on TV networks.

News consultants Research companies that advise stations about ways to improve the ratings of their news programs.

Noise Unwanted interference in a video or audio signal.

Orbital slot "Parking place" for a communications satellite in geosynchronous orbit.

Oscillation Vibration of a sound or radio wave.

Overbuild More than one cable system serving a community.

Package 1. News story that includes pictures of the newsworthy event, the natural sound, and a reporter's voice-over. 2. Series of theatrical movies made available by a distribution company for sale to cable and broadcast TV stations.

Panel method Research technique in which the same people are studied at different points in time.

Payola Bribes given to DJs to influence them to play particular records on radio stations.

Pay per transaction (PPT) System of videocassette rental in which the motion picture production company receives a portion of the rental fee whenever one of its movies is rented.

Pay per view (PPV) System in which cable subscribers pay a one-time fee for special programs such as movies and sporting events. *See also* Addressable converter.

PEG channel Cable station set aside for public, educational, or government use.

Peoplemeter In ratings, hand-held device that reports what TV show is being watched. Peoplemeters also gather demographic data about who is watching.

Persistence of vision Tendency of perceptual system to retain an image a split second after the image is removed from sight. Makes possible the illusion of motion in film and TV.

Phi phenomenon Tendency of perceptual system to "fill in blanks" between two light sources located close to one another. As one light blinks off and the other blinks on, the brain perceives the change as motion.

Photon Packet of light energy.

Pilot Sample episode of a proposed TV series.

Pirate station Unauthorized radio or TV station that generally broadcasts entertainment material.

Playback head In a tape recorder, the device that reproduces the signal stored on the tape.

Playlist List of records that a radio station plays. *See also* Format.

Plugola Gifts given to DJs for promoting a product on the air. Listeners are unaware that the DJ is being compensated for these mentions.

Preemption Show that a network affiliate refuses to carry.

Prime-time access rule (PTAR) In general, a regulation that limits the TV networks to three hours of programming during the prime-time period. Exceptions are made for news, public affairs, children's shows, documentaries, and political broadcasts.

Privatization Trend in which former public or state-owned broadcasting systems are becoming privately owned.

Promotional announcement (promo) Short announcement to remind viewers or listeners about an upcoming program.

Psychographic research Research that uses personality traits to segment the audience.

Psychographic variables Psychological factors that explain audience behavior.

Public service announcement (PSA) Announcement for charitable or other worthwhile endeavor presented free of charge by broadcasters.

Puffery Allowable exaggeration in advertising claims.

Pulse code modulation (PCM) Method used in digital recording and reproduction in which a signal is sampled at various points and the resulting value is translated into binary numbers.

Q score See TV Q.

Radio Act of 1927 Act that established the groundwork for modern broadcasting regulation.

Rating In TV, the percentage of households in a market that are viewing a station divided by the total number of households with TV in that market. In radio, the total number of people who are listening to a station divided by the total number of people in the market.

Recording head Device in a tape recorder that stores a new signal on the tape.

Retracing Scanning process that goes on inside a TV set.

Right of first refusal Network's contractual guarantee to prohibit a production company from producing a specific show for another client.

Rotation Mix or order of music played on a radio station.

Rough cut Preliminary rendition of an ad or a TV show produced so that viewers can get a general idea of the content.

Sampling 1. Selecting a group of people who are representative of the population. 2. In digital signal processing, selecting a number of points along an analog signal and converting the signal into binary numbers.

Satellite master-antenna television (SMATV) System used in apartment buildings in which a master receiving dish and antenna on top of the building pick up TV signals, which are then transmitted by wire to dwelling units.

Satellite news gathering (SNG) Use of specially equipped mobile units to transmit live and taped remote reports back to a local station.

Saturation Strength of a color as seen on color TV.

Scanner Radio monitor that is tuned to police and fire frequencies.

Scanning Technique by which the beam of electrons inside a TV camera traces its way across an image.

Sell-through A tape designed to be sold directly to consumers as opposed to video rental stores.

Sets in use In ratings, the number of radio or TV sets that are in operation at a given time.

Share In radio, the number of people who are listening to a station divided by the total number of people who are listening to radio at a given time. In TV, the total number of households watching a given channel divided by the total number of households using TV.

Sideband Signals above and below the assigned frequency of a carrier wave at a TV or radio station.

Signal-to-noise ratio Amount of desired picture or sound information that remains after subtracting unwanted interference.

Simulcast Radio program aired in both AM and FM at the same time on two different stations. Also a TV show, usually a concert, carried by an FM station at the same time it is being televised.

Single-system operator (SSO) Company that owns and operates one cable system.

Skip Tendency of radio waves to reflect off the ionosphere and back to earth, then back to the ionosphere, and so on. Makes possible long-distance radio transmission.

Specific-effects theory Theory that posits there are certain circumstances under which some types of media content will have a significant effect on some audience members.

Spot sets Segments in radio programming, such as commercials and promotions, which interrupt the normal programming content.

Stand-up Reporter standing in front of the camera providing an opening, bridge, or closing for a story. *See also* Package.

Station Organization that broadcasts TV or radio signals.

Step deal Contractual arrangement by which TV series are produced. Production proceeds in a series of defined steps, with the network having the option to cancel after each step.

Stimulation theory Theory suggesting that watching media violence will stimulate the viewer to perform aggressive acts in real life. Opposite the catharsis theory.

Storage instantaneous audimeter (SIA) Computer-assisted TV measurement device that makes possible overnight ratings.

Storyboard Drawings illustrating what a finished commercial or segment of a TV show will look like.

Stripping Scheduling the same show to run in the same time period from Monday through Friday.

Subscription television (STV) System that sends programs in scrambled form to TV sets equipped with decoders.

Subsidiary communications authorization (SCA) A service provided by FM stations using additional space in their channel to send signals to specially designed receivers.

Summative evaluation In corporate video, research that looks at the effectiveness of a completed program. *See also* Formative evaluation.

Superband That part of the cable signal that carries channels 23 to 69.

Superstation Local TV station that is distributed to many cable systems via satellite, giving the station national exposure.

Survey Research method that uses questionnaires or similar instruments to gather data from a sample of respondents.

Sustaining program Common in early radio, a program that had no commercial sponsors.

Switcher Device used to switch from one video signal to another. Can also be used to combine more than one video signal.

Synchronization pulse Signal that enables the output of two or more cameras and other video sources to be mixed together and also keeps the scanning process in the camera operating in time to coincide exactly with the retrace process in the TV receiver.

Syndicated exclusivity (syndex) FCC rule stating that local cable systems must black out programming imported from a distant station if it is being carried by a local station in the cable system's market.

Syndication Programming sold by independent companies to local stations and cable networks.

Target audience Specific group a radio station or TV program is trying to attract.

Target plate Mirrorlike device inside a TV camera that holds the image while the image is scanned by the electron beam.

Teaser Clever line used to introduce a newscast.

Teleconference Video link between individuals, frequently used for business conferences.

Teleports Facilities that provide uplinks and downlinks with communication satellites.

Teletext Cable service that offers text and graphics displayed on the screen.

Television receive-only earth station (TVRO) Home satellite dish that receives TV programming.

Tiering Process of selling cable subscribers increasing levels of service.

Timeshifting Recording something on a VCR to watch at a more convenient time.

Tip sheet Radio industry publication reporting current musical preferences across the country.

Total survey area (TSA) In ratings, a geographic area where at least some viewing of TV stations in a given market occurs. *See also* Area of dominant influence.

Tradeout Swapping advertising time for a product or service.

Transponder The part of a communication satellite that receives a signal from an earth station and retransmits it somewhere else.

Treatment Short narrative used to sell an idea for a TV show or series to a production company.

Trunk line Main cable lines connecting the headend to the feeder cables.

Turnkey automation Radio station that is fully automated.

TV Q Score that measures the popularity of TV celebrities.

Ultra high frequency (UHF) The portion of the electromagnetic spectrum that contains TV channels 14 to 83.

Underwriting Assisting a station in paying for a public radio or TV program in exchange for a mention on the air. Major corporations are the most frequent underwriters.

Uplink Ground source that sends signals to a communication satellite.

Uses and gratifications Research tradition that examines the reasons people use the media.

Velocity microphone Microphone that uses a thin metal ribbon and electromagnets to reproduce sound.

Vertical blanking interval Portion of the TV signal that occurs between fields; used to send teletext and closed-captioning.

Vertical integration Process by which a firm has interests in the production, distribution, and consumption of a product.

Very high frequency (VHF) The part of the electromagnetic spectrum that contains TV channels 2 to 13.

Video news release (VNR) In corporate video, a complete video package sent by a company to a news organization in an attempt to get broadcast time.

Videotex Two-directional information service linking a data bank with computer terminals via cable or telephone lines.

Video toaster A personal computer that generates special video effects.

Virtual reality Computer system that creates three-dimensional images that users interact with by means of special goggles and gloves.

Waveform Visual representation of a wave as measured by electronic equipment.

Wavelength Distance between two corresponding points on an electromagnetic wave.

Zapping Deleting the commercials when videotaping a program off the air for later viewing.

Zipping Fast-forwarding through the commercials when viewing a program recorded off the air.

PHOTO CREDITS

3 J.L. Atlan/Sygma. 5 ©1991 CNN, Inc. All Rights Reserved. 7 Shone/Gamma Liaison. 9 AP/Wide World Photos. 13 CBS Photography. 15 Brown Brothers. 16 Brown Brothers. 18 Brown Brothers. 23 Brown Brothers. 26 Brown Brothers. 29 CBS Photography. 30 David Sarnoff Library. 34 Brown Brothers. 36 Bettmann Archive. 38 Bettmann Archive. 43 Photo Works. 45 Hazel Carew/Monkmeyer. 46 Bettmann Newsphotos. 48 Courtesy of RCA/Thomson Consumer Electronics. 49 Brown Brothers. 50 Bettmann Archive. 53 Alfred Wertheimer/RCA Records/Photofest. 55 Photofest. 56 Courtesy CBS. 61 Courtesy CBS. 64 Photofest. 67 Kelly Mills/TBS. 69 Globe Photos. 72 Sygma. 77 Courtesy WCBS Newsradio 88. 78 *left*, Spencer Grant/Photo Researchers; *right*, Barbara Rios/Photo Researchers. 85 Courtesy Indianapolis Monthly. Photo by E. Anthony Valainis. 89 Courtesy KLOS Radio. 91 Lori Stoll/Capitol Press. 93 Courtesy WBGO. 103 Stuart Cohen/Comstock. 106 Tom Ballard/EKM Nepenthe. 111 Courtesy Hamilton Projects. 113 M. Brinton/CBS. 115 Bettmann Archive. 120 Arvind Garg/Photo Researchers. 124 MAJ/Picture Cube. 134 Courtesy Jerold Communications. 135 Bob Demyan/ESPN. 140 Eric Neurath/Stock, Boston. 143 Courtesy Blockbuster Entertainment Corporation. 149 Courtesy Aetna. 151 Bohdan Hrynewych/Stock, Boston. 154 Eric Neurath/Stock, Boston. 155 Courtesy of Video Arts. 156 Courtesy SmithKline Beecham. 162 Courtesy SmithKline Beecham. 168 Thomas Nebbia/Woodfin Camp & Associates. 171 J.L. Atlan/Sygma. 183 BBC Photograph Library. 184 Charles Marden Fitch/Taurus. 192 Capital Cities/ABC. 193 Minosa-Scorpio/Sygma. 199 Louie Psihoyos/Matrix. 200 Courtesy Universal Records. 213 Courtesy of WLUP AM 1000, Chicago. 215 Courtesy Radio Computing Services. 221 *top*, E. Adams/Sygma; *bottom*, Courtesy WCBS Newsradio 88. 222 Courtesy National Public Radio. 226 Gary Harper/KING 5 TV. 229 Courtesy of CBS. 231 Courtesy of CBS. 234 Kelly Mills/TBS. 235 ©1991 CNN, Inc. All Rights Reserved. 236 Gamma Liaison. 240 Kenneth Jarecke/Contact Press Images/Woodfin Camp and Associates. 241 Tom Ballard/EKM-Nepenthe. 246 ©CBS Inc. 1991. 249 Courtesy Burson-Marstellar. 252 Globe Photos. 253 Bob Nese/ABC. 257 Lifetime Television. 261 Courtesy NATPE International. 268 Steven LaBadessa. 273 Diego Goldberg/Sygma. 279 Courtesy Beyer Dynamic. 289 Courtesy Studer Revox. 296 Courtesy Grundig. 298 J.P. Laffont/Sygma. 301 Spencer Grant/Stock, Boston. 309 Courtesy Hughes Communicatons. 318 Courtesy Zenith Electronics Corporation. 333 Courtesy Nielsen Media Research. 337 Comstock. 340 Courtesy Nielsen Media Research. 351 Courtesy the Arbitron Company. 352 Courtesy the Arbitron Company. 354 Maureen Fennelli/Comstock. 365 Susan A. Anderson/Picture Cube. 369 Rick Kopstein/Monkmeyer. 376 AP/Wide World Photos. 380 Mark Antman/Image Works. 387 Courtesy NBC. 393 Photofest. 399 Russell Thompson/Omni-Photo Communications. 406 Dean Abramson/Stock, Boston. 413 Courtesy National Association of Broadcasters. 415 Courtesy Action for Children's Television. 420 Ken Hawkins/Sygma. 433 Mike Schwarz/Gamma Liaison. 443 Dennis Dunleavy/San Antonio Express News. 444 A. Tannenbaum/Sygma. 447 Courtesy Hawthorne Communications. 451 Patrick Forstier/Sygma. 455 © 1992 20th century Fox Film Corporation. 458 AP/Wide World Photos. 462 Jeffrey Markowitz/Sygma. 466 Photo Works. 471 Jacques Chenet/Woodfin Camp & Associates. 475 Courtesy NASA.

COLOR SECTIONS

DEVELOPMENTS IN PROGRAMMING: **page 1** *top*, E. J. Camp-Outline, Courtesy Multimedia Entertainment; *bottom*, Courtesy CBS, Inc., photo by Photo Works. **page 2** *top*, Gamma Liaison; *bottom*, ©1990 Capital Cities/ABC, Inc. **page 3** *top*, Courtesy of Twentieth Century Television, A Division of Twentieth Century Fox Film Corporation; *bottom*, Alliance Communications. **page 4** *top*, Courtesy Tokyo Broadcasting System, Inc.; *bottom*, Yves Forestier/Sygma.

DEVELOPMENTS IN TECHNOLOGY: **page 1** *top*, Courtesy Newtek; *bottom*, Courtesy Panasonic Company. **page 2** *top*, Courtesy Zenith Electronics Corporation; *bottom*, Courtesy Hughes Communications. **page 3** *top left*, Courtesy Sanyo Fisher Corporation; *top right*, Jacques Chenet/Woodfin Camp & Associates; *bottom*, Will Mosgrove/Apple Computer, Inc. **page 4** James King-Holmes/W. Industries/Science Photo Library/Photo Researchers.

INDEX

ABC, 32, 41, 58, 65, 70, 110, 116, 221, 247
Account executives, 108
ACT (Action for Children's Television), 64, 161
Addressability, 315
Advertising, on early radio, 24–26
Advertising regulation, 445–448
   and First Amendment, 445
   FTC rules for, 446–447
Aerial construction, 137
Affiliates, 112, 247
Aftermarket, TV, 255
AFTRA (American Federation of Televison
     and Radio Artists), 121
AGB (company), 342
Agenda setting, 385
All-Channel Receiver Bill, 58
Alternator, invention of, 17
AM, 40, 79, 94, 99, 278, 290, 292
   channels and classification on, 293–294
   and clear channels, 293
AM stereo, 40, 296–297
American Marconi Company, 19–21
Amortization, 266
Ampex Corporation, 57, 150
Amplifier, 288
Amplitude, 276–278
Analog signals, 275–276
Anchors, TV news, 242–243
Antitrust laws, 436
ARABSAT, 173
Arbitron, 341–343
Area of dominant influence (ADI), 347
Armed Forces Radio Service, 171
Armstrong, Edwin, 31
ASCAP, 28, 31, 431
Assignment editors, 240
AT&T, 14, 19–20, 24, 26
Attenuation, 292
Audience flow, 269
Audimeter, 339–340
Audio board, 289
Audio tape recording, 299
Audio technology, 274–300
   and signal processing, 278–299
   technical aspects of, 274–278
Audion, 18–19
Auditorium testing, 216, 355, 357, 358
Average quarter-hour persons, 349

Baird, John, 49
Ball, Lucille, 54–56
Barter syndication, 265–266
"Battlesphere," 476
BBC (British Broadcasting Corporation), 6, 171, 180
BBC World Service, 176–177
Bell, Alexander Graham, 14
Bell curve, 344–345
Berle, Milton, 52
Bertelsmann A.G. (company), 70
Beta tape format, 328
Betamax case, 142
Billboard, 216
Blacklist, 53–54
Blanket rights, 431
Blanking pulse, 307
Blockbuster Video, 143
BMI (Broadcast Music Inc.), 431
Branzburg v. Hayes, 442
Brinkley, David, 229, 237
Brinkley, John R., 31, 404
British Sky Broadcasting, 179
Brokaw, Tom, 242
Bumpers, 241
Burnout of hit songs, 355

Cable Communications Policy Act of 1984, 67, 403, 415
Cable network programming, 256–259
   and cable movies, 256
   and cable series, 256–257
   and motion pictures, 256
Cable signal transmission, 313–316
Cable system operator (CSO), 134
Cable television, 125–140
   basic services of, 127–130
   channel capacity of, 126
   economics of, 133–135
   finance of, 135–139
   growth of, 125–126
   history of, 58, 62–63, 67–68, 125–126
   and marketing, 131–133
   ownership of, 134–135
   and pay services, 130
   personnel of, 139–140
   and specialty services, 130–131
Cable testing, 359

INDEX

Call-outs, in research, 216, 355–356
Cameras in the courtroom, 443–445
Canada, electronic media in, 187–190
   economic support of, 188
   and laws and regulation, 189
   and programming, 189–190
   structure of, 188
Capital Cities Broadcasting, 70
Carlin, George, 434
Carnegie Foundation, 63
Carrier wave, 290
Categorical imperative, 464
Catharsis theory, 376
Cathode ray tube, 47
CBC (Canadian Broadcasting Company), 187–189
CBS, 27, 41, 51–52, 55, 110, 116, 247
Census, 344
Challenge programming, 269
Charge-coupled device (CCD), 308
Children's Television Act of 1990, 429
Chromakey, 311
Civil rights movement, 230
Clandestine radio stations, 170, 178
Claw, in movie projector, 305
Clear channels, 293
Cleese, John, 155
Clutter, advertising, 215
CNN, 4, 5, 8, 67, 178–179, 234–235, 238
Coaxial cable, 314
Codec, 162
Coincidental telephone interview, 338
College radio, 223
Color TV, history of, 55, 57
Commercial exploitation, 441
Commissions, 108
Communications Act of 1934, 29, 402, 421
Community stations, 76
Compact discs, 284
Comparative electronic media systems, 169, 180–194
Compensation, 248
Compulsory license, 260
Concept (sample script), 253
Confidence level, 361
Congress, U.S., 408–410
Conrad, Frank, 22–24
Consultants, 216, 366–367
Content analysis, 372
CONUS, 238
Converter box, 314
Cooperative advertising, 82
Cooperative Analysis of Broadcasting, 337
Copyright law, 429–432
Copyright Royalty Tribunal, 431
Corporate television, 150–167
   applications of, 153–159
     in education, 156–157
     in internal communication, 154–155

Corporate television (*Cont.*):
     in public relations, 157–158
     in sales, 158
     in training, 153
   careers in, 165–167
   defined, 150
   evaluating programs of, 165
   history of, 150–152
   networks, 161
   organization of, 159–161
   producing programs of, 163–165
   users of, 152–153
   video conferences, 162–163
Corporation for Public Broadcasting (CPB), 63, 258
Counterprogramming, 269
Courts, and media regulation, 410
Cronkite, Walter, 229, 233
Crossley, Archibald, 337
Cross-licensing agreement, 21–22
CRTC (Canadian Radio and Telecommunication Commission), 187
Cultivation analysis, 381–383
Cultural ethics, 464
Cumulative audience, 349
Cut, in special-effects video, 311

DAB (digital audio broadcasting), 43, 295
DAT (digital audio tape), 287–288, 299
Dayparts, 211, 351
DBS (direct-broadcast satellite), 72, 320
Defamation, 437–439
Deflection magnets, 307
De Forest, Lee, 17–19
Deutsche Welle, 177–178
Dichroic mirror, 307
Dielectric, 314
Digital compact cassette, 288
Digital signals, 275–276
Direct waves, 292
Disclaimer, 390
"Disneyland," 58
DuMont, Allen, 51
Duopoly, 181, 188

Edison, Thomas, 32
Effective radiated power (ERP), 294
Effects of broadcasting and cable, 370–395
   areas of research on, 374–395
   techniques of research on, 370–372
   theories of, 372–374
     hypodermic needle, 372
     limited-effects, 372–374
     specific-effects, 374
Egoism, in radio and TV, 463–464
Electrical transcription, 298

Electroencephalogram (EEG), 364
Electromagnetic spectrum, 290–295
Electron gun, 307
Equal employment opportunities, 435
Equal opportunities provision, 425–427
Equalizer, 288
Encoder, 308
ENG (electronic news gathering), 234–235
Erase head, 283
Escapism, 478
ESPN, 67
Ethics in radio and TV, 462–468
   practicing of, 465–466
   and real world, 466–467
   theories of, 462–464
Exit polls, 386
Expanded sample frame (ESF), 346
Externalities, 417

Facsimile technology, 274, 303–305
Fader, 289
Fair use doctrine, 431
Fairness doctrine, 427–429
False light, 441
Farnsworth, Philo, 49–52
Faulk, John Henry, 54
Federal Communications Commission (FCC), 29, 52, 57, 62, 89, 402–403, 405–408
   and cable, 424–425
   enforcement procedures of, 408
   and emerging technologies, 408
   and emerging technologies, 425
   and licensing, 406
   policy making of, 407
   role of, 421–424
Federal Radio Commision (FRC), 28, 402
Federal Trade Commission (FTC), 446–447
Feeder cable, 316
Fessenden, Reginald, 16–17
Fiber optics, 322–323
Fidelity, 274
Field experiments, 371
Field producers, 240
Fields, 310
Fill (listener call-ins), 216
Financial interest and syndication rules, 249
Fleming valve, 18
FM, 31, 40–41, 90, 94, 99, 278
   channels and classification of, 294
   receivers, 297
   sidebands, 294–295
Focus groups, 216, 359
Footprint, 318
Formative evaluation, 165
Fox Broadcasting Corporation, 68, 111–113, 179, 247, 249, 250, 457

Frame, 310
Franchise agreement, 134
Freeze frame, 325
Freeze on licensing of TV stations, 53
Frequency, 276–278
Frequency response, 278
Future file, 240
Future of broadcasting and cable, 472–478
   and emerging technologies, 473–477
   and industry outlook, 472–473
   and social concerns, 477–478

Galvanic skin response (GSR), 364
General Electric, 17, 19, 22, 24, 26
Georgia Pacific, 159
Gerbner, George, 381, 382
Globalization, 472
Golden mean, 464
Gorbachev, Mikhail, 4, 6, 7
Grazing, 325
Ground waves, 292

Hamlet, 154
Hammocking, 269
"Happy talk," 232–233
HBO, 67
HDTV (high-definition), 73, 309–311
Headend, 315
Height above average terrain (HAAT), 294
Helical scanning, 327
Herrold, Charles, 22
Hertz, Heinrich, 15, 29
Hertz (unit of measurement), 276
Holography, 476–477
Home video, 141–147
   and broadcasting, 144
   future of, 147
   and Hollywood, 143
   and pay cable, 146–147
   and video stores, 143
Homes passed, 132
Hooks, 356
Hooper, C. E., 338
Hoover, Herbert, 27
Hot clocks, 211–216
Households using television (HUT), 348, 351
Hue, 307
Huntley, Chet, 229, 233
Hypercard, 329
Hypodermic needle theory, 372

"I Love Lucy," 54–56
Inconoscope, 49
Indecency, 432–435

Independents, 113
Industrial television, 151
"Infotainment," 255
Information overload, 477
Instrumental viewers, 391
INTELSAT, 172
Interactive TV, 310
Interlace scanning, 306
International electronic media, 169–194
　history of, 170–173
　models of, 192–193
　radio, 173–178
　television, 178–180
International media systems, 169
International regulation, 448–450
INTERSPUTNIK, 172
Intrusion, 440–441
Invasion of privacy, 439–442
IQ and TV, 388
Isolation, 478
ITU (International Telecommunications Union), 448–449

Jenkins, Charles Francis, 49
Jennings, Peter, 242

Katz Communications, 108
Katz Radio Group, 85
KDKA, 23
Kenya, electronic media in, 190–191
Kinescope recording, 53, 325
Klapper, Joseph, 373
Koppel, Ted, 233

Laboratory experiments, 371
Lateral cut recording, 279
Laws, broadcasting and cable, 421–449
Lawson, Jennifer, 73
Learning and TV, 387–389
Least objectionable program, 248
Lenses, 302
Libel, 437
License agreement, 262
License fees, 249
Liner, 216
Live-assist production on radio, 202
Lobbyists, 412–414
Longitudinal studies, 372
Lotteries, 435
Lowband, 314
LPTV (low-power TV), 113–114, 321
Luminance, 307

McCarthy, Joseph, 54
Magid, Frank, 368
Mainstreaming, 382
Make-good, 119
Marconi, Guglielmo, 8, 15–16
Marconi Wireless Telegraph Company, 16
Market research, 357–364
Marketplace, 416–418
Markets, 346
Media Lab, 476
Metro ratings area, 346
Microphones, 279–281
Microwave relay, 320
Midband, 314
Miller v. California, 432
Minidisc, 288
Minidocs, 235
Mixing console, 289
MMDS (multichannel, multipoint distribution service), 141, 321, 425
Moderator, 359
Modulation, 275, 290
Monte Carlo method, 348
Morse, Samuel, 14
Morse code, 14, 17
Moscow Echo, 5
MSO (multiple-system operators), 68
MTV, 42, 67, 179, 206
Multichannel TV sound, 309
Multimedia, 473–474
Multipay household, 133
Multitrack recorders, 283
Murdoch, Rupert, 173, 179, 247
Murrow, Edward R., 54, 60, 229
Music licensing, 431
Music research, 355–357
"Must carry" rules, 127
Mutual Broadcasting System, 27

NABET, 121
National Advertising Review Board (NARB), 460
National Association of Broadcasters (NAB), 28, 411–413, 452–453, 460
National Association of Television Program Executives (NATPE), 261
National Cable Television Association (NCTA), 413
National Commission on the Causes and Prevention of Violence, 374
National Public Radio (NPR), 41, 80, 97
National representative, 107
National spots, 82
NBC, 26, 27, 32, 41, 46, 50, 52, 110, 116, 247
NET (National Educational Television), 63
Network ads (radio), 82
New World Information Order, 191–192

New York Times v. Sullivan, 438, 445
New York World's Fair of 1939, 46
News consultants, 235
News directors, 238–239
News producers, 239–240
NHK (Japan), 238
Nicholas II (Czar), 8
Nielsen, A. C., Inc., 338–340
Nipkow, Paul, 47
Noise, 274
Normal distribution, 344
NTIA (National Telecommunications and Information Agency), 411

"O & Os" (owned-and-operated stations), 112
Obscenity, 432–435
Off-net syndication, 262
One-way network, 161
Orbital slot, 318
Oscillation, 276
Overbuilds, 424

Pacifica, 434
Packages, of films for sale to networks, 241, 256
PAL, European TV system, 310
Paley, William S., 27–28
Panel method, 356, 372
Pay households, 133
Pay per transaction, 144
Pay per view, 133
Payola, 217
PEG channels, 415
People's Republic of China, electronic media in, 185–187
  financing of, 185–186
  programming of, 187
  regulation of, 186–187
  structure of, 185
Peoplemeters, 340–341
Peripherals, 141
Persistence of vision, 302–303
Pervasive presence theory, 401
Phi phenomenon, 302–303
Phonograph recording, 299
Photons, 302
Physiological measures, 364–366
Pilots, 253
PIP (picture-in-picture), 325
Pirate broadcasting, 178
"Pitching" a program, 253
Playback head, 275, 283
Political action committees (PACs), 413
Politics, and electronic media, 385–387
  impact of media on political system, 386–387
  and voting, 385–386

Preemptions, 248
Prime time, 337
Prime-time access rule (PTAR), 65, 249–250, 407
Privatization, 473
Profanity, 432–434
Profit margin, 80
Promotional announcements, 110
Prosocial behavior, 392–394
Psychographics, 363–364
Public access programming, 267
Public Broadcasting Service (PBS), 63, 258
Public radio, 96–98, 222–224
Public service announcements (PSAs), 110
Public television:
  audiences of, 259
  history of, 63, 70–71
  sources of programming for, 258–259
Pulse code modulation, 284–285

Radio:
  advertising on, 79–85
  careers in, 99–100
  competition in, 79–80
  economics of, 94–96
  FM and AM audiences of, 80
  formats on, 86–92
  history of, 14–44
    and early broadcasters, 22–24
    and early inventors, 15–19
    and growth of, 29–44
    and legal problems, 19–22
    and networks, 26–28
    in 1920s, 24–29
  organization of, 98–99
  ownership of, 85–86
  programming of, 86–92
  and promotions, 92–94
  salaries in, 100
  and types of stations, 79–80
Radio Act of 1912, 401
Radio Act of 1927, 401
Radio Beijing, 177
Radio broadcast data system (RBDS), 297
Radio Free Europe, 171, 174
Radio in the American Sector, 171
Radio Liberty, 171
Radio Martí, 172, 174
Radio Mocow, 5, 175–176
Radio programming, 200–225
  developing formats for, 207–209
  and format choice, 203–205
  and format design, 200–201
  and format evaluation, 216–217
  and listener characteristics, 209–211
  and music business, 205–206

Radio programming (*Cont.*):
   and networks, 220–221
   news and talk in, 217–219
   and production, 202
   and programming matrix, 201–202
   on public radio, 222–224
   by satellite, 221
   terminology in, 215–216
Radio Russia, 7
Radio spectrum, 290–293
Raised eyebrow, regulation by, 407–408
Ralph's Grocery Company, 153
Random sample, 345
Raster, 323
Rate card, 109
Rather, Dan, 242, 323
Ratings, 334–353
   history of, 326–343
   process, 343–353
   and ratings book, 349–350
   uses of, 334–336
RCA, 20–22, 24–26, 46, 50–52, 55
Real-time network, 161
Recording head, 275, 283
Red Channels, 53, 54
Red Lion Broadcasting v. FCC, 428
Registration, 308
Regulation of electronic media, 400–419
   history of, 401–405
   nexus of, 418
   rationale for, 400–401
   and regulatory forces, 405–419
Reporters, 240
Representation firms, 82
Resonance, 382
Retracing, 306
Retransmission consent, 127
RGB signal, 308
Ritualized viewing, 391
Robida, Albert, 46
Roosevelt, Franklin, 29, 32
Rosing, Boris, 47
Rotation, 355
Rough cuts, 358
RTNDA (Radio Television News Directors Association), 454

Sampling, 344–347
Sarnoff, David, 19, 25, 30, 46, 49
Satellites, 316–321
Saturation, 307
Scanning, 47, 306, 310
Scarcity theory, 400
Script preparation, 164
Section 315 of 1927 Radio Act, 425–427
Segue, 216

Self-regulation in broadcasting and cable, 452–462
   and citizens groups, 460–461
   codes of, 452–454
   and professional groups, 459
   standards of, 455–456
"Sell through" videotapes, 144
Semiautomation, and radio syndication, 202
"Sesame Street," 393–394
Sets (radios) in use, 338
Sex-role stereotyping, 383–385
Share of audience, 82, 338, 349, 350
SHF (super-high frequency), 318
Shield laws, 442–443
Signal-to-noise ratio, 275
Single-system operator (SSO), 68
Situational ethics, 464–465
Sixth Report and Order, of FCC, 55
"The $64,000 Question," 59
Sky waves, 292
Slander, 437
SMATV, 140–141
SNG (satellite news gathering), 234–235
Sony, 70, 142, 288
Soviet Union All-Union Radio, 170
Spots, on radio and TV, 81
State and local governments, 415–416
Station image, 362
Step deal, 253
Stereophonic sound, 40
Stimulation theory, 376
Storage instantaneous audimeter (SIA), 339
Storyboard, 164
Stripping, in TV scheduling, 263
Subscriber drop, 316
Subscription TV, 71
Sullivan, Ed, 52
Summative evaluation, 165
Superband, in cable TV, 315
Superstations, 67, 127
Surveys, 360, 371
Sustaining programs, in ratings, 337
Switcher, 311
Synchronization pulse, 307
Syndicated exclusivity, 260
Syndication, radio, 41
Syndication, TV, 247, 259–266
   and economics, 264–266
   market, 260–261
   types of, 261–265
Systematic interval selection, 345

Target audience, 86
Target plate, 307
Tartikoff, Brandon, 72
Teasers, in programming, 216, 241
Telegraphy, 14

Telephone, 14
Teleports, 320
Teletext, 72, 425
Television, 104–123
  business aspects of, 106–110
    and local advertising, 108
    and national spot advertising, 107
    and network advertising, 10
  and economics of local TV, 112–114
  history of, 46–73
    beginnings, 46–52
    growth period, 53–62, 57–62
    postwar period, 52–57
    and recent changes, 67–73
    stable times in, 62–66
  job outlook in, 120
  network, 110–112
  and salaries, 121
  and station organization, 117–120
  station ownership in, 114
  and station sales, 116
  and types of stations, 104–106
    commercial and noncommercial, 104
    VHF and UHF, 104–105
Television advertising, 389–391
Television channels, 312–313
  and station allocations, 312
  and width, 312
Television entertainment programming, 247–270
  costs of, 253
  independent producers of, 252
  local, 266–268
  network, 247–256
    process of, 249–251
    seasons in, 250–251
    strategies of, 268–270
Television news, 226–245
  audience for, 243
  history of, 227–235
  job outlook in, 243
  recent trends in, 236–238
  structure of, 238–245
Television violence, 429
Telstar, 317
Tianenmen Square, 186
Time Warner Inc., 70, 134, 472
Timeshifting, 142
Tip sheets, 216
Tolerated error, 361
Total survey area (TSA), 347
Traffic, 117
Transduction, 274–275, 303–305
Transponder, 318

Treatment (sample script), 164, 253
Trespass, 440–441
Trunk cable, 316
Turnkey automation and radio, 202
TV Martí, 172, 449
TV Q, 362
TVRO (television receive-only) Earth stations, 72, 140, 318

UHF band, 55, 57, 62, 63, 290, 312
U-Matic video cassette, 151
Underground plant, of cable, 137
United Fruit Company, 22
United Kingdom, broadcasting in, 180–184
  economics of, 181
  and law, 182–183
  and programming, 183
  structure of, 180–181
Uplinks, 318
Uses and gratifications, 391–392
USIA, 179
Utilitarianism, 463

"War of the Worlds," 32
Watergate, 231
Waveform, 276
WBAI, 434
WEAF, 24–26
Weighting, 348
Welles, Orson, 32
Wertham, Frederic, 375
Westinghouse, 22, 23, 26
Westwood One, 41, 221
WFAN, 204, 220
WGN, 67
White House, 410–411
Wireless Ship Act of 1910, 401
World Administrative Radio Conference (WARC), 449
Worldnet, 179
WTBS, 67

Yeltsin, Boris, 5, 6, 8
Young, Owen, 21

Zapping, in VCR recording, 146
Zipping, in VCR playback, 146
Zworykin, Vladimir, 49–50